THE PEPTIDES

Volume I

The Peptides

BY

EBERHARD SCHRÖDER

AND

KLAUS LÜBKE

HAUPTLABORATORIUM DER SCHERING AG
WEST BERLIN, GERMANY

Translated by Erhard Gross

NATIONAL INSTITUTES OF HEALTH
BETHESDA, MARYLAND

VOLUME I

Methods of Peptide Synthesis

1965

ACADEMIC PRESS New York and London

ACADEMIC PRESS INC.
111 Fifth Avenue, New York, New York 10003

United Kingdom Edition published by
ACADEMIC PRESS INC. (LONDON) LTD.
Berkeley Square House, London W.1

Library of Congress Catalog Card Number: 65-22760

PRINTED IN THE UNITED STATES OF AMERICA

PREFACE

In view of the time that has elapsed since the last work on polypeptides appeared, the publication of a new monograph seemed to be warranted, even more so since considerable progress has been made with regard to preparative techniques. We have thus set out to describe the current methods of peptide synthesis and their applications to the preparation of biologically active polypeptides.

The first volume of "The Peptides" bears the subtitle "Methods of Peptide Synthesis"; it presents a detailed description of protecting groups, individual amino acids, and coupling reactions. Problems of special interest are discussed in the closing chapters. An effort was made to stress problems which are of concern to the synthetic peptide chemist. Methods which have been little employed for the synthesis of biologically active polypeptides and those which are of interest only from a theoretical point of view have received correspondingly brief treatment. Owing to the detailed presentation of the material, a repetition of certain aspects could not be prevented in all instances.

The second volume of "The Peptides" is subtitled "Synthesis, Occurrence, and Action of Biologically Active Polypeptides." In it the synthesis of biologically active peptides and analogues is described. The origin, isolation, structural elucidation, and activity of naturally occurring biologically active peptides are discussed briefly.

Almost all of the approximately 3000 references were read in the original. Abstracting organs or brief English summaries were used only when papers were published in Japanese or Chinese. The authors endeavored to cover as completely as possible, at least from 1950 on, publications concerned with the synthesis of peptides. The literature was checked until close to the end of 1964. In order that the reader have a review of the entire literature on peptides, a complete bibliography has been included in each volume.

Synthetic peptide chemistry—now an integral part of preparative organic chemistry and biochemistry—has gone through periods of varying activity during the past sixty years.

The classic investigations of E. Fischer and T. Curtius had led to an understanding of the principles of protein structure and to the synthesis of peptides of at least nineteen amino acids. Following these investigations was a period of rather slow development. The next breakthrough came in 1932 with the use of the carbobenzoxy group by M. Bergmann and his school. This protecting group—still of great value—was used together with the acid chloride and azide methods for the synthesis of a variety of sterically homogeneous peptides. In collaboration with J. S. Fruton, J. P. Greenstein, K. Hofmann, E. L. Smith, V. du Vigneaud, and L. Zervas, M. Bergmann used these synthetic peptides to the elucidation of the substrate specificity of enzymes. The synthesis of higher peptides, however, was limited by the procedures available and the shortage of analytical control methods.

The period from 1944 to 1954 was governed by the development of analytical methods, modern separation techniques, and protein structure elucidation. The introduction of paper chromatography, preparative column chromatography, the more extensive use of electrophoresis and countercurrent distribution, and, finally, the determination of the structure of oxytocin by V. du Vigneaud and also by H. Tuppy and of insulin by F. Sanger led to the synthesis of peptides. After the synthesis of oxytocin great efforts were made to prepare other biologically active polypeptides. Peptide chemistry is still characterized by these objectives. In the course of only a few years a number of biologically active polypeptides has been synthesized in such quantities that it has become feasible to subject them to pharmacological and medical screening. These compounds are now beginning to be applied therapeutically. The synthesis of analogues of these peptides contributed to the understanding of the relationship between structure and action of biologically active polypeptides.

After J. S. Fruton's summary of 1949, reviews on the methods of peptide chemistry by W. Grassmann and E. Wünsch and by M. Goodman and G. W. Kenner were published in 1956 and 1957, respectively. The three volumes of "Chemistry of the Amino Acids" by J. P. Greenstein and M. Winitz, which were published in 1961, discuss not only techniques of peptide chemistry but also analytical methods, individual amino acids, and physicochemical problems.

We would like to thank the Schering AG for making this work possible. We are grateful to Dr. Gibian, with whom we started work in the peptide field several years ago, for his encouragement and support. We would also like to acknowledge the assistance of our co-workers. Their cooperation has enabled us in recent years to acquire most valuable practical experience. They have assisted us in reviewing the extensive literature, in writing the manuscript, and in composing the index.

Finally, we must express our gratitude to Dr. E. Gross who carried out the voluminous task of translating this monograph.

E. SCHRÖDER
K. LÜBKE

May, 1965

CONTENTS

V. THE SYNTHESIS OF CYCLIC PEPTIDES

VI. DEPSIPEPTIDES

VII. PEPTOIDS

VIII. THE PLASTEIN REACTION

IX. SOLID PHASE PEPTIDE SYNTHESIS

X. PROBLEMS OF RACEMIZATION

Contents of Volume II

Synthesis, Occurrence, and Action of Biologically Active Polypeptides

Nomenclature of Amino Acids and Peptides

The nomenclature that we have used for amino acids, peptides, and peptide hormones is based essentially on the proposals by:

(1) E. Brand and J. T. Edsall, *Ann. Rev. Biochem.* **16,** 224 (1947).
(2) E. Bricas and C. Fromageot, *Advan. Protein Chem.* **8,** 1 (1953).
(3) I. M. Goodman and G. W. Kenner, *Advan. Protein Chem.* **12,** 465 (1957).
(4) R. Schwyzer, *Chimia* (*Aarau*) **12,** 53 (1958).
(5) Committee on Abbreviations of the American Society of Biological Chemists, 1959.
(6) IUPAC, Section of Biological Chemistry, Nomenclature Commission 1960; cf. *J. Biol. Chem.* **237,** 1381 (1962).
(7) J. P. Greenstein and M. Winitz, "The Chemistry of the Amino Acids." Wiley, New York, 1960
(8) R. Schwyzer, J. Rudinger, E. Wünsch, and G. T. Young, Suggestions in the "Fifth European Peptide Symposium" (G. T. Young, ed.), p. 261. Macmillan (Pergamon), New York, 1963.

A. ABBREVIATIONS OF AMINO ACIDS

(1) Individual amino acids in the text are mentioned by their full names. Abbreviations are used only in tables, reaction schemes, etc., and for the presentation of peptides.
(2) Except for a few cases, the abbreviations consist of the first three letters of the trivial name (cf. Table on Nomenclature at the end of this section). If there is no trivial name, the chemical name is used in its Anglo-Saxon form and an abbreviation is formed from the first letters of the syllables. Structurally related amino acids have related abbreviations. The first letter is always capitalized, in the peptide chain likewise.
(3) The abbreviation represents the amino acid residue. The formulation of a free amino acid or of a free peptide is unequivocal only when it terminates in an H— at the amino group or an

—OH at the carboxyl group. The amino group always appears to the left, the carboxyl group to the right. Side chains are accommodated above or below the line.

$$\underset{}{NH\text{-}\overset{\overset{CH_3}{|}}{C}H\text{-}CO} = Ala \qquad H\text{-}HN\text{-}\overset{\overset{CH_3}{|}}{C}H\text{-}CO\text{-}OH = H\text{-}Ala\text{-}OH$$

H-Val-Leu-Ala-OH

Where the amino acid chain contains parts of an uncertain sequence the symbols are separated by a comma. The unknown amino acid sequence is put in parentheses:

H-Pro-Val-Leu-(Ala, Glu, Asp, Pro)-Arg-Gly-OH

(4) Additional designations are treated as follows:

(a) Amino acids without α-amino group or with an additional amino group in an unusual position are further characterized by the designation of this position:

β-Ala γ-Abu β,γ-Abu

(b) In the case of isomeric amino acids the iso compound, in analogy to Ileu, is abbreviated by adding the prefix I, the unbranched form, accordingly, by adding the prefix N:

Leu Ileu Val Nval

(c) Hydroxyamino acids carry the prefix "Hy" unless they have their own trivial name:

Hypro Hylys

The position of the hydroxy group can be specified by a preceding Greek letter: δ-Hylys

(d) No abbreviation has been introduced for "homo" (one more CH_2 group) and "nor" (one CH_2 group less):

Homoarg Norarg

(e) The substituent in *N*-substituted amino acids is placed before the amino acid in the form of its commonly used abbreviation separated by a period if no trivial name is used. The amino substitution can be designated clearly by the prefix N:

N-methylvaline = Me·Val or N-Me·Val

C-Substituted amino acids are always distinguished by a preceding C. The C atom will eventually have to be designated more closely:

C-Ph-Gly C^{α}-Me·Ala

(f) Cys stands for cysteine, $(Cys)_2$ for cystine. Cystine peptides are best expressed in two lines, e.g.:

cystinylbis-(valylleucine):

H-$(Cys)_2$-bis-(Val-Leu-OH) or H-$(Cys)_2$-$(Val\text{-}Leu\text{-}OH)_2$ or H-Cys-Val-Leu-OH
H-Cys-Val-Leu-OH

(5) For the designation of the optical activity the configuration is placed before the amino acid, separated by a hyphen.

The uncertainty about the configuration of an amino acid should be expressed by a question mark in the sequence of the peptide:

H-L-Me·Leu-OH   H-L-Ala-D-Val-OH   H-L-Ala-?-Leu-D-Val-OH

No designation always means the L-form.

Amino acids with two centers of asymmetry are expressed as follows:

D-Alloisoleucine   D-aIleu   L-Allohypro   L-aHypro

The amino acid is considered as a unit and will be preceded by all structural and configurational designations, e.g.:

allo = a   *erythro* = *e*   *threo* = *t*   *meso* = *m*

B. ABBREVIATIONS FOR THE BLOCKING GROUPS

Of the most widely known abbreviations for the carbobenzoxy group, Cbo- and Z-, Cbo will be used. Carbobenzoxy groups with substituents in the *p*-position will be designated by the substituents preceding them. Since for the *p*-phenylazocarbobenzoxy and the *p*-methoxyphenylazocarbobenzoxy groups no abbreviation derived from Cbo has been introduced, the present exclusively used designations PZ and MZ will be employed.

The abbreviations for *N*-acyl-, *N*-aryl-, and the carboxyl-protecting groups (cf. Tables II and III) are based essentially on those most commonly used in the literature. Protecting groups, derived from others by substitution, have no abbreviations of their own (e.g. OBzl and $OBzlNO_2$ for benzyl and *p*-nitrobenzyl esters).

C. PRESENTATION OF SUBSTITUTED PEPTIDES

(1) The abbreviated expression for peptide derivatives consisting of monoaminomonocarboxylic acids is unequivocal, e.g.:

Cbo-L-Val-L-Pro-OMe BOC-L-Val-L-Pro-OH H-L-Val-L-Pro-OBzl

Salt formation at the amino group:

H-Ala-OMe·HCl or $H_2^{\oplus}$-Ala-OMe $Cl^{\ominus}$

Salt formation at the carboxyl group:

Cbo-Ala-ONa or Cbo-Ala-$O^{\ominus}Na^{\oplus}$

Cbo-L-Phe-L-Ser-OH·dicyclohexylamine

but

Cbo-L-Phe-L-Ser-*O*-dicyclohexylammonium

Inner salt formation:

H_2^{+}-Ala-O^{-}

(2) Two different possibilities exist for bifunctional amino acids:

(a) One line expression:
The substituent of the ω-function is placed in parentheses directly behind the amino acid. The nature of the bond is clear from the abbreviation used and from the amino acid.

(b) Dual line expression:
It is clearer and will always be used when reaction schemes are presented. The substituent at the ω-function will be placed above, less frequently below.

Cbo-L-Ser(Bzl)-L-Arg(NO_2)-OMe or

```
       Bzl    NO2
        |      |
Cbo-L-Ser-L-Arg-OMe
```

Trit-L-Lys(BOC)-L-Asp(NH_2)-NH_2 or

```
        BOC    NH2
         |      |
Trit-L-Lys-L-Asp-NH2
```

(c) In addition to these abbreviation principles, the following more detailed formulations may be used equally well:

N^{α}-Trit-N^{γ}-BOC-L-Lys-L-Asp(NH_2)$_2$

(d) Salt formation at the ω-function can be expressed accordingly:

Cbo-L-Cys(Ag)-L-Phe-OMe	Ag | Cbo-L-Cys-L-Phe-OMe
H-L-Arg(HCl)-L-Glu-NH_2	$H_2^{\oplus}$ | H-L-Arg-L-Glu-NH_2 $Cl^{\ominus}$
Cbo-L-Glu(ONa)-L-Ala-OMe	ONa | Cbo-L-Glu-L-Ala-OMe

The salt-forming compound is more frequently placed behind the peptide separated by a period:

H-L-Val-L-Leu-OMe·picrate

H-L-Arg-L-Leu-OMe·2 HCl

(3) The following abbreviations result for trifunctional amino acids:

γ-Hylys(Cbo, γ-Me)-OMe or γ-Hylys(γ-Me)-OMe with Cbo written above Hylys, joined to it by a vertical line

N^{ϵ}-Cbo-*O*-Methyl-γ-hydroxylysine methyl ester

(4) In the case of tetrafunctional amino acids a clear expression in a single line is not possible. Therefore the following expression in several lines is proposed. It can also be used for cystine derivatives:

Cbo╲ ╱OMe Lan Tos╱ ╲OBzl	BOC╲ ╱OMe Dap Cbo╱ ╲OMe	Cbo╲ ╱OH $(Cys)_2$ Trit╱ ╲OMe

(5) α,ω-Peptides of monoaminodicarboxylic acids and diaminomonocarboxylic acids, respectively, are expressed as follows:

(a) In a single line:

Cbo-Glu-α-(Val-OH)-γ-(Leu-Ala-OH)

In two lines:

┌Leu-Ala-OH
Cbo-Glu-Val-OH

(b) In a single line:

α-(BOC-Phe)-ϵ-(Cbo-Ala-Val)-Lys-OEt

In two lines:

Cbo-Ala-Val┐
BOC-Phe-Lys-OEt

(c) In a single line:

α-(Cbo-Val)-ε-(Cbo-Glu-α-OMe-γ-)-Lys-OMe

In two lines:

```
          Cbo-Glu-OMe
               |
     Cbo-Val-Lys-OMe
```

D. CLASSIFICATION OF PEPTIDES

The peptides will be divided into two main groups: homomeric peptides and heteromeric peptides.

(1) *Homomeric Peptides*

In the group of homomeric peptides are all the compounds which on hydrolysis yield amino acids exclusively. On the basis of amino acid order and the type of linkages the following possibilities result:

(a) *Homodetic linear homomeric peptides:* Peptide bonds only between carboxyl and amino groups, irrespective of α- or ω-bond, e.g.:

H-Asp-Arg-Val-Tyr-Val-His-Pro-Phe-OH

(b) *Heterodetic linear homomeric peptides: O*-Peptides of serine, threonine, and other ω-hydroxy-α-amino acids or of *S*-peptides of cysteine:

```
Cbo-Val-Phe-O                  Cbo-Val-Phe-S
            |                              |
     H-Pro-Ser-Glu-OH               H-Pro-Cys-Glu-OH
```

(c) *Homodetic cyclic homomeric peptides:*

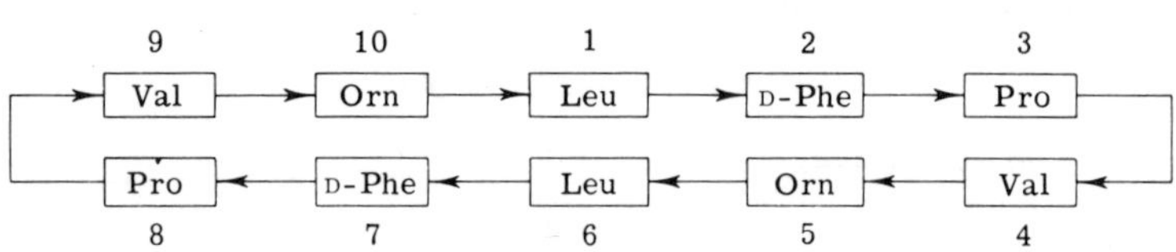

or

⌐Val-Orn-Leu-D-Phe-Pro-Val-Orn-Leu-D-Phe-Pro¬ (ring closed by a bracket beneath)

The formulation with boxes uses arrows to indicate the direction of the peptide bonds (CO → NH).

Amino acids in the ring should be numbered in such a way that the amino acid first in alphabetical order is assigned number 1 and additional numbering is given in clockwise direction. This mode of numbering cannot be applied to the schematic presentation of reaction routes in the frequently employed Schwyzer system, especially not to synthetic analogues of naturally occurring cyclopeptides. Here the linear amino acid sequence that has been cyclized will be used.

Amino acid side chains in cyclic peptides will be numbered 1′, 2′, 3′, etc., starting with the most distant amino acid.

The attachment of the side chain to the ring or an ω-bond in the cyclic peptide can be denoted by the symbols α,β,γ, etc.

However, it is clearer to use a formulation that corresponds to the expressions in several lines:

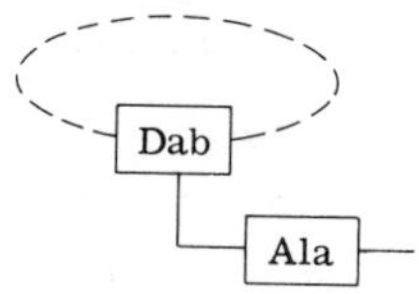

Side chain is in ω-bond

Side chain is in α-bond

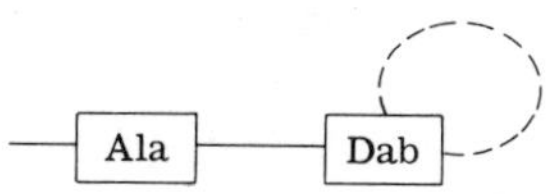

Ring contains an ω-bond

The linear sequence corresponding to a cyclopeptide can be denoted by the prefix "*seco*," e.g.:

seco-8,9-gramicidin S

= H-Val-Orn-Leu-D-Phe-Pro-Val-Orn-Leu-D-Phe-Pro-OH

The numbering will then change to the one for linear homodetic peptides. Diketopiperazines are not considered as cyclic peptides.

(d) *Heterodetic cyclic homomeric peptides:* The term "heterodetic" expresses the fact that cyclization has occurred through a hetero atom.

(i) *Cyclic disulfides:*

```
                    NH2 NH2
                     |   |
H-Cys-Tyr-Ileu-Glu-Asp-Cys-Pro-Leu-Gly-NH2
  |                     |
  S---------------------S
```

or

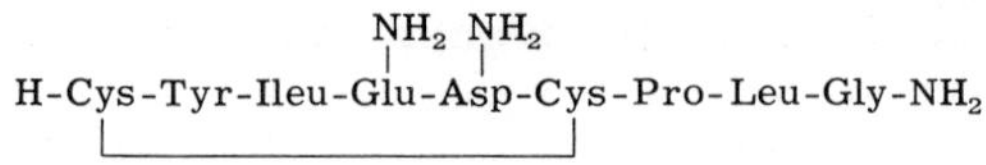

(ii) *Peptide lactones:*

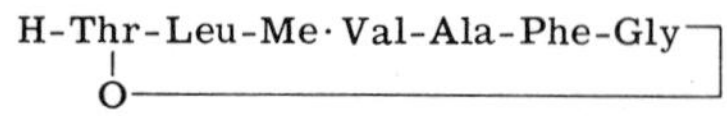

The O can be omitted.

(2) *Heteromeric Peptides*

Heteromeric peptides are compounds that contain in addition to amino acids a nonamino acid component. This component must be part of the compound, not only a temporary protecting group.

(a) Compounds of this type correspond to the proteids when the hetero component is at the beginning or at the end of the peptide chain or when it is linked to the chain by way of an ω-bond. We propose the term peptoids for this class of compounds. Chromo-, glyco-, lipo-, nucleo-, and phosphopeptides belong in the peptoids class; the peptide portion of the molecule, as in the case of the homomeric peptides, can be classified as linear, cyclic, homodetic, and heterodetic.

(b) If the hetero component is in the peptide chain, we are dealing with a heterodetic heteromeric peptide of the linear or cyclic form.

(3) For term depsipeptide see Chapter VI.

E. NOMENCLATURE OF SYNTHETIC ANALOGUES OF BIOLOGICALLY ACTIVE POLYPEPTIDES

If in a naturally occurring biologically active polypeptide an amino acid exchange has been effected by synthesis, the new amino acid will be placed ahead of the trivial name together with the position number.

Dab^1-Dab^9-bradykinin

This system has also been adopted for naturally occurring peptides that vary in amino acid composition:

Val^5-angiotensin and $Ileu^5$-angiotensin

Changes of functional groups are indicated in an analogous manner, e.g.:

$Asp(NH_2)^1$-Val^5-angiotensin

or

Val^5-angiotensin-Asp^1-β-amide

The elimination of an amino acid is denoted by placing the name of the amino acid with the prefix "des" in front of the compound:

Des-(Asp^{1})-Val^{5}-angiotensin II: H-Arg-Val-Tyr-Val-His-Pro-Phe-OH
2 3 4 5 6 7 8

The numbering system should not be changed here. This, however, may be done in a reaction scheme.

The denoting of eliminated amino acids by boxes is, especially in comparing several analogues, very instructive:

Des-(Tyr^{4})-Val^{5}-angiotensin II: H-Asp-Arg-Val-[ ]-Val-His-Pro-Phe-OH
1 2 3 4 5 6 7 8

If the peptide analogue contains an additional amino acid in the chain, the prefix "homo" will be attached to the trivial name and placed before the compound name:

Homo-(Tyr$^{4'}$)-Val^{5}-angiotensin

The addition of amino acids to the amino terminal or the carboxyl terminal is expressed in the following way:

Original sequence:

H-Asp-Arg-Val-Tyr-Val-His-Pro-Phe-OH Val^{5}-angiotensin II
1 2 3 4 5 6 7 8

Extended at the amino terminal by alanylglycine:

(Ala-Gly-Asp)^{1}-Val^{5}-angiotensin II

Extended at the carboxyl terminal by alanylglycine:

(Phe-Ala-Gly)^{8}-Val^{5}-angiotensin

Amino acids condensed to the amino terminal are denoted as Gly^{1_2} and Ala^{1_3}. Amino acids condensed to the carboxyl terminal are indicated as Ala^{8_2} and Gly^{8_3} or as Ala^{9} and Gly^{10}. In both cases one can switch to the numbering system 1–10 when a reaction scheme is to be presented.

TABLE ON NOMENCLATURE

I. Abbreviations of Amino Acid Residues

Alanine	Ala	Glycine	Gly
β-Alanine	β-Ala	Histidine	His
α-Aminobutyric acid	Abu	Homoarginine	Homoarg
γ-Aminobutyric acid	γ-Abu	γ-Hydroxylysine	Hylys
α-Aminoisobutyric acid (α-Methylalanine)	Iabu (C-Me·Ala)	Hydroxyproline	Hypro
		Isoleucine	Ileu
α-Aminoadipic acid	Aad	Isolysine (β-Lysine)	β-Lys
α-Aminopimelic acid	Api		
Arginine	Arg	Lanthionine	Lan
Aspartic acid	Asp	Leucine	Leu
Asparagine	$Asp(NH_2)$ NH_2–Asp (vertical)	Lysine	Lys
		Methionine	Met
Aspartic acid α-amide	$Asp\text{-}NH_2$ ($Asp\text{-}\alpha\text{-}NH_2$)	*N*-Methylvaline	Me·Val
		N-Methylleucine	Me·Leu
Cysteine	Cys	Norleucine	Nleu
		Norvaline	Nval
Cystine	$(Cys)_2$ Cys–Cys (vertical) Cys–S–S–Cys (vertical)	Ornithine	Orn
		Phenylalanine	Phe
		C-Phenylglycine	C-Ph·Gly
Cysteic acid	$Cys(O_3H)$ O_3H–Cys (vertical)	*N*-Phenylglycine	Ph·Gly
		Proline	Pro
Glutamic acid	Glu	Pyroglutamic acid	Pyroglu
Glutamine	$Glu(NH_2)$ NH_2–Glu (vertical)	Sarcosine	Sar (Me·Gly)

α,γ-Diaminobutyric acid	Dab	Serine	Ser
α-β-Diaminopropionic acid	Dap	Threonine	Thr
α,α'-Diaminoadipic acid	Daap	Tryptophan	Try
α,α'-Diaminopimelic acid	Dapi	Tyrosine	Tyr
Dihydroxyphenylalanine	Dopa	Valine	Val

II. Protecting Groups

A: *Acyl- and Alkyl-Type Protecting Groups*

Acetyl	CH_3CO—	Ac-
Benzoyl	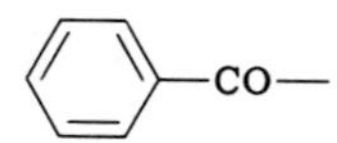	Bz-
Benzenesulfenyl	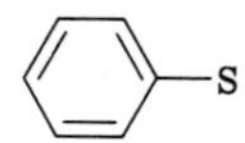	PhS-
Benzenesulfonyl	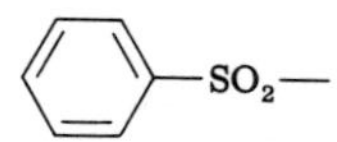	$PhSO_2$-
Benzyl	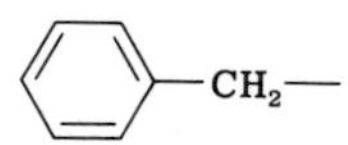	Bzl-
Benzylsulfonyl	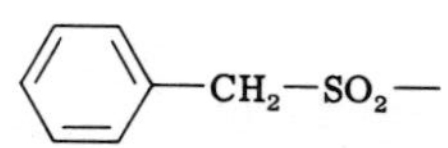	$BzlSO_2$-

TABLE ON NOMENCLATURE (Continued)

Benzylthiomethyl	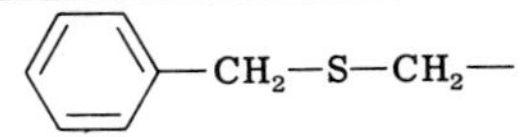	BTM-
2, 4-Dinitrophenyl	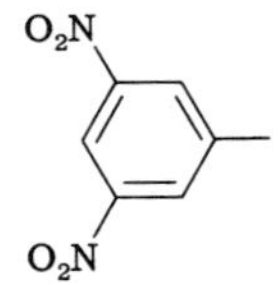	DNP-
Formyl	HCO—	Form-
o-Nitrophenylsulfenyl	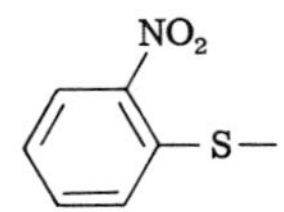	NPS-
Phthalyl	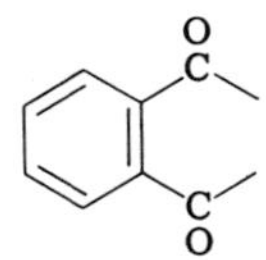	Phth-
Tetrahydropyranyl	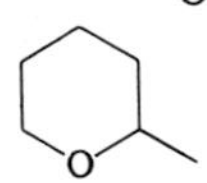	THP-
p-Toluenesulfonyl	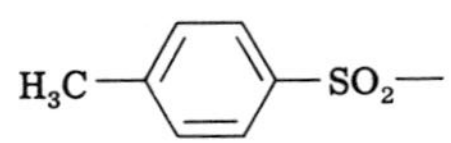	Tos-
Trifluoroacetyl	F_3C—CO—	TFA-
Triphenylmethyl	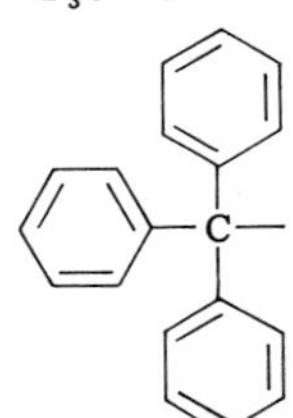	Trit-

B: *Urethan-Type Protecting Groups*

(*a*) *Carbobenzoxy and modified carbobenzoxy protecting groups*

Carbobenzoxy (Benzyloxycarbonyl)	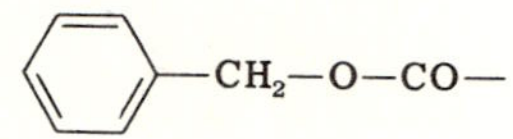	Cbo-	(Z-)
p-Bromocarbobenzoxy	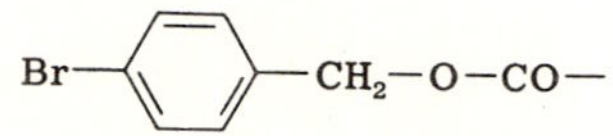	BrCbo-	(BrZ-)
p-Chlorocarbobenzoxy	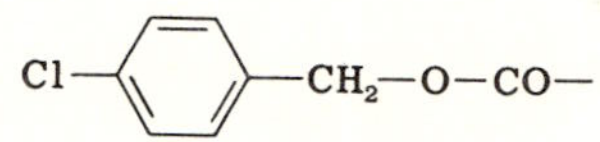	ClCbo-	(ClZ-)
p-Methoxycarbobenzoxy	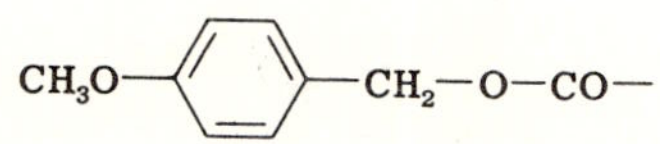	MeOCbo-	(MeOZ-)
p-Methoxyphenylazo-carbobenzoxy	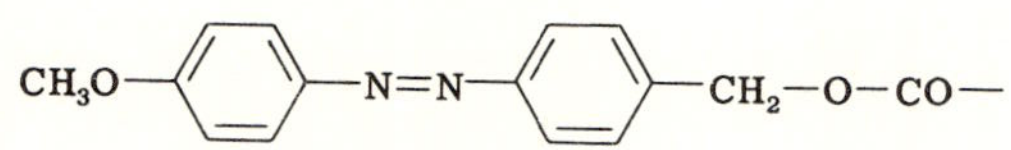	(MCbo-)	MZ-
p-Nitrocarbobenzoxy	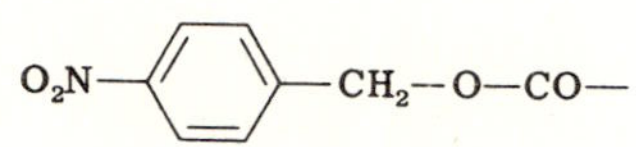	NO_2Cbo-	(NO_2Z-)
p-Phenylazocarbobenzoxy	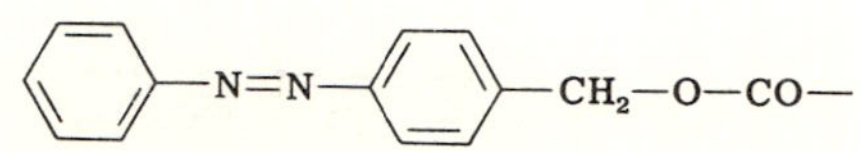	(PCbo-)	PZ-

TABLE ON NOMENCLATURE (Continued)

(*b*) *Other urethan-type protecting groups*

Allyloxycarbonyl	$CH_2{=}CH{-}CH_2{-}O{-}CO{-}$	AOC-
Benzylthiocarbonyl	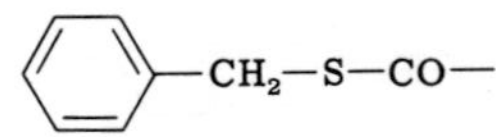	BTC-
tert-Butyloxycarbonyl	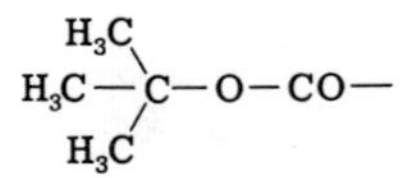	BOC-
Cyclohexyloxycarbonyl	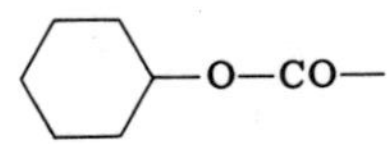	HOC-
Cyclopentyloxycarbonyl	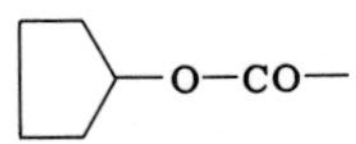	POC-
Phenylthiocarbonyl	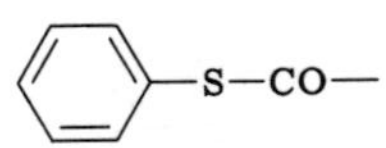	PTC-
Tolyloxycarbonyl	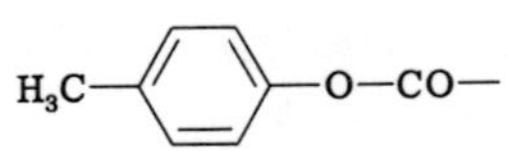	TOC-

C: *Carboxyl-Protecting Groups*

Amide	$-NH_2$	$-NH_2$
Ethyl ester	$-OC_2H_5$	$-OEt$
Azide	$-N_3$	$-N_3$
Benzyl ester	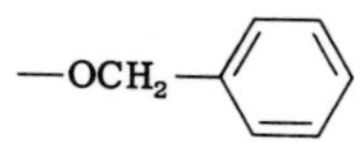	$-OBzl$
tert-Butyl ester	$-O-C(CH_3)_3$	$-O\,t\,Bu$
tert-Butyloxycarbonyl hydrazide	$-NHNH-OC-O-C(CH_3)_3$	$-NHNH-BOC$
Cyanomethyl ester	$-OCH_2CN$	$-OCH_2CN$

TABLE ON NOMENCLATURE (Continued)

p-Cyanophenyl ester	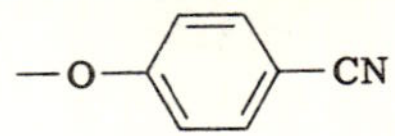	—OPhCN
Carbobenzoxy hydrazide	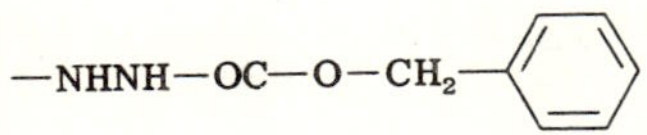	—NHNH—Cbo (—NHNH—Z)
Hydrazide	$—NHNH_2$	$—NHNH_2$
N-Hydroxysuccinimido ester	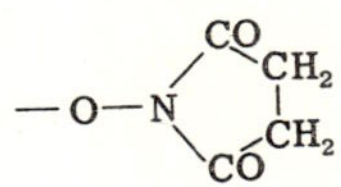	—ONSu
p-Methanesulfonylphenyl ester	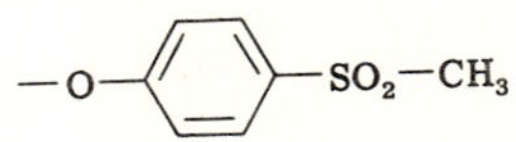	$—OPhSO_2Me$
p-Methoxybenzyl ester	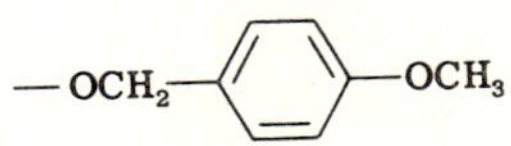	—OBzlOMe
Methyl ester	$—OCH_3$	—OMe
p-Nitrobenzyl ester	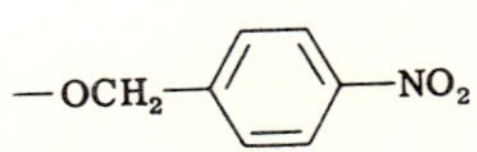	$—OBzlNO_2$

p-Nitrophenyl ester	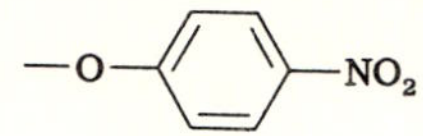	$—OPhNO_2$
p-Nitrothiophenyl ester	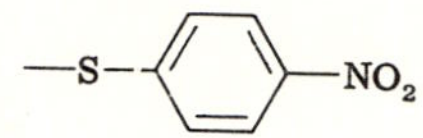	$—SPhNO_2$
Phenyl ester	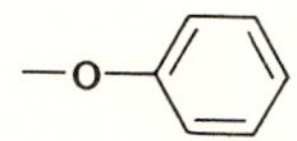	—OPh
Thioethyl ester	$—SC_2H_5$	—SEt
Thiophenyl ester	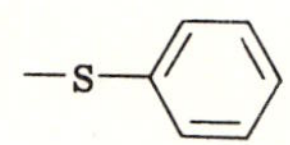	—SPh
2, 4, 5-Trichlorophenyl ester	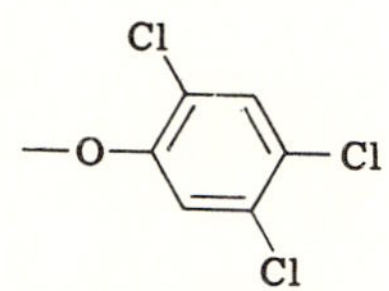	$—OPhCl_3$
Trityl hydrazide	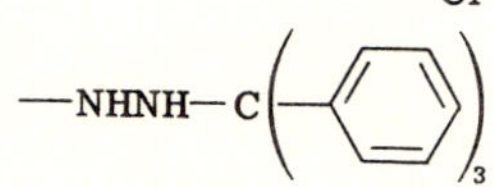	—NHNH—Trit
Vinyl ester	$—OCH{=}CH_2$	—OVi

INTRODUCTION

A peptide synthesis is characterized by the formation of an amide bond between the carboxyl group of one amino acid and the amino group of a second amino acid [1]:

$$H_2N-\underset{}{\overset{R}{\overset{|}{C}H}}-COOH + H_2N-\overset{R'}{\overset{|}{C}H}-COOH \xrightarrow{-\,H_2O} H_2N-\overset{R}{\overset{|}{C}H}-CO-NH-\overset{R'}{\overset{|}{C}H}-COOH \quad [1]$$

This reaction formally represents the elimination of water and requires "activation of the carboxyl group" by the introduction of a negative substituent X. The carbon atom is already positively charged in the polar form of the carboxyl group. Now it acquires an even higher charge and the addition of the amino nitrogen via its free electron pair is greatly facilitated [2]:

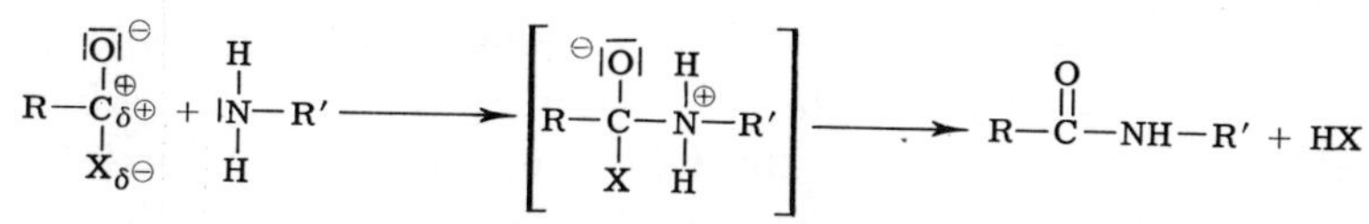
[2]

The peptide bond formation by "activation of the amino group" results also in a positively charged carboxyl carbon atom, since in the initial step the carboxyl group is added to the activating moiety. The reaction is favored by the resulting steric arrangement.

A large number of more or less different methods for activation are known. Relatively few of them have proved to be useful. These are sufficient, however, to form the peptide bond in almost all cases with satisfactory yield, and a minimum amount of racemization, and without the undesirable formation of by-products.

A peptide synthesis with free amino acids according to equation [1] cannot proceed unambiguously. The activated compound would react not only with the functional group of the second amino acid but also with its own amino group. An uncontrolled polycondensation would be the result. The synthesis of pure peptides requires the protection of all functional groups not expected to participate in a desired reaction [3]:

$$X{-}NH{-}\overset{\displaystyle R}{\overset{|}{CH}}{-}COOH + H_2N{-}\overset{\displaystyle R'}{\overset{|}{CH}}{-}CO{-}Y \xrightarrow{-H_2O} X{-}NH{-}\overset{\displaystyle R}{\overset{|}{CH}}{-}CO{-}NH{-}\overset{\displaystyle R'}{\overset{|}{CH}}{-}CO{-}Y \qquad [3]$$

The residues used for this blocking are called amino- or *N*-protecting groups (X) and carboxyl- or C-protecting groups (Y), respectively.

These residues, in addition to their function of protecting the amino and carboxyl groups, serve also another purpose. An amino acid is generally present as a zwitterion of different mesomeric forms [4]:

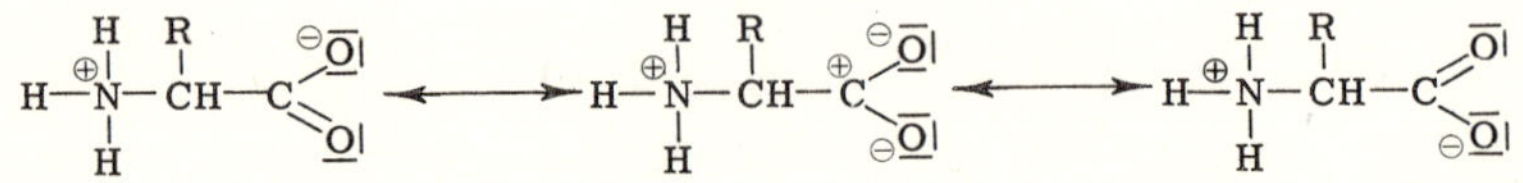 [4]

Activation of the carboxyl group would be difficult in this form and the addition of the amino nitrogen impossible. Furthermore, amino acids are poorly soluble in the organic solvents required as the medium for the majority of coupling reactions. Substitution of one of the functional groups deprives the amino acid of its ability to form an inner salt. The other functional group becomes available for reaction and better solubility in organic solvents is obtained. Properly chosen solvents will also eliminate the salt-forming properties and a peptide synthesis with a completely unprotected amino acid is possible. To continue the peptide synthesis, either the amino- or the carboxyl-protecting group must be selectively removed in order to obtain derivatives that can again enter into synthetic steps.

Additional functional groups in an amino acid, generally called ω-groups, that may give rise to side reactions must also be blocked. A protecting group must fulfill the following requirements, irrespective of its use as α-amino-, α-carboxyl-, or ω-protecting groups, before it can be employed in peptide chemistry. It must be possible to obtain the protected amino acid readily and, desirably, in quantitative yield. The blocking groups must retain their protecting properties, preferably under all coupling conditions. They must not give rise to side reactions. They must be readily and selectively removable so that other protecting groups for the same or other functional groups are not affected. For all practical purposes these conditions are not completely fulfilled. It will therefore be frequently necessary to arrive at compromises. The demand for protecting groups and combinations of protecting groups is, as a result of this, one of the main problems of peptide chemistry.

The importance of the protecting groups is stressed by the fact that they are discussed in detail together with the coupling methods. In the sections on amino acids and their peculiarities in peptide synthesis, especially multifunctional amino acids, possible combinations of protecting groups are described. Chapters V–X on cyclic peptides, depsipeptides, peptoids, and problems of racemization conclude this volume.

I. AMINO-PROTECTING GROUPS

An amino group can be acylated or alkylated. It is able to form Schiff's bases and accepts acids to produce salts. All these reactions represent a blocking or protection and may be employed to prepare *N*-protected amino acids. The following classification has resulted: acyl groups (RCO— and RSO_2—), urethan groups (ROCO— and RSCO—), alkyl and aryl groups (R—), arylidene groups (RCH=) plus the protection by salt formation. A survey of the diversity of possibilities for removing the individual blocking groups is presented in Table I at the end of Chapters I and II on amino- and carboxyl-protecting groups. The groups are arranged according to the method of removal. In discussing a protecting group, the preparation of the *N*-protected amino acid and the removal of the blocking group are important. Special problems in the application of these blocking groups during peptide synthesis will be covered.

The most important amino-protecting groups have been recently reviewed by Boissonnas (*286a*).

A. ACYL-TYPE PROTECTING GROUPS

In general, simple acyl groups are not suitable as amino-protecting groups. They form a new amide bond that would be difficult to cleave selectively in the presence of other peptide bonds. *N*-Benzoyl- and *N*-phenacetylamino acids were used earlier and made possible a systematic investigation of the formation of the peptide bond. The synthesis of free peptides, i.e., the removal of these groups, was possible only under special circumstances. Hillmann and Hillmann (*1005*) have investigated the stability of the acyl amide bond of acylglycylglycine versus the peptide bond by heating the compound in absolute alcoholic hydrochloric acid for several hours. Formyl-, acetyl-, chloroacetyl- and dichloroacetylglycylglycine can be cleaved with formation of glycylglycine ethyl ester hydrochloride. The benzoyl, phthalyl, trichloroacetyl, and phenacetyl compounds are stable. In the case of toluenesulfonyl-

and naphthalenesulfonylglycylglycine, only the peptide bond is cleaved, with formation of acylated glycine ester and glycine ester hydrochloride.

This indicates clearly that only those compounds are suitable as acyl-type protecting groups that can be removed by a special method. The formyl residue has been used; it can be hydrolyzed very readily under acidic conditions. The trifluoroacetyl group, due to the strong electronegative influence of the three fluorine atoms, can be removed by treatment with alkali. The phthalyl group is subject to hydrazinolysis. Sulfonic acid residues have also been used; they can be removed by reduction, with formation of mercaptans.

A rearrangement common to all acylamino acids, namely, the azlactone formation, limits severely the application of these residues as protecting groups, since the optical activity may be extinguished due to the formation of tautomers (see Chapter III,B,5). Phthalylamino acids as bisacyl derivatives and sulfonylamino acids do not undergo this reaction.

1. Formyl Group

Formylamino acids had been prepared by Fischer and Warburg (*723*) by heating amino acids in formic acid to 100°C, but their importance was not recognized. Fifty years later Hillmann and Hillmann (*1005*) found that they could be hydrolyzed very readily in acid. Formylamino acids are best prepared by the action of a mixture of formic acid and acetic anhydride on amino acids (*631, 640, 2420, 2579*). Formic anhydride, the possible intermediate, is most likely the acylating agent. To reduce the danger of racemization of optically active amino acids, temperatures of +5° to +15°C are recommended (*2073*). Peptides can also be formylated in this way (*2604*). If lysine is used as its formate the N^{α}-acyl derivative is formed. N^{ϵ}-Formyllysine can be obtained via the copper complex with ethyl formate (*1030*), or by the reaction of *p*-nitrophenyl formate with free lysine in aqueous solution (*1656a*).

Though formylamino acids are readily accessible by these methods, this protective group did not become of great importance in peptide chemistry for the synthesis of higher peptides. In most cases the peptide coupling proceeds with unsatisfactory results. The formyl group is not stable under the conditions of the acid chloride and phosphazo methods (*1233*). Racemization occurs very readily in methods proceeding through anhydrides (*2073*). This is caused by the formation of azlactones. Waley (*2414*), without experimental details, and Vogler and Lanz (*2386*) have described syntheses of optically active dipeptides using formylamino acids and the method of mixed anhydrides. The azide method has been used by King *et al.* (*1233*) but only to react formylglycine azide, obtained from formylglycine hydrazide, with aniline.

Sheehan and Yang (*2073*) have shown that the carbodiimide method gives good results with regard to yield and optical purity.

The cleavage of the formyl group can be readily achieved with a small excess of 0.5–1 *N* alcoholic HCl in 48 hours at room temperature or by boiling for 1 hour (*631, 2073*). The reaction time required for the deformylation varies with the individual amino acid. Under the conditions of deformylation, formylamino acids and peptides with free carboxyl groups may be esterified by the alcohol. Waley and Watson (*2420*) obtained H-L-Tyr-L-Lys(Cho)-OBzl in good yield from Form-L-Tyr(Ac)-L-Lys(Cbo)-OH by treatment with absolute benzyl alcohol/acetyl chloride for 40 hours. The *O*-acetyl group was eliminated. Sondheimer and Semeraro (*2163*) describe a series of side reactions during the deformylation of formylglutamine benzyl ester in methanolic hydrochloric acid. Transesterification of the benzyl ester and of the amide to the methyl ester takes place. To prevent side reactions of this kind the formyl cleavage may be carried out in dioxane/water as the solvent (*2073*).

The carbobenzoxy group is stable under the condition of formyl cleavage. It can be removed selectively by HBr in acetic acid (*2384, 2389, 2392*) while the formyl group is present, or by catalytic hydrogenation in the presence of formic acid (*1030*). The formyl group is stable to basic reagents. Formyl peptides can be obtained in good yield under the standard conditions of alkaline saponification (cf. page 55) of formyl peptide esters. Benzyl esters can be cleaved selectively by hydrogenolysis (*306*). According to recent studies by Losse and Nadolski (*1460a*), the formyl group can be removed quantitatively from *N*-formylamino acid esters and *N*-formyl peptide esters by catalytic hydrogenation in hydrogen chloride containing tetrahydrofuran, thereby forming amino acid and peptide ester hydrochlorides, respectively. Formylamino acids do not yield amino acid hydrochlorides, but only by-products. The hydrogenation of *N*-formylamino acid benzyl esters gave negative results.

The formyl group can also be removed in 2 hours at 60°C by oxidation with a two- to threefold excess of 15% H_2O_2 as oxidizing agent (*1463, 1464, 2718*). Presumably the reaction [1] proceeds via the carbamido acid.

$$\mathrm{H{-}\overset{\overset{\large O}{\|}}{C}{-}NH{-}\overset{\overset{\large R}{|}}{C}H{-}COOH \longrightarrow HO{-}\overset{\overset{\large O}{\|}}{C}{-}NH{-}\overset{\overset{\large R}{|}}{C}H{-}COOH \longrightarrow H_2N{-}\overset{\overset{\large R}{|}}{C}H{-}COOH + CO_2} \qquad [1]$$

No influence on the optical activity has been observed.

The formyl group has been used more recently by Hofmann *et al.* (*1025, 1030, 1037, 1043, 1044*) to protect the ε-amino group of lysine in the synthesis of partial sequences of ACTH, and by Vogler *et al.* (*2384, 2389, 2392*) for syntheses in the polymyxin field.

2. Trifluoroacetyl Group

The trifluoroacetyl group is an exceptional protecting group; it can be cleaved by very mild alkaline hydrolysis. It should therefore be a most welcome supplement to the well-known protecting groups that can be removed by hydrogenolysis or under acidic conditions. Although it was introduced into peptide chemistry 12 years ago by Weygand and Csendes (*2479*), it has not been widely used due to several disadvantages. It has recently become more important in the area of sequence analysis and for the investigation of steric homogeneity of peptides.

Trifluoroacetylamino acids are obtained by the action of trifluoroacetic anhydride on amino acids (*2479*). The reaction proceeds without racemization when 1.2 moles of trifluoroacetic anhydride are used in anhydrous trifluoroacetic acid as solvent at —10° to +10°C (*2480*). The use of 2 moles of trifluoroacetic anhydride in the absence of a solvent leads to the formation of the mixed anhydride [2] of the trifluoroacetylamino acid and trifluoroacetic acid. It rearranges with racemization either directly or via the symmetrical anhydride [3] to the azlactone [4] (2-trifluoromethyl-4-alkyloxazolone-6) (*2480, 2485, 2493*):

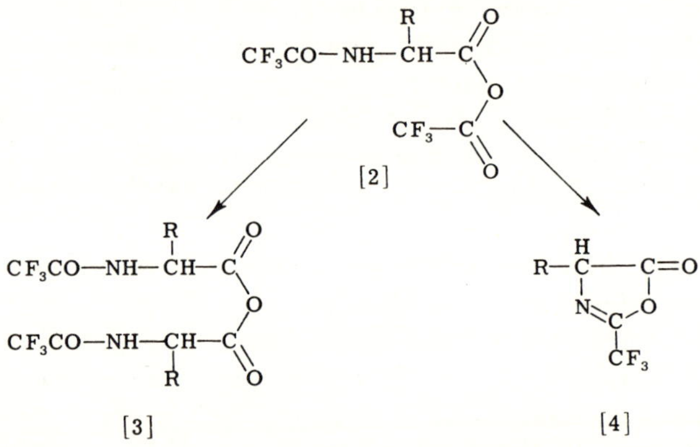

Aspartic acid and glutamic acid yield the trifluoroacetylated intramolecular anhydrides without racemization (*2490, 2496*).

Another possibility for the synthesis of trifluoroacetylated amino acids is the aminolysis of trifluoroacetic acid esters. Schallenberg and Calvin (*1923*) have used the ethyl thiolester and Weygand *et al.* (*2492, 2498*) the phenyl and methyl esters. The reactions must be carried out at pH values between 8 and 9 to prevent the hydrolysis of the trifluoroacetyl group. Trifluoroacetic acid imidazolide (*2187*) and trifluoroacetic acid pyrazolide (*1815*) have also been used for the introduction of this protecting group. They are applicable, however, to amino acid esters only.

The introduction of the trifluoroacetyl group causes difficulties for several amino acids. Serine and threonine lose water upon treatment with 1.2 moles of trifluoroacetic anhydride in trifluoroacetic acid to form the corresponding unsaturated acids [5] (*2497*):

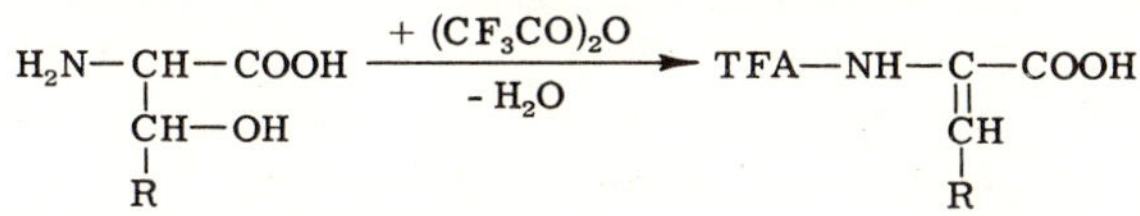

[5]

(R = H or CH_3)

N-Trifluoroacetylhydroxyamino acids are obtained in pure form by hydrolysis of the ester bond of the *N,O*-bis derivatives. These are prepared with 2.2 moles of trifluoroacetic anhydride. The acylation can be achieved more readily by treatment with trifluoroacetic acid ethyl thiolester (*1923*). The derivatives, however, are not very stable (cf. page 209). Tyrosine and tryptophan, when reacted in the usual manner, also give poor results. Their acyl derivatives can be obtained in high yield in ethyl ether (*2480*). The acylation of free lysine and ornithine with trifluoroacetic anhydride in trifluoroacetic acid yields the α-acyl derivatives exclusively. The ω-amino group, as the more basic group, is present in its ammonium form in the strongly acidic solution and cannot be acylated (*2480*). The ω-acyl derivatives (*1923*) can be obtained with trifluoroacetic acid ethyl thiolester. Subsequent reaction with trifluoroacetic anhydride in trifluoroacetic acid converts them to the bistrifluoroacetyl compounds.

Free peptides, in almost all cases, cannot be acylated by trifluoroacetic anhydride without side reactions. Acylation of the peptide bond and anhydride formation with the carboxyl group occur simultaneously. The peptide bond is weakened to the extent that it is partially cleaved [6] (*2483*). This cleavage is not encountered with peptide bonds of

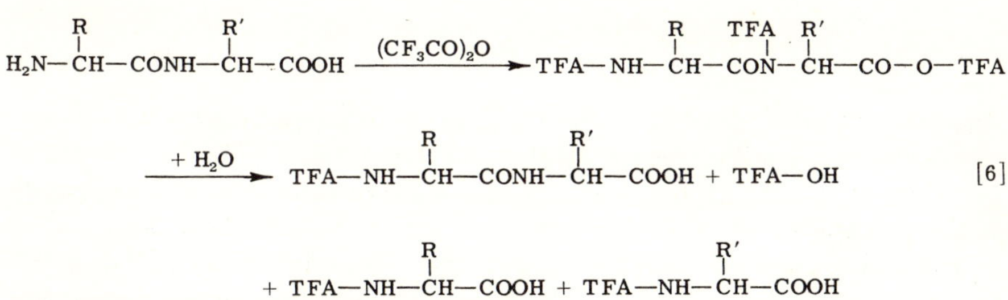

[6]

secondary amino acids, e.g., glycylproline.

Peptides with trifluoroacetylamino acids have been synthesized via the acid chloride method (*2490, 2493*), the azide method (*2496*), the phosphazo method (*2496*), the phosphorus oxychloride method (*2497*), and the cyanomethyl ester as activated ester (*2503*). Peptides are also formed via the intramolecular anhydrides of trifluoroacetylaspartic acid and -glutamic acid (*2478, 2490, 2496*). Due to the danger of azlactone formation, only the L-proline compound of the symmetrical trifluoroacetylamino acid anhydrides can be used for the preparation of optically pure peptides (*2490*).

The cleavage of the trifluoroacetyl group can be carried out in a weakly basic medium with 0.2 *N* sodium hydroxide in 10 minutes or with dilute ammonia solution (*2479*). Barium hydroxide (0.2 *N*) in water or in aqueous methanol has worked well. The barium ions can be readily precipitated with sulfuric acid (*2503, 2633a*). Trifluoroacetic acid can be removed with Amberlite IR-4B. So far as is known, racemization does not occur during this cleavage (*1923*).

It is a disadvantage of the trifluoroacetyl group that its selective cleavage in the presence of ethyl or methyl esters is restricted to rather special cases (*2496*). The conversion to the hydrazide is possible only for trifluoroacetyl peptide esters that are sterically hindered at the amino end (*2504*; cf. page 81). The difficulties during hydrazide formation can be prevented by using tritylhydrazine (*2500*). Trifluoroacetylamino acid azides are directly accessible from the acid chloride and the sodium or dicyclohexylammonium azide (*2496*). The free peptides can be obtained by simultaneous alkaline hydrolysis of the trifluoroacetyl group and, for example, the ethyl ester (*2481*).

The trifluoroacetyl group can also be cleaved by strong acids, e.g. boiling methanolic HCl (*2496*). These extreme conditions are contrasted by the ready acidic cleavage of trifluoroacetylserine and -threonine compounds that occurs under conditions as mild as those for the cleavage of the trityl group (see Chapter I,C,1). Weygand and Rinno (*2497*) assume that this is caused by an N → O acyl shift [7]:

$$\text{TFA—Ser—OH} \xrightarrow{H^{\oplus}} \text{H—}\overset{\text{TFA}}{\overset{|}{\text{Ser}}}\text{—OH} \xrightarrow{H_2O} \text{H—Ser—OH + TFA—OH} \qquad [7]$$

The ester bond formed is hydrolyzed in aqueous solution.

The trifluoroacetyl group has so far been used only by Weygand and Geiger (*2481*) in the case of glutathione to synthesize a biologically active peptide.

Trifluoroacetylamino acid esters, trifluoroacetyl dipeptides, and trifluoroacetyl peptide esters show high vapor pressures and can be readily sublimed in a high vacuum (*2484, 2491*). Thus amino acids of protein hydrolyzates can be isolated on a preparative scale by fractional distillation after their conversion to trifluoroacetylamino acid methyl esters (*2482*). Trifluoroacetylamino acid methyl esters and trifluoroacetyl dipeptide methyl esters (*1087a, 2406, 2474, 2475, 2492, 2677*) can be separated by gas chromatography. This offers new applications for the quantitative determination of amino acids and for sequential analysis. The gas chromatographic separation of diastereomeric trifluoroacetyl dipeptide methyl esters makes it possible to determine exactly the extent of racemization during peptide synthesis (*2474, 2495, 2502*).

The rate of hydrolysis of a number of trifluoroacetylamino acids catalyzed by acylase I has been investigated by Fones and Lee (*749, 750*).

For a summary of the possible applications of the trifluoroacetyl group in peptide chemistry, cf. Weygand (*2476*).

3. The Phthalyl Group

The use of phthalylamino acids in peptide chemistry was first described by Kidd and King (*1227, 1241*) and shortly thereafter by Sheehan and Frank (*2062*). Phthalylamino acids are actually α-(phthalimido)-carboxylic acids [8]. They are also called phthaloylamino acids, although in the earlier literature this name generally refers to *o*-carboxybenzoylamino acids [9] (*933*). Phthalylamino acids have been known

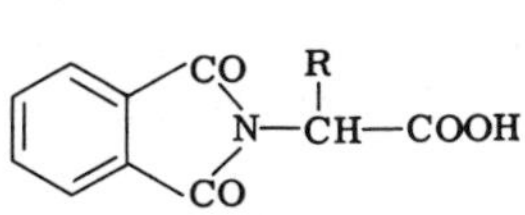

[8]

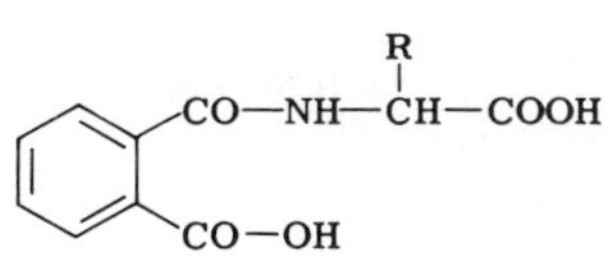

[9]

since the end of the last century (*617*), as well as their reaction with hydrazine to form phthalyl hydrazide [10] (2,3-dihydrophthalazinedione-2,4) (*1782*):

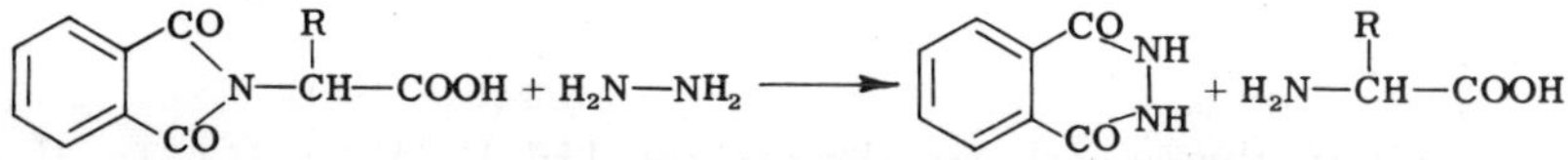

[10]

Ing and Manske (*1088*) used this reaction in 1926 for the synthesis of amines. Following the hydrazinolysis, they isolated the amines as hydrochlorides by the addition of hydrochloric acid. The addition of acid is required since phthalyl hydrazide—similarly to phthalimide—displays acidic properties. Therefore, on hydrazinolysis, the ammonium salt of the acid is formed first (*107*). A stronger acid precipitates the phthalyl hydrazide as an insoluble compound. It has been demonstrated that acetic acid is sufficient for this purpose (*848*). The most important method for the introduction of this protecting group remains the reaction of phthalic anhydride with amino acids or amino acid esters. With amino acids, the phthalylamino acids [11] are obtained directly:

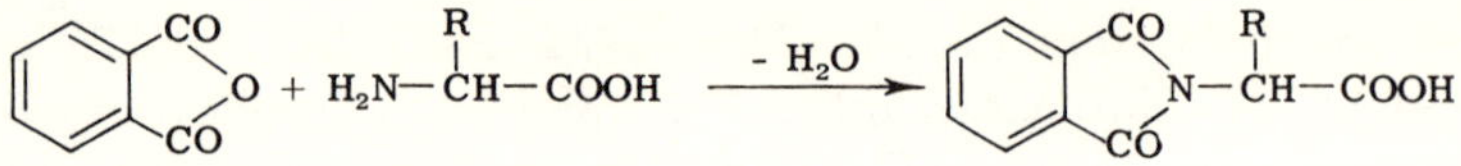
[11]

The reaction proceeds with elimination of water and can be carried out by simply fusing the components (*243, 730, 1227, 2062*). The originally suggested reaction temperature of approximately 180°C causes racemization in almost all cases (*1227, 2060*). This can be prevented as long as the temperature is kept below 150°C (*730, 2060*). High boiling solvents, such as dioxane (*2064*), *p*-cymene (*1601*), glacial acetic acid (*2423*), xylene (*2314*), and pyridine (*1241, 2262*), have been suggested to prevent local overheating. Certain amino acids, however, especially the bifunctional ones, racemize in these solvents also (*1241*). It has not been possible to convert tryptophan, tyrosine, serine, and taurine to phthalyl derivatives by these methods (*242, 2064*).

The methods leading to the formation of *o*-carboxybenzoyl derivatives as intermediates proceed under milder conditions. Amino acid esters form *o*-carboxybenzoylamino acid esters [12a] with phthalic anhydride, and *o*-carbethoxybenzoylamino acid esters [12b] with *o*-carbethoxythiobenzoic acid (*100, 101*):

[12a]

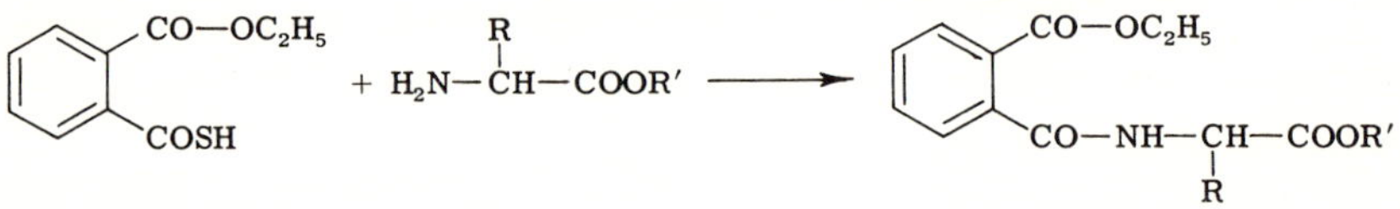
[12b]

Ring closure of the *o*-carboxy derivatives [12a] (R″ = H) to the phthalylamino acid esters can be achieved by treatment with thionyl chloride (*978*) or methanolic HCl (*1241*). Phthalylamino acid esters [13] can also be obtained from the carbethoxy compounds [12b] (R″ = C_2H_5) by treatment with alcoholic HCl. The phthalylamino acids [14] are obtained by ring closure and cleavage of the ester bond upon exposure to concentrated hydrochloric acid in glacial acetic acid (*2314*). The reaction of phthalic anhydride with amino acids and amino acid esters leads directly to the formation of phthalyl compounds when the reaction mixture is extensively heated in benzene, toluene, or aqueous dioxane in the presence of triethylamine. The elimination reaction can be accelerated by azeotropic removal of water (*308, 1015*). This treatment, however,

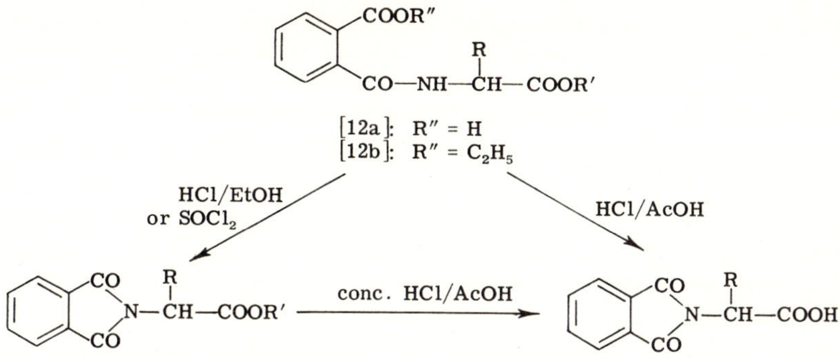

[13] [14]

causes partial formation of diketopiperazines. Alkaline hydrolysis of the phthalylamino acid esters is not possible. The exposure to alkali causes the opening of the phthalyl group at one of its imide bonds. The resulting *o*-carboxybenzene derivatives cannot be cleaved with hydrazine (*1354*).

Nefkens *et al.* (*1597, 1600*) have found that carbethoxyphthalimide [15] reacts with amino acids in aqueous alkaline solution at room temperature in a relatively short time to form phthalylamino acids. Sodium carbonate gives higher yields than sodium hydroxide. The reaction mechanism discussed assumes the opening of the five-membered ring [16], as under the conditions of alkaline hydrolysis. The phthalylamino

[15] [16]

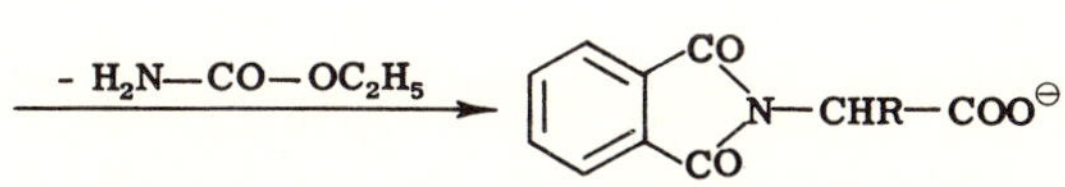

acid is formed with elimination of urethan. Phthalyl-DL-serine, N^{ϵ}-phthalyl-L-lysine, and *N,N′*-diphthalyl-L-cystine have been obtained for the first time with carbethoxyphthalimide. Attempts to prepare phthalyltryptophan were not successful. The mild reaction conditions and high yields of this method of phthalylation without racemization should be superior to the earlier procedures. Phthalylamino acids are also accessible from phthalic acid diphenyl ester and amino acids in phenol as solvent and at temperatures of 120°–130°C (*2489*).

The acid chloride method (*848, 1233, 2062, 2339*), the anhydride method (*214, 284, 1239, 1354, 1986*), *N,N′*-dicyclohexylcarbodiimide

(*2064*), *p*-nitrophenyl esters (*230, 265*), and intramolecular anhydrides of glutamic and aspartic acids (*1241, 1242*) have been employed for the synthesis of peptides with phthalylamino acids. The azide method was used with substituted hydrazides (*1024, 1027;* Chapter II,B,1).

If an amino acid ester is used for the synthesis of a peptide, the ester must be hydrolyzed under acidic conditions since the phthalyl group is not stable to alkali (*933, 2060*). The hydrolysis is normally conducted in acetone, acetone/water, or glacial acetic acid with 2 *N* HCl, 2 *N* H_2SO_4, or concentrated HCl by heating 0.5–2 hours (*1627, 2060, 2062, 2064*). No details have been reported for peptides of greater chain length. Cleavage of the phthalyl group may occur under more vigorous conditions. Fifteen percent of a free peptide was obtained (*2055*) in acetone/water with concentrated HCl by boiling for 3 hours. After alkaline hydrolysis, the *o*-carboxybenzoylpeptides may be converted to the phthalyl peptides by ring closure with thionyl chloride, via the mixed anhydride method, with phosphorus pentoxide or with *N,N'*-dicyclohexylcarbodiimide (*978, 1241*). Schumann and Boissonnas (*1986*) describe an alkaline hydrolysis in dioxane/sodium hydroxide without opening of the phthalimide ring. Hanson and Illhardt (*933*) have compared the alkaline and acidic hydrolyses for a number of cases. Under alkaline conditions only about 15–30% of the desired compounds is obtained, under acidic conditions 60–75%. The combination of phthalyl group and *tert*-butyl group is the most suitable one. The *tert*-butyl ester can be readily saponified under acidic conditions (*44, 48*).

The instability of the phthalyl group to alkali makes it also difficult to couple salts in aqueous alkaline solution (*2062*). An advantage may be offered by a reaction described by Yamashita and Yashiro (*2602*). The authors dissolved phthalylamino acid chloride in dioxane and added the aqueous solution of the amino acid in the presence of sodium or potassium acetate. The alkaline hydrolysis of the phthalimido group of phthalylamino acids and phthalyl peptides and their esters has been studied in the polarograph by Rudinger *et al.* (*1857*), who discuss a reaction mechanism. The phthalyl group is unstable under the conditions of enzymatic cleavage at pH values in the alkaline range (cf. *111*). Liberek (*1421b*) observed a base-catalyzed racemization, and suggests a reaction mechanism on the basis of carbanion formation at the asymmetric carbon atom.

The phthalyl group can be removed only by hydrazinolysis. The reaction is usually carried out with hydrazine hydrate in methanol or ethanol by refluxing for 1–2 hours (*848, 2060, 2062*) or standing at room temperature for an extended time (*230, 1241, 2055*). The use of *tert*-butyl alcohol (*2280*), dioxane (*230*), or aqueous sodium bicarbonate solution (*1241*) has also been described. The phthalazine is precipitated by acidifying with hydrochloric (*2055, 2060, 2062*) or acetic acid (*848*). It can also be removed as sodium salt with sodium carbonate solution (*2280*). The cleavage is normally carried out subsequent to saponification steps to prevent the simultaneous hydrazinolysis of ester bonds. Several cases of hydrazinolysis of phthalyl groups without attack of ester bonds have been reported (*894, 1236, 2055, 2072*). Phenyl esters as protection for carboxyl groups are cleaved preferentially with hydrazine hydrate, and phthalylamino acid hydrazides are obtained (*2373*). *tert*-Butyl esters are very stable to hydrazinolysis and only the phthalyl group will be cleaved (*48, 230, 2280*). Leach and Lindley (*1354*) have pointed out that the cleavage of the phthalyl group in peptides with C-terminal asparagine causes difficulties. No attack of peptide bonds during hydrazinolysis of the phthalyl group has been observed, but occurs under more vigorous reaction conditions (*25, 1622*). In addition to hydrazine hydrate, an excess of phenylhydrazine in the presence of tri-*n*-butylamine in ethanol has been suggested for the cleavage (*1984, 1985, 1986*). The method is supposed to offer the advantage of easier isolation of the peptides. Hydrazine acetate may be used in neutral solution under very mild conditions (*2005*). Removal of the phthalyl group is also possible in aqueous solution with copper acetate in the presence of basic copper carbonate (*848*). Experimental details have not been reported.

The phthalyl group is stable in liquid ammonia. In the presence of potassium, however, side reactions have been observed. They are apparently the result of an elimination of potassium phthalimide [17] (*76*):

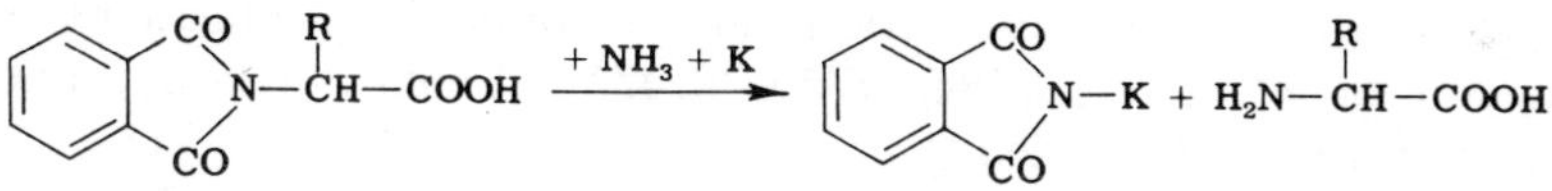

[17]

The phthalyl group was of no importance for the synthesis of biologically active polypeptides until about 3 years ago. It was recently used successfully by Beyermann and Bontekoe (*230*) and by Schröder (*1968*) for the synthesis of glucagon sequences. It also proved to be useful as ε-amino group protection for lysine in an α-MSH synthesis (*2005*).

4. The Toluenesulfonyl (Tosyl) Group

Fischer (*718*) described as early as 1915 the reductive cleavage of toluenesulfonic acid amides with hydroiodic acid and phosphonium iodide in glacial acetic acid at 60°C [18]:

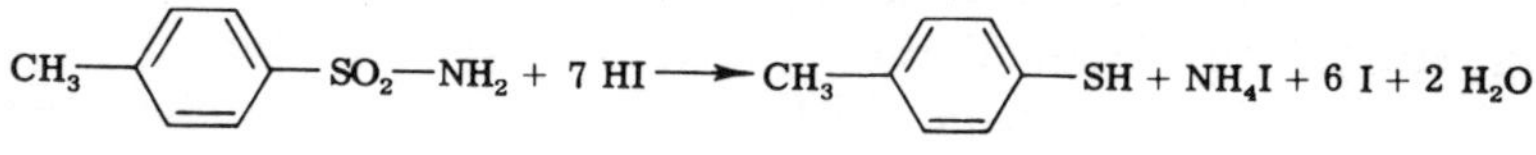

[18]

Tosylamino acids are used for the synthesis of *N*-methylamino acids (*720*; Chapter IV,A,3) since the amide hydrogen can be substituted very readily. Schönheimer (*1956*) demonstrated that peptide bonds will not be attacked during detosylation with hydroiodic acid. Thus the synthesis of free peptides is possible.

Tosylamino acids are prepared in alkaline solution from toluenesulfonyl chloride and the corresponding amino acid. Suitable reaction media are: ether/aqueous NaOH (*111, 720, 1478*), water/tetrahydrofuran in the presence of triethylamine (*2299*) or water alone. The amount of alkali required is added either at the beginning of the reaction (*2583*) or during the reaction, maintaining a constant pH value of about 9 (*1184*). The direct tosylation of tryosine leads to the formation of the *N,O*-bistosyl derivative (*718*). The monotosylamino acid can be obtained by reacting toluenesulfonyl chloride with tyrosine ester and saponification of the ester (*720*).

The tosyl group is almost exclusively removed by reduction with sodium in liquid ammonia. This method was first suggested by du Vigneaud and Behrens (*627*). The mechanism of the reaction has been discussed by Kovács and Ghatak (*1293a*). Under the same conditions the carbobenzoxy group and modified carbobenzoxy groups, the benzyl ester, and the S-benzyl group are removed. The *tert*-butyloxycarbonyl group, the formyl group, and the phthalyl group are stable (*301*). The behavior of methyl and ethyl esters is not quite clear (*2000*). The simultaneous cleavage of other protecting groups is of advantage only when a free peptide is to be synthesized. Examples of the synthesis of peptides using the N^{α}-tosyl protection can be found in publications of du Vigneaud *et al.* (*642, 1184, 1750, 2247*), Woolley (*2583*), and Rudinger *et al.* (*1057, 1854, 2652*).

The absolute stability of the tosyl group in the commonly used methods for the cleavage of other protecting groups, e.g. catalytic hydrogenation, HBr in acetic acid, alcoholic HCl, or alkaline saponification, is of great importance. It has been used frequently for the protection of the ω-amino group in the synthesis of biologically active polypeptides.

If the tosyl group is to be used for the synthesis of intermediates, other blocking groups that are also removed by sodium in liquid ammonia will have to be reintroduced afterwards. Examples are *N*-tosyl-S-benzyl derivatives. The mercapto group is again substituted by reacting it with benzyl chloride in liquid ammonia as solvent (*642, 1159*; cf. page 228).

Another disadvantage of the reduction with sodium in liquid ammonia is the difficulty of isolating the peptide in pure form. A large amount of inorganic salts must be removed; these are in part formed during the reaction, and in part are the result of the required destruction of the excess of metallic sodium by the addition of ammonium salts, such as ammonium chloride (*642*), ammonium iodide (*2247*), or ammonium acetate (*1847*), or of acetic acid (*2247*). It is therefore advantageous to remove the sodium by the direct addition of Dowex 50 (H^+ form) (*1516*; cf. *1202*). It is not possible to follow a general procedure for the reduction itself or for the subsequent desalting and purification steps. Different authors suggest different volumes of ammonia and amounts of sodium per mole of peptide. The reaction time and the duration of the blue color vary also from a few minutes to half an hour.

The choice of the isolation procedure is governed by the properties of the peptide. It is a very difficult one for peptides readily soluble in water (*2638*). Certain peptides can be precipitated from their aqueous solution with alcohol, since ammonium and sodium acetate are soluble in alcohol (*2247*). The isolation as insoluble salts, e.g. flavianate (*2638*), picrate (*627*), or barium salt (*2583*), is possible. More elaborate procedures must frequently be followed (*2583*). Incomplete reduction of the sulfonyl group causes the formation of sulfite; it is partially converted to sulfate by air oxidation (*1847*). Presently, the purification on ion exchangers should give the best results; it has been used exclusively in more recent investigations (*686, 1414, 1524, 1943*). Rudinger (*1847*) removed the metallic sodium with ammonium acetate. The sodium ions are replaced on a cation exchanger by ammonium ions. Sulfite and sulfate ions are precipitated by the addition of barium acetate. Again, a cation-exchange resin is employed to replace the barium ions by ammonium ions. The remaining salt is ammonium acetate; it is readily soluble in alcohol and does not interfere with the precipitation of the peptide; it can also be removed by lyophilization. Lithium, barium, and calcium have been tried for the reduction (*73, 1847*); especially calcium has the advantage that it can be removed as oxalate. The yields, however, are better when sodium is used.

Bajuz and Medzihradszky (*1519*) have reported on the stability of the peptide bond under the conditions of the reaction with sodium in liquid ammonia. Guttmann (*892*) has studied the influence of moisture on the cleavage of the peptide bond.

The tosyl group can also be removed with HBr in acetic acid (*2438, 2654*). Under the reaction conditions required—16 hours at 26°C—peptide bonds are no longer stable (*301*), and the method can be used only for tosylamino acids. More and more investigators are tending to use the *tert*-butyloxycarbonyl group and the formyl group in addition to the tosyl group as ω-protection.

Peptides with N^{α}-tosylamino acids have been synthesized by the acid chloride (*1367, 1750, 1956, 2247*), azide (*1956, 2583*), and tetraethylpyrophosphite (*111, 642*) methods. Syntheses via mixed anhydrides do not give satisfactory results (*111, 214a, 1005, 2298*) because of the high reactivity of the amide nitrogen (*2646, 2655;* cf. page 87). The ready decomposition of tosylamino acid chlorides in alkaline solution with formation of toluenesulfonamide, as observed by Beecham (*128, 129, 130, 131a*), is also the result of the active amino hydrogen [19]:

$$\text{Tos}-\overset{\text{H}}{\overset{|}{\text{N}}}-\text{CHR}-\text{COCl} + \text{NaOH} \longrightarrow [\text{Tos}-\bar{\text{N}}^{\ominus}-\text{CHR}-\text{COCl}]\,\text{Na}^{\oplus} + \text{H}_2\text{O}$$

$$\longrightarrow \text{Tos}-\text{NH}_2 + \text{O}{=}\text{CHR} + \text{NaCl} + \text{CO} \qquad [19]$$

Hillmann and Hillmann (*1005*) have described the great lability toward alcoholic hydrochloric acid of the peptide bond between glycine and the remainder of the peptide in tosylglycyl peptides. Several peptide ethyl ester hydrochlorides have been prepared with tosylglycine as "protecting group." The cleavage is more difficult for sterically hindered amino acids. Cleavage of other peptide bonds or racemization has not been observed after heating in alcoholic hydrochloric acid for 20 hours. Though this "protecting group" has not found practical application, its ready hydrolysis must be borne in mind since it can give rise to side reactions in syntheses with tosylglycine.

5. Aryl- and Alkylphosphoryl Groups

The investigations of *N*-phosphorylamino acids, conducted in connection with the exploration of phosphorylated compounds in biological syntheses (see Chapter VII,A), stimulated Cosmatos *et al.* (*529*) to examine the use of diarylphosphoryl- and dialkylphosphorylamino acids for the synthesis of peptides.

Dibenzylphosphorylamino acids (the authors suggest DBP as abbreviation) are obtained by acylating amino acid esters with dibenzylphosphoryl chloride in chloroform in the presence of triethylamine, followed by saponification in dioxane/1 *N* methanolic potassium hydroxide solution (cf. [20], page 17). The experiments were conducted with the *p*-nitro, *p*-bromo, and *p*-iodo derivatives since these compounds crystallize more easily.

Peptides were synthesized with *N,N'*-dicyclohexylcarbodiimide and via the mixed anhydride with diphenylphosphoric acid (see Chapter III,B,1,b,δ). The anhydride with ethyl carbonate gave rise to a cou-

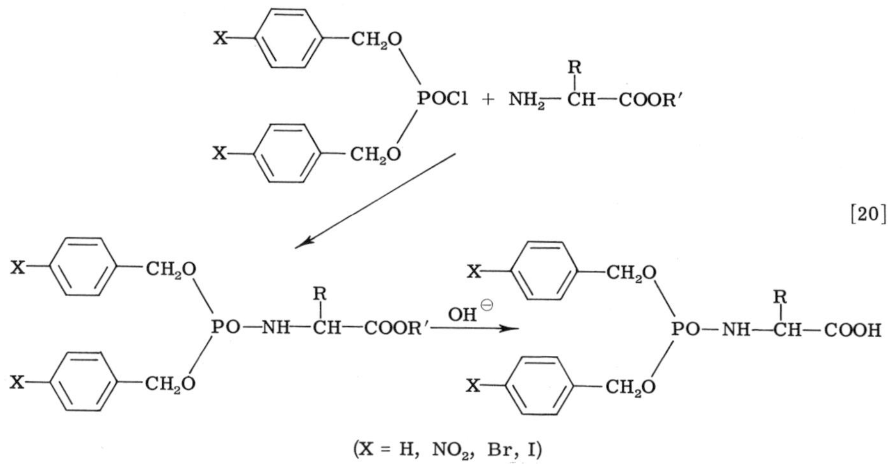

[20]

pling reaction of low yield only for glycine. Similar observations were made with the hydrazide; only the glycine esters reacted. The behavior of these *N*-protected amino acids is very similar to that of the tritylamino acids (see Chapter I,C,1). *N*-Diarylphosphoryl peptide esters can be very readily converted to hydrazides. The azide method can be employed in 60% acetic acid at 0°C despite the acid lability of the P—N bond. The hydrolysis of the ester bond has been described for amino acid derivatives only. The *N*-diarylphosphoryl protecting groups can be removed by catalytic hydrogenation in 80% ethanol in the presence of palladium black in the usual manner. The addition of a small amount of hydrochloric acid will dephosphorylate the phosphoryl compound formed. The free peptide will be obtained in one step [21] when the carboxyl group is protected by the benzyl residue:

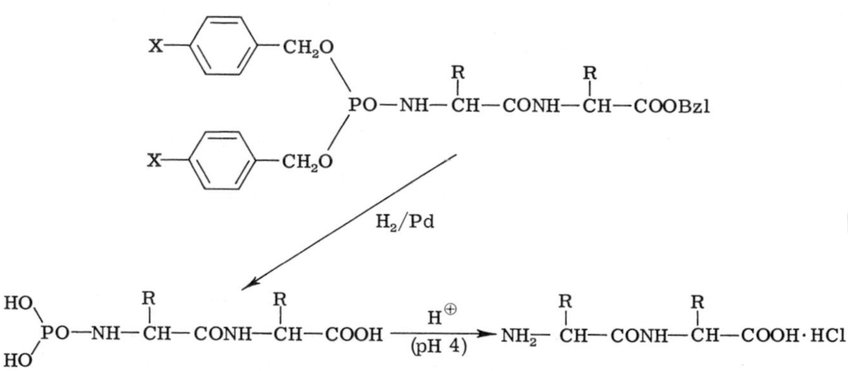

[21]

The elimination of the protecting group in chloroform with hydrogen bromide has also been described.

Efforts to use the diphenylphosphoryl group in the same way have failed (*529, 2040*). One phenyl group is removed during alkaline hydrolysis and also during catalytic hydrogenation. It is assumed that this hydrogenation proceeds by way of an intramolecular anhydride [22].

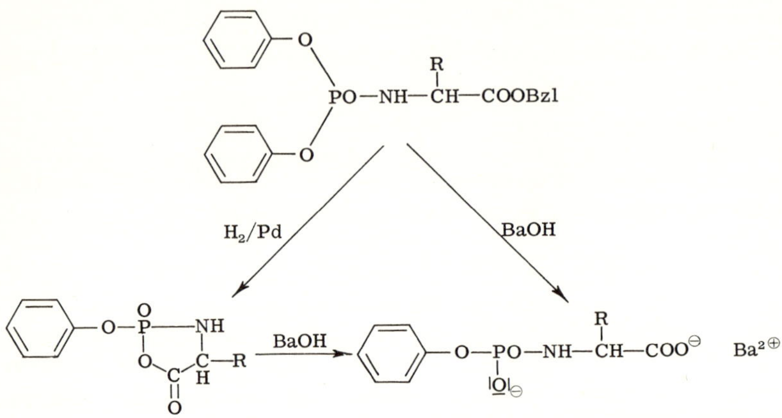

[22]

Zervas *et al.* (*2658, 2665*) has suggested the use of benzylphosphoryl groups for the synthesis of unsymmetrical cystine peptides. The principle observed is the simultaneous blocking of two amino acid or peptide residues by one protecting group. One benzyl group of dibenzylphosphorylamino acid esters can be cleaved selectively under mild conditions with sodium or barium iodide (*2660*). The resulting arylphosphorylamino acid ester can be coupled with a second amino acid derivative [23]:

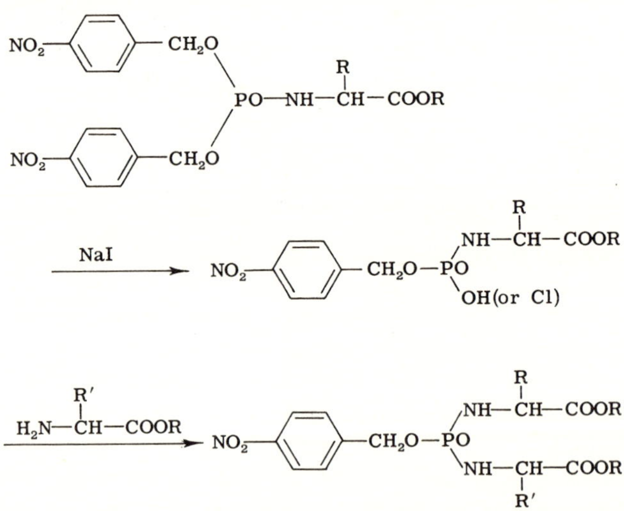

[23]

These types of protecting group are following an entirely new concept. No experience has yet been gained with higher peptides. It is therefore not possible to evaluate their importance for peptide chemistry.

6. Other Acyl-Type Protecting Groups

Other groups in addition to the acyl groups, which have been discussed in detail, have been used to protect the amino group, but are no longer of great importance.

a. Acetyl and Benzoyl Groups

The outstanding investigations of Curtius and his school on the use of azides for the formation of the peptide bond were done almost exclusively with benzoylamino acids and benzoyl peptides. The cleavage of this protecting group is possible only in special cases (*185, 498*).

Bergmann *et al.* (*187, 190, 191, 197*) were more successful in their use of the acetyl residue as amino-protecting group. Using symmetrical anhydrides or the azlactone method, they were able to synthesize peptides with such amino acids as phenylalanine or tyrosine. Due to side reactions this had not been possible by the classic methods of E. Fischer (see Chapter III,A). The acetyl group was removed with 1 *N* HCl in about 1 hour at 130°C; yields were approximately 50%.

In addition to the unfavorable cleavage, this procedure has the disadvantage that it is difficult to prepare acetylamino acids with acetic anhydride or acetyl chloride without racemization (*638a, 639*). Acetylation with ketene (*454*) is also accompanied by racemization. Acetylamino acid esters can be obtained with acetic anhydride without influence on the steric configuration (*192, 468*). Peptide esters or amides can be acetylated under very mild conditions with acetic acid *p*-nitrophenyl ester (*2005*) or thiolacetic acid (*894*). The acetyl residue has been used as N-protecting group for syntheses in the MSH and ACTH fields.

b. Benzylsulfonyl Group

The benzylsulfonyl group has been suggested by Milne and Peng (*1563, 1564*) as amino-protecting group since benzyl derivatives can be reduced relatively easily to toluene. Benzylsulfonylamino acids were prepared in aqueous alkaline solution from benzylsulfonyl chloride and the corresponding amino acids. Peptide syntheses with these derivatives, however, have been described in only a few instances using the acid chloride method.

The benzylsulfonyl group, like the toluenesulfonyl group, can be removed with sodium in liquid ammonia. It can also be reduced with

Raney nickel. The reaction can be carried out at normal pressure and room temperature. The nickel salts formed are removed from the aqueous solution with hydrogen sulfide. Removal of this group by catalytic hydrogenation in the presence of platinum oxide has not been possible.

c. Nitrophenylsulfenyl and Tritylsulfenyl Groups

Goerdeler and Holst (*812*) have prepared *o*-nitrophenylsulfenyl-amino acids by reacting the acid chloride with amino acids under neutral conditions [24]:

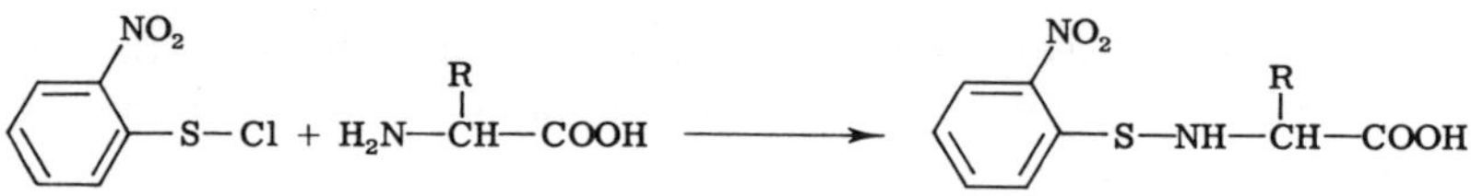

[24]

Amino acid esters can also be acylated in the presence of triethylamine. The disulfide [25] and the sulfonic acid thiolester [26] are formed as the main products in strongly alkaline solution:

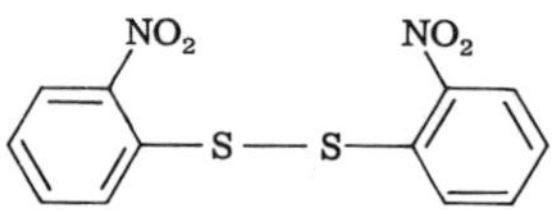

[25]

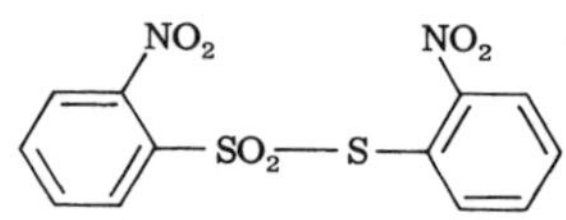

[26]

The *N*-protected amino acids can be easily purified via the dicyclohexylammonium salts. The salts are more stable than the free acids (*2659a*). Peptides were coupled via the anhydride and carbodiimide methods.

Zervas *et al.* (*2659a*; cf. *786a*) have described the tritylsulfenyl group. The reaction of amino acids with tritylsulfenyl chloride in aqueous alkali does not proceed in satisfactory yield. However, esters of amino acids (among them the *p*-nitrophenyl ester) can be acylated readily. The *N*-protected free acids are accessible via hydrolysis.

Both protecting groups can be removed conveniently by treatment with 2 equivalents of hydrogen chloride in ethanol or ether in several minutes. The *o*-nitrophenylsulfenyl group is selectively removable when the *tert*-butyl ester is present.

d. *o*-Nitrophenoxyacetyl and Chloroacetyl Groups

While the lactams of γ- and δ-amino acids can be obtained only under vigorous conditions, the corresponding lactam of *o*-aminophenoxyacetic acid is formed spontaneously. On the basis of this observation, Holley and Holley (*1050*) have used the *o*-nitrophenoxyacetyl and chloroacetyl residues as amino-protecting group.

To remove the *o*-nitrophenoxyacetyl group, *o*-nitrophenoxyacetyl-amino acids or peptides are reduced to the amino compounds. By heating them in water they decompose to *o*-aminophenoxyacetic acid lactam and amino acid or peptide [27]:

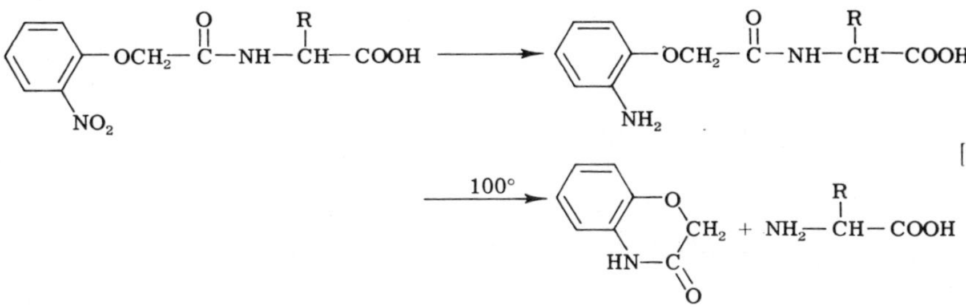

[27]

Following a similar reaction, the chloroacetyl group can be cleaved with *o*-phenylenediamine [28]:

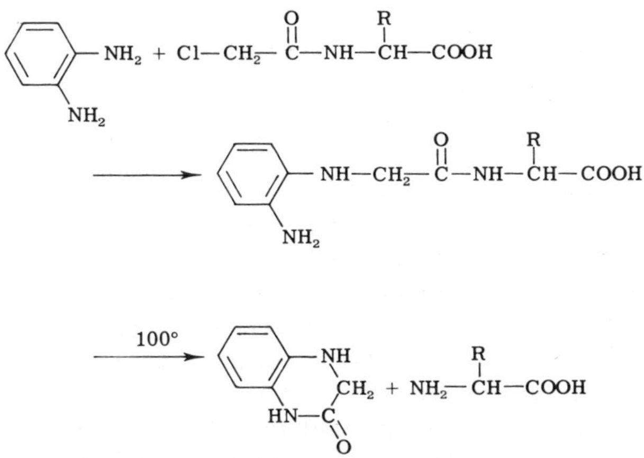

[28]

These protecting groups are introduced into amino acids or amino acid esters via the acid chloride by conventional procedures. A number of peptides have been prepared by the azide method and via mixed anhydrides. The protecting groups were removed subsequent to the hydrolysis of the ester bonds. Racemization was not observed when optically active amino acids were used. This cleavage reaction can also be applied as a method for the stepwise degradation of peptides (*1051*).

e. γ-Chlorobutyryl Group

Peter (*1714*) has described the use of the γ-chlorobutyryl group as amino-blocking group. For the removal of the protecting group the cyclic imino ether [29] is first formed with silver fluoborate. It is cleaved on coupling with acid chlorides.

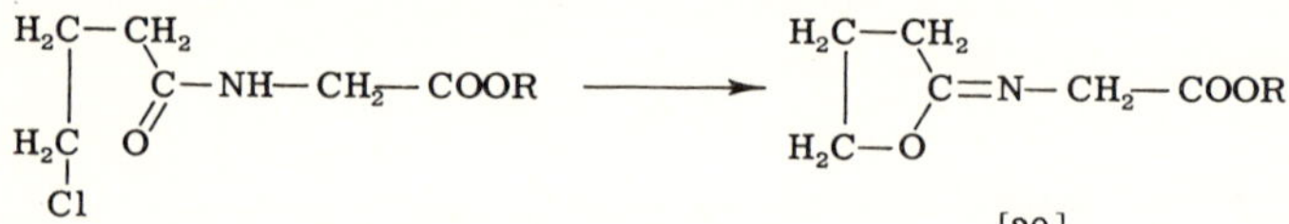

[29]

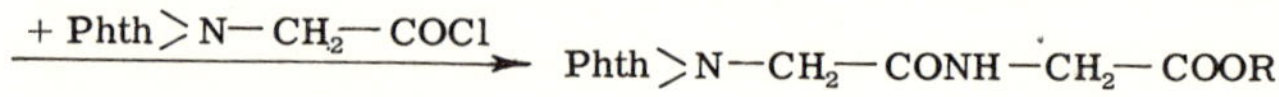

B. URETHAN PROTECTING GROUPS

Bergmann and Zervas (*193*) in 1932 suggested the use of the benzyloxycarbonyl group—frequently referred to as the carbobenzoxy group. The introduction of this amino-protecting group—it is still the most important one—marked the beginning of modern peptide chemistry. The outstanding properties of the carbobenzoxy group provided the stimulus for a systematic search for other urethan-type blocking groups. Certain advantages are offered by the *p*-chlorocarbobenzoxy group, the *p*-nitrocarbobenzoxy group, and the colored carbobenzoxy-like protecting groups. Attempts to use the allyloxycarbonyl, tolyloxycarbonyl, and phenyloxycarbonyl groups were unsuccessful. Efforts have been made to find a urethan-type blocking group that would make possible, with regard to removal, a differentiation between it and the carbobenzoxy group. It was obvious to investigate aliphatic groups; they are more stable to catalytic hydrogenation but can be cleaved more readily than the carbobenzoxy group under acidic conditions. Ethyl-, *sec*-butyl-, and diisopropylmethyloxycarbonyl were not suitable. Some selectivity was found for cyclopentyloxycarbonyl and to a higher degree for the cyclohexyloxycarbonyl group. The breakthrough came with the *tert*-butyloxycarbonyl group. It has become of almost the same importance as the carbobenzoxy group.

1. Carbobenzoxy Group

The carbobenzoxy group is of so great importance for the following reasons: ready accessibility of the *N*-protected amino acids, practically universal applicability in peptide synthesis, and, last but not least, numerous convenient ways of removal.

Carbobenzoxyamino acids [30] are obtained almost exclusively from carbobenzoxy chloride and the corresponding amino acid in aqueous alkaline solution (*193*):

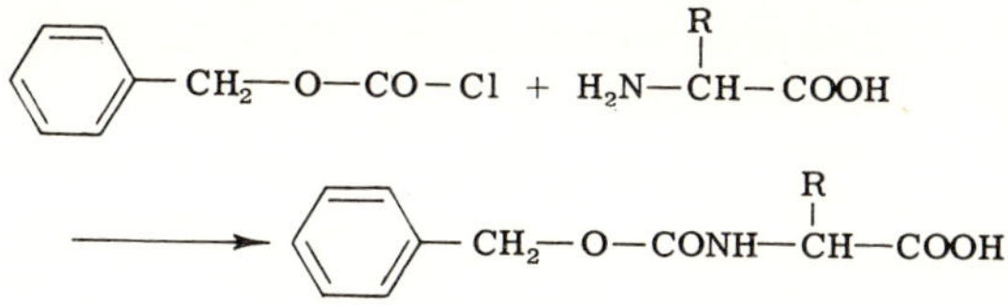

[30]

Sometimes the maintenance of an exact pH value during synthesis is required. This is true for S-benzylcysteine (*815*) or arginine (*2670*). Sodium bicarbonate is used as base. There are only a few cases where the reaction does not proceed satisfactorily; the hydroxyamino acids must be mentioned here. Yields of over 90% have, however, recently been reported for serine and threonine also (*85, 894, 1154*). For tyrosine the routine procedure leads to the formation of *N,O*-dicarbobenzoxy-tyrosine (*3, 1176*). The *N*-monocarbobenzoxy derivative can be obtained from the bis derivative by alkaline hydrolysis in 4 *N* sodium hydroxide (*593*), or by carbobenzoxylation of the amino acid ester and subsequent hydrolysis (*193*).

Acidification of aqueous alkaline solutions of carbobenzoxyamino acids or peptides sometimes gives rise to higher melting compounds. They are the result of a complex formation of the carbobenzoxyamino acid with its alkali salt (*415, 836, 850, 864*). The complex formation can be prevented by pouring the alkaline solution into an excess of acid at 0°C (*864*).

Tautomeric forms [31] of aromatic amino acids have been observed:

$$C_6H_5{-}CH_2{-}O{-}\overset{\displaystyle O}{\overset{\|}{C}}{-}NH{-}\overset{\displaystyle R}{\overset{|}{C}H}{-}COOH \rightleftharpoons C_6H_5{-}CH_2{-}O{-}\overset{\displaystyle HO}{\overset{|}{C}}{=}N{-}\overset{\displaystyle R}{\overset{|}{C}H}{-}COOH \qquad [31]$$

The carbobenzoxy group is used not only to block the α-amino function but also to protect simultaneously the α- and ϵ-amino groups of lysine (*201, 684, 1932*) and the α- and δ-amino groups of ornithine (*1178, 2251*). The preparation of these compounds follows in principle the same procedure as that for the simple amino acids (*111, 201, 204, 1178*). The N^{α},N^{G}-dicarbobenzoxy and N^{α},N^{G},N^{G}-tricarbobenzoxy derivatives of arginine are known (*2663, 2670*). The carbobenzoxylation of the guanido group requires strongly alkaline conditions (*2669, 2670*; see Chapter IV,B,5,a,γ).

The selective protection of the ω-amino function by the carbobenzoxy group has also been described. These compounds are obtained in the case of lysine (*1604, 1932*) and of ornithine (*958, 2251*) by carbobenzoxylation of the copper complexes (cf. page 152). Another synthesis proceeds via the *N*-carboxylic anhydrides, which are readily accessible from α-carbobenzoxyamino acid chlorides (see Chapter III,B,5,c). If the reaction is carried out with dicarbobenzoxylated diaminocarboxylic acids, cleavage of the *N*-carboxylic anhydrides [32] with water yields the ω-carbobenzoxyamino acids [33]. The corresponding esters [34] are obtained by treatment with alcohols (*204, 684, 1178, 2251*):

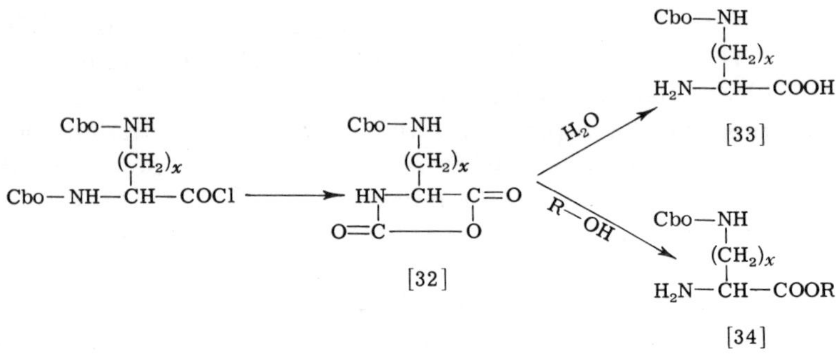

The guanido-protected derivative of arginine is also prepared from the N^{α},N^{G}-dicarbobenzoxy compound via the *N*-carboxylic anhydride (*2670*). This procedure had earlier served for the preparation of simple amino acid esters, especially benzyl esters (cf. page 59). The α-carbobenzoxy derivatives of diaminocarboxylic acids can be prepared via the ω-benzylidene compounds (*236*). The direct selective α-carbobenzoxylation is possible only for arginine (*2670*). The N^{α},N^{im}-dicarbobenzoxy derivative of histidine is known. N^{α}-Monocarbobenzoxyhistidine will be obtained from the N^{α},N^{im}-bis derivative on treatment with alkali (*1695, 1903*). The N^{im}-protected amino acid can be prepared in the form of its ester from dicarbobenzoxyhistidine ester by treatment with HBr in acetic acid (*1092, 1093;* cf. page 180). Peptides may be carbobenzoxylated according to established procedures in aqueous solution with magnesium oxide as base (*1750*).

The carbobenzoxyamino acid esters are of great importance as starting materials for the azide method. They may be prepared by esterification of carbobenzoxyamino acids or, better, by carbobenzoxylation of amino acid esters. This reaction may be carried out under a variety of conditions: in chloroform in the presence of triethylamine (*1052*), in ethyl acetate or in chloroform with magnesium oxide (*774, 2148*), in the heterogeneous phase of chloroform/water with magnesium oxide (*201*) or with sodium hydroxide and sodium carbonate (*86, 192*). It is frequently difficult to crystallize the carbobenzoxyamino acid esters, and they are used without further purification for the preparation of the hydrazides (*201, 2148*). During the carbobenzoxylation of amino acid esters, contaminated products are readily formed by the excess of carbobenzoxy chloride. Excessive reagent can be decomposed by the addition of pyridine. The resulting benzyl chloride must be removed before the hydrazide is formed (*1957*).

The carbobenzoxyaminodicarboxylic acid monoesters required for the synthesis of α- or ω-peptides can be prepared by selective esterification of carbobenzoxyaminodicarboxylic acids, by selective hydrolysis of carbobenzoxyaminodicarboxylic acid diesters, or from monoesters by the commonly employed procedures of carbobenzoxylation (see Chapter IV,C,1,α,β; 1,α,γ; 2,α,β; and 2,α,γ).

Carbobenzoxyamino acids can be used for the synthesis of peptides with practically no limitations. Most of the systematic investigations of coupling methods have used carbobenzoxyamino acids. Of special advantage, in contrast to the acyl-type blocking groups, is the low tendency to racemization. Due to this fact, it is possible to obtain the intramolecular anhydrides of carbobenzoxyglutamic acid and -aspartic acid in sterically pure form (*193*). Symmetrical or mixed anhydrides can also be used for peptide coupling without danger of azlactone formation (see Chapter X,B).

The only coupling method that may give rise to side reactions with carbobenzoxyamino acids is the acid chloride method, due to the ready formation of *N*-carboxylic anhydrides. Nevertheless many peptide syntheses were performed earlier with carbobenzoxyamino acid chlorides (*774, 938, 1451, 2251*). The *N*-carboxylic anhydride formation can be prevented by working at lower temperatures (cf. page 78).

The carbobenzoxy group is sensitive to alkali. MacLaren (*1489*) has observed that urea derivatives [35] or hydantoins [36] are formed, especially from carbobenzoxy peptide esters containing glycine as the second amino acid from the amino terminal (cf. *823, 1932, 2054a, 2419*):

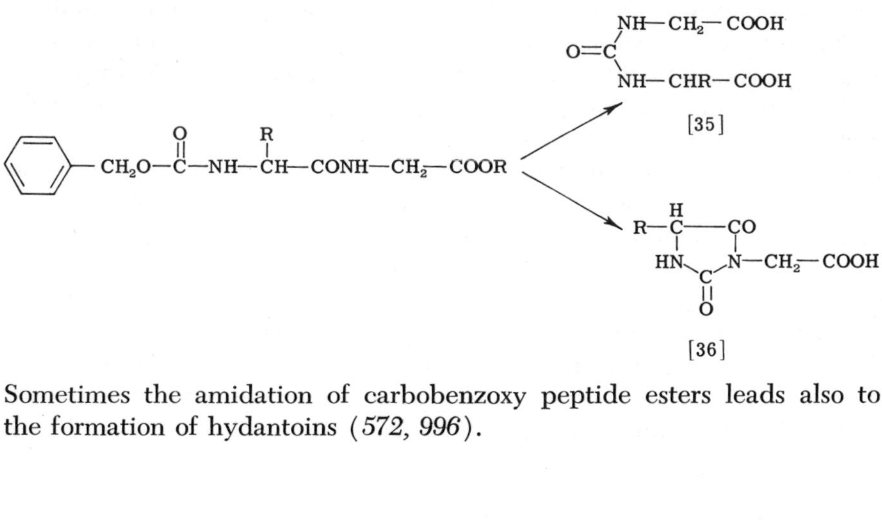

Sometimes the amidation of carbobenzoxy peptide esters leads also to the formation of hydantoins (*572, 996*).

The carbobenzoxy group can be removed under very mild conditions by catalytic hydrogenation. Presumably the reaction proceeds via the carbamic acid [37], forming toluene and carbon dioxide (*193*):

$$C_6H_5{-}CH_2{-}O{-}CO{-}NH{-}\overset{\displaystyle R}{\overset{|}{C}}H{-}COOH \longrightarrow C_6H_5{-}CH_3 + HO{-}CO{-}NH{-}\overset{\displaystyle R}{\overset{|}{C}}H{-}COOH$$

[37]

$$\longrightarrow C_6H_5{-}CH_3 + CO_2 + H_2N{-}\overset{\displaystyle R}{\overset{|}{C}}H{-}COOH$$

The additional formation of a minor amount of dibenzyl has been observed (*1697*). The following catalysts are used: palladium black (*108, 193, 745, 2663*), palladium on charcoal (*147, 2419*), or, according to Kuhn and Haas (*1307*), palladium on barium sulfate (*894, 899*). Platinum oxide (*301*) and ruthenium on charcoal (*1664*) are used less frequently.

The hydrogenation can be carried out in an open vessel. The end point will be recognized by the disappearance of carbon dioxide from the escaping hydrogen (*193, 1083, 2251*). The simultaneous hydrogenation of other groups, such as benzyl esters or nitro groups, cannot be controlled in this way. Erlanger and Brand (*684*) report for the hydrogenation of carbobenzoxy peptide benzyl esters that the reaction is not completed until 2 hours after the carbon dioxide evolution has ceased. If the hydrogenation is carried out in a closed system, care must be taken to remove the carbon dioxide formed (*1827, 2019a*). The hydrogenation can be controlled by the consumption of hydrogen, irrespective of the groups to be hydrogenated. No definite statements can be made on the amount of catalyst to be used. In general relatively large amounts of catalyst are used as compared to routine catalytic hydrogenations. Equal amounts of substance and catalyst and unusually long hydrogenation times are frequent (*108, 2251*). A wide variety of solvent combinations has been suggested. Their use is determined by each given situation, especially by the solubility and properties of the starting material and the end products. A mixture of methanol, acetic acid, and water is normally used to obtain free peptides from either carbobenzoxy peptides or their benzyl esters (*108, 1121, 2148, 2251*). The acetates are obtained initially, they are converted to the free peptides over potassium oxide *in vacuo*. Hydrogenation in aqueous ethanol is also possible (*1558*). Diketopiperazine formation has occasionally been observed for carbobenzoxydipeptide benzyl esters with proline and hydroxyproline (*1606, 2148*). If basic amino acids are present in the peptide, the equivalent amount of hydrochloric acid may be added to form the hydrochloride (*335, 2251*). Very poorly soluble compounds may be hydrogenated in glacial acetic acid (*2292*) or in the form of a suspension in dimethylformamide. Mixtures of water-saturated phenol with 2 *N* hydrochloric acid and ethanol have also served as a reaction medium at temperatures of 45°C (*334*). Hydrogenations are sometimes carried out in conventional solvents at higher temperature (*1938, 2104, 2620, 2626, 2637*) or under higher pressure (*2019a, 2071, 2104*). Peptide esters are obtained by hydrogenation in the presence of glacial acetic acid (*774*). However, the protonation of the amino group of dipeptide esters by acetic acid is frequently insufficient to prevent the formation of diketopiperazine (*771, 1483*). It is therefore better to carry out the hydrogenation in the presence of an equivalent amount of hydrochloric acid (*204, 1932*). Thus the stable hydrochlorides are obtained. Solvent combinations free of water, such as HCl in dioxane/ethanol or ethanolic HCl (*108*), have been described. The use of formic acid has been recommended for peptides containing the acid-labile formyl group (*1030*). Keil *et al.* (*1204*) have reported the hydrogenation of the phenyl residues of aromatic amino acids.

Frequently the catalytic hydrogenation fails in the presence of sulfur-containing amino acids (*291, 858, 895, 1354, 2119, 2508*). It has been described only for special cases (*583, 1414*). Sometimes the catalyst must be changed during the hydrogenation (*583*). Under certain circumstances, sulfur-free peptides cannot be catalytically hydrogenated, if they have been prepared via a sulfur-containing intermediate, e.g. the thiophenyl ester (*436*). The hydrogenation of cystine derivatives described by Berse *et al.* (*212*) is surprising. It proceeded without difficulty when the *p*-nitrocarbobenzoxy group was used instead of the carbobenzoxy group. Cystine can be reduced to cysteine during extended hydrogenation times, an observation already made by Bergmann and Michaelis (*188*). Hydrogen phosphide formed during coupling reactions via the phosphazo method can also give rise to difficulties by poisoning the catalyst in the subsequent catalytic hydrogenation (*814*).

Therefore many efforts have been made to develop other methods for the removal of the carbobenzoxy group. White (*2508*) described in 1934 the decarbobenzoxylation with hydrochloric acid, and shortly thereafter Harington and Mead (*937*) reported the cleavage with phosphonium iodide. The latter reaction, originally described as a reduction (*937*), is now believed to be an acid-catalyzed cleavage (*30, 147*). A modification of this method has been described by Waldschmidt-Leitz and Kühn (*2413*), who used hydrogen iodide in glacial acetic acid instead of phosphonium iodide.

The decarbobenzoxylation by hydrochloric acid has been investigated in great detail by several authors. White conducted the cleavage under very vigorous conditions, employing concentrated HCl at 70°C. Today 12 *N* hydrochloric acid is used at 37°C for 1–1.5 hours. Ester bonds present will be hydrolyzed (*1535, 1538, 1951*). Under these conditions the simultaneous cleavage of the carbobenzoxy and *tert*-butyloxycarbonyl groups is possible (*2022*). Ethanolic hydrochloric acid will cleave the carbobenzoxy group at 0°C in 30 hours. Esterification of the carboxyl groups takes place simultaneously (*108*). Other solvents that have been reported are glacial acetic acid (*149*)—no mention of the stability of the ester bond was made—and carbon tetrachloride in which the benzyl ester bond is not cleaved (*942*).

Vajda (*2350*) has investigated the stability of peptide bonds under the conditions of decarbobenzoxylation. He found that there is a considerable influence of the solvent. Under otherwise identical conditions, 30% of the peptide bonds of an acetyl dipeptide benzyl amide were cleaved in nitromethane, and only 9% in glacial acetic acid. Gawron and Draus (*785*) investigated the cleavage of a different peptide with ethanolic hydrochloric acid. The peptide bond is stable in 8 *N* solution after 4 hours at 45°C and in 2.2 *N* solution after 9 hours at 77°C. Cleavage occurred to a considerable extent in an 11.3 *N* solution at 45°C after 1 hour. By contrast, Goldschmidt and Jutz (*815*) have described the decarbobenzoxylation in boiling HCl-saturated ethanol within 2 hours.

The most important acid-catalyzed cleavage of the carbobenzoxy group is probably the reaction with hydrogen bromide. Glacial acetic acid is normally used as solvent (*147, 149, 296*). The use of nitromethane (*30, 152, 2032*) and the passing of hydrogen bromide through chloroform (*236*), ethyl acetate, or ether (*1104*) have also been described. The reaction mechanism has been discussed by Ben-Ishai and Berger (*149*) and by Albertson and McKay (*30*; cf. *1815, 2594*). At least 2 equivalents of hydrobromic acid are required, since in this reaction benzyl bromide and the amino acid or peptide hydrobromide are formed [38]:

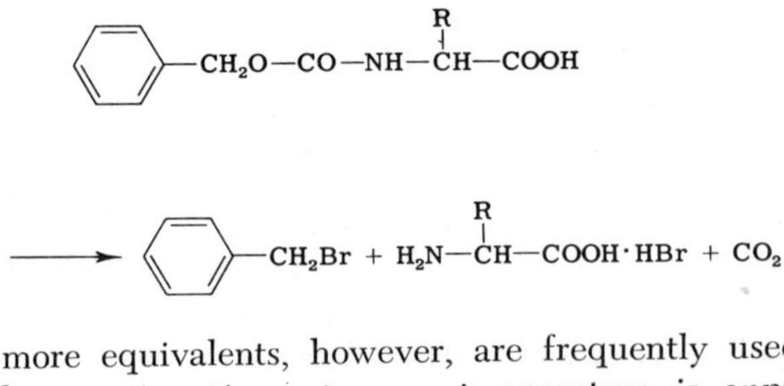

[38]

Three or more equivalents, however, are frequently used (*296, 1133, 1615a*). The reaction time at room temperature is approximately 30 minutes. It may be—especially for higher peptides—considerably longer (*490b, 1133*), and the temperature may have to be raised to about 60°C (*806, 1441*). Cleavages at 0°C have been reported for special cases at very long reaction times (*291a, 895*). Brenner and Curtius (*344a*) as well as Zahn *et al.* (*2619, 2633a*) have occasionally used liquid hydrogen bromide for the decarbobenzoxylation at temperatures as low as −80°C.

Side reactions have been observed for some amino acids. The methyl group in methionine is displaced by the benzyl group with formation of S-benzylhomocysteine (*30, 785, 1487;* cf. page 242). This side reaction can be prevented by the addition of methyl ethyl sulfide (*490b, 894, 895*). Carbobenzoxy-L-methionine *p*-nitrophenyl ester has been decarbobenzoxylated with HBr in ethyl acetate or ether (*1104*). *O*-Acetylation of serine has been observed occasionally (*596a, 1615a, 1617a*). It does not occur when an HBr/glacial acetic acid solution with a 30% water content (*894*) or ethyl acetate or trifluoroacetic acid (*894, 1104*) is used as solvent. Tryptophan-containing peptides may give poor results. Sometimes the solutions are intensely colored (*291*) and the yields are low (*1487*). The side reactions are caused by the bromine formed during the reaction. Phenol (*1605, 2594*) or diethyl phosphite (*291, 895*) is added to prevent them. The reaction products after cleavage with HBr frequently show a high content of bromine that cannot be removed by recrystallization or precipitation (*273, 1971*).

The reaction with hydrogen bromide is a specific decarbobenzoxylation that can be applied even in the presence of the formyl (*2384, 2389, 2392*) or trityl group (*236*). Boissonnas and Preitner (*301*) have found that under the conditions of complete decarbobenzoxylation (1 hour, 20°C) the allyloxycarbonyl group is cleaved to the extent of 40%, while only 1% of the *sec*-butyloxycarbonyl group is affected. The carbethoxy group is stable. Cleavage of peptide bonds occurs only on heating to 100°C and amounts to about 50% after 1 hour. ω-Acid amides are not hydrolyzed under the standard reaction conditions (*296*). Taschner and Kupryszewski (*2267*) have investigated in detail this acidolysis of peptide bonds in glacial acetic acid. The peptide bonds are cleaved to the extent of 70–95% after 8 hours at 100°C when a 3% solution of HBr in glacial acetic acid is used. For some peptides this cleavage could be observed at room temperature. Hydrogen chloride is less effective than hydrogen bromide.

The tosyl and phthalyl groups are among the amino-protecting groups that cannot be removed under the conditions of decarbobenzoxylation (*48, 301*). The *tert*-butyloxycarbonyl group will be cleaved (*49*). Methyl and ethyl esters are stable and, provided the exposure time to the reagent is not too long, the benzyl ester will also survive (cf. page

61). Poroshin *et al.* (*1754*) have investigated the stability of various esters to hydrogen bromide. The extent of cleavage increases in this order: Methyl ester → *p*-nitrobenzyl ester → cyclohexyl ester → benzyl ester.

Weygand and Steglich (*2499*) describe a decarbobenzoxylation by boiling in trifluoroacetic acid. It proceeds with carbon dioxide evolution and formation of trifluoroacetic acid benzyl ester. The latter very readily benzylates aromatic amino acids. This side reaction can be prevented by the addition of such compounds as phenol, resorcinol, or anisole that are very easily substituted. The reaction time varies from 12 to 150 minutes, depending on the amino acid used. Yields are 40–98%. Prolonged action of the reagent will also cleave *O*-benzyl ethers (e.g. *O*-benzyl-tyrosine) as well as benzyl esters. According to Khosla and Anand (*1224b*), decarbobenzoxylation occurs under even milder conditions (3 hours, 38°C). The benzyl, *p*-nitrobenzyl, and methyl esters are cleaved in addition to the carbobenzoxy group, if a mixture of trifluoro-acetic acid and hydrochloric acid is applied (4 hours, 40°C). The carbo-benzoxy group can be readily removed from pyroglutamic acid (*1975*).

The reductive cleavage of the carbobenzoxy group with sodium in liquid ammonia has been described by Sifferd and du Vigneaud (*2119*). In contrast to the catalytic hydrogenation, dibenzyl is formed as the main product. The formation of toluene to a smaller extent has been observed (*1697*). The cleavage is carried out in the same way as that of the toluenesulfonyl group (cf. page 14). The same difficulties in purification and isolation are encountered. Therefore the method is applied to the synthesis of free peptides only where ω-amino groups are protected by the tosyl group (e.g., ACTH, bradykinin, kallidin), or where mercapto groups carry a benzyl protection (e.g. oxytocin or vasopressin).

Birkhofer *et al.* (*243*) have described a decarbobenzoxylation by reduction with triethylsilane. It has been applied successfully to carbo-benzoxy-S-benzylcysteine, with preservation of the S-benzyl protection.

In addition to the protection of amino groups, phenolic hydroxyls (see Chapter IV,F,1,b,γ) and the mercapto function (see Chapter IV, G,1,e) can also be blocked by the carbobenzoxy group.

2. Modified Carbobenzoxy Groups

a. Substituted Carbobenzoxy Groups

α. The p-Chloro- and p-Bromocarbobenzoxy Groups. p-Chlorocarbo-benzoxyamino acids were prepared for the first time by Boissonnas and Preitner (*301*). They were investigated in detail by Kisfaludi and

Dualszky (*1247*) several years later. The synthesis from *p*-chlorocarbobenzoxy chloride and the amino acid in aqueous alkaline solution follows the standard procedure. *p*-Chlorocarbobenzoxy chloride is prepared from *p*-chlorobenzyl alcohol and phosgene at —15°C in ether. It remains as an oil after the evaporation of the solvent (*301*). Asparagine is chlorocarbobenzoxylated with sodium carbonate as base, and aspartic acid with magnesium oxide. Amino acid esters can be acylated in heterogeneous phase, in chloroform/water with magnesium oxide, sodium bicarbonate, or sodium carbonate as base. All derivatives show excellent crystallization properties. Crystalline products are obtained in many cases where the carbobenzoxyamino acids are known only as oils. The melting points are up to 30° higher than those for the corresponding carbobenzoxy compounds (*1247*).

Peptide syntheses have been described by the carbodiimide method, by the mixed anhydride method with ethyl chlorocarbonate in chloroform or with *o*-phenylene chlorophosphite in toluene, and via activated esters (cyanomethyl ester) (*1246, 1247*). This protecting group has been used for the synthesis of naturally occurring polypeptides, e.g. glutathione (*1246*) and partial sequences of ACTH (*98, 1249*). The cleavage of the *p*-chlorocarbobenzoxy group by catalytic hydrogenation is slower than that of the carbobenzoxy group (*301*). The reaction with hydrogen bromide in glacial acetic acid was used in the synthesis of glutathione (*1246*). It is also slower than the corresponding decarbobenzoxylation (*301*). The *p*-chlorocarbobenzoxy group can be readily removed in liquid ammonia (*301, 1246*).

Channing *et al.* (*482*) have described the *p*-bromocarbobenzoxy group. It shows the same behavior as the *p*-chloro compound with regard to the preparation of the acid chloride, the reaction with amino acids, and the coupling of *N*-protected amino acids. Higher melting points and the better crystallization properties are stressed for several glycine derivatives, which have been compared with the corresponding carbobenzoxy derivatives.

β. p-Nitrocarbobenzoxy Group. *p*-Nitrocarbobenzoxy chloride was prepared from *p*-nitrobenzyl alcohol and phosgene. Thiele and Dent (*2307*) used chloroform as solvent and carried out the reaction in a sealed tube at 60°–65°C. The compounds can be more readily obtained, according to Carpenter and Gish (*463*), in dioxane at room temperature. *p*-Nitrocarbobenzoxy chloride is a stable crystalline compound.

The blocked amino acids can be prepared by the standard procedures. The acid chloride is added in dioxane solution (*463*). The derivatives of serine, threonine, and tyrosine are best obtained in a dioxane-water mixture in the presence of equimolar amounts of alkali and sodium bicarbonate (*804*). The *N,O*-bis-*p*-nitrocarbobenzoxy derivative of tyrosine and the N^{α},N^{G}-bisacyl compound of arginine are readily formed (*804*). The N^{α}-mono derivative can be obtained by acylation at pH 10, the derivative protected at the guanido group only from the bis compound via the *N*-carboxylic anhydride (*805*). *p*-Nitrocarbobenzoxy-nitro-L-arginine (*215*) and bis-*p*-nitrocarbobenzoxy-L-lysine (*804*) are also known. Histidine (DL-form) forms an N^{α}-amino derivative (*804, 2067*). The formation of di-*p*-nitrophenyl carbonate as a by-product of the acylation of amino acids is sometimes observed. This compound will also be formed on treatment of the acid chloride with aqueous alkali (*463*). Most *p*-nitrocarbobenzoxyamino acids are readily crystallizing compounds. Another advantage over the carbobenzoxy group is the presence of the nitro group. Its ultraviolet absorption at 265 mμ can be used for analytical purposes (*463, 478*).

Peptide syntheses that have been described preferred the acid chloride method for the coupling (*463, 634, 805, 1483*). This protecting group has also been used for the synthesis of sequences of biologically active polypeptides (*213, 634, 1854*) and of peptolides (*2086*). The cleavage of the *p*-nitrocarbobenzoxy group has so far been described only with hydrogen bromide in glacial acetic acid (*634*) and by catalytic hydrogenation (*463, 478, 1483*). *p*-Aminotoluene (*463*) is one of the reaction products of the hydrogenation. It can be removed very readily *in vacuo* in the form of its volatile acetate (*463, 478*). It came as a surprise that the blocking group can also be removed from cystine derivatives by catalytic hydrogenation (*212, 213*).

γ. *Colored Carbobenzoxy Groups.* Schwyzer *et al.* (*2037*) have introduced as colored protecting groups of the urethan type the *p*-(*p′*-methoxyphenylazo)-benzyloxycarbonyl residue [39] (abbreviated MZ) and the *p*-(phenylazo)-benzyloxycarbonyl residue [40] (abbreviated PZ).

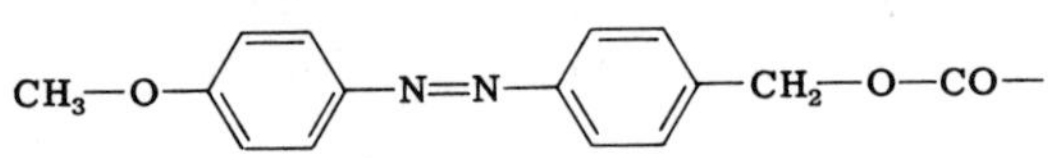

[39]

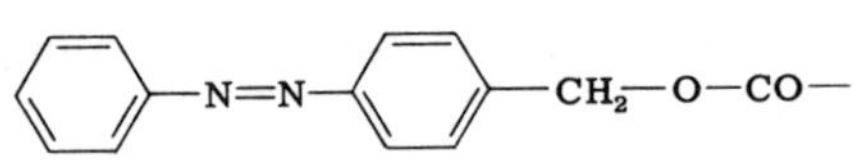

[40]

The reaction of amino acids with the acid chlorides follows the standard procedure. Suitable solvent systems are: ether-water, dioxane-water, or acetone-water (*2037*). The acid chlorides are prepared from the corresponding substituted benzyl alcohols and phosgene in dioxane. *p*-(*p'*-Methoxyphenylazo)-benzyl alcohol is accessible by diazotization of *p*-aminobenzyl alcohol, reaction of the diazonium salt with phenol, and methylation with dimethyl sulfate. *p*-(Phenylazo)-benzyl alcohol is prepared from *p*-aminobenzyl alcohol and nitrosobenzene.

In addition to a series of simple amino acids, the *p*-(*p'*-methoxyphenylazo)-benzyloxycarbonyl derivatives of L-arginine (acylation in a mixture of sodium hydroxide and sodium bicarbonate), L-aspartic acid β-methyl ester, L-glutamic acid γ-methyl ester, and N^{im}-benzyl-L-histidine have been prepared (*2037*). N^{ϵ}-Protected lysine was synthesized via the copper complex, the corresponding methyl ester via the *N*-carboxylic anhydride prepared from the N^{ϵ}-protected compound and phosgene (*2033*). Due to their color it is possible to locate and determine quantitatively the protected amino acids or peptides on chromatograms and during column chromatography or countercurrent distribution (*2037*).

p-(*p'*-Methoxyphenylazo)- and *p*-(phenylazo)-benzyloxycarbonyl derivatives have been described by Schwyzer *et al.* in a series of publications on syntheses of sequences of naturally occurring polypeptides [ACTH (*1173, 2021, 2022, 2024*), tyrocidin A (*2032, 2038*), gramicidin S (*2033*)]. The colored protecting groups have also been used for the synthesis of peptolides (*2004*). The applications demonstrated that these protecting groups have another great advantage in addition to color: their derivatives show excellent crystallization properties (*2032, 2033*).

Cleavage can be achieved by catalytic hydrogenation. Three moles of hydrogen are taken up and aromatic amines are formed as reaction products (*2037*) [41]:

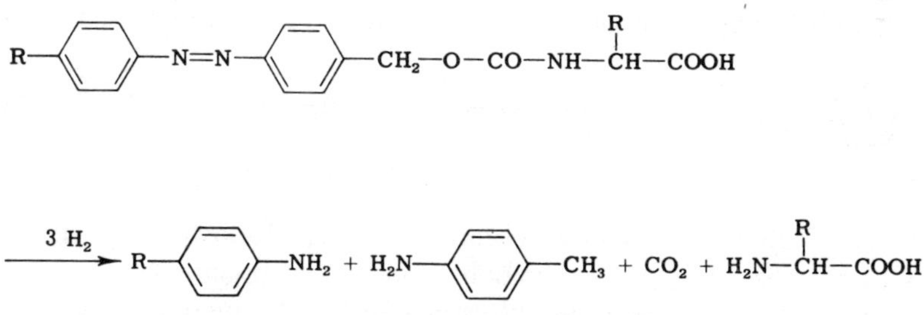

[41]

(PZ: R = H; MZ: R = CH_3O)

The hydrogenolysis is very fast, even in neutral solution (*2000*).

The removal of the blocking groups with HBr in glacial acetic acid requires longer reaction times than the corresponding decarbobenzoxylation. The *p*-(*p'*-methoxyphenylazo)-benzyloxycarbonyl group can be removed in 1–1.25 hours at 40°C (*2032, 2033*). Crystalline phenylazo or *p*-methoxyphenylazobenzyl bromide is formed during the reaction. For the isolation of the peptide, the solvent is evaporated and the residue shaken in an aqueous-organic solvent system. The hydrobromide goes into the water phase, while benzyl bromide is dissolved in the organic phase (*2037*). Cleavage with sodium in liquid ammonia is also possible (*2000*).

δ. *p-Methoxycarbobenzoxy Group.* McKay and Albertson (*1487*) obtained for the first time *p*-methoxycarbobenzoxyamino acids [43] by the reaction of isocyanatocarboxylic acid esters [42] with anisyl alcohol and subsequent hydrolysis:

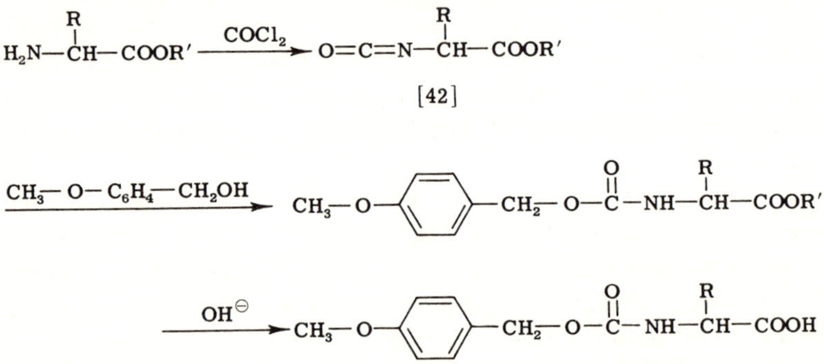

[42] [43]

These compounds can be more readily obtained, according to Weygand and Hunger (*2487*), in a synthesis analogous to that of the *tert*-butyloxycarbonylamino acids (see Chapter I,B,3,a). The amino acids are allowed to react with *p*-methoxycarbobenzoxy azide [44] or with *p*-methoxycarbobenzoxy *p*-nitrophenyl ester [45]. The azide is prepared

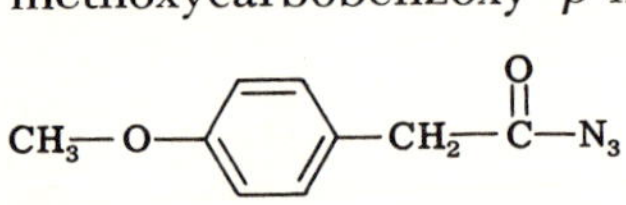

[44]

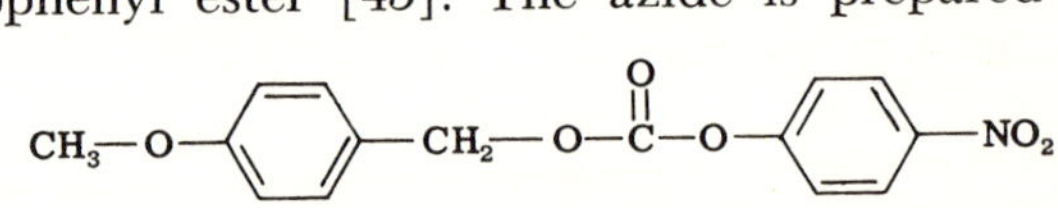

[45]

from anisyl alcohol and phenyl chlorocarbonate, followed by hydrazinolysis of the resulting *p*-methoxybenzylphenyl carbonate and reaction with nitrous acid. The *p*-nitrophenyl ester can be obtained from *p*-nitrophenyl chlorocarbonate and anisyl alcohol. Both compounds are allowed to react with amino acids at room temperature for 1–2 days. Dioxane/water is used as solvent and magnesium oxide as base. The *N*-protected amino acids must be precipitated from the alkaline solution with an ion exchanger and taken up in ethyl acetate, since the protecting group is not stable to acid (*2487*).

Peptides with *p*-methoxycarbobenzoxyamino acids have been synthesized via the mixed ethyl- and isobutylcarbonic acid anhydride (*1487, 2488*), the carbodiimide method (*2488*), and the phenyl thiolester (*2488*). The protecting group can be removed by treatment with hydrogen chloride or hydrogen bromide in nitromethane (*1487*), or by boiling in glacial acetic acid for 2 hours. The mildest conditions are offered by the treatment with trifluoroacetic acid—0.5 hour at 0°C. Anisole must be added to prevent side reactions by the initially formed *p*-methoxybenzyl cation. The carbobenzoxy group and the benzyl ester are completely stable under these conditions (*2487, 2488*). The *p*-methoxycarbobenzoxy group can also be removed by catalytic hydrogenation (*2487*).

b. Carbobenzoxy-Like Protecting Groups

α. Allyloxycarbonyl Group. The allyloxycarbonyl residue, a carbobenzoxy-like protecting group [46], has been investigated in detail by

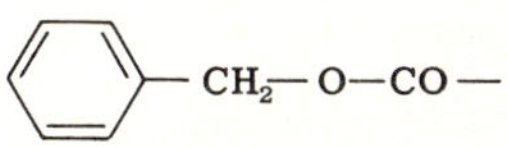

Carbobenzoxy

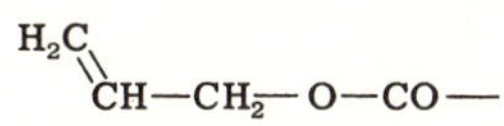

[46]

Stevens and Watanabe (*2213*). It is interesting from a theoretical point of view that certain analogies to the carbobenzoxy group have been found (*301, 2213*). This group has not yet been applied to peptide synthesis.

β. Phenoxy-, Tolyloxy-, and Naphthyloxycarbonyl Groups. For the synthesis of phenoxycarbonyl chloride [47] and tolyloxycarbonyl chloride [48], the corresponding anhydrous sodium phenolates must be reacted with phosgene. Naphthyloxycarbonyl chloride [49] can be obtained by

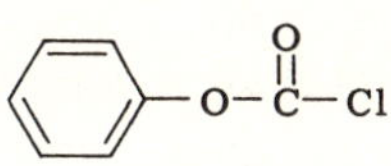

[47]

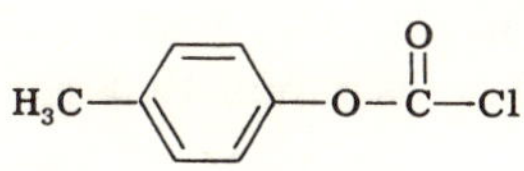

[48]

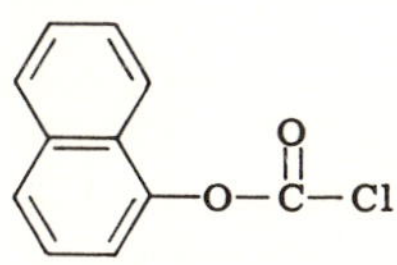

[49]

reacting α-naphthylcarbinol with phosgene (*482*). The *N*-protected amino acids are prepared by the standard procedure (*301, 482*). Cleavage may be carried out by very long catalytic hydrogenation or by treatment with sodium in liquid ammonia (*301*). Conditions for the removal with HBr in glacial acetic acid are so vigorous that a cleavage of peptide bonds cannot be prevented (*301*). Therefore these groups are not suitable for peptide synthesis.

3. Aliphatic Urethan Groups

a. *tert*-Butyloxycarbonyl Group

The *tert*-butyloxycarbonyl group is the only one of the urethan-type protecting groups offering a differentiation to the carbobenzoxy group that has become of importance.

Diketopiperazine formation had been observed during the cleavage of isopropyloxycarbonylglycyl-DL-phenylalanine with HBr in glacial acetic acid (*30*), but not in the case of the carbobenzoxy derivative. This observation and the discussion of the possible reaction mechanism [50] led McKay and Albertson (*1487*) to the discovery of the *tert*-butyloxycarbonyl group:

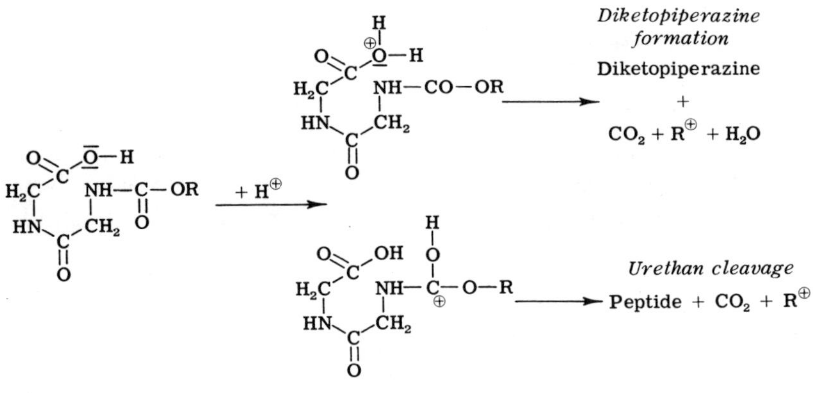

[50]

Anderson and McGregor (*49*) described the group only a short while later. They assumed that the olefin would be formed as a very volatile product in the reversed reaction of the proton-catalyzed esterification with isobutylene [51]:

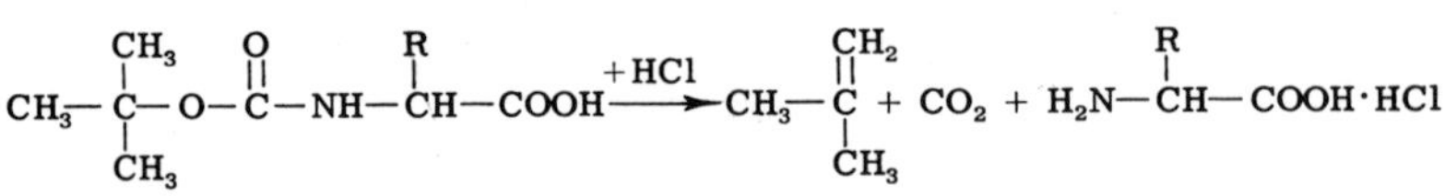

[51]

Difficulties were initially encountered with the synthesis of the *N*-protected amino acids. *tert*-Butyloxycarbonyl chloride decomposes above +10°C. Its preparation from sodium *tert*-butylate and phosgene in *n*-butane proceeds with a yield of only 20% (*490a*). McKay and Albertson have therefore described a synthesis of *tert*-butyloxycarbonylamino acids in which α-isocyanatocarboxylic acid esters are allowed to react with *tert*-butanol. The resulting esters are hydrolyzed subsequently (*1487*) (reaction scheme analogous to that on page 34). The vigorous conditions required for the synthesis of the isocyanates from amino acid esters and phosgene (*823*) and the additional ester hydrolysis are a disadvantage. Nevertheless the method is applied occasionally, especially when a subsequent hydrazine formation is intended (*894*).

A much simpler approach to these compounds is offered by the aminolysis of suitable *tert*-butyloxycarbonyl aryl esters. While the yields are low for the phenyl ester and the phenyl thiolester (*1487*), satisfactory results can be obtained with the *p*-nitrophenyl ester (*49*) [52]:

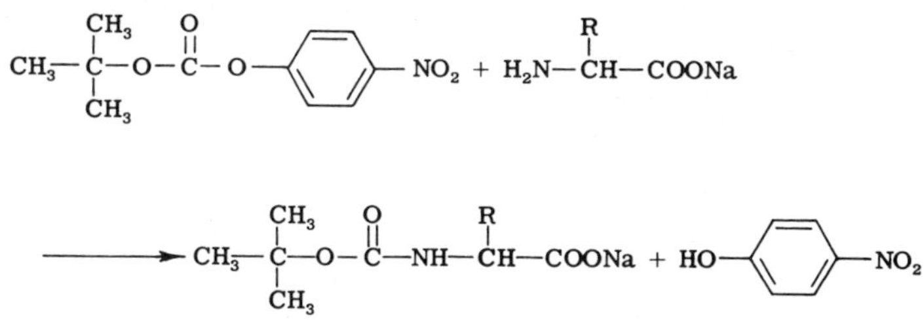

[52]

The *tert*-butyloxycarbonyl phenyl ester has become readily accessible through the investigations of Carpino (*464*). The corresponding *p*-nitrophenyl ester [53] can be obtained by the same procedure (*49*). *tert*-

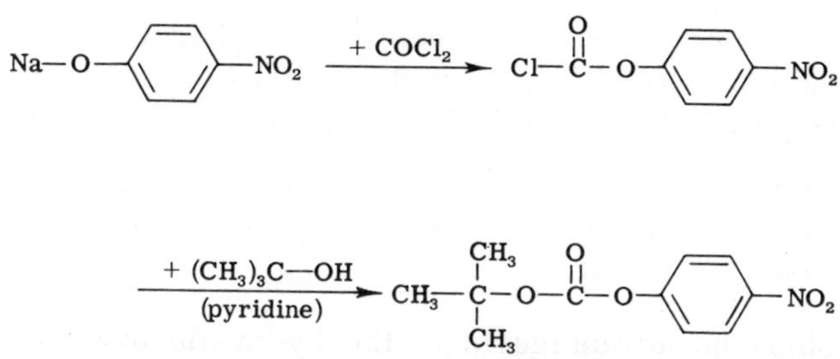

[53]

Butyloxycarbonylamino acids are prepared by keeping the starting materials a long time at room temperature or by boiling them briefly in a mixture of *tert*-butanol and water. Sodium hydroxide or sodium carbonate is added for salt formation and to bind the resulting nitrophenol (*49*).

It is not possible to convert all amino acids to the protected derivatives by this method. Schwyzer *et al.* (*2036*) have therefore suggested the use of *tert*-butyloxycarbonyl azide. This reagent has also been described by Carpino *et al.* (*465, 465a*), and is prepared from the phenyl ester via the hydrazide. It is a stable compound and can be distilled *in vacuo* without decomposition. The azide can also be obtained from the more readily accessible *p*-nitrophenyl ester. Acylations are carried out in dioxane or dioxane/water. Magnesium oxide or sodium hydroxide is used as base. The reactions are carried out at 40°–50°C and may require 24 hours for completion (*866, 2036*).

Amino acid esters or peptide esters can be protected with the *tert*-butyloxycarbonyl group by reacting them in pyridine with *tert*-butyloxycarbonyl *p*-nitrophenyl ester (*2020*) or, better, with the azide (*1103, 2036*). *tert*-Butyloxycarbonylamino acids cannot be esterified by an alcohol in the presence of an acid since the protecting group is very labile to acid (*49, 1487*). Esterification with diazomethane is possible (*2022*).

Klee and Brenner (*1255*) have described *tert*-butyloxycarbonylimidazole [54], which can be prepared from *tert*-butanol and carbonyldiimidazole. The reaction with ethanol ($R = C_2H_5$) consumes 2 moles of alcohol, and diethyl carbonate [55] is formed. In the case of *tert*-butanol [$R = (CH_3)_3C$], *tert*-butyloxycarbonylimidazole is the final product:

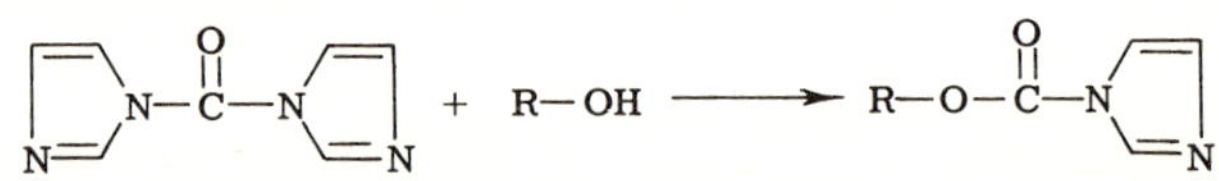

[54]

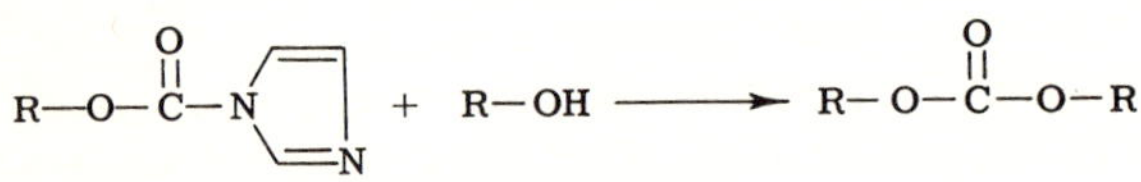

[55]

The *N*-protected amino acids are formed in low yield by heating the reagent with sodium salts of amino acids to 100°C. *tert*-Butyloxy-carbonyl hydrazide [56] is formed in 80% yield with anhydrous hydrazine:

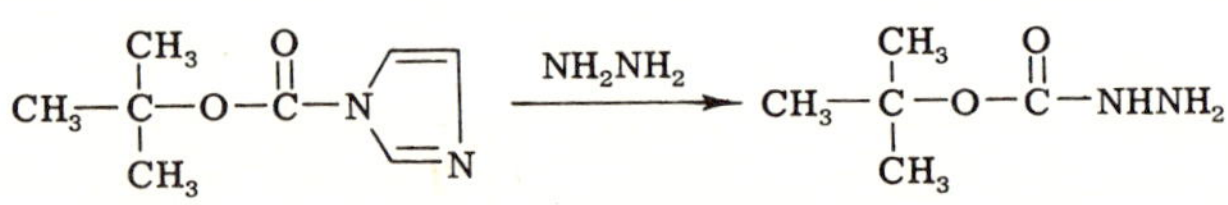

[56]

This reaction offers no advantages over the hydrazide formation from phenyl or *p*-nitrophenyl esters.

Most amino acids can be converted to their *tert*-butyloxycarbonyl compounds. Difficulties are encountered only with aspartic and glutamic acids. More recently, however, derivatives of these two amino acids have also been described (*1974, 1975*). The N^{α}-*tert*-butyloxycarbonyl-N^G,N^G-dicarbobenzoxy derivative (*866*), the N^{α}-*tert*-butyloxycarbonyl-N^G-tosyl derivative (*1785*) and the N^{α}-*tert*-butyloxycarbonyl-N^G-nitro derivative (*1020b*) of arginine have been described. Hydroxyamino acids can be acylated only in poor yields. The *tert*-butyloxycarbonylamino acid hydrazides are advantageous for subsequent coupling reactions. They can be prepared from the esters that have been obtained by acylation (*1103*).

tert-Butyloxycarbonylhistidine hydrazide can be prepared by the same procedure (*1972*).

This blocking group is of great importance for the protection of the ω-amino group. The α-carbobenzoxy-ε-butyloxycarbonyl derivative of lysine and the ε-*tert*-butyloxycarbonyllysine methyl ester are known (*2022*).

The instability to acid of the *tert*-butyloxycarbonyl group should be borne in mind during the preparation of *N*-protected amino acids, since they are frequently precipitated by acid from alkaline solutions. The same is true for the synthesis of peptides. The basic components are best removed with citric acid (*2033, 2036*). Low temperatures (—5° to —30°C) should be maintained under the strongly acidic conditions of azide formation. This can be achieved by applying the azide method as modified by Honzl and Rudinger (*1058*). The use of anhydrous solvents makes it possible to maintain such low temperatures (cf. page 84).

The cleavage of the *tert*-butyloxycarbonyl group occurs upon relatively short action of acids (*49, 1487*). At least 2 moles of acid are required to complete the cleavage (*49*). It may therefore be assumed that not isobutylene but *tert*-butyl chloride is formed as the cleavage product. Hydrogen chloride is used predominantly as acid in such solvents as benzene, diethyl phosphite (*49*), or ethyl acetate (*2020, 2033*). The cleavage is carried out in 15–30 minutes at room temperature and under moderate cooling. Trifluoroacetic acid proved to be useful for higher peptides; the reaction time at room temperature is up to 1 hour (*1172, 2019*). The cleavage can be conducted without difficulty so that carbobenzoxy groups are not affected (*2038*). Both protecting groups can be removed simultaneously with HBr in glacial acetic acid. The *tert*-butyloxycarbonyl group is stable to sodium in liquid ammonia and to catalytic hydrogenation (*49*). Thus the selective cleavage of the carbobenzoxy group is possible (*2022*). The synthesis of *N*-protected peptide acids has been carried out, employing benzyl esters which can be cleaved by catalytic hydrogenation. This combination is especially useful in cases where alkaline ester hydrolysis, e.g., for peptolides, is not possible (*802*). The *tert*-butyloxycarbonyl group cannot be cleaved by heating in 75% acetic acid. Detritylation of peptides, however, may occur under these conditions (*2022, 2038;* cf. page 155). The *tert*-butyloxycarbonyl group has been used for the synthesis of protected hydrazides (see Chapter II,B,1,b).

The *tert*-butyloxycarbonyl group, though known for only a relatively short time, has become second in importance to the carbobenzoxy group, for example, for the synthesis of sulfur-containing peptides (*1470, 1904,*

2020) or, in combination with the carbobenzoxy group, as ω-protection in the synthesis of peptides with diaminocarboxylic acids (e.g., ACTH sequences or eledoisin).

b. OTHER ALIPHATIC URETHAN GROUPS

Many aliphatic urethans have been investigated for their applicability as protecting groups. Only the *tert*-butyloxycarbonyl group has become of importance. Others that have been described are the diisopropylmethoxycarbonyl [57] (*1487*), *sec*-butyloxycarbonyl [58] (*301*), isopropyloxycarbonyl [59] (*30*), and ethoxycarbonyl [60] (*301*) groups.

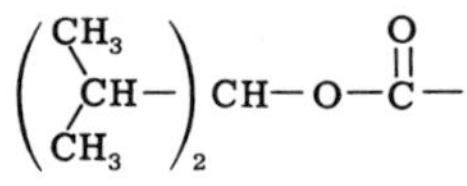

[57]

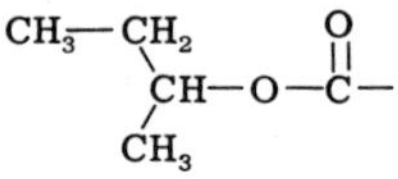

[58]

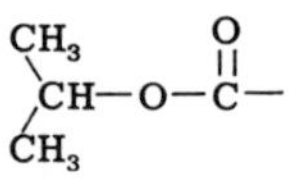

[59]

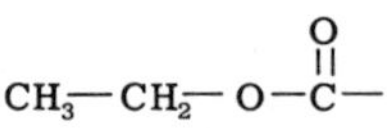

[60]

All these groups can be introduced via the corresponding alkyl chlorocarbonate by established procedures (*301, 1487*). Their cleavage presents difficulties and cannot be carried out without side reactions in the majority of cases.

Diketopiperazine formation has been observed for isopropyloxycarbonyl dipeptides and tripeptides on treatment with HBr in glacial acetic acid. One peptide bond was cleaved in the case of a tripeptide (*30;* cf. page 36).

Three days are required to cleave the diisopropylmethoxycarbonyl group with hydrogen bromide in nitromethane (*1487*).

The *sec*-butyloxycarbonyl and ethoxycarbonyl groups are cleaved with HBr in glacial acetic acid. High temperatures (100°C) are required and the reaction times are 10 minutes and 1 hour, respectively (*301*). Cleavage of peptide bonds occurs to a considerable extent under these conditions. The peptide bonds of methoxycarbonyl peptides are cleaved preferentially. This reaction can be used for the determination of the N-terminal amino acid in structural studies of peptides and proteins (*488*).

Kader and Stirling (*1164*) have used β-(*p*-toluenesulfonyl)-ethyloxycarbonyl chloride [61] to prepare *N*-protected amino acids, and applied them to the synthesis of peptides via the *p*-nitrophenyl esters. The advantage of this urethan-type blocking group is its easy removal by

$$CH_3-C_6H_4-SO_2-CH_2-CH_2-O-\overset{\overset{\displaystyle O}{\|}}{C}-Cl$$

[61]

alkali. It was demonstrated in one case that the cleavage proceeds under conditions mild enough that ester bonds are not attacked. The synthesis of *N*-protected peptides from benzyl esters by catalytic hydrogenation is possible.

4. Hydroaromatic Urethan-Type Protecting Groups

Efforts to use hydroaromatic protecting groups did not meet with great success.

a. Cyclopentyloxycarbonyl Group

The removal of the cyclopentyloxycarbonyl group (*1487*) with HBr in glacial acetic acid is more difficult than that of the carbobenzoxy group. A selective cleavage, however, is not possible. Only by catalytic hydrogenation has it been possible to differentiate between the two groups (*1487*). This was demonstrated by the synthesis of a tyrocidin A sequence (*2038, 2236*).

b. Cyclohexyloxycarbonyl Group

The cyclohexyloxycarbonyl group can also be removed with HBr in glacial acetic acid (*1487*). Here the circumstances are somewhat more favorable. The differences are great enough to permit the selective cleavage of the carbobenzoxy group. This was applied to the protection of the ϵ-amino group during the synthesis of cysteine-containing lysine peptides (*1345, 1983*). The N^{α}-carbobenzoxy protection was removed at room temperature by the standard procedure with 1.5 *N* HBr in glacial acetic acid for 2 hours. After the completion of all coupling steps the N^{ϵ}-cyclohexyloxycarbonyl protection was also removed by treatment with 1.5 *N* HBr in glacial acetic acid, but at 60°C for 1 hour.

Debabov and Shibnev (*575a*) succeeded in synthesizing several proline-containing peptides by using this protecting group in combination with the catalytically removable benzyl ester. However, the conditions required for the cleavage with hydrogen bromide in glacial acetic acid, of the higher peptides particularly, led to partial hydrolysis of peptide bonds.

5. Thiourethans

The aryl- and alkylthiocarbonyl compounds [62] also belong to the urethan-type protecting groups. Ehrensvärd (*655*) and somewhat later

[62]

Lindemann *et al.* (*1432*), intrigued by the stability of phenylthiocarbonyl amides toward strong acids and their ready alkaline hydrolysis, studied the application of phenylthiocarbonylamino acids—also referred to as *N*-(phenylmercaptoformyl)-amino acids—to peptide synthesis.

Phenylthiocarbonyl chloride will acylate only amino acid esters. The reaction is carried out in organic solvents, such as ether, ethyl acetate, (*655*) or chloroform (*1432*), in the presence of a second equivalent of the amino acid ester (*655*) or, better, a tertiary base such as triethylamine (*1432*). Amino acids, in the form of their salts, react in aqueous solution, aminolysis of the thiolester bond takes place, and substituted urea derivatives are formed [63] (*1272*):

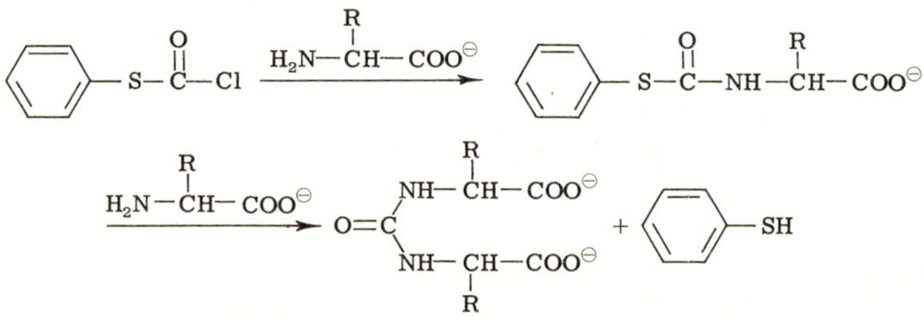

[63]

The *N*-protected amino acids are obtained by acid hydrolysis of the esters (boiling for 10 minutes with concentrated HCl/glacial acetic acid 1:1) (*655, 1432*). Phenylthiocarbonylamino acid esters, due to their aminolytic lability, are decomposed by hydrazine hydrate to form carbonylbis-amino acid hydrazides and *S,S*-diphenyl dithiocarbonate (*1273*).

Only the acid chloride method has so far been used for synthesis of peptides (*655, 1432*). Phenylthiocarbonyl-L-glutamic acid and acetic anhydride give phenylthiocarbonyl-L-glutamic acid anhydride without racemization (*1432*). Aminolysis with amino acid esters yields glutamyl α-peptides (*1273*).

The removal of the blocking group from amino acids and their esters can be carried out with lead tetraacetate in 70% ethanol or with 0.1 *N* aqueous alkali in the presence of lead hydroxide or lead carbonate (*655*). Insoluble lead phenylmercaptide is precipitated during this reaction. According to Lindenmann *et al.* (*1432*), however, hydantoins [64] and not the free esters are formed from phenylthiocarbonyl peptide esters:

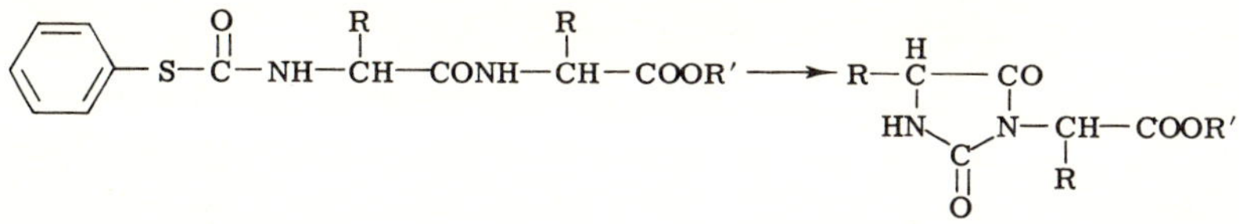

[64]

Phenylmercaptan is apparently eliminated and an isocyanate formed which closes the ring to form the hydantoin. Kollonitsch *et al.* (*1271, 1273*) have therefore described the oxidative cleavage of the phenylthiocarbonyl group. They use 2.5 moles of perbenzoic acid at —5°C in acetic acid/benzene/tetrahydrofuran, dioxane, or dioxane with 2% water [65]:

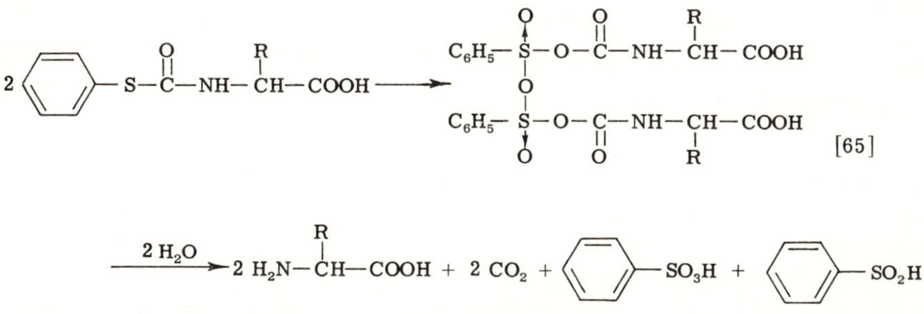

[65]

A number of dipeptides have been prepared by this procedure. The disadvantage of aminolysis of the thiolester bond in aqueous solution can be overcome by using thiolesters with a less pronounced tendency to aminolytic cleavage.

Kollonitsch *et al.* (*1271, 1272*) report the use of methyl-, ethyl-, butyl-, and benzylthiocarbonyl chlorides that have all been prepared from the corresponding mercaptan and phosgene. The butyl and benzyl derivatives are especially suitable for peptide synthesis. The *N*- protected amino acids can be prepared without difficulty in a Schotten-Baumann reaction. They show better crystallizing properties than the corresponding methyl- or ethylthiocarbonyl compounds.

Peptide syntheses have been reported, employing the acid chloride, mixed anhydride, phosphazo (*1272*), and carbodiimide (*942, 1272*) methods. The cleavage proceeds best with perbenzoic acid in dioxane or in dioxane/tetrahydrofuran at 0°C. The presence of a small amount of water prevents the cleavage of peptide bonds (*1272*). The protecting group is stable to acids, including HBr in glacial acetic acid (*1271, 1272*). It is labile under alkaline conditions. Esters can therefore be hydrolyzed only with acids.

Phenyl- and ethylthiocarbonylamino acids have also been used for polymerizations (*919*).

C. ALKYL-TYPE PROTECTING GROUPS

Monoalkylation normally does not provide complete protection of the amino function. It can still be acylated although with greater difficulty. The amino group is fully blocked in dialkylated derivatives. The dibenzyl derivatives have been used; they are suitable since they can be cleaved by catalytic hydrogenation. Monobenzyl derivatives have also been described, but their application is very limited. Only the triphenylmethyl (trityl) group has become important for the protection of the amino function. It provides complete protection through steric hindrance. The main advantage lies in the mild conditions for its removal.

1. Triphenylmethyl (Trityl) Group

The preparation of *N*-protected amino acids and their use in peptide synthesis do not make the trityl group the ideal type of amino protection. It is for its ready cleavage that it is chosen again and again for special purposes. It can be removed selectively even in the presence of the *tert*-butyloxycarbonyl group.

Simple tritylamino acids and trityl peptides have been known for some time. Systematic studies, however, have been done only during the past ten years by Amiard *et al.* (*35, 37, 38*), Zervas and Theodoropoulos (*2668*), and Stelakatos *et al.* (*2203*). Trityl chloride is used exclusively as the alkylating agent. The easiest tritylation is that of amino acid esters. It is usually carried out at 0°C in chloroform as solvent and in the presence of a tertiary base, e.g. triethylamine (*37, 236, 291, 1459, 2668*). The yields are very good. Difficulties, however, are encountered with the hydrolysis of the esters. The tritylglycine ethyl ester and the tritylalanine methyl ester are the only esters that can be hydrolyzed at room temperature in the usual manner with a small excess of 1 *N* alcoholic KOH, giving yields of 92 and 85%, respectively. The tritylphenylalanine methyl ester is cleaved to the extent of only 10% under analogous conditions (*2668*). Conditions such as boiling for 5 minutes in 20% solution of potassium hydroxide in propylene glycol have been reported for the tritylated esters of other amino acids. It has been claimed and substantiated for the hydrolysis of N^{α},N^{im}-ditritylhistidine that racemization does not occur (*37*). Trityl-L-serine was obtained without racemization by hydrolysis of its ester (*37, 490b, 891, 2067*). The hydrolysis of tritylarginine methyl ester, according to Boissonnas *et al.* (*291*), proceeds also without racemization when carried out with a tenfold excess of potassium hydroxide by boiling for 15 minutes in methanol/water. Attempts by Haas (*904*) to hydrolyze tritylnitroarginine methyl ester in the same way were unsuccessful. The trityl group is in danger of being cleaved during hydrolysis in aqueous alcoholic sodium hydroxide (*1006*). These difficulties are not encountered with trityl dipeptide esters (*2668*). The trityl group is therefore frequently introduced into peptide esters that have been synthesized with another protecting group (*291, 2022, 2027*).

The tritylamino acids can be obtained by catalytic hydrogenation if the benzyl esters are used instead of the methyl or ethyl esters (*230, 2203*). The reaction must be stopped after the uptake of 1 mole of hydrogen, otherwise the hydrogenolysis of the trityl group will proceed. Methanol or ethyl acetate (*2017a, 2203*) and cyclohexane (*38*) have been used as solvents. Dioxane is especially suitable; the hydrogenation stops almost completely in this solvent after 1 mole of hydrogen has been taken up (*1980*). The hydrogenolysis of benzyl esters can also be improved by the addition of an equivalent amount of triethylamine (*38*).

It is also possible to tritylate amino acids in the form of their salts in aqueous-organic solution. The competing reaction of the hydrolysis of trityl chloride to triphenylcarbinol is a disadvantage. This hydrolysis occurs almost exclusively when triethylamine or pyridine is used as base. Somewhat more satisfactory results are obtained with diethylamine in isopropanol/water. Triphenylcarbinol and trityldiethylamine are isolated as by-products. Therefore the yields are frequently less than 50% (*2203, 2668*). The tritylamino acids are best isolated as diethylammonium salts (*2668*). The free acids are obtained by adding the equivalent amount of hydrochloric acid (*2203*). Amiard *et al.* (*38*) have used the difficult hydrolysis of tritylamino acid esters for the synthesis of tritylaminodicarboxylic acid monoesters. One equivalent of alkali hydrolyzes selectively the γ-ester bond of tritylglutamic acid diesters. Tritylglutamic acid α-monomethyl and α-monoethyl esters have been prepared in this way. N^{ϵ}-Trityllysine can be obtained by selective detritylation of the ditrityl derivative (*35, 236*). This procedure can be applied to N^{α},N^{ϵ}-ditrityllysyl peptides (*35*; see Chapter IV,B,1,b,δ). Certain trifunctional amino acids will yield the ditrityl derivatives by the usual tritylation. The N^{α}-trityl group can be removed selectively from *N,S*-ditrityl cysteine (*39, 2668*) by boiling with 50% acetic acid (*39;* see Chapter IV, G,1,c). N^{α},N^{im}-Ditritylhistidine (*37, 2203*) can also be detritylated selectively. The *O*-protected derivative of tyrosine has been obtained by the same approach from *N,O*-ditrityltyrosine benzyl ester (*2203*).

The influence of the trityl group makes the synthesis of peptides with tritylamino acids more difficult. Tritylamino acid esters, with the exception of the glycine derivative, form no hydrazides (*1006, 1096,*

2668). Iselin (*1096*) tried to prevent these difficulties by selective tritylation of amino acid hydrazides, but did not succeed. Trityl peptide esters, however, can be converted to the hydrazides. The instability of the trityl group to acid does not interfere with the azide coupling; it is entirely possible to use it at sufficiently low temperatures (*1096*). Efforts to obtain derivatives suitable for coupling reactions from amino acid *p*-nitrophenyl esters by tritylation were also unsuccessful. Steric hindrance makes the formation of the peptide bond impossible (*786a*).

The anhydride method is applicable to tritylglycine and tritylalanine (*37, 1006, 1121*). It fails with other amino acids such as phenylalanine, leucine, and asparagine. The steric influence of the trityl group diminishes so much in trityl dipeptides that the mixed anhydride method is again applicable (*2668*). There are no difficulties with the carbodiimide (*37, 230, 291, 1459, 2203*) and phosphazo (*2203*) methods or when peptide bonds are to be formed via the cyanomethyl ester (*2018, 2031*). The acid chloride method has also been employed (*490*). Tritylamino acid chloride hydrochloride is formed during the reaction with phosphorus pentachloride (*2668*) [66]:

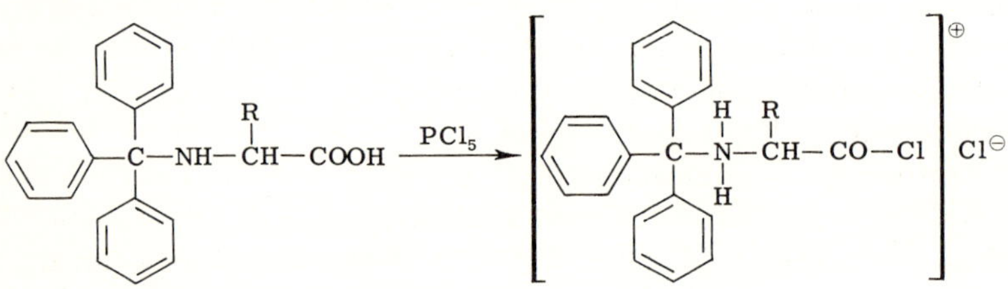

[66]

Hydrochlorides are also formed on other occasions. They are hydrolyzed very readily in water (*2668*).

The trityl group is normally removed by acid under mild conditions. Standing of trityl peptides or trityl peptide esters for 10 minutes in 50% acetic acid or boiling in glacial acetic acid is sufficient (*37, 291, 236, 2017a*). The cleavage poses no problems for higher peptides. An undecapeptide was detritylated in 10 minutes in 80% acetic acid at a temperature of 100°C (*295*). The *tert*-butyloxycarbonyl group is stable under these conditions (cf. page 155). This selective detritylation was used by Schwyzer *et al.* (*2022*) for the synthesis of an ACTH sequence. The protected peptide was treated with 75% acetic acid for 30 minutes at 30°C. The removal of the trityl group can be carried out in a few minutes with a small excess of hydrogen chloride or aqueous hydrochloric acid in ethanol (*1121, 2668*), acetonitrile (*2031*), acetone (*2369*), or ethyl acetate/acetonitrile (*2018*). Cleavage of peptide bonds, however, may occur in 10% ethanolic hydrochloric acid (*784*). Trityl ethyl ether is formed in addition to the peptide or peptide ester hydrochloride during the detritylation under anhydrous conditions, triphenylcarbinol in the presence of water (*2668*) [67]:

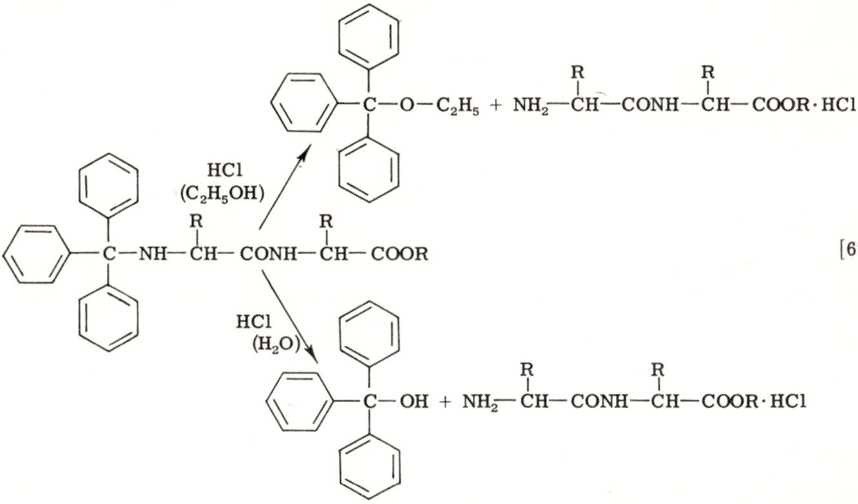

[67]

Trifluoroacetic acid proved to be useful for the detritylation of higher peptides (exposure of the blocked peptide to the reagent at 0° to —10°C for a few minutes). In pure trifluoroacetic acid the cleavage is brought about by the slow addition of water; the triphenylcarbinol precipitates. Trifluoroacetic acid can be used in aqueous acetic acid (*2027, 2029, 2030*). The trityl group has been frequently used to prepare linear derivatives for the synthesis of cyclic peptides. The acid hydrolysis can be carried out under conditions mild enough that sensitive ester groups, such as the cyanomethyl ester, are not affected (cf. page 104).

The trityl group can also be removed by catalytic hydrogenation (*2668*); triphenylmethane is formed. The reaction has not been often used.

The trityl group is not the first among the protecting groups in peptide chemistry. It has become an important one, however, in spite of all its disadvantages and the many difficulties encountered with it.

2. Mono- and Dibenzyl Groups

The blocking of the amino function with the benzyl residue presents a situation entirely different from that of the large trityl group. Monobenzyl- and dibenzylamino acids form hydrochlorides very readily. They are poorly soluble in cold water and thus very well suited for isolation and purification. They are different in this respect from other *N*-protected amino acids. The hydrochlorides are readily soluble in chloroform (*2370*).

Normally the *N,N*-dibenzylamino acids are used in peptide synthesis. They are prepared from amino acids and benzyl chloride in boiling aqueous alcohol in the presence of alkali. By this procedure *N,N*-dibenzylamino acid benzyl esters are formed; they are subsequently hydrolyzed (*2370, 2371*). The monobenzyl derivatives are prepared from *N*-benzylideneamino acid esters by catalytic hydrogenation followed by hydrolysis (*2370*), or from dibenzylamino acids by selective hydrogenolysis of one benzyl group (*903*). The reaction of dibenzylamine with a cyanohydrin and the hydrolysis of the nitrile group (*41*) yields only DL-compounds. Frankel *et al.* (*763*) have prepared *N*-benzyl-DL-aspartic acid and derivatives of it by adding benzylamine to maleic acid [68]:

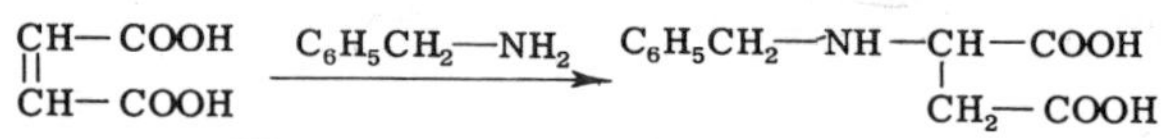

[68]

N-Benzylasparagine (*763*) has been prepared from maleic acid monoamide by an analogous reaction. β-Asparaginyl peptides are obtained from maleic anhydride by opening the anhydride with amino acid esters and the addition of benzylamine (*1440*). It is remarkable that this reaction yields β-peptides exclusively (see Chapter IV,C,2,a,α_5). *N*-Monobenzyl- and *N,N*-dibenzylamino acids form acid chlorides with phosphorus pentachloride. They are present as hydrochlorides (*2370*). These derivatives have been used in peptide syntheses in the usual manner. Dibenzylamino acids have been coupled with amino acid esters by the mixed anhydride method (*1741, 1742, 2372*). The secondary amino group can be acylated. *N*-Benzyl-DL-aspartic acid and acetic anhydride form *N*-benzyl-*N*-acetyl-DL-aspartic acid anhydride (*1442*).

The protecting group can be removed by catalytic hydrogenation. The monobenzyl derivatives are formed when dibenzyl derivatives are hydrogenated by the standard procedure using palladium black or, better, palladium on barium sulfate as catalyst in methanol/hydrochloric acid (*903*). The hydrogenolysis must be carried out in acetic acid or aqueous ethanol at 70°–80°C to complete the cleavage of the second group.

N-Monobenzyl- and *N,N*-dibenzylamino acids are not of importance for the synthesis of peptides. The *N*-benzylation is used only for nitrogen protection in the imidazole ring of histidine.

3. Trialkylsilane as Alkyl-Type Protecting Group

Amino acid esters can be converted to trialkylsilylamino acid esters by treatment with trialkylchlorosilane (*249, 1880*). Free amino acids yield trialkylsilylamino acid trialkylsilyl esters [69] (*1877, 1878, 1881*), which can also be obtained with silazanes (*250, 1879, 1881*):

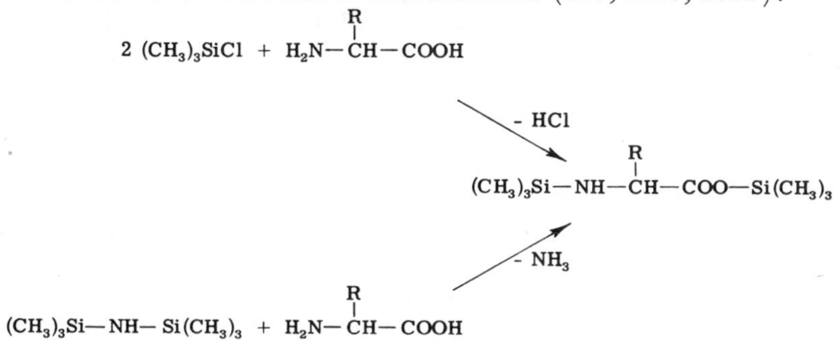

[69]

The silylated amino acids can be coupled with *N*-protected amino acids using the imidazolide or mixed anhydride method (*246, 251*). The silyl residues are hydrolyzed during isolation and the *N*-protected peptide acids are obtained directly [70]:

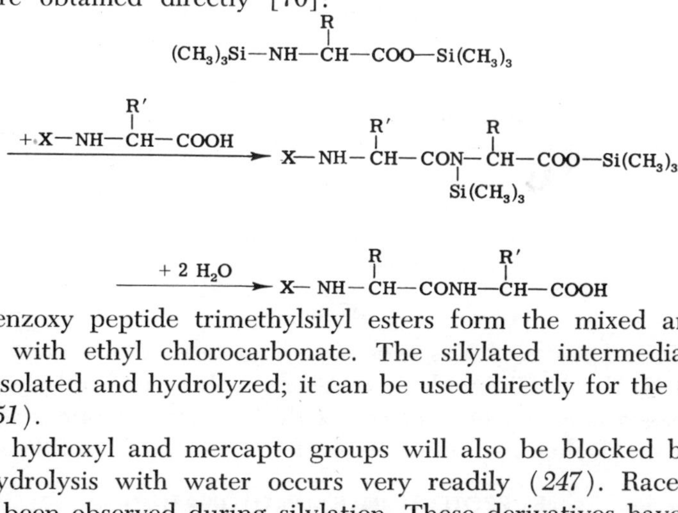

[70]

Carbobenzoxy peptide trimethylsilyl esters form the mixed anhydride directly with ethyl chlorocarbonate. The silylated intermediate need not be isolated and hydrolyzed; it can be used directly for the coupling step (*251*).

Free hydroxyl and mercapto groups will also be blocked by silylation. Hydrolysis with water occurs very readily (*247*). Racemization has not been observed during silylation. These derivatives have not yet been used.

D. ARYLIDENE-TYPE PROTECTING GROUPS

Schiff's bases are normally so unstable that they cannot be used as protecting groups. The N^{ϵ}-benzylidene derivative of lysine has been described; it can be applied conveniently for N^{α}-carbobenzoxylation (*236*). Hydrazides have also been isolated occasionally as benzylidene or isopropylidene derivatives (*642, 2173, 2719;* cf. page 81). Attempts to apply α-benzylideneamino acid *p*-nitrophenyl esters to peptide synthesis did not meet with success (*786a*). Several carbonyl derivatives that form stable compounds with amino acids have been described more recently. They are suitable as amino-protecting groups for peptide syntheses.

Sheehan and Grenda (*2066*) have used the 1-hydroxy derivative of benzaldehyde. The Schiff's base is stabilized by a hydrogen bond [71].

[71]

The ready cleavability of these protecting groups is a great advantage. They can be removed with 1 equivalent of acid at 0°C. There is little tendency to racemization, since the formation of azlactones is not possible. Several peptide syntheses using the carbodiimide method have been reported. The mixed anhydride method is also applicable.

The reaction of 1,3-dicarbonyl compounds, such as benzoylacetone, acetylacetone, or cyclopentan-2-one-1-carboxylic acid ethyl ester, with amino acids in an alcoholic solution of alkali leads to the formation of enamines, which are also stabilized by a hydrogen bond [72]. Dane

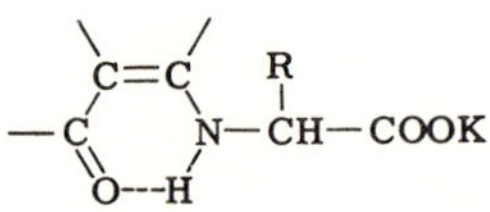

[72]

et al. (*563*) have used this type of amino protection and synthesized peptides by the carbodiimide method or via the cyanomethyl ester. The protecting groups can be removed with 2 *N* hydrochloric acid or 2 *N* acetic acid. Alkaline hydrolysis yields the *N*-protected peptide acids.

E. PROTECTION BY PROTONATION

Protonated amino groups are of importance for the synthesis of peptides in only a few special cases. This is in contrast to the protection of the carboxyl group by salt formation. Protonation, however, becomes essential for the prevention of side reactions during the synthesis of amino acid and peptide derivatives. Esters, due to ready diketopiperazine formation or polycondensation, can be obtained in a stable form only as salts. This is very important for activated esters of amino acids and peptides that are required as starting material for cyclization.

Amino acid and peptide chlorides can be obtained only as salts. One amino group in diaminocarboxylic acids can be protected selectively for certain reactions by salt formation. Under neutral conditions these amino acids are present as inner salts. It is therefore possible to N^{ω}-acylate lysine with carbobenzoxy phenyl ester or trifluoroacetic acid ethyl thiolester. The α-amino group, however, will be acylated by trifluoroacetic anhydride in trifluoroacetic acid because the ω-function is protonated.

Protection by protonation is of great value for the selective protection of the strongly basic guanido function of arginine. Here it makes possible the synthesis of peptides, irrespective of the presence of arginine residues in the carboxyl or amino component. The cleavage of guanido-protecting groups is difficult for higher peptides. Protonation, therefore, has been used frequently in the synthesis of biologically active polypeptides.

II. CARBOXYL-PROTECTING GROUPS

The prime objective of the substitution of the carboxyl group is not protection but rather the release of the amino acid from its zwitterion state. It also frequently facilitates the isolation of the reaction products. Protection is emphasized only in cases where one carboxyl group in amino-dicarboxylic acids is to be activated.

The most important and normally easiest type of substitution is esterification. The methyl and ethyl ester, should be mentioned first. They are supplemented very prominently by the benzyl ester, which can be hydrogenated catalytically, and by the *tert*-butyl and *p*-methoxybenzyl esters. They are both convenient for their ready hydrolysis in acids.

Another possibility for carboxyl substitution is the formation of amides or hydrazides. There are limitations to the use of amides. Their cleavage with preservation of the peptide bond, as in the case of the acyl-type amino-protecting groups, is possible only in exceptional cases. Acid hydrazides do not offer sufficient protection for the carboxyl group. The hydrazide group can also be acylated during peptide synthesis. This undesirable side reaction can be prevented by blocking the hydrazide function with an amino-protecting group. The protection of the carboxyl group by salt formation is becoming increasingly important, especially for the synthesis of higher peptides.

The carboxyl-protecting groups will be treated, like the amino-protecting groups, according to their chemical nature. The protecting groups are again compiled (in Table I) according to cleavability.

A. CARBOXYL PROTECTION BY ESTERIFICATION

1. Methyl and Ethyl Esters

There is little difference between the methyl and the ethyl ester with regard to preparation and application to the synthesis of peptides.

The esterification of an amino acid can be carried out with an excess of the corresponding alcohol in the presence of an acid as catalyst [1]:

$$H_2N-\underset{R}{\overset{|}{C}}H-COOH + HO-CH_3 \xrightarrow[-\ H_2O]{(HCl)} H_2N-\underset{R}{\overset{|}{C}}H-COO-CH_3 \cdot HCl \qquad [1]$$

The amino acid esters are obtained as ammonium salts. It is therefore necessary to use at least 1 equivalent of acid. Hydrochloric acid and thionyl chloride are used almost exclusively as catalysts. It is possible to use the thionyl chloride method (*348, 349*; cf. *1315*) with a lower excess of alcohol, since water is not formed during the reaction [2]:

$$H_2N-\underset{R}{\overset{|}{C}}H-COOH + HO-CH_3 + SOCl_2 \longrightarrow H_2N-\underset{R}{\overset{|}{C}}H-COO-CH_3 \cdot HCl + HCl + SO_2 \qquad [2]$$

The reaction is carried out by mixing the alcohol with 1.1 moles of thionyl chloride at $-10°C$ and adding the amino acid. Esterifications with dimethyl sulfite in the presence of toluenesulfonic acid are known (*2290*). It is likely that in the case of thionyl chloride the initially formed chlorosulfinic acid ester [3] represents the esterifying reagent:

$$CH_3OH + SOCl_2 \longrightarrow CH_3-O-SO-Cl + HCl \qquad [3]$$

The reaction conditions for an esterification differ significantly from amino acid to amino acid. Standing at room temperature (*2251*) for varying periods or brief heating (*478, 2364*) is sufficient for simple amino acids. The application of these techniques to amino acids with long side chains frequently leads to the formation of amino acid hydrochlorides only (*1115*). In such cases, the esterification can be achieved by extended heating to higher temperatures (*1115, 2364*). It is sometimes necessary to repeat the operation to complete the reaction (*259, 296*). Dipeptides and tripeptides may be esterified by the same procedure (*715*). These methods furnish the amino acid methyl ester and amino acid ethyl ester hydrochlorides in good yield and with a high degree of purity. Other procedures are therefore of importance only for special cases, e.g. amino acid *N*-carboxylic anhydrides are cleaved with methanol to yield methyl esters. Bergmann *et al.* (*204*) prepared N^{ϵ}-carbobenzoxylysine methyl ester by this method and Zervas *et al.* (*2670*) N^{G}-carbobenzoxyarginine methyl ester. The direct esterification of ω-carbobenzoxydiaminocarboxylic acids may give rise to side reactions, due to the relatively great acid lability of the ω-group. It has nevertheless been described several times (*686, 1119, 2099*). Taschner and Wasielewski have reported a method for the transesterification of *N*-protected amino acids with ethyl or methyl formate or the corresponding acetate (*2276, 2279, 2282*). They used sulfuric acid or toluenesulfonic acid as catalyst. Only in the case of phenylalanine were they successful in esterifying the free amino acid with ethyl acetate (*2279*). When this method was applied to aminodicarboxylic acids with perchloric acid as catalyst, only the ω-esters were obtained (*241*). They can be separated from the unchanged acid on Amberlite IR-4B. Transesterifications are not possible with phenyl acetate or cyanomethyl acetate.

The acid-catalyzed esterification of free amino-dicarboxylic acids under sufficiently vigorous conditions leads to the formation of diethyl or dimethyl esters (*714, 834, 849, 1424, 1950*). The ω-esters can be prepared selectively at room temperature and in relatively short reaction times. The reaction conditions must be strictly observed (*195, 931, 1560*; see Chapter IV,C,1,a,β and 2,a,β). The α-ethyl and α-methyl esters are accessible by alcoholysis of the *N*-protected intramolecular anhydrides. The cleavage depends upon the nature of the *N*-protecting group. Mixtures with varying contents of α- and ω-esters are obtained. They can be separated by differences in p*K* values (*1369*) or by differences in solubility of the dicyclohexylammonium salts (*2481, 2486, 2496;* cf. page 184).

N-Protected amino acid methyl or ethyl esters are obtained by attaching an amino-protecting group to the esterified amino acid. Frequently the procedure is reversed, i.e. *N*-substituted amino acids are esterified with thionyl chloride or hydrogen chloride as catalyst (*1315, 2251*). Many *N*-protecting groups and the amide group of glutamine and asparagine are attacked under the conditions of an acid-catalyzed esterification. Taschner and Wasielewski (*2276, 2277*) therefore studied a series of acid chlorides for their applicability as catalysts. Under mild reaction conditions, yields were over 80% with thionyl chloride, sulfuryl chloride, acetyl chloride, phosphorus pentachloride, phosphorus pentabromide, and tetrachlorosilane. The mildest conditions for protecting groups labile to acids are presently offered by the esterification with diazomethane (*204, 692, 725, 1033, 1636, 1844, 2022, 2160, 2654*). This method can be applied to *N*-protected glutamine and asparagine. Diazoethane has been described for the synthesis of ethyl esters (*108*).

The ammonium form of an ester is not capable of adding to the carboxyl C atom. The use of the free ester is therefore usually required for a peptide coupling reaction. Free amino acid esters are liberated from their hydrochlorides by aqueous alkali and potassium carbonate and can be extracted with ether or ethyl acetate (*714, 2251*). They may also be obtained by the addition of a base, such as triethylamine or ammonia, to the hydrochloride in an organic solvent, e.g. chloroform (*95, 1007*) or methanol (*478*). The ammonium chloride is precipitated with ether and the free ester remains after the evaporation of the solvent. The hydrochloride has also been removed with sodium methylate/methanol (*719, 722, 1625, 2029*). The reaction with dicyclohexylamine in the absence of a solvent and the subsequent distillation of the free ester have been described (*2496*). The esters are normally very unstable compounds. Aminolysis leads to the formation of the corresponding diketopiperazines [4] (*714*). This reaction is faster for amino acids with short side chains (*8*). Dipeptide esters also form diketopiperazines (*715*):

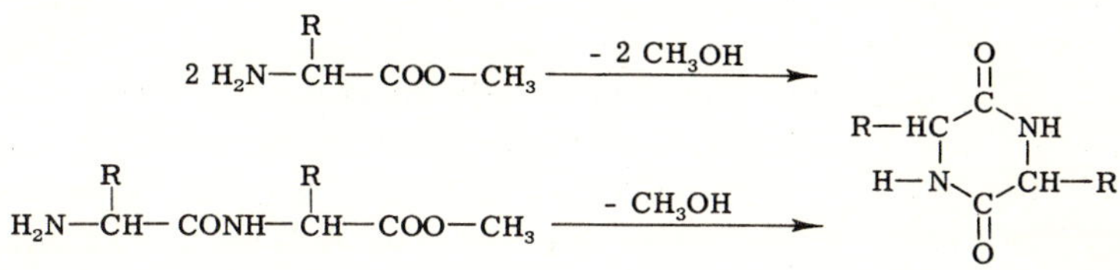

[4]

The free esters of higher peptides, however, are relatively stable compounds. Some of them have been crystallized (*685, 2022*). Free esters are normally not isolated. An equivalent amount of a tertiary base, usually triethylamine or tri-*n*-butylamine, is added to the hydrochloride in a suitable solvent, such as tetrahydrofuran, dimethylformamide, or chloroform. This mixture is used directly for the coupling reaction. The precipitated trialkylammonium hydrochloride is sometimes removed by filtration. Ion-exchange resins have been used for the binding of the acid, especially in cases of higher peptide esters (*296, 297*). Sometimes the activated carboxyl component and the amino acid ester hydrochloride are dissolved together and the ester is liberated by the subsequent addition of a base (*493, 2060*).

The cleavage of methyl or ethyl esters after the peptide synthesis is done in most cases by alkaline hydrolysis. An immense number of applications has been reported. There are reports in abundance on useful solvents and solvent combinations. Methanolic or ethanolic potassium hydroxide solution (*2637*) or aqueous organic mixtures, e.g. methanol/water and ethanol/water (*336*) or acetone/water (*274, 683, 1606, 1932*), have been used. Dioxane and dioxane/acetone/water (*296, 639a, 683, 1183, 1625, 2251*) have been shown more recently to be useful. They are excellent solvents for higher peptides (*1625*). The use of dimethylformamide, dimethyl sulfoxide (*232, 2225*), and pyridine (*296*) has been described. Peptide esters have been hydrolyzed in heterogeneous phase, e.g. in ethyl acetate/aqueous sodium hydroxide (*894*).

The hydrolysis is normally carried out with a small excess of alkali for 0.5–1.5 hours at room temperature. Reaction times of up to 24 hours however, may be sometimes required (*685, 1009, 2052, 2386*). Lower temperatures have been reported for the extended hydrolysis times (*1113*). The excess of alkali may be as high as tenfold (*685, 686, 2027, 2029*) and temperatures of up to over 40°C have been reported (*336, 2029, 2302, 2628*). The hydrolysis of tritylamino acid esters by brief boiling in 20% alkaline propylene glycol represents an extreme case (*37*). Peptide bonds are normally not attacked during alkaline hydrolysis, seryl peptides are an exception (*894, 2637*). There is no difficulty with the addition of the base in small portions (*894, 949*; cf. *692*). *O*-Substituted serine does not show these side reactions (*1090*).

The hydrolysis becomes increasingly difficult with the growing peptide chain (*683*). The reaction conditions required are dangerous for the stability of *N*-protecting groups and chances are enhanced for racemization. There are certain advantages to the acid hydrolysis, which is normally conducted in aqueous hydrochloric acid. Phthalyl or carbobenzoxy peptide esters can be hydrolyzed in such solvents as acetone or dioxane in 0.5–1.5 hours at room temperature (*1183, 2060*). Carbobenzoxy peptide esters are hydrolyzed and decarbobenzoxylated in 12 *N* HCl at 39°C (*1538, 1951*). The cleavage of the ester bond is facilitated in peptide esters where the amino group is not protected (*478, 2017*). Rather vigorous reaction conditions, if required for the acid hydrolysis of methyl or ethyl esters, may lead to the cleavage of peptide—especially ω-acid amide bonds. The hydrolysis of the ω-acid amide bond has been used with advantage in several cases. It makes possible the removal of temporary blocking groups from the ω-carboxyl function during peptide synthesis (*2016, 2017*). The acid hydrolysis is sometimes carried out with hydrogen bromide (*42*). Taschner *et al.* (*2268, 2269, 2282*) have studied it in detail in glacial acetic acid as solvent. Yields of 80% are obtained when 1.5 moles of hydrogen bromide are used per mole of ester. The reactions were carried out at room temperature for 4–6 days. The phthalyl and tosyl groups and peptide bonds are not attacked under these conditions. The ester bond in carbobenzoxy peptide esters can be hydrolyzed selectively with toluenesulfonic acid (*2263*).

Taschner (*2263*) and Taschner and Liberek (*2270, 2271*) have carried out nonhydrolytic cleavages by heating *N*-protected amino acid or peptide esters with lithium halides in pyridine. Yields in this halolytic reaction increase in the order: lithium chloride → lithium bromide → lithium iodide. Methyl esters are cleaved more readily than ethyl esters. The following reaction mechanism has been proposed [5]:

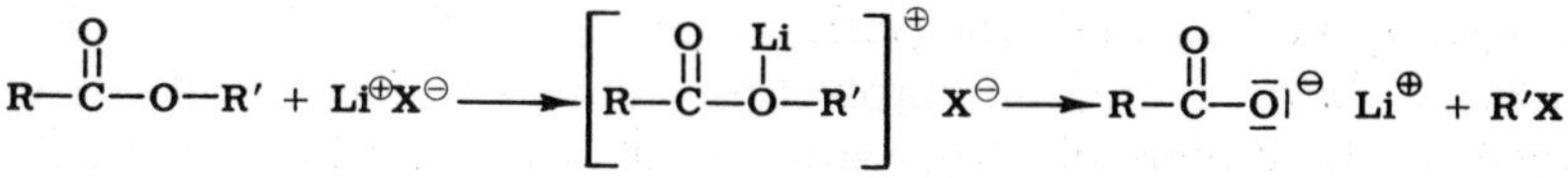 [5]

Kloss and Schröder (*1263a*) have reported the enzymatic hydrolysis on a preparative scale of *N*-protected or free peptide esters with chymotrypsin or trypsin. The esterase activity of the enzymes is widely independent of the C-terminal amino acid. Peptide esters with terminal D-amino acids and ω-esters will not be hydrolyzed. Proteolytic cleavage is minimal under the conditions of saponification (cf. *1470*).

Nefkens (*1598, 1600a*) has described the phthalimidomethyl ester, a modified methyl ester that can be prepared from *N*-chloromethylphthalimide and carbobenzoxyamino acids, followed by catalytic hydrogenation [6]:

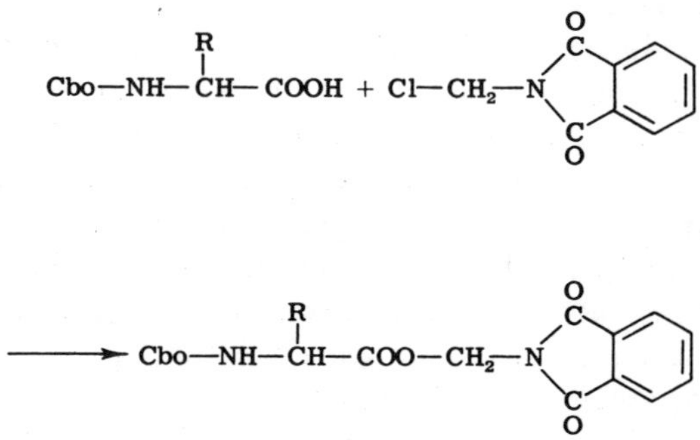 [6]

N-Protected peptides can be esterified by the same procedure. It is relatively easy to hydrolyze this ester with acid. Conditions are: hydrogen chloride in dioxane or ethyl acetate for 16 hours or hydrogen bromide in glacial acetic acid for 10–15 minutes. Furthermore, the ester can be hydrolyzed with hydrazine hydrate for 3 hours in ethanol at room temperature, whereby the free acid is formed. Racemization has not been observed during the synthesis of the ester.

2. *tert*-Butyl Esters

An important breakthrough came with the introduction of the *tert*-butyl ester (*43, 48, 2272, 2280*). It can be hydrolyzed very readily in acid, while more vigorous reaction conditions are required for methyl and ethyl esters.

Several methods are available for the synthesis of amino acid *tert*-butyl esters. The esterification with isobutene must be mentioned first. It is catalyzed by sulfuric acid and can be carried out with free amino acids in dioxane (*1862*) or with *N*-protected amino acids in methylene chloride (*48, 1977*). The carbobenzoxy group (*48*), removable by catalytic hydrogenation, and the phthalyl group (*1977*), cleavable by hydrazinolysis, have served as protecting groups. Transesterification takes place on treatment of free amino acids with *tert*-butyl acetate. Better results are obtained, however, with the *N*-protected amino acids (*2265, 2280*). Perchloric acid is used as catalyst for this transesterification; sulfuric acid or toluenesulfonic acid is not as good. The reaction of *N*-protected amino acids with *tert*-butanol and phosphorus oxychloride in the presence of an excess of pyridine also leads to the formation of *tert*-butyl esters. Carbobenzoxy- and phthalylamino acids can be esterified by this method, but not tosylamino acids. Racemization may occur (*2264*). Another approach, although of theoretical interest only, is the esterification of α-halogen carboxylic acids with isobutene, reaction of these esters with sodium azide, and reduction to the amino acid *tert*-butyl esters (*2394*).

The free *tert*-butyl esters of most amino acids are stable and distillable liquids. They are not subject to self-condensation (*48*), even if kept at room temperature (for the autocondensation of glycine *tert*-alkyl esters; cf. *2395*). This is another advantage of the *tert*-butyl esters in addition to their possible acid hydrolysis. They are very stable to hydrazinolysis and aminolysis (*48*) and much more difficult to hydrolyze by alkali than the methyl or ethyl esters. The variations in the synthesis of peptides with aminodicarboxylic acids have thus been greatly extended. It is not safe to assume that aspartic acid β-*tert*-butyl esters will always be stable to hydrazine and alkali (*2017a*). Suitable derivatives for peptide coupling at the α-carboxyl group have been obtained as ω-*tert*-butyl esters (*1173, 1974, 1975, 2007, 2019, 2598, 2598a*). The reverse is true for α-*tert*-butyl esters for coupling reactions to be carried out at the ω-carboxyl group (*2274, 2281, 2283*). The esters are so stable to alkali that it is possible to hydrolyze a nitrile to the acid amide while the *tert*-butyl ester is present (*1419*). The synthesis of the *tert*-butyl esters of arginine, N^G-substituted arginine, histidine, and tryptophan has not yet been reported. During the esterification of serine and threonine with isobutene, simultaneous etherification of the hydroxyl function takes place (*228*). This does not pose a problem since the *tert*-butyl ethers can be cleaved just as readily as the esters. Quite the contrary is true in peptide synthesis and it may be advantageous to use *O*-protected hydroxyamino acids (*230, 457, 1962*).

The *tert*-butyl esters of amino acids are also known as crystalline salts. Roeske (*1862*) and Taschner *et al.* (*2265, 2280*) have prepared the chlorides and Anderson and Callahan (*48*) the phosphites of several *tert*-butyl esters of amino acids. For the prevention of hydrolysis, especially when strong acids are used, it is important to observe low temperatures and to eliminate an excess of acid during the salt formation (*1977, 2394*).

The cleavage of the *tert*-butyl esters can be carried out with hydrogen chloride in ethyl acetate (*1977*) or methylene chloride (*2283*), with toluenesulfonic acid in benzene (*48*), or with trifluoroacetic acid (*1173, 2024*). The latter reagent has been frequently used for the synthesis of sequences of biologically active peptides. Of advantage is the simultaneous removal of the *tert*-butyloxycarbonyl group. The *tert*-butyl ester can be cleaved with HBr in glacial acetic acid as long as the *N*-protecting groups present are stable to acid (e.g., the phthalyl group) (*48*). The combination of trifluoroacetyl group and *tert*-butyl ester—selective cleavage of the trifluoroacetyl group with alkali—is also known (*48*).

The *tert*-butyl ester, although relatively new as carboxyl-blocking group in peptide chemistry, has become very important. Today it is one of the most useful protecting groups.

3. Benzyl Esters

Bergmann *et al.* (*204, 205*), in addition to the carbobenzoxy group, introduced to peptide chemistry the benzyl ester. It can readily be cleaved by catalytic hydrogenation and was for long the only carboxyl-protecting group, besides the methyl and ethyl esters, that could be removed selectively.

The preparation of amino acid benzyl esters by the reaction of amino acid chloride hydrochlorides (*938, 1876*) or amino acid *N*-carboxylic anhydrides (*204, 683, 684, 1059*) with benzyl alcohol is not of great importance. Several N^{ω}-protected diaminocarboxylic acid benzyl esters, however, are accessible only through the *N*-carboxylic anhydrides (see Chapter III,B,5,c).

The esterification in benzyl alcohol with hydrogen chloride as catalyst, which is the standard method for the synthesis of methyl and ethyl esters, requires several repetitions of the procedure (*1606*) and the yields are frequently poor. Nevertheless several benzyl esters, e.g., those of proline (*1606*) and hydroxyproline (*8*), are still obtained by this method. Peptides can be esterified more easily by this procedure (*1035, 2619*). The use of thionyl chloride as catalyst has also been described (*1786, 2099*). Better results are obtained when the esterification is carried out in a benzene solution and the water is removed by azeotropic distillation (*1539*). Other catalysts, however, are used more frequently. Erlanger and Hall (*690*) esterify for 4 hours at 90°–95°C in the presence of polyphosphoric acid. The amino acid benzyl esters are either isolated directly from the reaction mixture or precipitated by hydrogen chloride as hydrochlorides after the aqueous alkaline solution has been extracted with ether. Miller *et al.* (*1558, 1561*) have described benzenesulfonic acid as catalyst and Izumiya and Makisumi (*1116*) toluenesulfonic acid. The esters may again be obtained as hydrochlorides or as sulfonates directly. Peptides can also be esterified by this method (*1035, 2054a, 2302*). The high temperature (110°–120°C) that must be maintained during the reaction is disadvantageous. It is therefore better to use as solvent, e.g., benzene (*542, 1117, 2670*), toluene (*178*), or carbon tetrachloride (*236, 494*). The esterification is favored by the presence of benzyl borate. It is highly reactive with water and the equilibrium is constantly shifted in the desired direction (*1274*). It is advantageous during the esterification in the presence of benzene- or toluenesulfonic acid to remove the water formed by azeotropic distillation (*1474*). The esterification of serine and threonine may be accompanied by a side reaction and lead to the formation of benzyl ethers (*1940*). The application of sulfonyl chloride has its limitations. The esterification is successful only when the amino acids or peptides are used as toluene sulfonates in *sym*-tetrachloroethane as solvent (*2278*). Amino acid benzyl esters are also accessible in the presence of toluenesulfonic acids by heating to 100°C in a five- to tenfold excess of dibenzyl sulfite (*2290*). The method is not applicable to alanine, lysine, and proline.

The ω-benzyl esters of aminodicarboxylic acids are obtained by selective esterification of the unprotected amino acid in the presence of 1 equivalent of toluenesulfonic acid (*498*) or in the presence of hydrogen iodide (*931*) or hydrogen bromide (*259*). The reaction of the copper complex with benzyl chloride and potassium hydrogen carbonate has also been described (*931*). The α-benzyl esters of aminodicarboxylic acids can be prepared by partial acid hydrolysis of the free diesters with hydroiodic acid (*436, 1889*). *N*-Protected aminodicarboxylic acid monoesters are accessible from the free monoesters by the introduction of the amino-protecting group. Partial hydrolysis of *N*-protected dibenzyl esters leads to the formation of ω-esters (*178, 1750*). The steric hindrance of the protecting group permits the selective hydrolysis of the ω-ester group of trityl compounds. The α-benzyl esters, especially of carbobenzoxy compounds, are obtained by cleavage of the intramolecular anhydrides (*205*). They can be readily isolated in a pure state in the form of their dicyclohexylammonium salts (*1260*).

N-Protected amino acid benzyl esters of simple amino acids are rarely prepared, since their amino-protecting groups would have to be removed to use them as carboxyl-protected derivatives. They are accessible by esterification of the *N*-protected amino acids with catalytic amounts of toluenesulfonic acid (*149, 596*) or sulfuryl chloride (*2278*) and also with the equivalent amount of thionyl chloride (*1315*) in benzene or *sym*-tetrachloroethane. Other methods are the reaction with diazophenylmethane (*1911a, 2163*) or of silver salts of *N*-protected amino acids with benzyl halides (*759*).

Amino acid or peptide benzyl esters are used in peptide chemistry similarly to the methyl and ethyl esters. Their cleavage can be achieved by catalytic hydrogenation and leads to the formation of *N*-protected peptide acids when amino-protecting groups stable to hydrogenolysis are present. Free peptides are obtained in the presence of groups that can be cleaved by hydrogenolysis. The hydrogenolysis is carried out in the same way as that of the carbobenzoxy group (cf. page 26). The catalytic hydrogenation of benzyl esters in ethanol may be accompanied by transesterification and formation of the ethyl ester. This can be prevented by the use of *tert*-butanol as solvent (*152, 542*).

It is not possible to cleave benzyl esters selectively by catalytic hydrogenation in the presence of the carbobenzoxy group (*966*). Benzyl esters may be hydrogenated selectively in the presence of the trityl protection, although this group is also prone to hydrogenolysis (cf. page 47). The carbobenzoxy group can be removed selectively by phosphonium iodide in glacial acetic acid at 50°C (*937, 938*), or by hydrogen bromide at room temperature in 15–30 minutes either in glacial acetic acid (*149*) or in nitromethane (*30*). A more or less pronounced cleavage of the benzyl ester cannot always be prevented (*125, 1248, 2637*). The action of 30% HBr in glacial acetic acid (15 minutes) is sufficient to lead to the formation of monoester and free acid in the case of glutamic acid and aspartic acid dibenzyl esters. The cleavage of the ω-monoesters is faster than that of the α-esters. It is worthwhile noting that considerable amounts of the diester are formed on treatment of L-glutamic acid α-benzyl ester with hydrogen bromide in glacial acetic acid. Peptide bonds are also in danger of being cleaved (*2637*). The excessive action of HBr/glacial acetic acid at room temperature (*966*) or the passing of hydrogen bromide through a solution of the carbobenzoxy peptide benzyl ester in glacial acetic acid at 60°C (*147*) caused the complete hydrolysis of the benzyl ester, with formation of the free peptide hydrobromide.

The benzyl ester, as a protecting group that can be removed reductively, is also cleaved by sodium in liquid ammonia (*971, 1828*; cf. *2119*). This reaction offers another possibility for removing it simultaneously with the carbobenzoxy group and eventually other protecting groups.

It was demonstrated in several cases that *N*-protected peptide benzyl esters can be converted with hydrazine hydrate to the hydrazides (*686, 1902, 1912, 2637*). The alkaline hydrolysis of benzyl esters is less important. The conditions are the same as for methyl or ethyl esters (*236, 806, 1605, 1606, 1863, 1991, 2007, 2300*).

4. Modified Benzyl Esters

The *p*-nitro- and *p*-methoxybenzyl derivatives are known, as in the case of the substituted carbobenzoxy groups, as modified benzyl esters.

a. *p*-Nitrobenzyl Ester

The combination of the carbobenzoxy group with the benzyl ester is very well suited for the synthesis of smaller peptides. Difficulties must be faced only during the synthesis of the intermediates. The required selective cleavage of the carbobenzoxy group with HBr in glacial acetic acid is not always assured in the presence of the benzyl ester. Certain advantages are offered by the *p*-nitrobenzyl ester, which has been described by Schwyzer and Sieber (*2032*). Schwarz and Arakawa (*1991*) treated simple carbobenzoxy dipeptide *p*-nitrobenzyl esters for 1 hour with HBr in glacial acetic acid. No free peptide, or only an insignificant amount, was found. The extent of cleavage of the ester is small, even after 2 hours of reaction time, amounting to 15%. The ready crystallizability of the *p*-nitrobenzyl esters (*1991, 2032*) offers a further advantage.

The amino acid *p*-nitrobenzyl esters are prepared by boiling the carbobenzoxyamino acids and *p*-nitrobenzyl bromide or chloride in ethyl acetate while a tertiary base such as triethylamine is present. The carbobenzoxy group is removed subsequently by treatment with HBr in glacial acetic acid (*292, 1991, 2032, 2163*). They are also accessible from the free amino acids by esterification with *p*-nitrobenzyl alcohol in the presence of *p*-toluenesulfonic acid in carbon tetrachloride as solvent (*2103, 1975*). A further synthetic method consists in the reaction of carbobenzoxyamino acids or carbobenzoxypeptides (such as sodium- or trialkylammonium salts) with *p*-nitrobenzyl tosylate in acetone–dimethylformamide (boiling for ½ hour). Yields are above 70%; racemization is not observed. Tritylamino acids can also be esterified in this manner. No side reactions were noted for hydroxyamino acids with the hydroxyl group (*2306b*). The *p*-nitrobenzyl esters can be cleaved by catalytic hydrogenation. During the reaction *p*-aminotoluene is formed; it is not known to what extent its presence causes difficulties in the isolation and purification (cf. the cleavage of the *p*-nitrocarbobenzoxy group, page 32). The alkaline hydrolysis has also been reported (*1104*). The *p*-nitrobenzyl ester has been used many times for the synthesis of biologically active polypeptides.

b. *p*-Methoxybenzyl Ester

Weygand and Hunger (*2487*) have described the *p*-methoxybenzyl ester. It is not possible at this time to evaluate what advantages it offers. The ester can be cleaved by catalytic hydrogenation and by acid hydrolysis under very mild conditions. This means that a selective cleavage of amino-protecting groups is possible only for the trifluoroacetyl residue. There are so far no reports on its stability to hydrazine that would make possible a combination with the phthalyl group.

5. Carboxyl Protection and Simultaneous Activation

Ester bonds may be cleaved by hydrazinolysis, ammonolysis, or aminolysis. These properties are used in peptide chemistry for the synthesis of hydrazides and amides and in the case of suitable esters for the formation of peptide bonds. Certain conditions permit the use of activated esters as carboxyl-protecting groups. Esters with a higher degree of activation than, for example, methyl or ethyl esters are at an intermediate level between activated esters and carboxyl-protecting groups. They are not sufficiently activated to give rise to side reactions during peptide synthesis. Their main advantage is the ready amide or hydrazide formation. This would be difficult to achieve, especially for higher peptides, with the commonly used methyl or ethyl ester. The phenyl esters and ethyl thiolesters are useful for this purpose.

a. Use of Truly Activated Esters

Normally the amino-protecting group is removed first from the activated ester of an *N*-protected amino acid or an *N*-protected peptide. The resulting free ester is used for the peptide synthesis by established procedures. The method employed must provide a higher degree of activation than the activated ester or it will not be possible to prevent side reaction (cf. page 101).

b. Phenyl Esters

Kenner (*1212*) has suggested the use of the phenyl ester as carboxyl-protecting group. It can be prepared—like many aryl esters—from *N*-protected amino acids and diphenyl sulfite or triphenyl phosphite (cf. page 99). It has also been obtained via acid chlorides and by the anhydride method (*265*). The esterification of *N*-protected peptides results in racemization. Intramolecular anhydrides of aspartic acid and glutamic acid are cleaved with phenol. The removal of the amino-protecting group is required in each case. The ester bond can be hydrolyzed in aqueous acetone at pH 11, or at pH 7.5 by boiling in aqueous dioxane in the presence of imidazole. Racemization was observed after boiling with mineral acids in aqueous dioxane (*1212*). Klieger and Gibian (*1260*) noted the ready hydrazinolysis and amide formation of the phenyl esters. (Cf. page 103 for the use of phenyl esters for peptide coupling reactions.)

c. Ethyl Thiolester

The ethyl thiolester displays properties quite similar to those of the phenyl ester. As an activated ester it requires rather vigorous reaction conditions (see Chapter III,B,2,b,β). Little is known about its application as carboxyl-protecting group. Its use was first discussed by Wünsch and Wieland (cf. *2263*). The oxidative cleavage offers an advantage and may be carried out, as in the case of the phenylthiocarbonyl group, with peracetic or perbenzoic acid. Thiolesters can also be cleaved with salts of heavy metals. Klieger and Gibian (*1259*) demonstrated the facile hydrazinolysis and amide formation in the case of glutamic acid. It is selective in the presence of methyl esters. Only the thiolester bond of carbobenzoxyglutamic acid γ-methyl ester α-ethyl thiolester is subject to ammonolysis in liquid ammonia.

B. CARBOXYL PROTECTION BY HYDRAZIDE OR AMIDE FORMATION

1. Substituted Hydrazides

The carboxyl protection rendered by the esters discussed in the previous section also offered activation to a certain degree. The carboxyl-protecting groups to be considered now already contain the hydrazide group. The hydrazide group itself is not suitable for carboxyl protection. An amino group cannot be acylated selectively in its presence (*2637*). The hydrazines used are therefore blocked with an amino-protecting group to prevent the side reactions. This approach makes feasible the formation of hydrazides, and the azide method can be employed in the case of very long peptides, for example, where hydrazinolysis is normally not possible. It can also be used in the presence of the phthalyl group, which is subject to hydrazinolysis of its own or in combination with the trifluoroacetyl group, which would be cleaved due to the strongly basic nature of hydrazine. The guanido group in peptides with arginine (*890*) or nitroarginine (*292*) is the cause of side reactions during hydrazinolysis. Again, the azide method can be employed only when protected hydrazides are used. The required use of an additional amino-protecting group is a disadvantage. The group must be chosen in such a way that the selective removal of any other protecting group is guaranteed. The cleavage of the hydrazide bond with formation of the free acid is not possible. This type of carboxyl protection is therefore suitable only for the synthesis of intermediates where chain extension at the carboxyl group by the azide method is contemplated.

a. Carbobenzoxy Hydrazides

Hofmann *et al.* (*1024, 1027*) described as early as 1952 the combination of the phthalyl, tosyl, and carbobenzoxy groups as amino protection with carbobenzoxyhydrazine as carboxyl protection. They have made use of the variations offered by this combination for synthesis. *N*-Protected peptide hydrazides—the hydrazide function [7] was blocked by the carbobenzoxy group—were prepared from phthalylglycine chloride and carbobenzoxyhydrazine. The phthalyl group was removed and the resulting N^1-glycyl-N^2-carbobenzoxyhydrazine coupled with the *N*-protected amino acid. *N*-Protected peptide hydrazides [8] can be obtained by catalytic hydrogenation when the phthalyl or tosyl group is used as amino protection. New coupling reactions can be carried out at the amino group after the hydrazinolysis of phthalyl peptide carbobenzoxy hydrazides [9]. The catalytic hydrogenation of carbobenzoxy peptide carbobenzoxy hydrazides yields free peptide hydrazides [10] that can be used as starting material for polycondensations or cyclizations (*1493*) (cf. page 66).

Carbobenzoxyhydrazine is obtained by partial hydrazinolysis of dibenzyl carbonate (*1024*) or by carbobenzoxylation of hydrazine hydrate (*309*). The bis derivative is formed to some extent in both cases. It can be removed by fractional crystallization together with hydrazine hydrochloride. An improved synthesis uses carbobenzoxy *p*-nitrophenyl ester and an equivalent amount of hydrazine hydrate. The reaction is carried out at room temperature in the absence of solvents (*1979*).

b. *tert*-Butyloxycarbonyl Hydrazides

Boissonnas *et al.* (*292*) and Schwyzer *et al.* (*2038*) introduced *tert*-butyloxycarbonylhydrazine at practically the same time. They applied it in combination with the carbobenzoxy group as amino protection to the synthesis of bradykinin and a tyrocidin sequence.

N-Protected amino acids or peptides are coupled with *tert*-butyloxycarbonylhydrazine by the anhydride method (*292, 890*) or with carbodiimide in methanol (*2003*). The carbobenzoxy group can be removed by catalytic hydrogenation. The cleavage of the *tert*-butyloxycarbonyl group is carried out in the usual manner with hydrogen chloride or trifluoroacetic acid (cf. page 39). The solution resulting from cleavage with trifluoroacetic acid can be used directly for the azide formation (*2038*). Reaction times are 45–60 minutes.

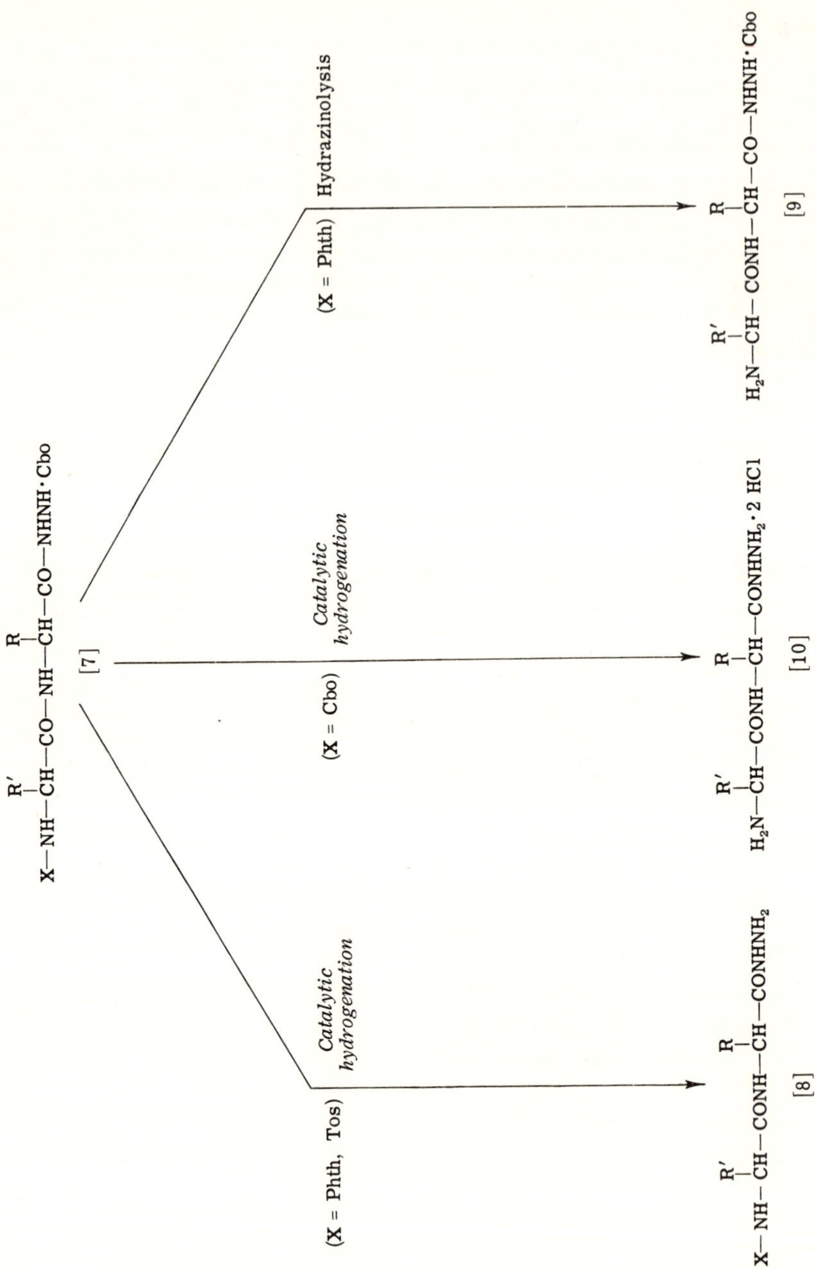
X—NH—CH(R′)—CO—NH—CH(R)—CO—NHNH·Cbo
[7]
(X = Phth, Tos)
Catalytic hydrogenation
(X = Cbo)
Catalytic hydrogenation
(X = Phth)
Hydrazinolysis
X—NH—CH(R′)—CONH—CH(R)—CONHNH₂
[8]
H₂N—CH(R′)—CONH—CH(R)—CONHNH₂·2 HCl
[10]
H₂N—CH(R′)—CONH—CH(R)—CO—NHNH·Cbo
[9]

tert-Butyloxycarbonylhydrazine can be prepared, according to Carpino (*464*), from *tert*-butyloxycarbonyl phenyl ester and hydrazine hydrate.

c. Trityl Hydrazides

Weygand and Steglich (*2500*) have used tritylhydrazine in combination with the trifluoroacetyl group. They were not so much interested in carboxyl protection as in attempting to make trifluoroacetylamino acids suitable for the azide method. The *N*-protected amino acid trityl hydrazides were prepared by the carbodiimide and phosphite methods or via the cyanomethyl esters. The trityl group can be removed with 1 *N* anhydrous hydrogen chloride. The trifluoroacetyl group is cleaved in the usual manner with alkali by heating in alcohol with an equivalent amount of 2 *N* sodium hydroxide. The free amino acid trityl hydrazides are normally not crystalline. They can be used without isolation for peptide synthesis.

Tritylhydrazine is prepared by tritylation of hydrazine hydrate with trityl chloride in ether (*2500*). Bistritylhydrazine is formed as a by-product; it is insoluble in ether and can be readily removed.

d. Phenyl Hydrazides

The phenyl group cannot be removed and the conversion to free hydrazide is not possible when phenylhydrazine is used as carboxyl protection (*1207a, 1562, 2413*). The free carboxyl compound is accessible by oxidative cleavage of the phenylhydrazine group with copper II salts or with manganese dioxide in aqueous acetic acid.

2. Amides

The chemical difference between an acid amide bond and a peptide bond is rather insignificant. In general it is not possible to protect a carboxyl group reversibly by amide formation. A discussion of acid amides as carboxyl protection at this point is justified for these reasons: a number of biologically active polypeptides are α-amides (α-MSH, eledoisin, oxytocin, vasopressin) or ω-amides (oxytocin, vasopressin, glucagon). It is known that biologically active polypeptides in some cases (e.g., angiotensin, α-MSH) upon ω-amide formation do not lose their biological activity, or only to a small extent.

The amide formation and the reactivity of ω-amides of aminodicarboxylic acids may cause difficulties. The minor differences in chemical behavior as compared to peptide bonds must not be ignored during peptide synthesis. They frequently give rise to side reactions.

a. Synthesis of Amides

Amides are prepared by ammonolysis of amino acid or peptide esters. Methyl or ethyl esters serve almost exclusively as starting material. Benzyl esters are used less frequently (*1806, 2604*). It is possible to use compounds with free (*257, 478, 2150, 2152*) or protected amino functions (*133, 204, 291, 774, 775, 1788, 1863, 2150, 2152*). Methanol or ethanol saturated with ammonia is normally used as solvent for the reaction. Hydantoins may be formed during the amide reactions of carbobenzoxy dipeptide esters (*583, 774;* cf. page 25). Normally methyl and ethyl esters do not undergo ammonolysis in liquid ammonia, although it has been reported on various occasions (*2369, 2717*). Amides are always obtained in liquid ammonia from activated esters such as the ethyl thiolester or phenyl ester (*1259*). The ammonolysis of peptide bonds does not occur unless extreme conditions are employed: 47 hours at 120°C in liquid ammonia in a sealed tube (*479*). Amides are sometimes prepared from *N*-protected acids and ammonia by a common coupling method; the mixed anhydride method (*150, 296, 306, 469, 1225, 1301, 1805, 2263*) and the acid chloride method (*133, 1354, 2163, 2490*) have been used. The synthesis of amides is also possible from *N*-protected amino acid phenyl esters by fusion with ammonium acetate (*2489*).

b. Special Problems of Acid Amides

Acid amides can be prepared from nitriles by partial hydrolysis, e.g., with HBr in glacial acetic acid. It is therefore not possible to decarbobenzoxylate a cyanomethyl ester. An *O*-substituted glycolic acid amide [11] would be formed (*837*):

$$\text{Cbo-peptide-OCH}_2\text{CN} \xrightarrow{\text{HBr/glacial acetic acid}} \text{H-peptide OCH}_2\text{CONH}_2\cdot\text{HBr} \qquad [11]$$

This reaction has been used in several cases for the synthesis of ω-amides from the corresponding ω-nitriles. The partial hydrolysis can also be carried out for 24 hours at 25°C with 1 *N* sodium carbonate in acetone in the presence of hydrogen peroxide (*1419*).

The reverse reaction, the dehydration of an amide to the nitrile, may occur as a side reaction when the carbodiimide method is applied to *N*-protected asparagine. The synthesis of the α-*p*-nitrophenyl ester (*273*) and also peptide coupling reactions (*296*) give low yields for the same reason (see Chapter IV,C,2,a,δ).

The ω-amide bond and the ω-peptide bond are less stable than the α-peptide bond. This lability gives rise to side reactions under certain circumstances. ω-Amides are partially hydrolyzed (*2016, 2017*) during the acid hydrolysis of esters. The acid-catalyzed esterification may cause alcoholysis of the amide bond. There are considerable differences in the behavior of different compounds. The β-amide bond of tosylasparagine is more stable than the γ-amide bond of tosylglutamine (*2654*). Alcoholysis was also observed during the acid hydrolysis of the formyl group in methanol (*2163*). The decarbobenzoxylation with HBr in glacial acetic acid proceeds without complication, showing that the former reaction is not a hydrolysis. This lability is also the cause of the hydrazinolysis of the amide bond that can be sometimes observed during the preparation of α-hydrazides (*1491*). Another side reaction is the facile imide formation of asparagine and glutamine derivatives (*2160*; cf. page 188). Chibnall and Westall (*489*) have described the reaction of the amide group of glutamine with nitrous acid, which may cause side reactions during an azide coupling. Wieland (cf. *2263*) has discussed this reaction as a possible reversible carboxyl protection. Data on this application are not available.

Akabori *et al.* (*27, 2104*) have described the xanthyl residue as amide-blocking group to prevent side reactions at the ω-amide function. It can be introduced readily with xanthydrol in glacial acetic acid and will be cleaved by HBr in glacial acetic acid. It is stable to catalytic hydrogenation. No general procedure is presently known for using an ω-amide as reversible protection of the carboxyl function. Schwyzer *et al.* (*2016, 2017*) and Hofmann *et al.* (*1037, 1040, 1041, 1045*), however, have used ω-amides as carboxyl protection in syntheses of angiotensin, α-MSH, and ACTH sequences. Cleavage was achieved in all cases by acid hydrolysis under vigorous conditions.

C. CARBOXYL PROTECTION BY SALT FORMATION

The protection of the carboxyl group by salt formation represents the least elaborate approach. It can be achieved readily and gives practically quantitative yields. The salt need not be isolated. The free acid for the next coupling step can be obtained by acid treatment at the completion of the peptide synthesis. Unfortunately there are preparative difficulties that prevent a general application of this method.

Amino acids in the form of their salts are normally soluble only in water. The coupling reaction has therefore to proceed in water and the hydrolysis of the activated compound is likely to occur. The difficulties can to some extent be overcome by working in aqueous-organic solvent mixtures and by maintaining fairly low temperatures under not too basic conditions (*2626, 2628*). It is of great advantage to keep the pH constant with the help of an automatic titrator (*899, 902*). The salts of higher peptides, especially the salts with organic bases, are frequently soluble in dimethylformamide, thus eliminating the possibility of hydrolysis.

The isolation of the reaction products is difficult since the starting carboxyl component and the end product are present as acids. It is therefore advantageous to couple a short carboxyl component with a long peptide salt. The end product, in contrast to the starting acid, is no longer soluble in such solvents as ether, ethyl acetate, or methanol, and purification becomes feasible (*225, 272, 469, 2386, 2628*). Compounds that crystallize easily can be purified by several recrystallizations. Less favorable circumstances may require a countercurrent distribution (*1228*) or preparative electrophoresis (*590*) after the amino-protecting group has been removed.

The following methods have been used with salts as coupling components: the acid chloride method, the mixed anhydride method, and the azide method. Activated esters have recently been employed more frequently. Several examples in the following compilation demonstrate the numerous possibilities for variation.

Mixed Alkyl Carbonic Anhydrides: Sodium hydroxide and triethylamine are preferred for the salt formation. The mixed anhydrides are prepared by standard procedures in tetrahydrofuran (*225, 272, 274, 469, 590, 1538, 1807, 2386, 2620, 2628*), in dioxane (*1354, 1890, 1986*), in dimethylformamide (*2386, 2629*), or in toluene (*1844*). The amino component is usually added in aqueous solution wherein the temperature must not rise above 0°C (*2626, 2628*). The chances for an opening of the active intermediate at the wrong anhydride bond are greater in aqueous alkaline solution than in a pure organic phase (*677, 678, 679*).

Acid Chloride: A peptide synthesis via the acid chloride method corresponds to a Schotten-Baumann reaction. The acid chloride is dissolved in dioxane (*1184, 1459, 2339*) or benzene (*1184*) or is used directly (*208, 938, 1750, 2247, 2659*). It is added to the aqueous solution of the triethylammonium salt (*938, 1459*), the sodium salt (*208, 2339, 2659*), or the magnesium salt (*1184, 1750, 2247*). Extensive hydrolysis was observed for phthalylvaline chloride and phthalylleucine chloride (*1459*).

Azide Method: The salts are formed with triethylamine. The solution of the azide in ethyl acetate is prepared according to the standard procedure. It is combined under vigorous shaking or stirring (*1022, 1951, 2497*) with the solution of the amino component in water or in an organic solvent, e.g., dimethylformamide (*1228*). The azide solution can also be added in several portions (*1043*).

Activated Esters: Among the activated esters the *p*-nitrophenyl ester has been used most frequently for the coupling reaction with amino acid salts. The use of thiolesters has also been reported; they offer certain advantages since they are relatively stable to hydrolysis (*996*). The amino component is again added as sodium or triethylammonium salt. Suitable solvent combinations are chloroform/ethyl acetate (*902*) and aqueous mixtures with tetrahydrofuran or dioxane (*899, 902*). Low pH values should be maintained to prevent the hydrolysis of the esters (*899, 902*).

The protecting groups are not equally well suited for peptide syntheses with salt components. Side reactions by partial ring opening are likely for the phthalyl group (cf. page 12). Tosylamino acid chlorides are decomposed in aqueous alkaline solution, because their amide nitrogen is activated (cf. page 16). Coupling reactions with salts are becoming increasingly important despite these complications. The yields are sometimes low, but this is compensated by the fact that no esters need be hydrolyzed. The final step in the synthesis of higher biologically active peptides is frequently carried out by a salt coupling reaction, since an ester hydrolysis of higher peptides is always accompanied by difficulties.

Weygand and Steglich (*2501*) and Weygand and Kaelicke (*2489;* cf. *2711*) have described a fundamentally different approach to the application of free amino acids for the synthesis of peptides. Most amino acids are soluble in glacial acetic acid. Their zwitterion nature is repelled by complex formation with the solvent. Activated esters, especially the phenyl thiolester, have been used in the coupling reaction. The carbobenzoxy and phthalyl residues served as *N*-protecting groups. The reaction is carried out at higher temperatures and requires several hours for completion. Acetylation is favored when sterically hindered amino acids are used. Diketopiperazine formation was observed for dipeptides. Racemization did not occur in a single case including *N*-protected dipeptide phenyl thiolesters.

TABLE I Stability and Cleavability of

Cleaving reagent and standard conditions		Acyl-type protecting groups		
		Form	TFA	Phth
HBr	In AcOH, $MeNO_2$, EtOAc, $CHCl_3$, Et_2O, etc. (30-60 min, room temp.)	Stable[a]	–	Stable
	In AcOH, etc. (1-2 hr, up to 60°C)	–	–	–
HCl	1 *N* alcoholic HCl (1-2 days, room temp.; or 1-hr boiling)	*Cleavable*	–	Stable
	10 *N* HCl or 10 *N* HCl in Me_2CO, dioxane, etc. (1-2 hr, 100°C)	Cleavable	Cleavable	Stable
TFA-OH	Anhydrous or 90% (30-120 min, 0°C to room temp.)	–	Stable	Stable
AcOH	Glacial or aqueous AcOH (up to 10 min, 100°C)	Stable	Stable	Stable
KOH or NaOH	In MeOH, EtOH, dioxane, Me_2CO, etc., also with H_2O (up to 2-hr, room temp.)	Stable	Cleavable	Side reaction
NH_4OH	0.1-1 *N* aqueous solution (up to 30 min, room temp.)	Stable	Cleavable	Side reaction
H_2/Pd	In MeOH, AcOH, also with H_2O or 1 *N* HCl (room temp.)	Stable	Stable	Stable
Na	In liquid NH_3 (no generally applicable standard conditions)	–	–	Side reaction
NH_2NH_2	In MeOH, EtOH, etc. (up to 2-hr, boiling; or 1-2 days, room temp.)	Stable	Side reaction	*Cleavable*

Amino- and Carboxyl-Protecting Groups

	Urethan-type protecting groups				
Tos	Cbo	ClCbo, BrCbo, NO_2Cbo	PZ, MZ	MeOCbo	BOC
Stable	*Cleavable*[b]	*Cleavable*	*Cleavable*	Cleavable	Cleavable
Cleavable	Cleavable	Cleavable	Cleavable	Cleavable	Cleavable
Stable	Limited stability	–	–	Cleavable	*Cleavable*
Stable	*Cleavable*	Cleavable	Cleavable	Cleavable	Cleavable
Stable	Stable	Stable	Stable	*Cleavable*	*Cleavable*
Stable	Stable	Stable	Stable	–	Stable
Stable	Stable	Stable	Stable	Stable	Stable
Stable	Stable	Stable	Stable	Stable	Stable
Stable	*Cleavable*	*Cleavable*	*Cleavable*	Cleavable	Stable
Cleavable	Cleavable	Cleavable	Cleavable	–	Stable
Stable	Stable	Stable	Stable	–	Stable

TABLE I

Cleaving reagent and standard conditions		Alkyl-type protecting groups: Trit	Alkyl-type protecting groups: Bzl, N-, O-, or S-	OMe, OEt
HBr	In AcOH, $MeNO_2$, EtOAc, $CHCl_3$, Et_2O, etc. (30-60 min, room temp.)	Stable[a]	Stable	Stable
	In AcOH, etc. (1-2 hr, up to 60°C)	–	–	Limited stability
HCl	1 *N* alcoholic HCl (1-2 days, room temp.; or 1-hr boiling)	*Cleavable*	Stable	Stable[a] or cleavable[a]
	10 *N* HCl or 10 *N* HCl in Me_2CO, dioxane, etc. (1-2 hr, 100°C)	Cleavable	–	*Cleavable*
TFA-OH	Anhydrous or 90% (30-120 min. 0°C to room temp.)	*Cleavable*	Stable	Stable
AcOH	Glacial or aqueous AcOH (up to 10 min, 100°C)	*Cleavable*	Stable	Stable
KOH or NaOH	In MeOH, EtOH, dioxane, Me_2CO, etc., also with H_2O (up to 2-hr, room temp.)	Stable	Stable	*Cleavable*
NH_4OH	0.1-1 *N* aqueous solution (up to 30 min, room temp.)	Stable	Stable	Cleavable
H_2/Pd	In MeOH, AcOH, also with H_2O or 1 *N* HCl (room temp.)	Cleavable	Cleavable[a]	Stable
Na	In liquid NH_3 (no generally applicable standard conditions)	–	*Cleavable*	Side reaction
NH_2NH_2	In MeOH, EtOH, etc. (up to 2-hr, boiling; or 1-2 days, room temp.)	Stable	Stable	Stable[a]

[a] Generally stable; cleavage occurs only under specialized conditions.
[b] The cleavage methods most important for the removal of a given protecting group are in italics.

(Continued)

Esters					
O*t*Bu	OBzlOMe	OBzl	$OBzlNO_2$	ω-NH_2	Peptide bonds
Cleavable	Cleavable	Limited stability	Stable	Stable	Stable
Cleavable	Cleavable	Limited cleav-ability	–	Limited stability	Tendency to cleavage
Cleavable	Cleavable	Side reaction	–	Limited stability	Stable
Cleavable	Cleavable	Cleavable	–	Limited *cleav-ability*	Tendency to cleavage
Cleavable	*Cleavable*	Stable	Stable	Stable	Stable
Stable	–	Stable	Stable	Stable	Stable
Limited stability	–	Cleavable	Cleavable	Stable	Stable
Stable	–	–	–	Stable	Stable
Stable	Cleavable	*Cleavable*	*Cleavable*	Stable	Stable
Side reaction	–	Cleavable	Cleavable	Stable	Stable
Stable	–	Stable[a]	–	Limited stability	Stable

III. FORMATION OF THE PEPTIDE BOND

A. CLASSICAL METHODS

Initial attempts to join amino acids through amide bonds go back to the last century; these are of historical interest only. The methods, however, that led to the formation of the first well-defined peptides, the so-called "classical methods of peptide synthesis," will be described briefly (cf. Fruton, *772;* Goodman and Kenner, *829;* Greenstein and Winitz, *861*).

Free dipeptides were synthesized for the first time by Fischer and his co-workers by acid or alkaline hydrolysis of diketopiperazines. While acid hydrolysis gave good results for diketopiperazine itself, substituted diketopiperazines had to be cleaved by alkali. Racemization was observed. Unsymmetrical diketopiperazines yielded a mixture of the two possible peptides. Diketopiperazines are obtained in a simple way by cyclic condensation of amino acid esters. Ester condensation can also lead to the formation of linear chains. A modification of this reaction is frequently used today. This method of activated ester takes advantage of esters that can readily be cleaved by aminolysis.

Curtius explored the possibility of forming peptides in a Schotten-Baumann reaction from amino acids or their esters and acylated amino acid azides. Fischer described at about the same time the use of carbethoxyamino acid chlorides. It was not possible to remove this blocking group; Curtius was faced with the same situation in the case of the benzoyl group. Twenty years later Bergmann and his school prepared all peptides by one of these two methods. The azide method is still very important for the synthesis of peptides, whereas the acid chloride method has been losing more and more of its significance.

The α-halogen acyl halide method of Fischer represents the classical approach to the synthesis of peptides. An α-halogen carboxylic acid chloride is reacted with an amino acid or a peptide or their esters to form an α-halogen acyl derivative. Exchange of the halogen by ammonia gives a free peptide or its ester. The synthesis of the octadecapeptide

H-L-Leu-(Gly)$_3$-L-Leu-(Gly)$_3$-L-Leu-(Gly)$_9$-OH by repetition of this reaction was successfully carried out as early as 1907 (*716*).

The ready formation of azlactones from acylamino acids and their cleavage with amino acid esters offer another route to synthetic peptides, but the method is not of interest since racemization is likely to occur due to the possible tautomerism.

Leuchs discovered the ready conversion of Fischer's carbethoxyamino acids to *N*-carboxylic anhydrides. Their cleavage with amino acids leads also to the formation of peptides. This method is used widely for the synthesis of polyamino acids, and occasionally for the synthesis of defined peptides.

B. PEPTIDE SYNTHESIS BY ACTIVATION OF THE CARBOXYL GROUP

1. Activation by Anhydride Formation

The activation of a carboxyl group by anhydride formation is more pronounced when the second acid displays strong electron-attracting properties. This is clearly demonstrated by the mixed anhydrides with inorganic acids. Organic acids, especially carboxylic acids as anhydride components, may disproportionate to form symmetrical anhydrides. As a consequence of the "wrong opening of the anhydride" it is also possible that the acid used for the anhydride formation enters into the amide formation [cf. the excellent review by Albertson (*29*); cf. Wieland (*2521*)].

a. Mixed Anhydrides with Oxygen-Free Inorganic Acids

α. Acid Chloride Method. N-Protected amino acid chlorides are prepared with a large excess of thionyl chloride in indifferent solvents such as benzene (*1459*) or without a solvent (*805, 1184, 1459, 1956*). Phosphorus pentachloride is used more frequently in benzene (*1459, 2062*), ether (*199, 208, 215, 938, 958, 1483, 1625, 2443*), or chloroform (*958, 2299*). Other functional groups must be protected. Protonation offers sufficient protection for the guanido group of arginine (*805*). Tosyl- (*214, 958, 1184, 1750, 2299*), trifluoroacetyl- (*2490, 2496*), and phthalyl-amino acids (*386, 979, 1459, 2060, 2062*) yield stable and often easily crystallizing acid chlorides. Side reactions [1] must be expected when tosylamino acid chlorides are used in combination with a salt coupling reaction (*128, 129, 130*):

$$\text{Tos—NH—}\underset{}{\overset{\text{R}}{\overset{|}{\text{CH}}}}\text{—CO—Cl} + \text{OH}^{\ominus} \longrightarrow \text{Tos—NH}_2 + \text{H—}\overset{\text{R}}{\overset{|}{\text{C}}}\text{=O} + \text{CO} + \text{Cl}^{\ominus} \qquad [1]$$

The extent of side reaction depends upon the amino acid used and the basicity of the reaction mixture. The use of protecting groups of the urethan type is most disadvantageous. It is difficult to crystallize the acid chlorides formed. Some are very unstable (*208, 1243*) and readily form *N*-carboxylic anhydrides. This rearrangement can be largely prevented by working at low temperatures (*199, 208, 215, 463, 938, 1625, 1771*). The isolation of the acid chloride is sometimes omitted and its solution used directly for the coupling reaction (*1844*). Carbobenzoxypyroglutamic acid (*799*) forms a stable acid chloride. The ω-acid chlorides of carbobenzoxyaminodicarboxylic acid α-esters are frequently used (*205, 303, 636, 639a, 847, 937, 1861, 2173*). The bis acid chlorides of these amino acids are also known (*847, 2198*). Tritylamino acid chlorides have been prepared despite the great acid lability of the protecting group (*2668*); due to the alkyl-type protection they are present as hydrochlorides.

An additional mole of base must be added to coupling reactions with amino acid esters to bind the hydrochloric acid released. The acylation has been carried out in many cases, especially according to earlier reports, in the presence of 2 moles of amino acid ester (*184, 1606, 2148*). It is more rational, however, to add a second mole of base. The hydrochloric acid can also be extracted from an organic solution of the two coupling components by shaking with aqueous alkali (*957, 1020, 1159, 2189, 2202*).

Acid chlorides are relatively stable under the conditions of a Schotten-Baumann reaction; they are, therefore, well suited for coupling reactions with salts. Phthalylvaline chloride and phthalylleucine chloride, however, cause difficulties in the course of the reaction. The hydrolysis is faster than the acylation (*1459*). The high degree of activation of acid chlorides makes possible a satisfactory coupling with *N*-methylamino acids; this is true also in cases where other methods give only poor yields (*183, 208*). Very few *N*-protected peptide chlorides have been reported (*1459*). The free ω-carboxyl groups of peptides with glutamic or aspartic acid have been converted to acid chlorides; they have been used for the synthesis of ω-amides (*938, 1533*). Acid chlorides have been employed with great success in the synthesis of peptolides (*2086, 2091*). The reaction of *N*-protected amino acid chlorides with alcohols is sometimes used for the synthesis of esters (*2119, 2339, 2654*). *N*-Unsubstituted amino acid chlorides or peptide chloride hydrochlorides (*351, 1382*) are best suited for polycondensations (*1443*); they proved to be of value for the synthesis of cyclic peptides and peptolides.

Coupling reactions with acid chlorides as intermediates have recently been added to the classical methods for the synthesis of acid chlorides. Unsymmetrical dichloroalkyl ethers [2] form acid chlorides with *N*-protected amino acids on heating to 70°–80°C in ethyl acetate or without solvent (*992, 1743, 1744, 1812*):

$$R{-}COOH + R'{-}CCl_2{-}O{-}R'' \longrightarrow R{-}COOCl + HCl + R'{-}CO{-}OR'' \qquad [2]$$

N-Protected peptide esters are formed directly when this reaction is carried out in the presence of amino acid esters or their hydrochlorides. Suitable ethers are dichloromethyl methyl ether (*1812*), dichloroethyl ethyl ether (*992*), or ethyl chlorovinyl ether (*992*). Rieche and Gross (*1812*) followed this procedure and obtained *N*-carboxylic anhydrides from carbobenzoxyamino acids when they worked without solvents.

Bossard *et al.* (*310*) demonstrated that the acid chloride formation with thionyl chloride is catalyzed by dimethylformamide. The actual chlorinating agent is *N,N*-dimethylchloroformimidium chloride [3]. This reagent can be prepared as a crystalline moisture-sensitive compound from dimethylformamide and thionyl chloride, phosgene, phosphorus pentachloride, or oxalyl chloride (*70, 71, 311*):

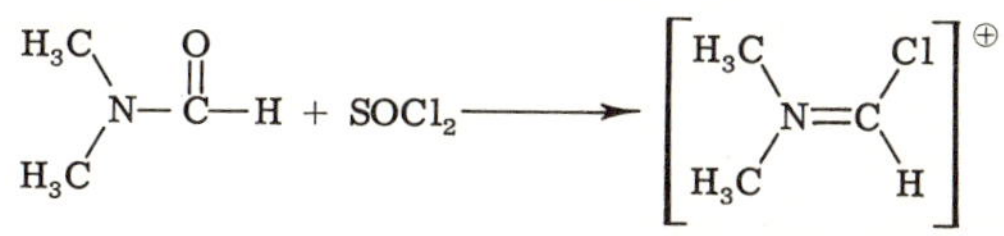

[3]

According to Zaoral and Arnold (*2649*), the compound is suitable for peptide synthesis. The *N*-protected amino acid or peptide acid is combined with reagent [3] at −5° to −10°C in chloroform, methylene chloride, or dimethylformamide. Without isolation, the acid chloride is added to the amino component and 2 equivalents of base. The synthesis of the dimethylchloroformimidium chloride and the formation of the acid chloride can also proceed in a one-step reaction: the *N*-protected amino acid is dissolved in dimethylformamide and 1 equivalent of thionyl chloride is added at −15° to −30°C (*1980*).

β. Azide Method. A number of complications may be encountered with the azide method. Yet it is regarded very highly among the methods for the synthesis of peptides, mainly because racemization during coupling reactions has not been so far observed (*1021, 1033*). The reaction proceeds via several steps: The azide [5] must be prepared via the hydrazide from an *N*-protected amino acid or peptide ester [4] by treatment with nitrous acid, and is then allowed to react with the amino component:

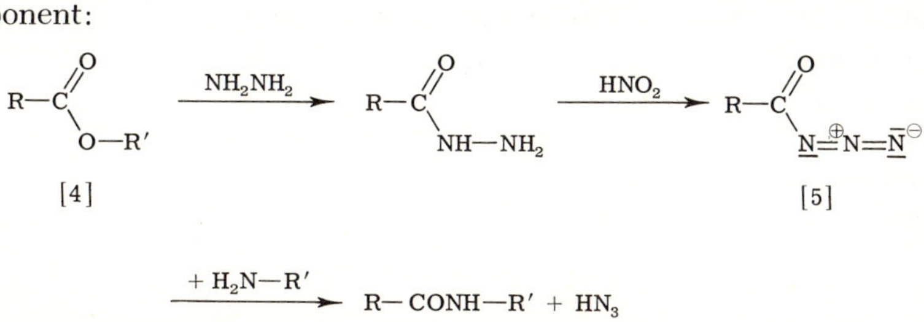

Salt formation with hydrazoic acid liberated during the coupling reaction has not been observed. This is a great advantage over the analogous coupling reaction with acid chlorides.

β_1. *Synthesis of N-protected amino acid or peptide hydrazides.* The most common approach to the synthesis of an *N*-protected amino acid or peptide hydrazide is the hydrazinolysis of a suitable ester. Methyl and ethyl esters are preferred. Benzyl esters are used less frequently (*1191, 1912, 2134*). The hydrazides are very readily accessible from esters that are to a certain degree activated. They make possible a selective hydrazinolysis, e.g., in the presence of a methyl ester (*1259, 1260*). The *tert*-butyl esters are not subject to hydrazinolysis. This is important for the synthesis of monohydrazides of *N*-substituted aminodicarboxylic acids. *N*-Protected monohydrazides of monoaminodicarboxylic acids with the unsubstituted second carboxyl group are known (*972, 1246, 1843, 2173*); coupling reactions with these derivatives, however, are not satisfactory (*1246, 1891*).

The hydrazinolysis of methyl and ethyl esters is normally carried out in alcoholic solution. Brief standing of the alcoholic solution with hydrazine hydrate is sufficient in simple cases. An excess of hydrazine hydrate is used to prevent the formation of symmetrical bishydrazides (*2583*). Amino acids with long and branched side chains and higher peptides require longer reaction times, higher temperatures, and a larger excess of hydrazine hydrate. The use of a twentyfold excess at room temperature (*692*) and boiling for 40 hours with a threefold excess (*2029*) have been described. Anhydrous hydrazine was used in extreme cases (*2419, 2620*). The hydrazinolysis is carried out in dimethylformamide (*803a, 2224*) or in hydrazine hydrate without solvent (*1957*). Ferren *et al.* (*713*) have observed that the hydrazinolysis of simple carboxylic acid ester is exceptionally fast in *n*-butanol. This alcohol has, therefore, also been described as solvent (*1491*). The hydrazinolysis is sometimes carried out at low temperatures (*894, 1904b*). Tritylamino acid esters may be so sterically hindered as not to form hydrazides at all (*1006*). Trityl ethylglycinate is hydrolyzed under extreme conditions, e.g., boiling with hydrazine hydrate in propanol (*1006*). Zervas and Theodoropoulos (*2668*) prepared the hydrazide from the methyl ester by keeping it for 2 days in a mixture of dioxane and methanol with a large excess of hydrazine hydrate. Trityl peptide esters can be converted to hydrazides without difficulty. Iselin (*1096*) tried to obtain tritylamino acid hydrazides by selective tritylation of the α-amino group of amino acid hydrazides; the results were not very satisfactory. Selective acylations of amino acid hydrazides have been reported by Curtius and Levy (*556*). The reaction with acetic anhydride or benzoyl chloride yields the N^{α},N'-bisacyl derivative. The acylation with an acid azide is claimed to lead to the formation of the N^{α}-monoacylated product. This observation could not be fully confirmed by Zahn and Schnabel (*2637*).

Esters of higher peptides frequently undergo the hydrazinolysis more readily than an alkaline hydrolysis (*1059*). Another advantage of the *N*-protected amino acid and peptide hydrazides is their easy crystallizability. Therefore the crude products of *N*-protected amino acid or peptide ester are frequently converted to the hydrazides. The purification of the hydrazides causes no difficulties. According to Zahn and Schnabel (*2637*), crude hydrazides are relatively unstable. If circumstances are very unfavorable, the hydrazide can be isolated as the isopropylidene (*692, 2719*) or benzylidene derivative (*2173*). The alkylidene residue need not be removed prior to the azide formation. It is unstable under the required acidic reaction conditions (*2719*).

In some cases the hydrazinolysis of *N*-protected amino acid or peptide esters leads to the formation of undesired reaction products. Racemization was reported for serine-containing peptides (*1946;* cf. *2005*). Side reactions are caused by the guanido group in arginine-containing peptides (*772*). Benzyl mercaptan may be lost during the formation of hydrazides of S-benzylcysteine derivatives (*1491*) (see Chapter IV,G, 1,a,α). However, the formation without difficulty of hydrazides of these derivatives has also been described. Phthalylamino acid esters or phthalyl peptide esters cannot be converted to hydrazides. The hydrazinolysis of the phthalyl group is faster than the hydrazide formation. Trifluoroacetyl derivatives are cleaved under the basic conditions of hydrazinolysis. Sterically hindered amino acids or peptides, however, yield 40–70% of hydrazide, depending on the reaction conditions. The free amino acid or peptide hydrazides are formed as by-products. Dipeptides usually form diketopiperazines (*2504*).

ω-Amide bonds may also cause side reactions. They are cleaved more readily by hydrazine than α-amides or peptide bonds. This reaction seems to be dependent upon the *N*-protecting group. MacLaren *et al.* (*1491*) obtained Tos-Glu($NHNH_2$)-Cys(Bzl)-$NHNH_2$ from Tos-Glu(NH_2)-Cys(Bzl)-OEt by reacting it with hydrazine hydrate in *n*-butanol for 18 hours at 40°C. The corresponding carbobenzoxy compound did not react to form the bishydrazide. Anhydrous hydrazine and temperatures of up to 125°C are required for the hydrazinolysis of α-amide and peptide bonds (*25, 1622*). Side reactions, therefore, need not be expected under the normal conditions of ester hydrazinolysis.

In cases where the normal hydrazide formation is impossible, the azide can be obtained directly from an *N*-protected amino acid chloride and sodium or dicyclohexylammonium azide (*724, 2496*). It is more advantageous, however, to use a hydrazine carrying an amino-protecting group as a substituent. The *N′*-protected hydrazide displays the characteristics of a carboxyl-protecting group and the chain can be extended at the amino terminal. The hydrazide protection can be removed selectively and a coupling reaction conducted via the azide. This method has been applied to the synthesis of a number of naturally occurring peptides (*292, 890, 2038, 2236*).

Little is known of the possible hydrazide formation from *N*-protected amino acids or *N*-protected peptides by acylation of hydrazine. Only Klee and Brenner (*1255*) have used the imidazolide method for this purpose. The method is more of theoretical interest since it uses an intermediate that is already activated for coupling.

β_2. *Azide coupling.* The azide formation takes place only in strongly acidic solutions. Aqueous acetic acid, as used originally by Curtius and Levy (*556*), is in many cases insufficient. Therefore mixtures of acetic acid with hydrochloric acid or hydrochloric acid alone are frequently used. Upon the addition of sodium nitrite, the azide is precipitated as an oil or sometimes in the crystalline form. The nitrite is added at low temperatures to prevent side reactions. The azide readily decomposes in aqueous solution (*972*) and is therefore extracted directly with an organic solvent, such as ethyl acetate or ether. Acid is removed by washing with hydrogen carbonate solution. The excess of nitrite can be destroyed by the addition of an aqueous solution of ammonium sulfamate (*2017a*). The solution is dried prior to the reaction with the amino acid ester. The azide can be added to the solution of the amino component in an excess of hydrogen carbonate without neutralization, if prolonged standing of its solution is to be avoided (*1912*). If hydrazides are poorly soluble in glacial acetic acid/hydrochloric acid, this solvent system can be replaced by dimethylformamide/hydrochloric acid. The azide solution is then neutralized with triethylamine and combined directly with the amino component (*890, 894, 896*), or the azide is precipitated with water after adding potassium carbonate to the solution (*1132*). Trifluoroacetic acid (*803a, 2038*) may also be employed as solvent. Poorly soluble hydrazides can be used in suspension (*2636, 2637*). Histidine-containing peptides, due to the basic nature of the imidazole ring, can be extracted from aqueous alkaline solutions only (*724, 1052*). Under these conditions, however, the azide group may react with the imidazole nitrogen in an undesirable fashion (*1538, 1950*). Merrifield and Woolley (*1538*) therefore, add the amino component as hydrochloride to the azide solution while it is still acidic. They extract both reactants simultaneously with ethyl acetate after the addition of potassium carbonate. Schnabel and Zahn (*1945*) have observed nitration of the aromatic nucleus in tyrosine-containing peptides. An excess of nitrite may introduce the nitroso group at the indole nitrogen in tryptophan-containing peptides (*337*). Peptides with acid-sensitive protecting groups, such as the trityl (*1096*) or *tert*-butyloxycarbonyl group (*1103, 2019*), can also be reacted despite the strongly acidic reaction conditions as long as certain precautionary measures are observed. All operations must be carried out in a rapid sequence and at low temperatures. A temperature of —10°C (*1096*) was reported for the trityl group, —20°C for the *tert*-butyloxycarbonyl group. Free peptide hydrazides serve as starting material for polycondensations (*1027, 1493*) or cyclizations (*2003, 2072,* cf. *2031*). It is important in these cases that the nitrous acid reacts faster with the hydrazide group with azide formation, than with the amino group with diazotization (*1493*).

Side reactions associated with the azide method were observed very early (cf. *1942, 1942b*). They are to some extent the result of the rearrangement of the azide to isocyanate [6]:

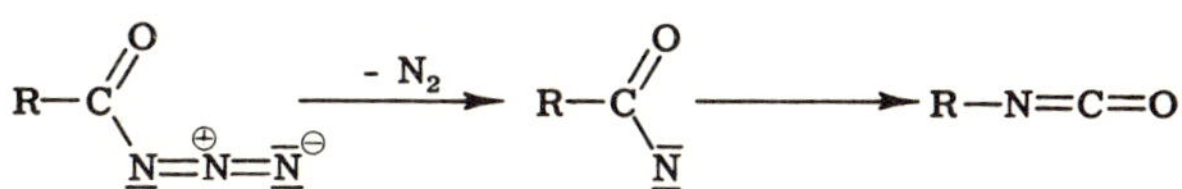

[6]

The reaction with the added amino component makes possible the formation of urea derivatives [7]:

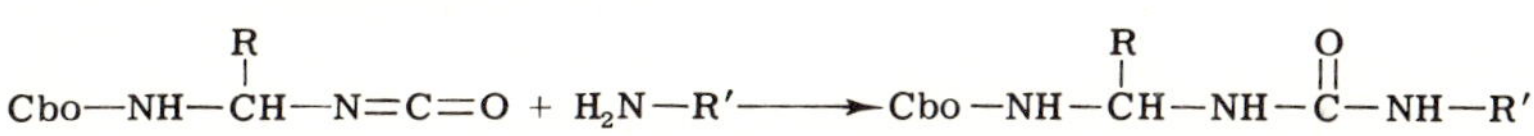

[7]

This side reaction had been observed by Curtius (*552, 553*), and was later confirmed by many authors (*1009, 1033, 1059, 2322, 2419*). The isocyanate formation in serine-containing peptides may lead to the formation of cyclic urethans [8] (*769, 949*):

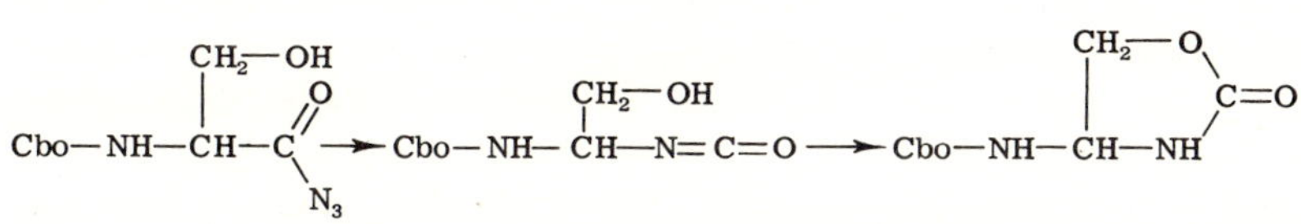

[8]

The urethan is formed as the main product upon heating an azide solution in the presence of an alcohol (*553, 555*). The formation of the ester with the alcohol used as solvent has also been described (*2620*). Coupling reactions with azides have nevertheless been carried out in alcoholic solutions (*288, 896, 2616*). The isocyanate formation was also observed by Heyns *et al.* (*1002*) during cyclizations. A carbobenzoxy tripeptide azide yielded the carbobenzoxy dipeptide amide on catalytic hydrogenation under conditions observing the dilution principle. The isocyanate had been hydrogenated and was subsequently hydrolyzed. Temperatures of about 0°C are insufficient to suppress these side reactions (*2322, 2637*). Even an isolated crystalline azide rearranges slowly to the isocyanate at 0°C (*2019*). Schwyzer and Kappeler (*2019*) were able to follow with ease the formation of isocyanate by infrared spectroscopy. While the azide has a characteristic band at 4.75 μ, the isocyanate shows a sharp band at 4.5 μ. N^1,N^2-Diacylated hydrazines may be formed as long as the azide formation is incomplete and the reaction mixture contains still unchanged hydrazide (*2126*).

The formation of acid amides is another side reaction frequently encountered. It was first observed by Prelog and Wieland (*1771*) and later confirmed by a number of authors (*344, 687, 1853*). This amide formation seems to occur very readily with carbobenzoxy-S-benzylcysteine azide (*972, 1491, 1863*). It is not the result of a rearrangement of the azide but, according to Honzl and Rudinger (*1058*), another product of the reaction of the hydrazide with nitrous acid. The fact that the reaction of an acid chloride with sodium azide shows no amide formation at all is cited as supporting evidence. The following course of the reaction has been discussed [9]:

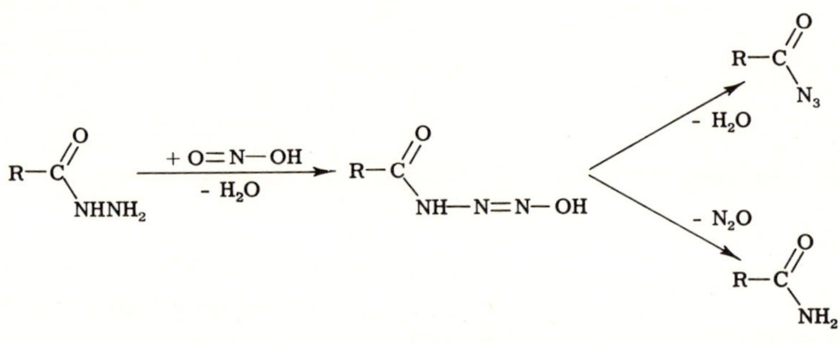

[9]

Honzl and Rudinger (*1058*) have studied in the case of carbobenzoxy-S-benzylcysteine hydrazide the influence of temperature, acid concentration, and water content on the amide formation. Higher temperatures favor the formation of the amide. Unreacted hydrazide can be recovered if the acid concentration is too low. The kind of acid used exerts an influence in addition to its concentration. Under otherwise identical conditions, less amide is formed in perchloric acid than in hydrochloric acid. The presence of larger quantities of water favors the amide formation. No amide was formed, however, in a homogeneous phase of tetrahydrofuran/water/hydrogen chloride with a water content of 5% or less if the temperature was kept below —15°C. These results led to the discovery of a new method that uses *tert*-butyl nitrite or nitrosyl chloride instead of sodium nitrite. Now the formation of the azide is possible in aqueous-organic phase, e.g. tetrahydrofuran/concentrated hydrochloric acid, or in pure organic phase, e.g. tetrahydrofuran or ether and gaseous hydrogen chloride. The azide formation and the subsequent coupling reaction can be carried out at temperatures of −10° to −20°C without difficulty. This modification, in addition to the prevention of side reactions, offers great advantages when acid-labile protecting groups are present.

β_3. *Synthesis of peptides by oxidation of hydrazides.* Wolmann *et al.* (*2577, 2578*) have developed an interesting coupling method based on the observation that glutamic acid γ-hydrazide reacts with 2 equivalents of *N*-bromosuccinimide with elimination of nitrogen and formation of pyroglutamic acid. *N*-Protected amino acid or peptide hydrazides are combined with the amino component in tetrahydrofuran, dimethylacetamide, or aqueous dioxane, and an oxidizing agent is added at 0°C in the presence of 2 equivalents of a tertiary base. Iodine, hydrogen peroxide, silver oxide, etc., can be used instead of *N*-bromosuccinimide. Peptides have been isolated in yields of over 90%. In the absence of an amino component, bis-(*N*-acylamino-acyl) hydrazide will be isolated as the reaction product. Cbo-Gly-L-Phe-Gly-OEt was checked for racemization; 1.1% of the DL-form was found to be present. Medzihradszky-Schweiger (*1522*) has discussed a possible reaction mechanism, assuming the formation of a diazonium salt [10]:

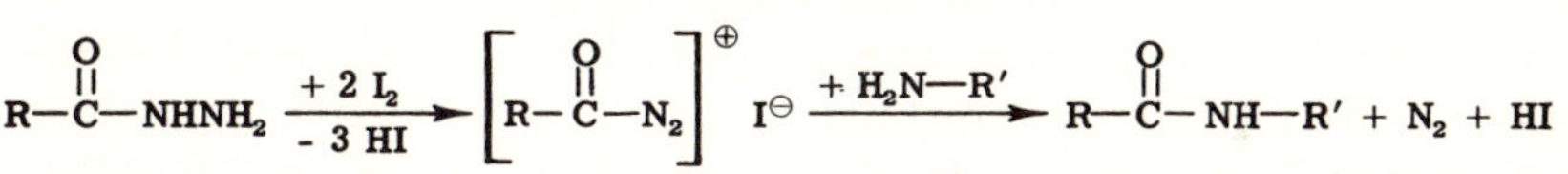
[10]

b. Mixed Anhydrides with Oxygen-Containing Inorganic Acids

Anhydrides with oxygen-containing inorganic acids have an electron distribution such that the wrong opening of the anhydride bond and undesirable side reactions occur very rarely. The mixed anhydrides of phosphoric acid, sulfuric acid, sulfurous acid, and carbonic acid have practically not been used for coupling reactions. The mixed anhydride accessible via phosphorus oxychloride was more successful. The coupling reactions via mixed anhydrides with monoesters of carbonic acid, however, has become very important.

α. *Anhydrides with Carbonic Acid Monoesters.* The anhydrides with monoesters of carbonic acid are outstanding among the mixed anhydrides and are normally referred to as the "anhydride method." The procedure was submitted for publication with only a day's difference by Wieland and Bernhard (*2524*) and by Boissonnas (*284*), and a short time later by Vaughan (*2362*). The method has the advantage that no undesirable by-products are formed. The aminolytic cleavage of the anhydride yields, in addition to the peptide, only carbon dioxide and alcohol [11]:

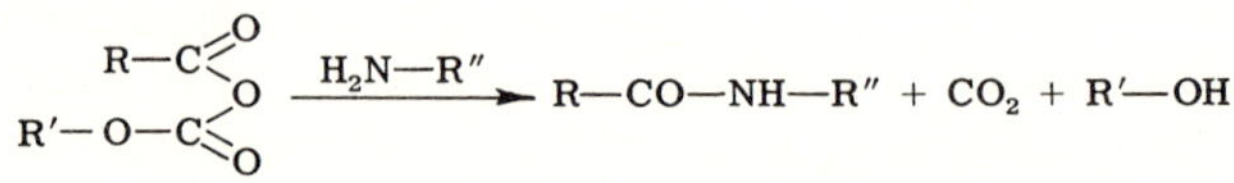

[11]

Investigations on the applicability of various alkyl or aryl esters of carbonic acid showed that the methyl and phenyl esters gave poorer results (*2524*). Otherwise only a minor influence of the alkyl group was noted (*2367*). Therefore the ethyl ester, being readily accessible, and the isobutyl ester are used most frequently. The *sec*-butyl ester (*436, 1854, 2362*) offers certain advantages, as in the case of mixed anhydrides with carboxylic acids (see Chapter III,B,1,c,α), through its branched side chain (*2367, 2524*). The mixed anhydrides formed with benzyl chlorocarbonate (*2656*) or octyl chlorocarbonate (*435*) are rather unusual. A thiolester of carbonic acid, phenylthiocarbonyl chloride, was also investigated for the formation of mixed anhydrides (*61*).

The anhydrides are usually formed at temperatures of 0°C or lower. Several cases of temperatures of +10°C and higher have been reported (*272, 692, 1028, 1089*). No mention was made of a disproportionation that, according to Wieland and Bernhard (*2524*), occurs above 0°C. Triethylamine is normally used as base to bind the hydrochloric acid liberated during the anhydride formation. Tri-*n*-butylamine (*217, 291, 685, 1028, 1064, 1089, 2417*), tri-*n*-propylamine (*1934*), and *N*-ethylpiperidine (*1854, 2652*) have also been used. Tri-*n*-butylamine is claimed to be particularly useful due to the ready solubility of its salt (*284*). It is more difficult, however, to remove it after the coupling reaction (*29*). Tetrahydrofuran is the solvent of choice. Dioxane (*217, 692, 1028, 1033, 1089*), toluene (*436, 1121, 1844, 2359, 2567*), chloroform (*37, 236, 436, 573, 1491, 1930, 2298, 2359*), ethyl acetate (*1381, 1605*), or dimethylformamide (*1104, 1121, 1192*), and mixtures of these solvents have also been used. Albertson (*29*) cautions on the use of dimethylformamide since reaction with the alkyl chlorocarbonate is a strong possibility. Alcohols cannot be used since the danger of *O*-acylation exists. *tert*-Butanol, however, can be employed (*2534*). Reaction times for the anhydride formation vary from a few minutes (*2292*) to half an hour (*685, 692, 2386*). Vaughan (*2363*) reports maximum reaction times of 5–10 minutes at −5°C. The amino component, as ester or amide, is normally added in an anhydrous solvent. Esters may be added in aqueous solution (*572, 590*) to hold down the solvent volumes for poorly soluble amino acid or peptide esters. The method has also been applied to coupling reactions with salts (see Chapter II,C).

Peptide formation via mixed anhydrides is relatively fast even at low temperatures. Sometimes heating or refluxing seems to be required for the completion of the reaction (*20, 297, 2359, 2363, 2364*). No comments are made on the thermal decomposition of the mixed anhydrides.

Coupling reactions with mixed anhydrides are free of complication for most amino acids, irrespective of their use as carboxyl or amino component. Hydroxyl groups and the imidazole ring need not be protected. Difficulties have been reported for the synthesis of asparaginyl peptides (*296, 1354;* see Chapter IV,C,2,a,δ). Most tritylamino acids are sterically hindered. Only tritylglycine and tritylalanine can be coupled (*37, 1006*). Ressler (*1805*) has reported that carbobenzoxyglutamic acid with 1 equivalent of isobutyl chlorocarbonate forms the mixed anhydride only at the α-carboxyl group.

Acyl imides [12] [13] (*1286, 1930*) are sometimes formed when *N*-protected glycine is used as carboxyl component. Zaoral and Rudinger

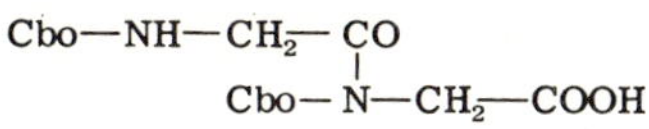

[12]

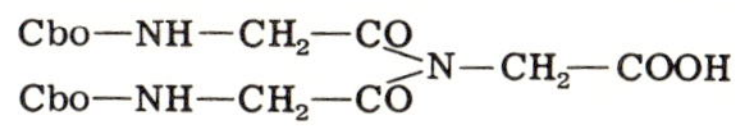

[13]

(*2655*) have studied this side reaction in detail for tosylglycine. Its amide hydrogen can be substituted very readily. The mixed anhydride of tosylsarcosine with *sec*-butyl carbonate and aniline form the anilide in 80% yield. The analogous reaction with tosylglycine yields only 15% of the anilide. Formula [14] shows how the isolated by-products have been formed.

Cleavage of the anhydride in a different manner and formation of urethans have rarely been observed (*120, 803a, 1233*). Battersby and Robinson (*120*) have discussed the possibility of urethan formation by alkyl chlorocarbonate that has not been consumed for the formation of the anhydride. Wieland *et al.* (*2534*) report that the disproportionation can be prevented by working at low temperatures, but not the undesirable opening of the anhydride.

Racemization during coupling reactions with *N*-protected amino acids is primarily dependent upon the kind of amino-protecting group used. It is favored by acyl-type blocking groups such as formyl or trifluoroacetyl that form azlactones very readily. The optically active C-terminal amino acids in *N*-protected peptide acids are racemized in almost all cases. Vaughan (*2363*) has studied in detail the coupling of Cbo-Gly-L-Phe-OH with glycine ethyl ester under various reaction conditions. The strongest influence is exerted by the solvent. Total racemization was observed in chloroform, none in tetrahydrofuran; it amounted to 10% in dioxane and rose in a toluene/chloroform mixture with increasing

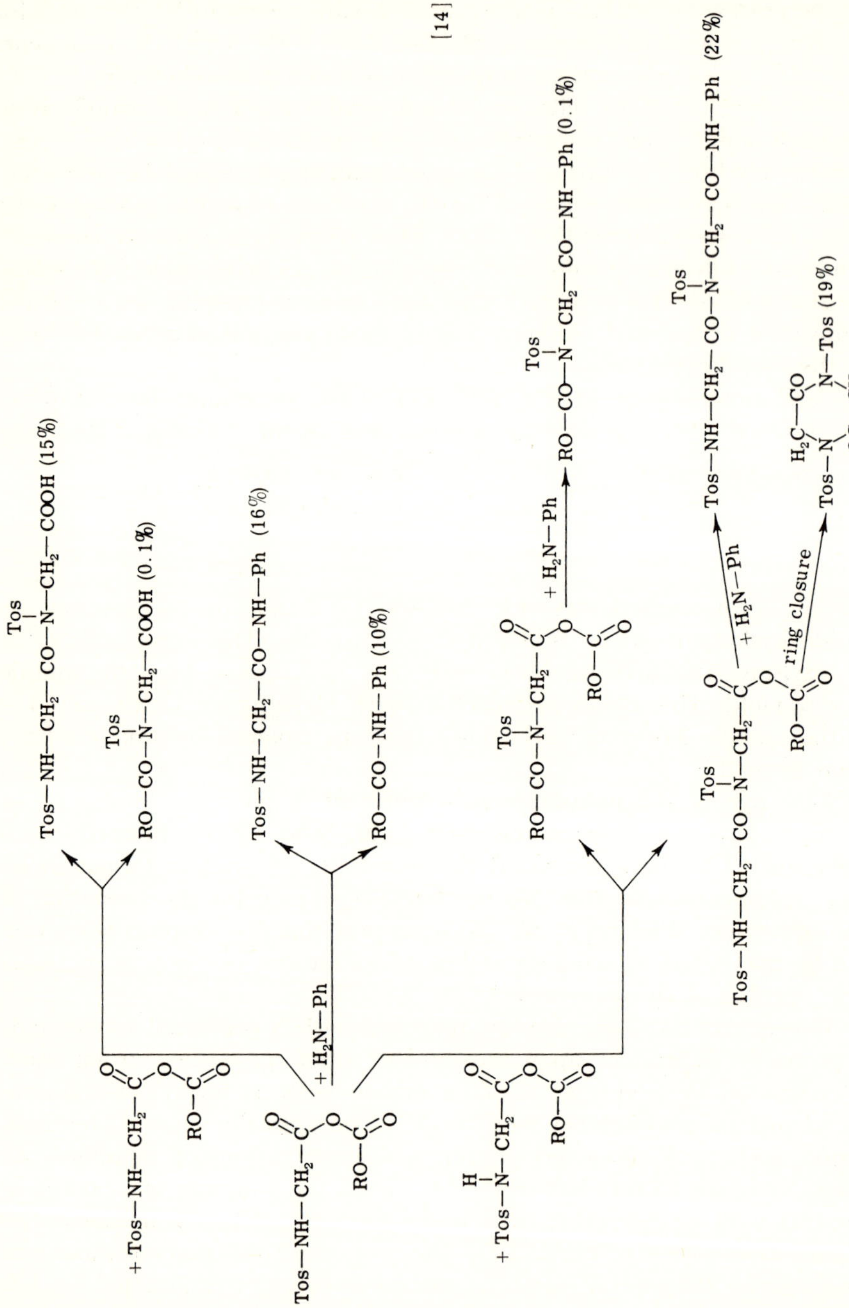

[14]

amounts of the latter. Furthermore, it is influenced by the time permitted for the formation of the mixed anhydride; the longer it is the higher the degree of racemization. On the other hand, the temperature during the coupling with the amino acid ester is without influence. Determann and Wieland (*590a*) confirmed these observations. The authors noted in the same case that racemization does not occur upon formation of the anhydride in tetrahydrofuran at −15°C and coupling within 10 minutes. Chances for racemization are greater during coupling reactions with salts (*2364*). Diastereoisomers have been separated by countercurrent distribution in the system ethyl acetate/1 *N* phosphate buffer. Ingle (*1089*) found racemization of the amino component during the coupling of carbobenzoxyglycine with L-tryptophan in aqueous alkaline solution. This would not normally be expected.

From observations during syntheses of biologically active polypeptides, several authors have confirmed that other methods, especially azides, yield better products. Vogler and Lanz (*2386*), however, obtained the same rotation values when they compared the anhydride, azide, and carbodiimide methods as applied to tri-, tetra-, and pentapeptides in the polymyxin series. The anhydride method was used where the danger of racemization is greatest, namely, for coupling reactions with salts. Vaughan and Eichler (*2364*) have also reported the synthesis of optically pure peptides by coupling reactions with *N*-protected dipeptide acids.

Mixed anhydrides with monoesters of carbonic acid can be used for the synthesis of cyclic peptides (see Chapter V,A). Polycondensations by catalytic hydrogenation of mixed anhydrides of carbobenzoxy dipeptides are also possible (*428*).

β. Anhydrides with Carbonic Acid. The mixed bisanhydrides of *N*-protected amino acids with carbonic acid [15] disproportionate very rapidly, even at 0°C, to form the symmetrical anhydrides (*2524*). Mixed monoanhydrides [16], however, are accessible at very low temperatures (−70°C) (*357*). They have been frequently used for the synthesis of *N*-protected amino acid esters. Peptide syntheses are possible in prin-

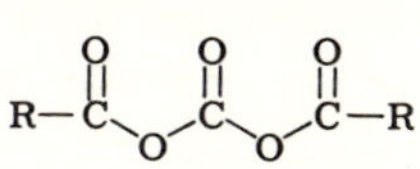

[15]

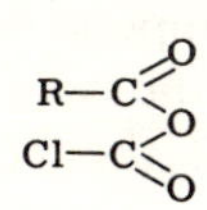

[16]

ciple. It was shown, however, that complete racemization occurred during the reaction of salicoyl-L-phenylalanine with glycine methyl ester (*357*.) The question remains open as to what extent the acyl protection of the amino group is responsible.

The synthesis of a mixed anhydride by Liwschitz *et al.* (*1436, 1439*) from benzylaspartic acid with an excess of phosgene without the addition of a base is in sharp contrast to these observations. It was carried out at room temperature or by heating to 60°C in dioxane whereby only the mixed anhydride at the α-carboxyl group was formed (cf. *29*).

γ. *Phosphorus Oxychloride Method.* First attempts to prepare mixed anhydrides [17] from 3 moles of an *N*-protected amino acid and 1 mole of phosphorus oxychloride for the synthesis of peptides failed (*2524*). The formation of the monoanhydride [18] however, was successful. It can be reacted under anhydrous conditions with the amino component in the form of an amino acid or peptide ester. It is essential to work at temperatures of —15°C or lower to prevent disproportionation (*2532*).

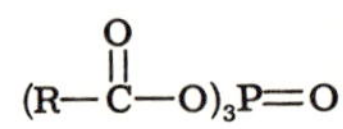

[17]

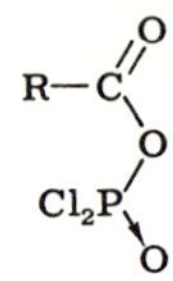

[18]

If carbobenzoxyglycine is used as carboxyl components, the initially formed mixed anhydride acylates the imino nitrogen of a second mole of carbobenzoxyglycine to form *N*-(carbobenzoxyglycyl)-*N*-carbobenzoxyglycine [19]. This structure was confirmed by catalytic hydrogena-

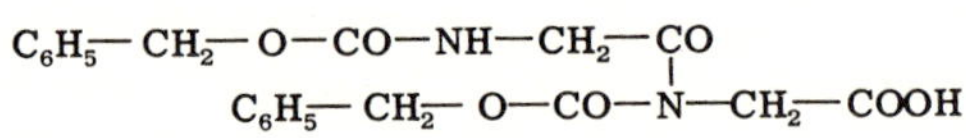

[19]

tion, which gave glycylglycine, and by the reaction with HBr in glacial acetic acid, resulting in the cleavage of only one carbobenzoxy group and the formation of *N*-glycyl-*N*-carbobenzoxyglycine (*2532*). This side reaction can be prevented by carrying out the activation of the carboxyl component in the presence of the amino component. Therefore phosphorus oxychloride is allowed to react with the mixture of the carboxyl and amino components; the tertiary base can be added prior to or together with the acid chloride. The reaction can also be induced by the addition of the base, preferably triethylamine or pyridine, to the mixture of the carboxyl component, amino component, and phosphorus oxychloride (*1491, 2029, 2032, 2533*).

Tetrahydrofuran and chloroform are preferred as solvents. Wieland *et al.* (*2534*) have used *tert*-butanol as solvent. Due to steric hindrance it is supposedly not subject to *O*-acylation. It must be pointed out, however, that *tert*-butyl esters have been prepared by *O*-acylation of *tert*-butanol by means of phosphorus oxychloride (*2264*). Wieland and

Heinke (*2532*) reported there to be no need for special protection of hydroxyl groups in the carboxyl or in the amino component. Weygand and Rinno (*2497*), however, were not able to couple trifluoroacetylserine successfully, while other trifluoroacetylamino acids gave good results.

This method is used for the synthesis of peptides less frequently than the alkyl carbonic acid anhydride method, but quite useful for the synthesis of *p*-nitrophenyl or thiolphenyl esters (*2533*).

δ. *Anhydrides with Phosphoric Acid Esters.* Sheehan and Frank (*2063*) prepared *N*-protected amino acid dibenzyl phosphoric acid anhydrides from phthalyl- or carbobenzoxyglycine chloride and the silver salt of phosphoric acid dibenzyl ester. Katchalski and Paecht (*1179, 1180*) reversed the procedure and reacted the silver salt of carbobenzoxyamino acids with the monochloride of phosphoric acid dibenzyl ester [20]:

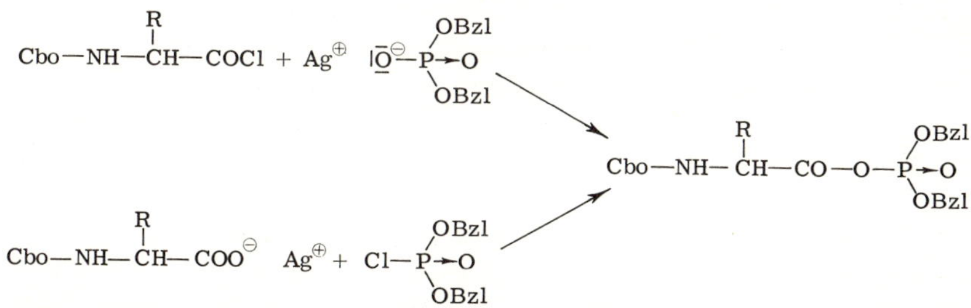

[20]

Chantrenne (*483*) and Wieland and Bernhard (*2524*) have used the acid chlorides of monophenyl and diphenyl esters of phosphoric acid for the formation of anhydrides. Cramer and Gärtner (*538, 539*) have reported the synthesis of a mixed anhydride from an *N*-protected amino acid and an enolic ester of phosphoric acid [phosphoric acid diethyl ester (α-ethoxy-β-carbethoxyvinyl ester)] [21]:

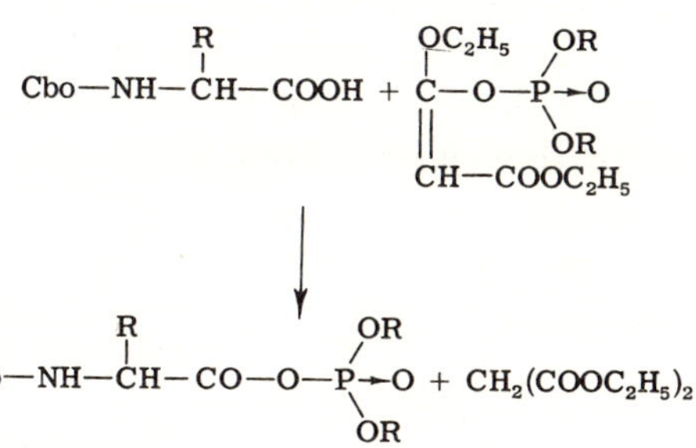

[21]

Disproportionation to the symmetrical anhydrides was not observed, although this synthesis had to be carried out at a temperature of +70°C. Cosmatos *et al.* (*529*) have used *N*-(dibenzylphosphoryl)-amino acid diphenylphosphoric anhydrides for the synthesis of peptides.

The peptide coupling reaction can be carried out with amino acid esters or, in aqueous alkaline solution, with the alkali salts of amino acids. No racemization was observed for dipeptides. Reports on the synthesis of higher peptides are not available. The yields of over 80% for this anhydride cleavage are remarkable (*2063*) since they were obtained under physiological conditions (pH 7.4 and room temperature).

ϵ. *Anhydrides with Unsubstituted Phosphoric Acid.* Katchalski and Paecht (*1179, 1180*) obtained mixed anhydrides of amino acids with phosphoric acid by reacting carbobenzoxyamino acid dibenzylphosphoric acid anhydrides with hydrogen bromide in carbon tetrachloride. These anhydrides condense very readily in aqueous solution and yield polymers. In aqueous ammonia amides are formed. The same compounds have been described by Bentler and Netter (*153*). Their starting material was α-azidocarboxylic acid dibenzylphosphoric acid anhydrides. The azido group was reduced with Raney nickel/hydrogen. The benzyl ester groups were cleaved by catalytic hydrogenation in the presence of palladium on charcoal. Mixed anhydrides with a protection at the amino group are accessible from phthalylamino acid chlorides and the monosilver salt of phosphoric acid (*459*).

ζ. *Anhydrides with Phosphorous and Arsenious Acid.* Mixed anhydrides with mono- and diesters of phosphorous acid and diesters of arsenious acid have also been used for the formation of anhydrides. Normally the acid chloride is added to the mixture of the carboxyl and amino components and the reaction proceeds via an activation of the amino group (see Chapter III,C,3). Therefore these methods will be discussed together with other types of activation of the amino group.

η. *Thionyl Chloride as Anhydride-Forming Reagent.* Yields were far below 50% when peptides were synthesized via anhydrides with thionyl chloride in coupling reactions with salt components (*2303, 2524*). The assumption that the actual acylating agent is the symmetrical anhydride formed by the disproportionation of the mixed anhydride is likely to be correct, since sulfur dioxide was liberated during the anhydride formation [22]:

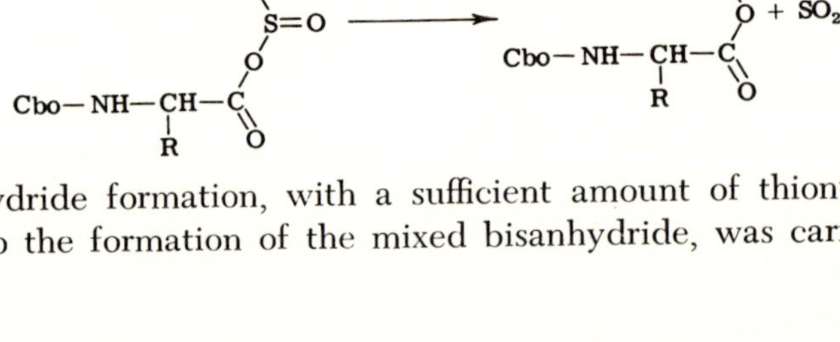

[22]

The anhydride formation, with a sufficient amount of thionyl chloride to lead to the formation of the mixed bisanhydride, was carried out at

$-10°C$. Complete racemization was observed under these conditions (*2303*).

θ. Anhydrides with Sulfuric Acid. The use of sulfuryl chloride as anhydride-forming reagent gave no satisfactory results (*2524*). Kenner (*1210*), Kenner and Stedmann (*1213*), and Clayton *et al.* (*497*) have described an interesting coupling reaction with the mixed anhydride of free sulfuric acid. The complex [23] (cf. page 79) is accessible from dimethylformamide and sulfur trioxide. It reacts with the sodium or lithium salt of an *N*-protected amino acid in dimethylformamide as solvent to form the mixed anhydride [24]. The complex is very unstable and was never isolated. Its solution is used directly for the coupling reaction:

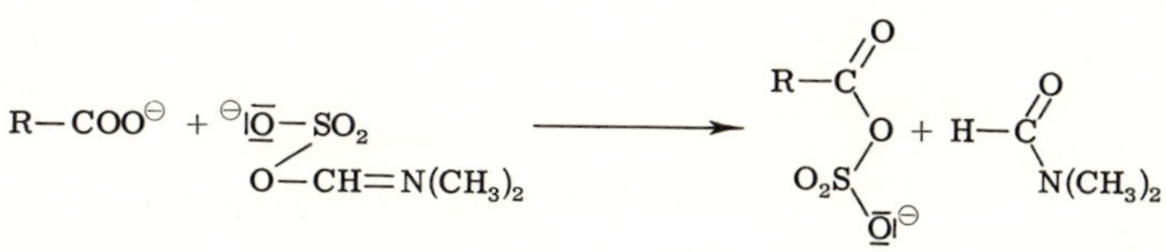

[23] [24]

In contrast to other mixed anhydrides, the *N*-protected amino acid sulfuric acid anhydride is present as its anion. It is therefore soluble in water and very well suited for coupling reaction with salts in aqueous alkaline solution.

It was shown for Cbo-Gly-L-Phe-OH that the mixed anhydride is sterically stable in dimethylformamide solution. Racemization occurs on hydrolysis with aqueous alkali. A coupling reaction with glycine in aqueous alkaline solution yielded a peptide corresponding in rotation and melting point to that prepared by the azide method and subsequent alkaline hydrolysis of the ethyl ester (*1213*). Other investigations have shown that the danger of racemization can be reduced by working in aqueous dimethylformamide and using magnesium carbonate as base. Sterically pure products were obtained under anhydrous conditions. Hydroxyl groups need not be protected when they are present in the amino component; they must be protected in the carboxyl component (*497*).

The studies on mixed sulfuric acid anhydrides have been reviewed by Kenner (*1211*).

c. Mixed Anhydrides with Organic Acids

α. Organic Carboxylic Acids as Anhydride-Forming Reagents. Due to their more symmetrical form and in contrast to the mixed anhydrides of *N*-protected amino acids with inorganic acids [25], it is more likely that the anhydrides with carboxylic acids [26] will be cleaved in a different manner. Wieland *et al.* (*2537*) were able to use mixed anhy-

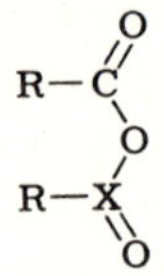

[25]

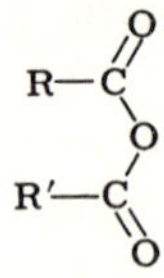

[26]

drides formed from sodium salts of carbobenzoxyamino acids and benzoyl chloride for the synthesis of peptides (*2553*). These mixed anhydrides are of low stability and disproportionate very readily to the symmetrical anhydrides (*2537*). The removal of benzoic acid after the synthesis of peptides was not always easy. The syntheses were all performed via coupling reactions with salts in aqueous solution. The reaction with amino acid esters is more convenient.

Emery and Gold (*677, 678, 679*) were studying the aminolysis of mixed anhydrides at the time of the first publications by Wieland. They found that the following factors influenced the course of the cleavage of the anhydride: Under otherwise identical conditions the carbonyl C atom with the lowest electron density will acylate the nucleophilic amine. If one anhydride component is sterically hindered, the other will react preferably to form the amide. This is true, however, only under anhydrous reaction conditions. Discrepancies and even the reverse have been observed in the presence of water [27]:

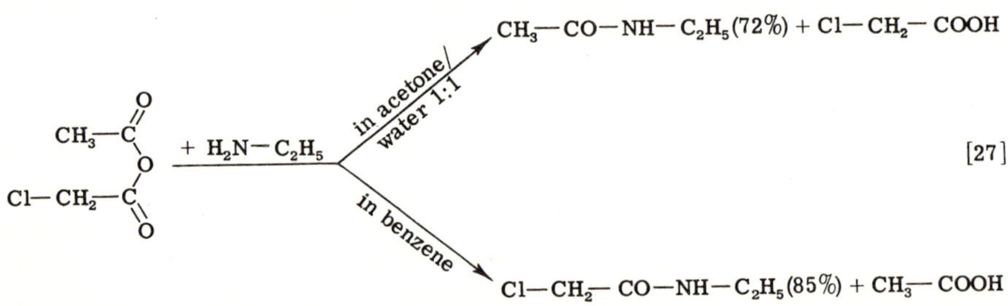

[27]

Wieland and Stimming (*2554*) have shown an increase in the yield of acetylhydroxamic acid in the reaction of hydroxylamine with mixed anhydrides of acetic acid in going from propionic to butyric and benzoic acid as anhydride component. According to Kenner (*1211*), the basicity of the amine may also influence the course of the reaction, as was concluded also by Vaughan and Osato (*2366*) from an extensive series of experiments. Carbobenzoxyglycine was allowed to react with a number of carboxylic acid chlorides to form the mixed anhydrides. The yields of the cleavage with aniline were compared. Sterically hindered, aliphatic carboxylic acids, such as diethylacetic acid, trimethylacetic acid (pivalic acid), or isovaleric acid, gave very satisfactory results. Aromatic acids, however, especially substituted compounds, gave only contaminated products in low yield. Losse and Demuth (*1455*) have confirmed this steric effect for mono-, di-, and triphenylacetic acid as anhydride component. Weygand and Reiher (*2496*) have reported the partial opening in the undesired manner of the mixed anhydride from trifluoroacetylglycine and benzoyl chloride. The results were much better under anhydrous conditions than in aqueous solvents. Low temperatures are also more favorable.

In addition to the mixed anhydrides with benzoic acid (*1806, 2419*), those with isovaleric acid (*775, 1191, 1806*) and pivalic acid (*1367, 2648*) are sometimes described. Acylation of the tosylamino group (cf. page 87) in tosylglycine was not observed with mixed anhydrides of pivalic acid (*2648*). Carbobenzoxyasparagine can be coupled by this method in yields of 75–80%. Other procedures normally yield less than 50% (see Chapter IV,C,2,a,δ). Losse and Demuth (*1455*) have prepared mixed anhydrides of *N*-protected amino acids with carboxylic acids by reacting them with ketene, e.g., diphenylketene, that could be used for the synthesis of peptides.

The ready accessibility of trifluoroacetylamino acid trifluoroacetic acid anhydride from 2 moles of trifluoroacetic anhydride and 1 mole of amino acid prompted Weygand *et al.* (*2480, 2485, 2493*) to test its applicability to peptide synthesis.

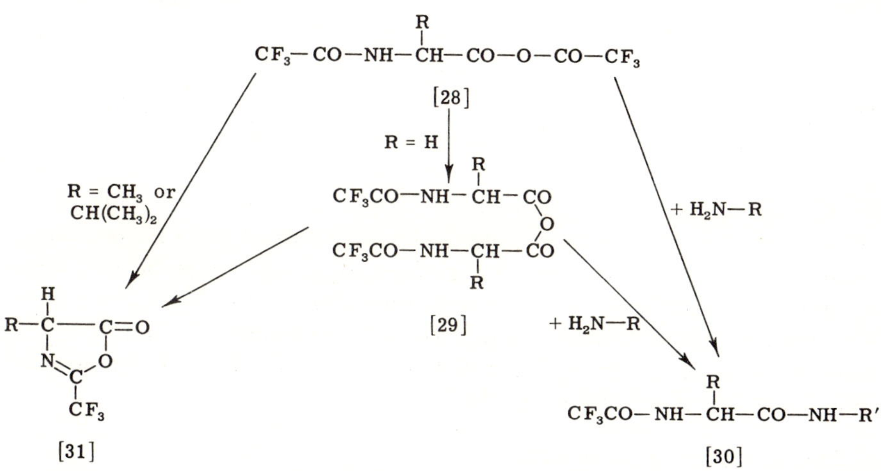

The symmetrical anhydride [29] is formed from the mixed anhydride [28] upon the action of triethylamine if glycine is the amino acid (R = H). The structure of the symmetrical anhydride was confirmed by its reaction with aniline which gave the anilide [30]. The preparation of the symmetrical anhydride is more difficult when R = CH_3 or $CH(CH_3)_2$ (alanine or valine). Here, formation of the azlactone, 2-trifluoromethyl-4-methyloxazolone-5 [31], is predominant and accompanied by racemization. The unsymmetrical anhydride of trifluoroacetyl-L-proline and trifluoroacetic acid disproportionates slowly to form the optically active symmetrical trifluoroacetyl-L-proline anhydride and trifluoroacetic anhydride (*2490*). Aspartic acid and glutamic acid form the trifluoroacetylated intramolecular anhydrides without racemization (*2490*, *2493*). Disproportionation to the symmetrical anhydrides was observed during the reaction of the mixed anhydrides with benzylglycinate. The resulting trifluoroacetic anhydride or free trifluoroacetic acid present caused a transesterification of the benzylglycinate and formation of benzyltrifluoroacetate.

β. *Sulfonic Acids as Anhydride-Forming Reagents.* Sokolowska *et al.* (*2157*, *2158*, cf. *2524*) made the benzenesulfonyl chloride method applicable to peptide synthesis. Brewster and Ciotti (*359*) have carried out esterifications with benzenesulfonyl chloride in pyridine. They postulate a reaction mechanism via the symmetrical anhydride. Nevertheless the yields are higher than 50%, because the acid released after acylation reacts with the sulfonyl chloride still present and forms the mixed anhydride again [32]:

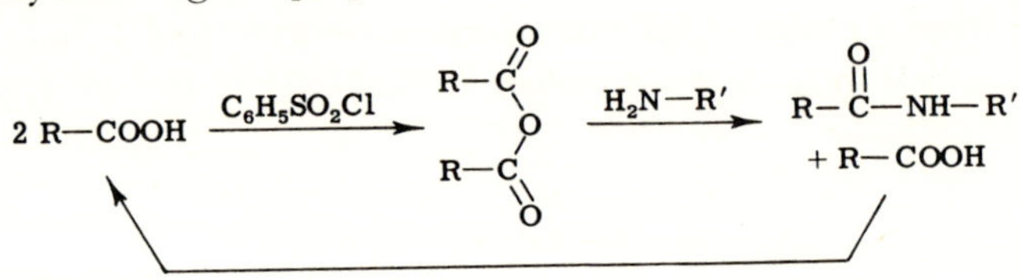

[32]

The method gives good results with tosyl- and phthalylamino acids. Side reactions occur with carbobenzoxyamino acids, but have not been studied in detail.

Theodoropoulos and Gazopoulos (*2303*) have used toluenesulfonyl chloride for the formation of anhydrides. In contrast to benzenesulfonyl chloride, it is also applicable to carbobenzoxyamino acids. Several cases of undesired anhydride cleavage with formation of tosylated compounds have been reported. Complete racemization was observed when dipeptide acids were used as carboxyl components in tetrahydrofuran or acetonitrile. Yields of about 50% were obtained.

d. Symmetrical Anhydrides

The most important methods for the synthesis of symmetrical anhydrides of *N*-protected amino acids are the disproportionations of mixed anhydrides (*2158*, *2490*) or the reaction with carbodiimide (*338*,

1987, 1988). These anhydrides have the advantage that the undesired cleavage cannot occur. Their application is not very economical, however, with maximum yields of only 50%.

Evidence has been provided that several mixed anhydride methods actually proceed through symmetrical anhydrides. A similar intermediate was postulated for the carbodiimide method. The symmetrical anhydride is constantly regenerated, thus assuring good yields. Weygand and Reiher (*2496*) have reported a peptide synthesis via the symmetrical trifluoroacetylamino acid anhydrides that is of theoretical interest. These anhydrides rearrange in absolute alcohol in the presence of 1 equivalent of dicyclohexylamine to form trifluoroacetyl dipeptide dicyclohexylammonium salts. The formation of pyroglutamic acid from the trifluoroacetylglutamic acid anhydride and dicyclohexylamine (cf. page 193) represents an analogous case. Presumably an acyl imide [33] is formed and reacts with the solvent with elimination of trifluoroacetic acid ethyl ester.

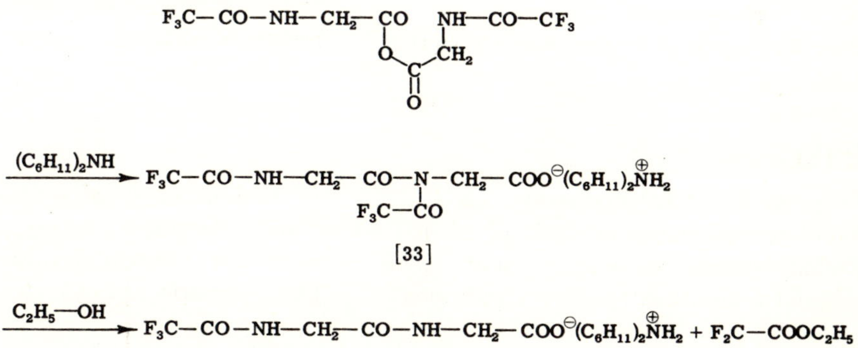

[33]

e. Intramolecular Anhydrides of Glutamic Acid and Aspartic Acid

The aminolysis of the intramolecular anhydrides of glutamic acid and aspartic acid is not uniform and both the α- and ω-isomers must be expected in peptide syntheses. The cleavage of these anhydrides with alcohols is applied mainly to the synthesis of starting materials where a relatively simple procedure for the separation of the mixtures is available (see Chapter IV,C,1,a,α, and 2,a,α).

2. Activated Esters

The synthesis of peptides via the aminolysis of esters has long been known, but a systematic search for suitable "activated" esters was not started until about 1950. Wieland *et al.* (*2548*) have described the synthesis of peptides via the phenyl thiolesters. Bodanszky (*265*) was able to show later that this type of activation is not restricted to the thiol group. *O*-Aryl esters, especially those with a negative substituent at the aromatic nucleus, are also suitable. Schwyzer *et al.* (*1995, 2011, 2014*), at about the same time, prepared activated alkyl esters by the introduction of negative substituents in methyl esters. The *p*-nitrophenyl esters are favored—next to the frequently used 2,4,5-trichlorophenyl hydroxysuccinimido and cyanomethyl esters—among the large number of tested aryl and alkyl esters and aryl and alkyl thiolesters. The esters can be arranged with respect to their activating properties in approximately the order (*265, 840, 841, 2520*) [34]:

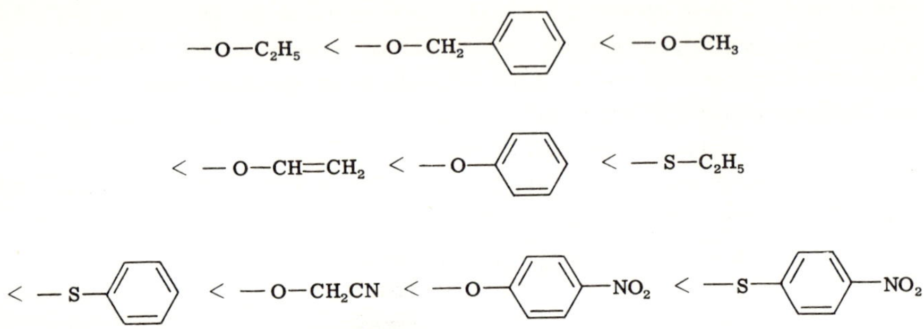

[34]

a. *O*-Esters

α. Aryl Esters. Bodanszky (*264, 265*) and Bodanszky *et al.* (*282*) have studied the aminolysis of phenyl esters and nitrophenyl esters of phthalylglycine by ethylglycinate. Their results were obtained under otherwise identical reaction conditions [35]. The *p*-nitrophenyl ester was

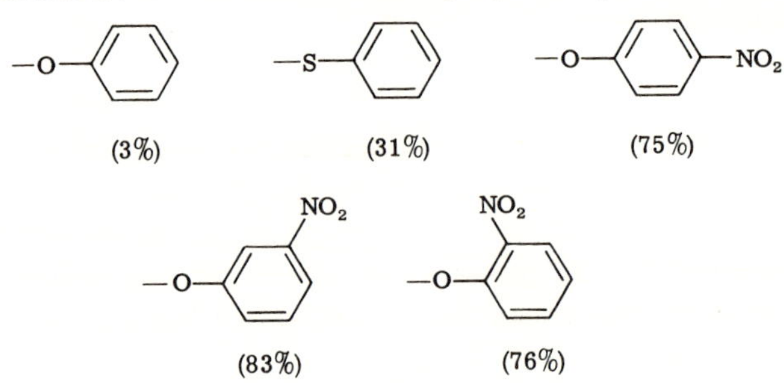

[35]

preferred for its better crystallizability.

Bodanszky (*265*) prepared *N*-protected amino acid *p*-nitrophenyl esters via the acid chloride method by fusing the components. *o*-Nitrophenol reacts only at higher temperatures. 2,4-Dinitro- and 2,4,6-trinitrophenol are sterically hindered and do not react. Ethyl-*p*-nitrophenyl carbonate was isolated as a by-product of the alkylcarbonic acid anhydride method. Other approaches to the synthesis of *N*-protected amino acid *p*-nitrophenyl esters are the anhydride method via phosphorus oxychloride (*2533*), diphenylketene (*676*), or ethoxyacetylene (*269*), and the carbodiimide method (*273, 674, 1871*). The yields for an esterification with *N,N'*-dicyclohexylcarbodiimide can be improved if pyridine is used as the solvent. Buzas *et al.* (*453*) have claimed, therefore, a somewhat different reaction mechanism [36], that does not proceed via the symmetrical anhydride as intermediate as it occurs in peptide coupling reactions (see Chapter III,B,3,a):

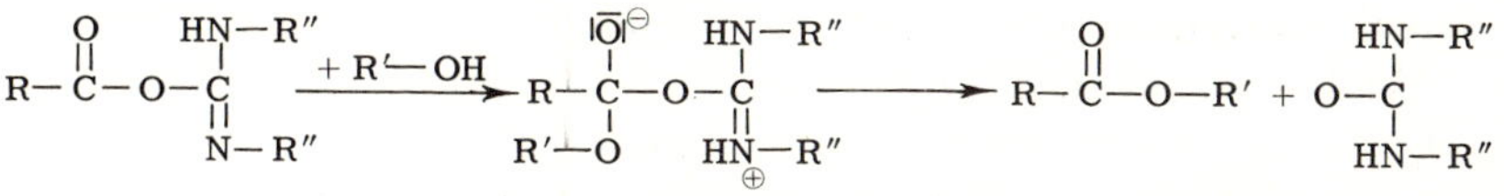
[36]

The reaction of *N*-protected amino acids with di-*p*-nitrophenyl sulfite in pyridine or ethyl acetate with 2 equivalents of pyridine at room temperature offers another convenient synthesis. Di-*p*-nitrophenyl sulfite [37] can be readily prepared from *p*-nitrophenol and thionyl chloride:

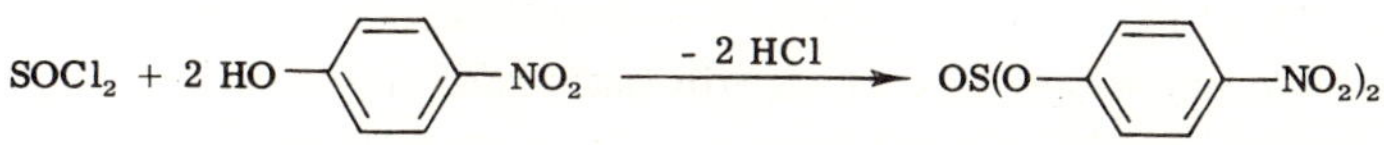
[37]

Iselin *et al.* (*1100*) have postulated the following reaction mechanism for the esterification. The diphenyl sulfite [38], in the presence of a base, is attached to the carbonyl compound [39]. The mixed anhydride with the half ester of sulfurous acid [40] is formed with elimination of a

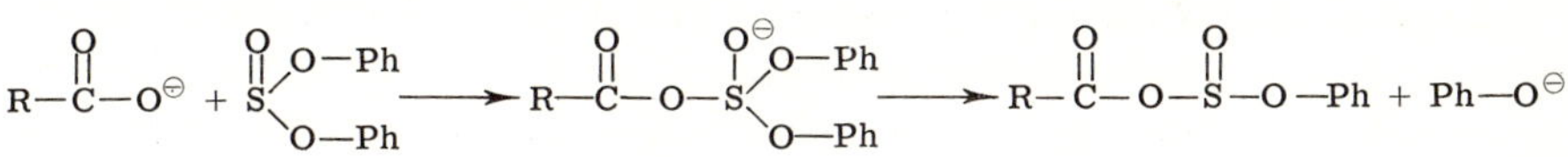

[38] [39] [40]

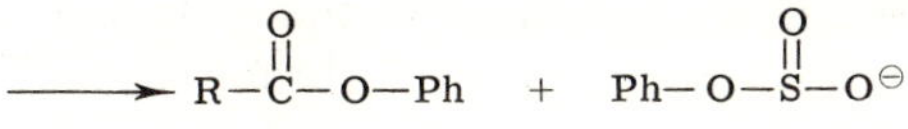

[41] [42]

phenyl cation. It reacts with the phenolate to form the ester [41]. The half ester of sulfurous acid [42] is present as anion. It decomposes to phenolate and sulfur dioxide. It was shown for ethyl-*p*-nitrophenyl sulfite that aryl-alkyl sulfites transfer only the aryl residue. Accordingly, an unsymmetrical diaryl sulfite, e.g., phenyl-*p*-nitrophenyl sulfite, reacts to form exclusively the *p*-nitrophenyl ester (*1100*). The elimination of the nitrophenolate anion is favored by the electron-attracting nitro group. The reaction of diaryl sulfates with carboxylic acids has not yet been conducted successfully.

Like the diaryl sulfites, the triaryl phosphites may be used for the preparation of *N*-protected amino acid aryl esters (*1079*). They are synthesized by reacting phosphorus trichloride with phenol. They were first used by Farrington *et al.* (*704*) for the synthesis of aryl thiolesters. Presumably a similar reaction mechanism must be formulated. The structure and number of other reaction products, however, have not been established (*704*). Two *p*-nitrophenyl residues are transferred in the presence of pyridine. Mixed aryl-alkyl phosphites and unsymmetrical diaryl phosphites show the same behavior as the corresponding sulfites. Wieland *et al.* (*2534*) have used di-*p*-nitrophenyl carbonate for the synthesis of *p*-nitrophenyl esters. Good yields were obtained by refluxing for 3 hours in ethyl acetate in the presence of pyridine.

Certain amino acids do not readily form the *p*-nitrophenyl esters. *N*-Protected glutamine and asparagine p-nitrophenyl esters were obtained only in low yield. Irrespective of the method, elimination of water from the amide group must be accepted as a side reaction (see Chapter IV,C,I,a,ϵ and 2,a,δ). Lactam formation was observed in the case of carbobenzoxynitroarginine (*280*). *p*-Nitrophenyl esters of N^{α}-tosylamino acids are not known (*214*). Nor have *N*-protected serine or threonine *p*-nitrophenyl esters been synthesized. Only the 2,4-dinitrophenyl esters of carbobenzoxy-L-serine and carbobenzoxy-L-threonine have been more recently described (*1499a, 1829a*).

Peptide syntheses with *p*-nitrophenyl esters are carried out in, for example, ethyl acetate or tetrahydrofuran—depending on the solubility—at room temperatures or slightly higher temperatures. Dimethylformamide is sometimes added. A number of naturally occurring polypeptides have been synthesized via *p*-nitrophenyl esters exclusively (*267, 273, 277*), demonstrating clearly the advantages of this method. Khurgin and Dimitrieva (*1226*) have investigated the influence of chain length on the rate of aminolysis. It is more strongly influenced by the chain length of the carboxyl component than of the amino component. The aminolysis is strongly catalyzed by the addition of imidazole (*1515a*). The *p*-nitrophenyl ester method can also be applied to coupling reactions with salts (*283, 996*). Guttmann and Boissonnas (*899*) subject the esters to aminolysis in a mixture of dioxane and aqueous sodium hydroxide. A constant pH value (automatic titrator) prevents the undesired alkaline hydrolysis.

The aminolysis of the *p*-nitrophenyl esters proceeds without racemization. Bodanszky and Birkhimer (*268*), however, have shown a time-dependent change in optical rotation at 20°C for solutions of carbobenzoxyamino acid *p*-nitrophenyl esters in the presence of only 1% of triethylamine. While glutamine and asparagine derivatives are stable, the changes in rotation are fast for carbobenzoxy-S-benzyl-L-cysteine *p*-nitrophenyl ester, for example. Therefore an excess of triethylamine for the liberation of the amino acid esters should be avoided. The alkaline hydrolysis of Cbo-Gly-L-Phe-$OPhNO_2$ gives the racemic acid. Cbo-Gly-L-N-Me·Phe-$OPhNO_2$, on the other hand, shows only very little racemization. An oxazolone derivative has been discussed as a possible intermediate (*839*). Cbo-Gly-L-Pro-$OPhNO_2$ prefers to form a carbobenzoxydiketopiperazine (*838*).

The preparation of *p*-nitrophenyl esters of *N*-protected peptides may be accompanied by racemization, irrespective of the method employed (*889, 1102, 1467*). Optically pure nitrophenyl esters of peptides can be prepared by conventional techniques only when the carboxyl-terminal amino acid cannot be racemized (glycine or proline) (*1102, 2027*). In all other cases, free amino acid *p*-nitrophenyl esters can be obtained by removing the blocking group from the *N*-protected amino acid *p*-nitrophenyl esters [43]. They react with *N*-protected amino acids or *N*-protected peptides to form the *N*-protected peptide *p*-nitrophenyl esters [44] (*832, 1080, 1102, 1134, 1467*). Goodman *et al.* (*834, 837*) did not even isolate the activated esters, but allowed them to react immediately with the amino components:

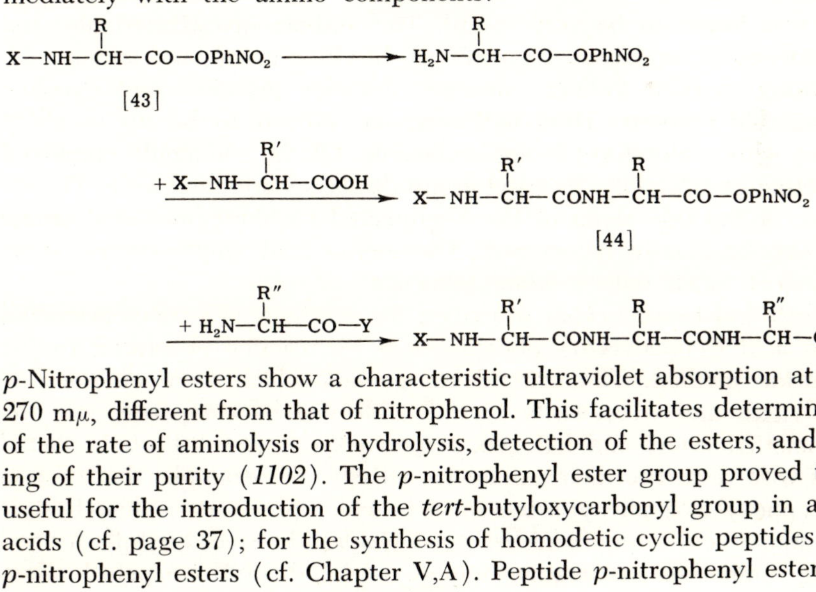

p-Nitrophenyl esters show a characteristic ultraviolet absorption at 268–270 mμ, different from that of nitrophenol. This facilitates determination of the rate of aminolysis or hydrolysis, detection of the esters, and testing of their purity (*1102*). The *p*-nitrophenyl ester group proved to be useful for the introduction of the *tert*-butyloxycarbonyl group in amino acids (cf. page 37); for the synthesis of homodetic cyclic peptides with *p*-nitrophenyl esters (cf. Chapter V,A). Peptide *p*-nitrophenyl esters are suitable for polycondensations (*589a*).

Iselin *et al.* (*1100*) have described a number of other carbobenzoxyglycine aryl esters. Aryl sulfites or aryl phosphites were used for the esterification. The *p*-methanesulfonylphenyl and *p*-cyanophenyl esters (*1100, 2693, 2695*) of various *N*-protected amino acids and peptides were prepared. The latter esters have apparently not yet been used for the synthesis of peptides. The *p*-methanesulfonylphenyl ester was used to activate derivatives for cyclizations; in contrast to most other activating esters, it is stable to catalytic hydrogenation. It was therefore, possible, to remove the carbobenzoxy group by catalytic hydrogenation from a derivative prior to cyclization (*2031*).

Kupryszewski (*1309, 1311*) has described the use of 2,4,6-trichlorophenyl esters of carbobenzoxy-, phthalyl-, tosyl-, and benzoylamino acids for the synthesis of dipeptides. He showed that the aminolysis of phthalylglycine 2,4,6-trichlorophenyl ester with glycine methyl ester is faster than that of the corresponding *p*-nitrophenyl ester. The trichlorophenyl ester gave under otherwise identical conditions a yield of 58%, while only 37% was obtained with the *p*-nitrophenyl ester. The activated esters are accessible via the phosphorus oxychloride method (*1314*). The same method was applied successfully to the synthesis of pentachlorophenyl esters (*1312*).

Pless and Boissonnas (*1738b;* cf. *1738a*) have published a detailed report on their investigations of the influence of substitution of a series of phenyl esters on the aminolytic properties. The 2,4,5-trichlorophenyl ester was found to be very useful. The authors investigated also the dependence of aminolysis upon solvent, blocking group, and amino acid. According to their findings, dioxane, aqueous pyridine, and pyridine are suitable solvents. Dimethylformamide proved to be an excellent solvent while chloroform is not applicable. Of the commonly employed protecting groups, only the trityl group displays steric hindrance. The influence of the side chain of the *N*-protected trichlorophenylated amino acid can be clearly recognized. The amino acid employed as amino component exerts only a minor influence.

Barth and Losse (*113a*) described the synthesis of various protected amino acid-4-(phenylazo) phenyl esters by reaction according to the anhydride method. The colored compounds which were obtained in good yields and were easily crystallizable can be compared with the cyanomethyl esters and halogen, respectively, nitro-substituted, phenyl esters with regard to their aminolysis activity. For example, the reaction of -4-(phenylazo) phenyl esters or -4-(4′-chlorophenylazo) phenyl esters with amino acid esters furnishes good yields. The free dyestuffs formed during aminolysis are readily soluble in tetrahydrofuran/petrol ether, ethyl acetate/petrol ether, or ether/petrol ether and, therefore, can be easily separated from the peptide derivative.

Phenyl esters occupy a position intermediate between activated esters and carboxyl-protecting groups (*1212, 1260*). They react under very mild conditions with reagents of strong aminolytic action, such as ammonia and hydrazine. The synthesis of a peptide has also been described (*1261*; cf. *265*). Heating for 48 hours to 45°C in tetrahydrofuran was required. The phenyl ester of trifluoroacetylglycine reacts under vigorous conditions (20 minutes at 150°C) with leucine in phenol/triethylamine to form the peptide derivative (*2498*). Trifluoroacetic acid phenyl esters can be used for the synthesis of trifluoroacetylamino acids (*2498*).

β. Alkyl Esters. Methyl and ethyl esters are normally not sufficiently activated to be useful for the synthesis of peptides. They have nevertheless been employed on occasion for polycondensations (*352, 435, 1177*) and for the synthesis of cyclic (*400*) and linear peptides (*1177*). The trifluoroacetyl protecting group can be introduced under very mild conditions with methyl or ethyl trifluoroacetate (*2482, 2492, 2503*). The aminolysis of the methyl esters is greatly enhanced by the presence of imidazole (*2557*) [cf. (*1796*) for kinetic studies of the aminolysis of simple alkyl esters].

The introduction of negative substituents increases, as in the case of the aryl esters, the activation of alkyl esters. This was systematically studied primarily by the group of Schwyzer (*1099, 2008, 2009, 2012, 2014*). Benzoylglycine was chosen for the investigation of a series of substituted methyl esters. The cyanomethyl ester proved to be best (cf. *1573*).

Cyanomethyl esters can be readily prepared by the reaction of *N*-protected amino acids with chloroacetonitrile in the presence of triethylamine (*2009*). The reaction must be carried out at room temperature if an excess of chloroacetonitrile is used as solvent. Partial to complete racemization was observed in reactions of carbobenzoxyamino acids with chloroacetonitrile at higher temperatures (*1099, 1422a*). The following reaction conditions are recommended when ethyl acetate is used as solvent: boiling for 3–5 hours or standing for 60 hours at room temperature. The danger of racemization remains under the conditions of boiling.

The most favorable conditions for the aminolysis of cyanomethyl esters are described in the detailed publications. High molar concentrations and 0.05–0.1 equivalent of acetic acid, to act as catalyst, gave the best results (*1099, 2008, 2012*). The aminolysis is carried out at room temperature and proceeds without racemization (*1099, 2012*). As in the case of *p*-nitrophenyl esters (cf. page 100), the aminolysis of cyanomethyl esters is catalyzed by the addition of imidazole, pyrazole, triazoles, etc. (*235a*).

N-Protected dipeptides have also been converted to activated esters without racemization. Amino acid and peptide cyanomethyl esters are synthesized from *N*-protected derivatives by removing with acid the trityl, formyl, or *tert*-butyloxycarbonyl group (*837, 1466, 2018*). The acid concentration should be such as to prevent the hydrolysis of the cyano group to the acid amide group [45]:

$$R{-}CO{-}O{-}CH_2{-}CN \xrightarrow{+\,H_2O} R{-}CO{-}O{-}CH_2{-}C{\lt}^{O}_{NH_2} \qquad [45]$$

This hydrolysis does not permit the use of the carbobenzoxy group and its removal with hydrogen bromide in glacial acetic acid (*837*). Cyanomethyl esters have rarely been used for the synthesis of higher biologically active polypeptides. They have played, however, an important role in elucidating more basic problems, e.g. the synthesis of cyclic peptides (*2018*).

Efforts to use tetrahydropyranyl esters for activation did not meet with success (*1101*). The formation of a new center of asymmetry proved to be a disadvantage [46]:

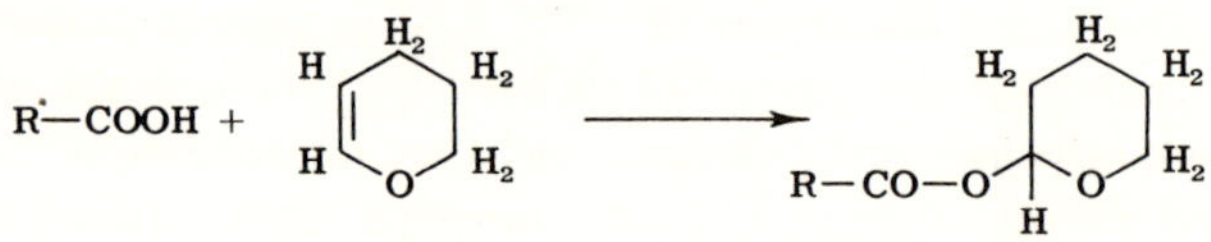

[46]

During aminolysis an aldehyde [47] is formed; it may cause side reactions:

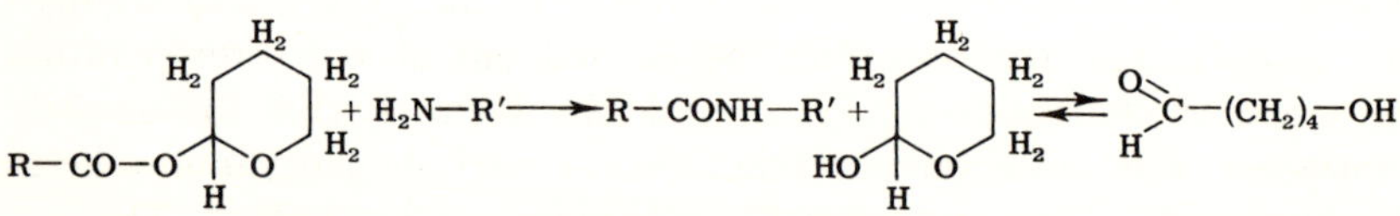

[47]

The vinyl esters of *N*-protected amino acids have been prepared by transesterification with vinyl acetate in the presence of palladium chloride (cf. *2109, 2110*). They are supposedly suitable for coupling reactions with sterically hindered amino acids. Acetaldehyde is released during the aminolysis. It can be trapped with, for example, malonic ester. Mainly trifluoroacetylamino acid vinyl esters have so far been described. They react with amino acid esters with peptide bond formation at room temperature for 10 hours or on heating to 65°C for 40 minutes. It was shown by gas chromatography that racemization occurs to a small extent during the esterification or the coupling reaction of the synthesis of TFA-L-Val-L-Val-OMe (*2502*). [Cf. (*266*) for aminolysis of propargyl esters.]

Nefkens and Tesser (*1599*) have prepared hydroxyphthalimido esters [48] from carbobenzoxy amino acids and *N*-hydroxyphthalimide via the carbodiimide method. These esters are activated and react with amino acid esters at 0°C within a few seconds to form dipeptide derivatives in yields of 40–80%:

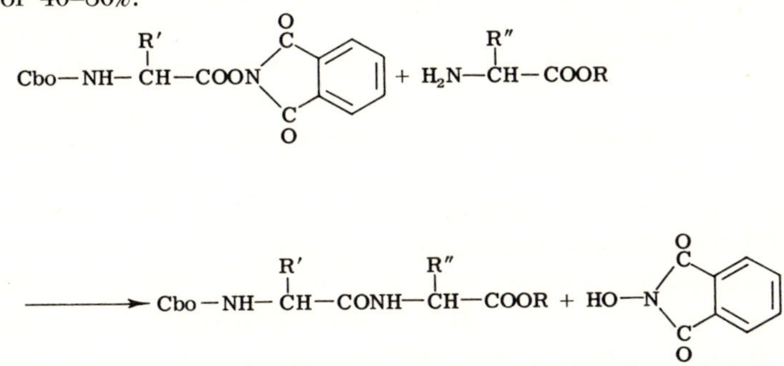

[48]

Racemization has so far not been observed. *N*-Hydroxyphthalimide can be synthesized by briefly heating hydroxylamine hydrochloride/triethylamine with carbethoxyphthalimide in absolute ethanol.

Anderson *et al.* (*53a,b*) have suggested for application to peptide synthesis an analogous ester, namely, the hydroxysuccinimido ester. The solubility in water of *N*-hydroxysuccinimide offers the advantage that the coupling products can be conveniently purified. This activated ester has quickly become very useful for the synthesis of higher peptides (*1020c*).

b. THIOLESTERS

α. Aryl Thiolesters. Aryl thiolesters of *N*-protected amino acids are prepared via the mixed anhydride method using, for example, mixed alkylcarbonic acid anhydrides (*996, 1059, 2548*), mixed sulfuric acid anhydrides (*704*), and phosphorus oxychloride as anhydride-forming reagent (*2533*). According to Wieland and Heinke (*2533*), it is advantageous in the application of the phosphorus oxychloride method to add reactants to the *N*-protected amino acid in this order: thiophenol, phosphorus oxychloride, base. Pyridine, as well as triethylamine, is a suitable base and serves at the same time as solvent. The cleavage of the intramolecular anhydride of carbobenzoxyglutamic acid with thiophenol may yield the α- or the γ-esters, depending on the reaction conditions (*1893, 2559*). *N,N′*-Dicyclohexylcarbodiimide can also be used for the synthesis of *N*-protected amino acid aryl thiolesters (*1871, 1456,*

2501). Kenner *et al.* (*704, 705, 1214, 1215*) have used tri-(*p*-nitrophenylthiol) phosphite for the synthesis of *p*-nitrophenyl thiolesters. The phosphite was prepared from phosphorus trichloride and sodium *p*-nitrophenolate. The esters were obtained by allowing the phosphite to react with, for example, the lithium salts of carbobenzoxyamino acids or peptides at 20°C. Better results were obtained by the direct reaction with *N*-protected amino acids or peptides in pyridine (20 hours, 18°C) or in dimethylformamide (20 minutes, 85°C) under an atmosphere of nitrogen (*704, 705*).

Cbo-Gly-L-Phe-SPhNO$_2$ was chosen by several authors to study the extent of racemization during the synthesis of *N*-protected dipeptide *p*-nitrophenyl thiolesters. The crude product of e.g., Cbo-Gly-L-Ala-SPhNO$_2$ prepared via the lithium salt with tri-(*p*-nitrophenylthiol) phosphite gave the Cbo-tetrapeptide (30% racemate) when reacted with H-L-Phe-Gly-OH (*704*). Judging from the experience with the corresponding *p*-nitrophenyl esters, it is likely that the racemization occurred during the synthesis of the activated ester. Farrington *et al.* (*705*) obtained optically inactive Cbo-Gly-DL-Phe-SPhNO$_2$ in a synthesis via the mixed ethylcarbonic acid anhydride. Wieland and Heinke (*2533*), however, were able to isolate an ester in 65% yield that was optically pure to the extent of 90%. They employed the same mixed anhydride in tetrahydrofuran. Sterically uniform products, though in yields of only 40%, are obtained via the phosphorus oxychloride method. The Cbo-dipeptide acid (45%) was recovered in its optically active state. A partially racemized ester was obtained when the mixed sulfuric acid anhydride in dimethylformamide was used for the acylation (*704*).

Farrington *et al.* (*704*) rate the *p*-nitrophenyl thiolester as one of the most highly activated esters. For the example

$$\text{Cbo—Gly—Y} + \text{H—Ala—OH} \longrightarrow \text{Cbo—Gly—Ala—OH}$$

the following differences in the relative rates of reaction [49] were given:

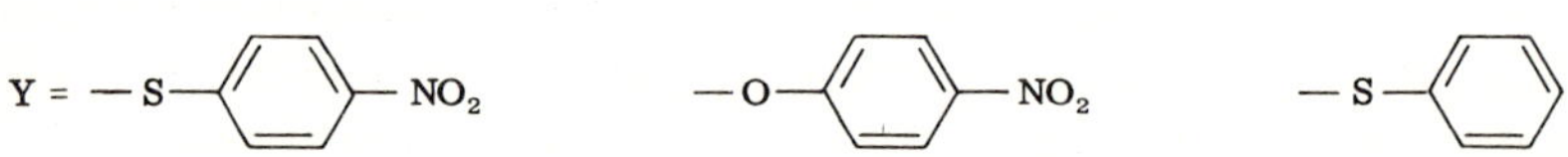

(140) (16) (1)

[49]

Thiolesters may be allowed to react with amino acid esters in indifferent solvents or alcohols to form peptides (*2533, 2548*). Reactions with free amino acids or peptides are also feasible. Thiophenol formed in these reactions can be oxidized with hydrogen peroxide to the disulfide to be extracted from alkaline solution. Thiolesters are relatively stable to hydrolysis and alcoholysis (*2533, 2548*). This proves to be a great advantage for coupling reactions with salts (*705, 2548*). They are indeed so stable to alkaline hydrolysis that they can be compared with, for example, methyl or ethyl esters.

Weygand *et al.* (*2473, 2489, 2501, 2711*) reacted alkyl and aryl thiolesters of *N*-protected amino acids and *N*-protected peptides with free amino acids and peptides in hot or boiling glacial acetic acid (cf. page 71).

Free amino acid or peptide thiolesters can be prepared, in the form of their hydrochlorides or hydrobromides, either from amino acid chloride hydrochlorides and thioalcohols (*2538, 2547*) or, better, from the corresponding carbobenzoxyamino acid thiolesters or carbobenzoxy peptide thiolesters by cleavage of the protecting group with hydrogen bromide in glacial acetic acid (*1214, 2533*). Amino acid phenyl thiolesters may be condensed with *N*-protected amino acids or peptides to form *N*-protected peptide phenyl thiolesters. The acylation of dipeptide phenyl thiolesters via the anhydride method is faster than the self condensation. Diketopiperazine formation was observed to a considerable extent only with proline-containing peptides and when the more active *p*-nitrophenyl thiolester was used (*2533*). Amino acid and peptide aryl thiolesters, when used under suitable conditions, proved to be useful for polycondensations (*2525, 2547*) and for the synthesis of cyclic peptides (*1214, 1215, 2533*).

N-Protected peptide phenyl thiolesters react very rapidly with a slight excess of hydrazine hydrate to form the corresponding hydrazides in high yield. Half molar amounts of hydrazine hydrate gave the symmetrical *N*-protected bispeptidyl hydrazides (*2533*). Racemization was not observed during the coupling of *N*-protected peptide phenyl thiolesters with amino acid esters in organic solvents or with amino acids in aqueous alkaline tetrahydrofuran at 60°C. The presence of imidazole accelerates the rate of aminolysis of phenyl thiolesters (*2529*). [Cf. (*1500*) for trichlorophenyl thiolesters of phthalylamino acids.]

Aryl thiolesters are not of importance for the synthesis of higher peptides. This may be due to the inconveniences associated with the handling of mercaptans and, perhaps more so, to the complications encountered during catalytic hydrogenations (*436*).

β. Alkyl Thiolesters. There are few reports on alkyl thiolesters as active esters. They can be considered as active esters (*1259*) only in reactions with compounds of strong aminolytic properties, e.g. ammonia or hydrazine. Their application to peptide synthesis requires rather vigorous reaction conditions (*1995, 2498*). The trifluoroacetic acid ethyl thiolester was used for the introduction of the trifluoroacetyl group (see Chapter I,A,2). The carboxymethyl thiolester (*S*-acylthioglycolic acid), however, displays the properties of an activated ester (*1995, 2011*). Its synthesis via the mixed anhydride method makes use of *N*-protected amino acids or peptides and thioglycolic acid. Peptide carboxymethyl thiolesters have been used for the synthesis of cyclic peptides (*2018*). The thiolesters were prepared from carbobenzoxy compounds by elimination of the protecting group with hydrogen bromide in glacial acetic acid (*1996*).

The alkyl thiolesters can be arranged with regard to increasing reactivity in the following order (*1995*) [50]:

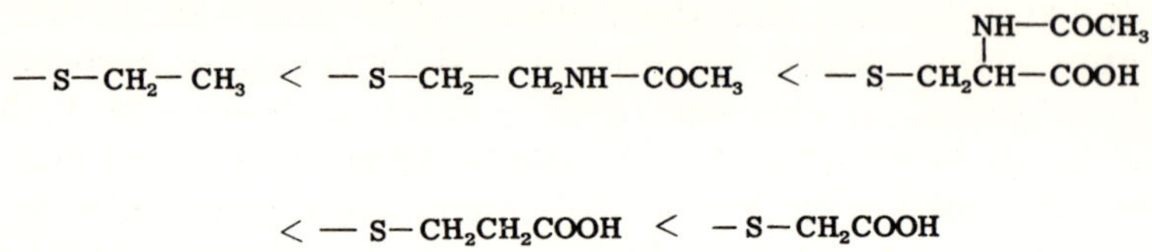

[50]

Thiolactones, e.g. homocysteine thiolactone (*141*) or cysteine thiolactone (*728, 729, 558*), are also activated thiolesters and can be cleaved by aminolysis. Some form of activation is already present in the thio acids. They are capable of acylation (*543*).

3. Activation by a C—N— or C—C— Multiple Bond

a. Carbodiimide

Sheehan and Hess (*2068*) introduced in 1955 a new method for the formation of peptide bonds using *N,N'*-dialkylated carbodiimides, preferably *N,N'*-dicyclohexylcarbodiimide. The application of this reagent became in a short time one of the most important coupling methods. The preparative procedure is simple and the fact that there is very little tendency to racemization offers a great advantage. The reaction can be expressed in the following equation [51]:

$$R{-}COOH + H_2N{-}R' + R''{-}N{=}C{=}N{-}R'' \longrightarrow R{-}CONH{-}R' + R''{-}NH{-}CO{-}NH{-}R'' \quad [51]$$

Its mechanism, however, is not yet fully understood. Khorana (*1223*) has suggested an activation of the carboxyl group initiated by the addition of its proton to the double bond of the carbodiimide [52]:

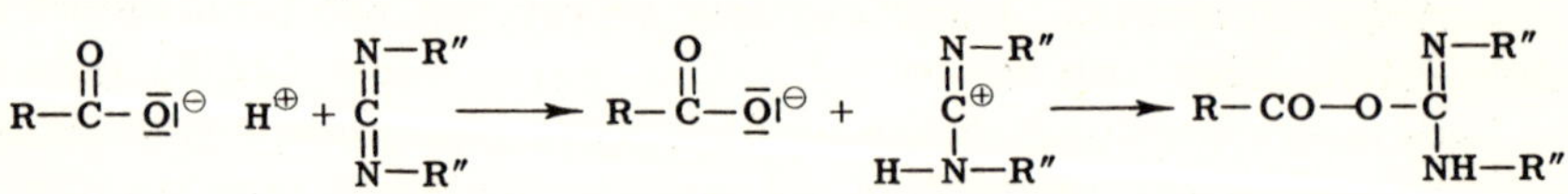

[52] [53]

The resulting compound [53] may react in two different ways. O → N acyl migration (reaction a) and formation of acylurea [54] is one

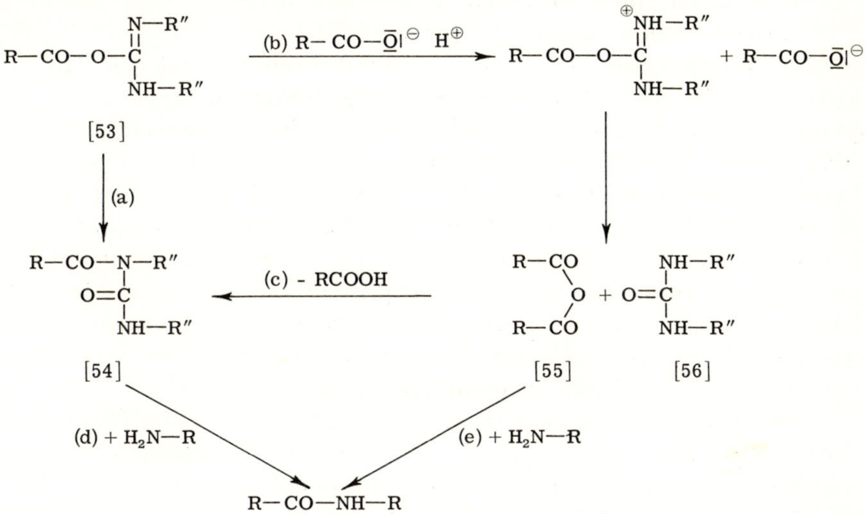

alternative if there are no further protons available. This course of the reaction was confirmed by Schüssler and Zahn (*1988;* cf. *1249a, 1987*) for the reaction of carbobenzoxyglycine with *N,N′*-dicyclohexylcarbodiimide in the presence of triethylamine (low concentration of protons). The acylurea was isolated in a yield of 40%. In the presence of further acid a second addition of a proton takes place (reaction b), and the O → N acyl migration will be prevented. Stabilization occurs by way of formation of the symmetrical anhydride [55] and disubstituted urea [56]. Several authors have described the synthesis of symmetrical anhydrides, using *N,N′*-dicyclohexylcarbodiimide (see Chapter III,B,1,d). The intramolecular anhydrides of glutamic acid and aspartic acid are also accessible by this procedure. Smith *et al.* (*2154*) have discussed an alternate formation of the acylurea (reaction c). It has a chance to occur when carbodiimide and carboxylic acid are kept for a long time at room temperature and no other reactants are present.

The formation of peptide bonds, according to Khorana (*1224*), may proceed either from the acylurea (reaction d) or from the symmetrical anhydride (reaction e). Diehl (*602*), however, claims that the acylurea is not an activated compound. Schüssler and Zahn (*1988*) have been able to prove that the reactivity of symmetrical anhydrides of carbobenzoxyamino acids and carbobenzoxy peptides is sufficient for peptide coupling reactions under the conditions of the carbodiimide method. In cases where symmetrical anhydrides do not react, a reaction cannot be brought about with carbodiimide either (*2077*). There remains a discrepancy in the reaction mechanism via a symmetrical anhydride, i.e. no racemization has been observed in coupling reactions with acylamino acids carrying, for example, the formyl (*2073*) or trifluoroacetyl group (*2497*). Racemization might be expected from the intermediate anhydride on the basis of the ready azlactone formation, since acetyl- and formylamino acids have been used directly for the synthesis of azlactones (*1633, 2117*; see Chapter III,B,5,a). Anderson and Callahan (*47*) have demonstrated an influence of solvent and temperature on the racemization for the reaction of *N,N′*-dicyclohexylcarbodiimide with Cbo-Gly-L-Phe-OH and ethylglycinate. The DL-form was isolated in a yield of 7–8% from the reaction in tetrahydrofuran at room temperature, while only 0.5% was found when the reaction was performed at 0°C. The racemic mixture was isolated in a yield of 12% from a reaction carried out in methylene chloride at room temperature (cf. *2141*).

The undesired formation of acylurea (*978, 1104, 1538, 2017a, 2055, 2064, 2616, 2617*) is frequently observed. It is dependent upon a number of factors. Low temperatures reduce its formation (*978, 1950*); therefore the reaction is normally carried out at a temperature of 0°C (*626, 1193, 2073*) or lower (*978, 2497, 2638*), rarely at higher temperatures (*291, 2300*). The formation of acylurea is also reduced significantly in methylene chloride (*1535, 1539, 2064*) or acetonitrile (*152, 894, 2064, 2617*). Other solvents that have been used are tetrahydrofuran (*236, 626, 1538*), dimethylformamide (*272, 274, 291, 1104, 2386*), and chloroform (*1903*), or various combinations of them. Aqueous tetrahydrofuran (90%) is supposedly very good (*225, 226, 231, 232, 233, 234, 1525*). Pyridine can also be used as solvent (*288, 291, 978, 2019a*) despite the fact that the presence of bases, especially triethylamine (*1988, 2638*), favors the formation of acylurea. The other partner of the reaction, the added amino component, may also act as base. Zahn and Diehl (*2617*) have therefore suggested that it be added over a longer period of time.

In the preparative procedure both the *N*-protected carboxyl and amino components are dissolved, whereby salt formation may occur. This does not necessarily prevent the coupling reaction (*1104*). The carbodiimide is added to the mixture, sometimes in several portions (*1523*). When glutamine or asparagine is used as carboxyl component, γ- or β-cyano derivatives may be formed as the result of an elimination of water (see Chapter IV,C,1,a,ϵ and 2,a,δ). Sometimes the by-product of the reaction, *N,N′*-dicyclohexylurea, causes difficulties during the isolation or crystallization of peptides. Therefore other carbodiimides with more favorable solubility properties have also been investigated (*2061, 2069*). Carbodiimides and the resulting urea with a tertiary amino group are soluble in water as ammonium salts. *N*-(Cyclohexyl)-*N'*-(*p*-diethylaminocyclohexyl)-carbodiimide [57] may serve as an example. Compounds with quaternary amino groups, *N*-cyclohexyl-*N'*-[2-morpholinyl-(4)-ethyl]-carbodiimide metho-*p*-toluenesulfonate [58], are well suited. They can be applied to coupling reactions in purely aqueous solutions (*2069*).

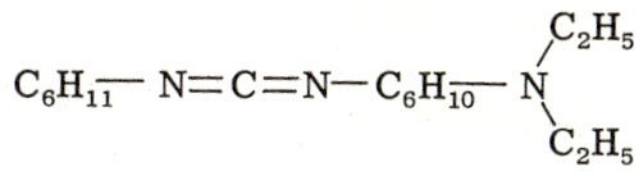

[57]

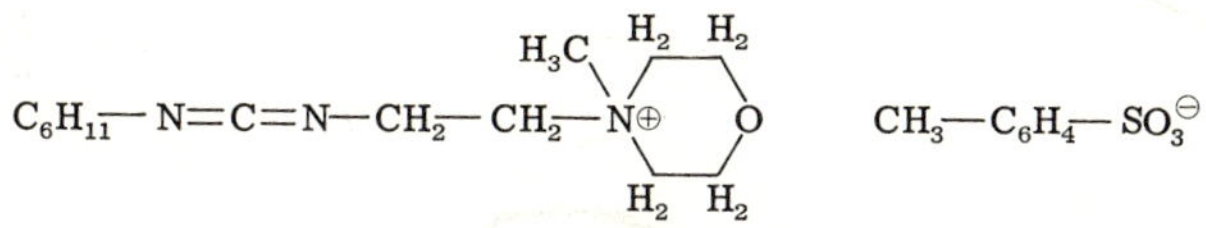

[58]

N,N'-Dicyclohexylcarbodiimide—in addition to its use for the formation of peptide bonds—is frequently applied to esterifications, especially to the synthesis of activated alkyl and aryl thiolesters. Peptides have been cyclized with *N,N'*-dicyclohexylcarbodiimide and also with water-soluble carbodiimides (*1284a, 1285, 2071, 2545*). Propiothiolactone was also synthesized with the aid of *N,N'*-dicyclohexylcarbodiimide (see Chapter IV,G,1,g,β).

b. Cyanamides

Losse and Weddige (*1461, 1462*) discovered a reaction analogous to the carbodiimide method when they allowed *N,N*-dialkyl cyanamides [59] to react with *N*-protected amino acids. The initially formed activated adduct [60] reacts with an amino component with formation of the peptide bond.

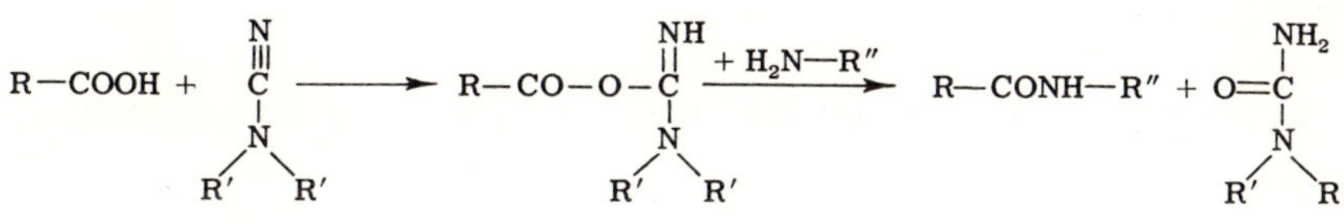

[59] [60]

(R' = benzyl, phenyl, or ethyl)

c. Ethoxyacetylene

Arens (*65*) and Panneman *et al.* (*1690*) have described the use of ethoxyacetylene as coupling reagent for the synthesis of peptides. The reaction [61] can be expressed summarily in the following equation:

$$R{-}COOH + H_2N{-}R' + HC{\equiv}C{-}OC_2H_5 \longrightarrow R{-}CONH{-}R' + CH_3{-}COOC_2H_5 \qquad [61]$$

Its mechanism should depend on the conditions employed. It is possible to obtain symmetrical carboxylic acid anhydrides with ethoxyacetylene (*66*), and a mechanism via the symmetrical *N*-protected amino acid anhydrides ([63], route a) has been discussed. Route b, however, is more

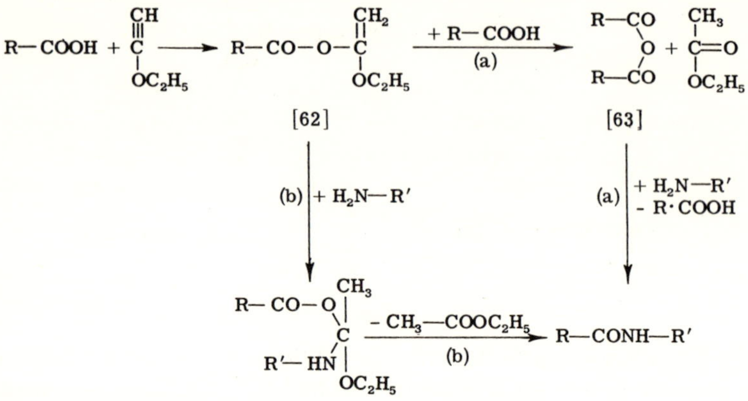

[62] [63] [64]

likely, since Sheehan and Hlavka (*2070*) isolated the *N*-protected amino acid ethoxyvinyl ester [62], in addition to the desired peptide, from a coupling reaction with ethoxyacetylene. This ester is an activated derivative. It reacts with amino acid esters to form peptides in high yield. Addition of the amino component to the vinyl ester [64] has been discussed as reaction mechanism (*1690*). The reaction certainly takes a different course when the hydrochloride (*992, 1690*) is used instead of the free amino component. The first step is the addition of hydrogen chloride to the triple bond. The ethyl chlorovinyl ether [65] reacts with the carboxyl component with formation of the acid chloride (cf. page 78):

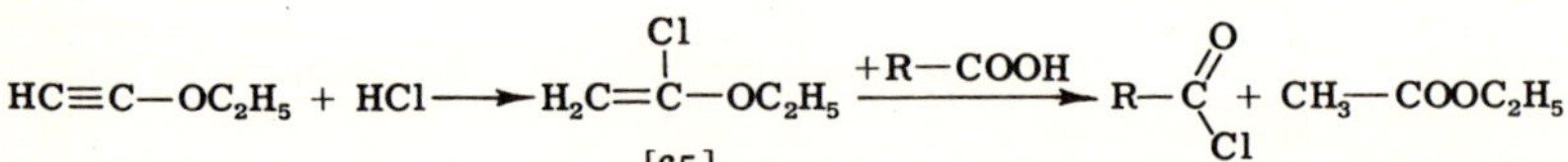

[65]

The reaction requires a four- to fivefold amount of ethoxyacetylene. The reagent serves also as solvent. A better solvent is ethyl acetate with a water content of 0.5%. Racemization may occur when the amino component enters the coupling reaction with *N*-protected peptides as hydrochloride. No racemization, not even with peptides as carboxyl components, was observed when the free amino component was used and the reaction was carried out in ethyl acetate that contained a trace of water.

Panneman *et al.* (*1690*) have applied this method to the synthesis of a number of partial sequences of ACTH (up to hexapeptides). Yields compare favorably with those obtained by other methods. Ethoxyacetylene has also been used for the synthesis of cyclic peptides (see Chapter V,A).

d. KETENIMINES

The structural relationship of the nitrogen analogues of ketenes, the ketenimines [66], to carbodiimides prompted Stevens and Munk (*2212*) to investigate their applicability as coupling reagents for the synthesis of peptides. The diacyl amide [67] was isolated as the main product

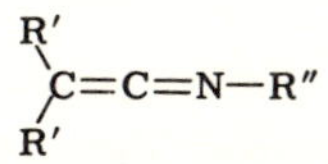

[66]

of a reaction with diphenylketene-*p*-tolylimine ($R' = C_6H_5$, $R'' = CH_3-C_6H_4$). It corresponds to the acylurea of the carbodiimide method. However, its reaction with an amino component showed that it is an activated derivative, which is not the case for the acylureas:

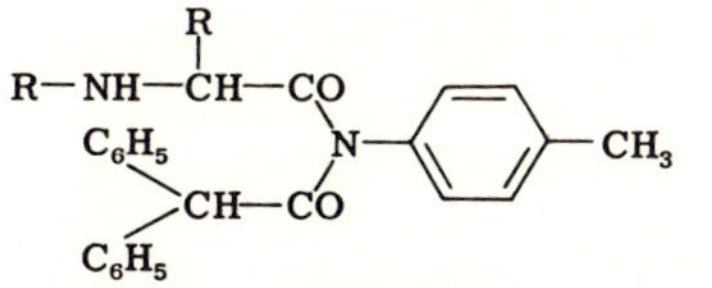

[67]

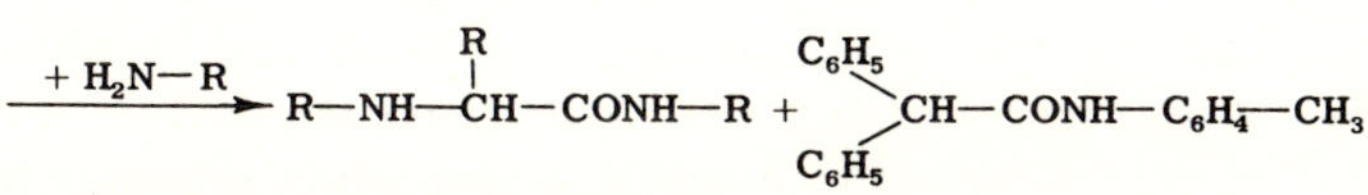

e. ISOXAZOLIUM SALTS

Woodward *et al.* (*2581*) have used isoxazolium salts to activate the carboxyl group. This activation, as in the case of the ethoxyacetylene method, proceeds via the vinyl ester intermediate. Woodward and Olofson (*2580*; cf. *1661*) have suggested the following reaction mechanism. The isoxazolium cation [68] and the carboxyl anion form initially a ketimine [69], followed by the addition of the carboxyl group and formation of an isoimide [70]. A rearrangement involving an acyl

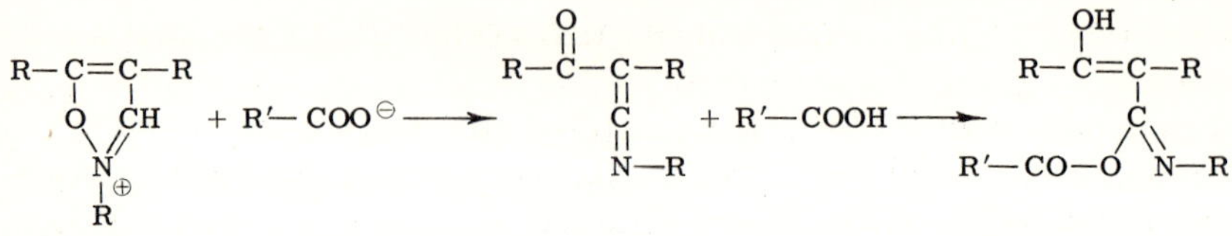

[68] [69] [70]

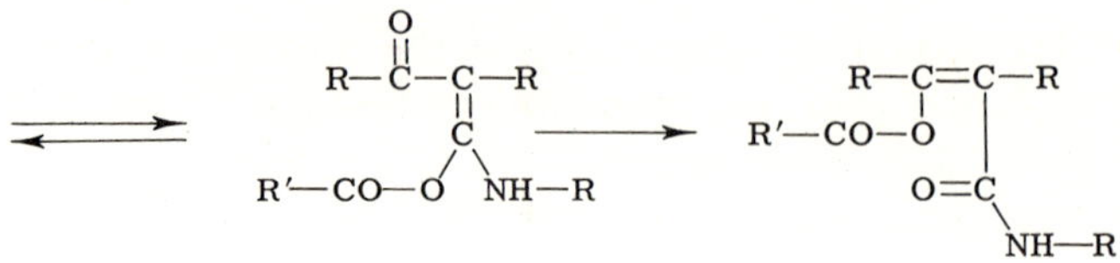

[71]

migration to the actual activated derivative, namely, the vinyl ester [71], was suggested on the basis of infrared and ultraviolet spectroscopic data.

A suitable isoxazolium salt is *N*-ethyl-5-phenylisoxazolium-3′-sulfonate [72] (*1662, 2581*). In the preparative procedure the *N*-protected

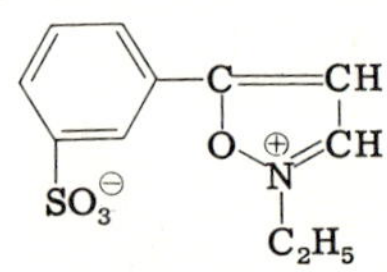

[72]

carboxyl component, triethylamine, and coupling reagent are combined in acetonitrile or nitromethane and kept at 0°C for some time; the amino component is then added. Good yields, even with asparagine or glutamine, and the fact that hydroxyamino acids can be used without protection of the hydroxyl function have been stressed as advantages. Racemization was observed to a minor extent when peptides were used as carboxyl components (*2581*). The method has been used for the synthesis of naturally occurring polypeptides (*1322, 1411*).

4. Heterocyclic Amides as Activated Compounds

The acylation of an amino group by an acyl amide represents a transamidation and should be formulated, therefore, as an equilibrium [73]:

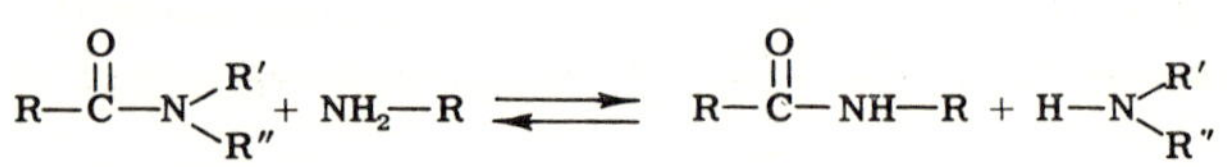 [73]

Several acyl amides with the proper residues R′ and R″, however, are known to undergo the reaction exclusively in the desired direction. All these compounds have in common a nitrogen-containing heterocyclic ring system as amide component. Staab (*2181*) has reported in a review on the preparation and properties of these acyl amides. The compounds that have been studied can be arranged by their rate of hydrolysis in the order illustrated [74] (the figures in parentheses represent $K \times 10^5$

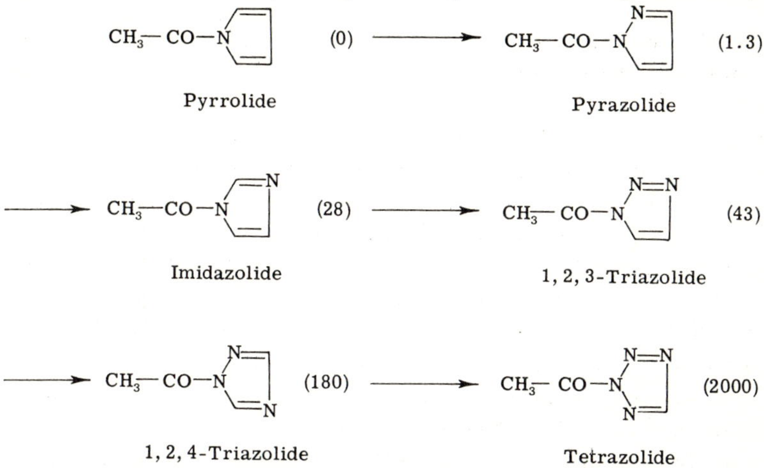

[74]

sec^{-1}). Imidazolides, pyrazolides, and 1,2,4-triazolides belong to this class of compounds and have been used in peptide chemistry.

a. Imidazolide Method

The transfer of acyl residues from derivatives of N^{im}-acylhistidine to amino groups, and thus the possibility of carboxyl activation by imidazolide formation has been described by several authors (*1285, 1695, 1903*). Sheehan *et al.* (*2067*) found that the intramolecular amide of nitrocarbobenzoxy-L-histidine (cf. page 177) reacts with benzylamine to form nitrocarbobenzoxy-L-histidine benzylamide. Staab *et al.* (*2180, 2182, 2183, 2184, 2185, 2187, 2188*) have studied the acyl imidazolides in detail. They can be synthesized from acid chlorides and 2 equivalents of imidazole. A more elegant approach uses the carboxylic acids directly and allows them to react with *N,N*-carbonyldiimidazole [75] (*2182*). *N,N′*-Carbonyldiimidazole can be readily prepared by reacting phosgene with imidazole (*1700, 2188a*).

Paul and Anderson (*1700*) assume that the formation of the imidazolide proceeds via a mixed anhydride [76] that has not yet been isolated:

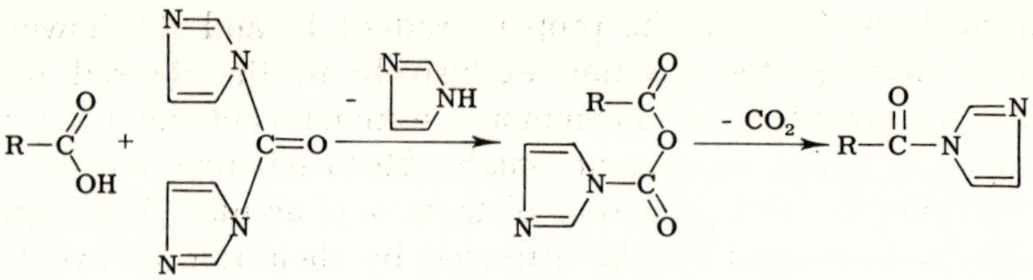

[75] [76]

The most important reactions of the acyl imidazolides are acylations, i.e. ester or amide formation, and the synthesis of acid anhydrides (*2188*).

Anderson and Paul (*51*) were the first to use *N,N'*-carbonyldiimidazole for the formation of peptide bonds. This very fast coupling method, its high yields, and the convenient preparative procedure make it one of the most elegant in peptide chemistry. *N*-Protected amino acids or peptides in different solvents, such as tetrahydrofuran, dimethylformamide, or methylene chloride, are combined with an equivalent amount of *N,N'*-carbonyldiimidazole. The reaction is carried out at room temperature (*1700*) or, better, at lower temperatures (*1700, 1701*). The amino component will be added as soon as the carbon dioxide evolution has ceased or somewhat later. The reaction may be carried out with free amino acid or peptide esters or with their hydrochlorides (*1700*). It is important that moisture be excluded completely during the imidazolide formation. An excess of carbonyldiimidazole should be prevented, since it reacts very readily with the amino component to form substituted ureas [77]:

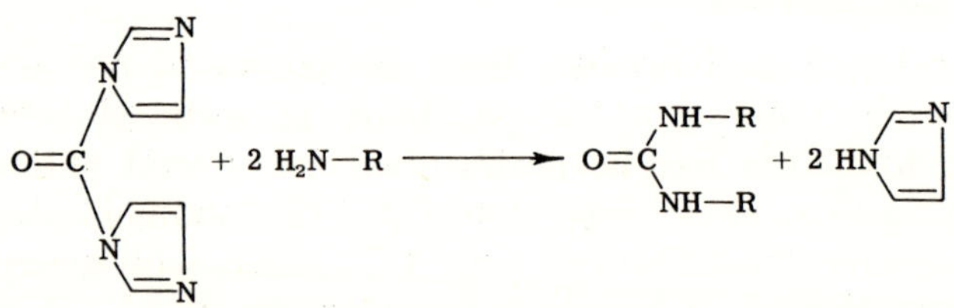

[77]

The extent of racemization was studied in the classic example of the coupling of Cbo-Gly-L-Phe-OH with ethylglycinate. Five percent of the racemic mixture was formed in tetrahydrofuran at room temperature, and less than 0.5% when the reaction was carried out in dimethylformamide at −10°C (*44, 1700*).

The fact that coupling reactions with salts can be carried out in aqueous alkaline solution—although with much lower yields—proves that the *N,N'*-carbonyldiimidazole is the moisture-sensitive compound, not the acyl imidazole (*44*). This directed attention to *N,N'*-thiocarbonylimidazole. It can be prepared in an analogous fashion from imidazole and thiophosgene (*1813, 2186*). *N,N'*-Carbonyldiimidazole will be hydrolyzed immediately in aqueous solution. Staab and Walther (*2186*) report a half-life of more than 7 hours for the thio compound in 2% aqueous tetrahydrofuran. Formation of the acyl imidazolides is brought about by heating the reactants. Peptide coupling reactions have not yet been reported. Wieland and Vogeler (*2556*) have suggested the use of thionylimidazole for the synthesis of peptides. The coupling may be carried out in a one-step reaction by combining the *N*-protected amino acid and the amino acid ester hydrochloride with a solution of the diimidazole. The process is brought to completion by the addition of triethylamine.

Paul and Anderson (*1701*) have synthesized Ileu^5-angiotensin II-Asp^1-β-amide, using the imidazolide method exclusively. They were able to isolate almost all of the intermediates in the crystalline state and found that the yields were higher than with other methods (ethyl dichlorophosphite, carbodiimide, mixed anhydride, azide). The method has also been applied to the synthesis of sequences of other naturally occurring polypeptides (*1045, 1170, 1172*).

Carbonyldiimidazole can also be used for the synthesis of hydrazides (*2180, 2182*). The use of anhydrous hydrazine is not essential. Hydrazine hydrate can also be employed, since the hydrazinolysis of the imidazolide is much faster than its hydrolysis. The formation of ester bonds is especially important for the synthesis of peptolides. Amino-protecting groups have also been introduced with the aid of imidazolides. Staab and Walther (*2187*) have described the trifluoroacetylimidazolide, and Klee and Brenner (*1255*) the *tert*-butyloxycarbonylimidazolide. The aminolysis of esters (*1515a, 2529, 2557*) and peptide coupling reactions with tetraethyl pyrophosphite (*50*) are favored by the presence of imidazole. Imidazolides of phosphoric acid have been used as coupling reagents (cf. *540*).

b. Pyrazolide Method

Ried and Schleimer (*1817*) have synthesized pyrazolides of *N*-protected amino acids by reacting the acid hydrazides with 1,3-dicarbonyl compounds [78]:

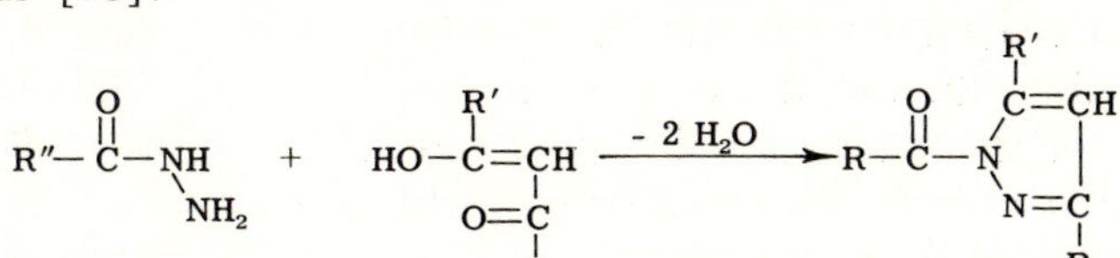

[78]

The reaction with acetylacetone ($R = R' = CH_3$) was carried out in boiling ethanol. In another synthetic approach, the acid is activated via a mixed anhydride and coupled with dimethylpyrazole (*1815*). The derivatives required for these syntheses, e.g. hydrazides, can be readily activated or they are already activated (mixed anhydrides). The method follows synthetic pathways that in the majority of cases are not really needed.

The pyrazolides, in contrast to the imidazolides, are rather stable to water. Their isolation is easy and they can be recrystallized from aqueous ethanol (*1817*). Accordingly, more vigorous reaction conditions are required for the aminolysis, e.g. allowing them to react with free amino acid esters without solvent by heating for 1 hour on a water bath (*1814, 1817*), or heating or boiling the reactants in ethanol or tetrahydrofuran (*1815*). In several cases the substituted pyrazole released during the reaction was removed by steam distillation (*1814, 1817*). The extraction with aqueous acid from the organic solution of the peptide offers milder conditions. DL-Amino acids or derivatives with carboxyl-terminal glycine were used as starting material for the synthesis of peptides by this method. No statement can be made, therefore, with regard to racemization.

Ried and Franz (*1815*) have described a synthesis of glutathione, but otherwise the method has not been applied to the synthesis of biologically active polypeptides.

c. Triazolide Method

Beyermann and Maassen van der Brink (*235*) have recommended the use of *N,N'*-carbonyl-di-1,2,4,-triazole for the synthesis of peptides. The compound is accessible by the reaction of phosgene with triazole. This reaction is slower than the formation of the imidazolide. Its aminolysis, however, is markedly enhanced. The synthesis of Cbo-Gly-L-Phe-Gly-OEt with carbonylditriazole proceeded without racemization. The synthesis of the same peptide carried out under identical conditions with carbonyldiimidazole gave the racemic mixture to the extent of 17% (room temperature, triethylammonium chloride not removed).

5. Activation by the Oxazole Ring System

Different reactants form, via the two functional groups of an amino acid molecule, different derivatives of oxazole ([79], [80], [81]). The compounds of the oxazolidinedione-2,5 type [81] may be considered as intramolecular anhydrides of carbamic acid derivatives of amino acids. The analogues of these *N*-carboxylic anhydrides, the intramolecular anhydrides of thiothiolcarbamic acids [82] and of *N*-phosphorylamino acids

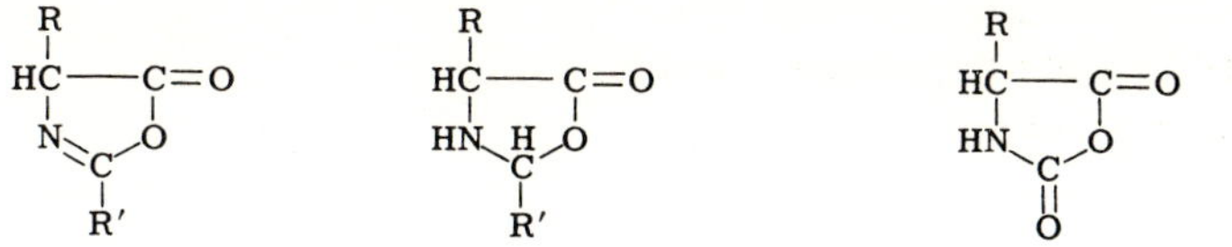

Oxazolone (Azlactone) [79] Oxazolidinone-5 [80] Oxazolidindione-2, 5 [81]

[83] are known. These compounds can be cleaved by aminolysis. All of them may serve as starting material for the synthesis of peptides. They

[82] [83]

have one disadvantage in common. The participation of the amino group in the ring system limits the number of amino-protecting groups that are applicable.

a. Oxazolone (Azlactone)

Oxazolones are accessible from acylamino acids by the elimination of water. Acetic anhydride is used predominantly as water-eliminating agent [84]:

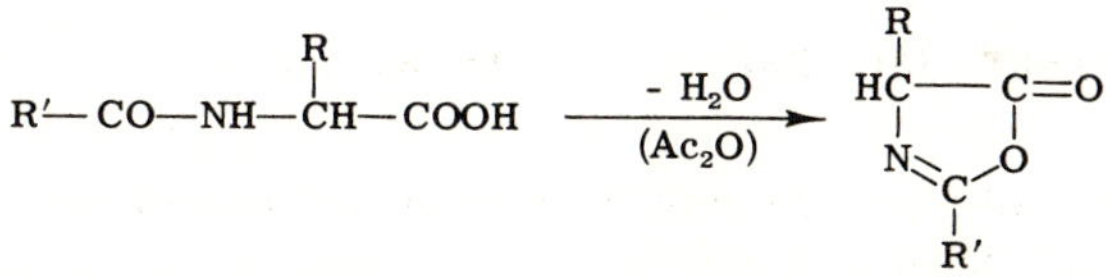

[84]

N,N'-Dicyclohexylcarbodiimide (*1633, 2117*) or ethyl chlorocarbonate (*1634*) can also be used for the elimination of water. The aminolysis of an oxazolone leads to the formation of *N*-acylamino acyl amides or peptides [85]:

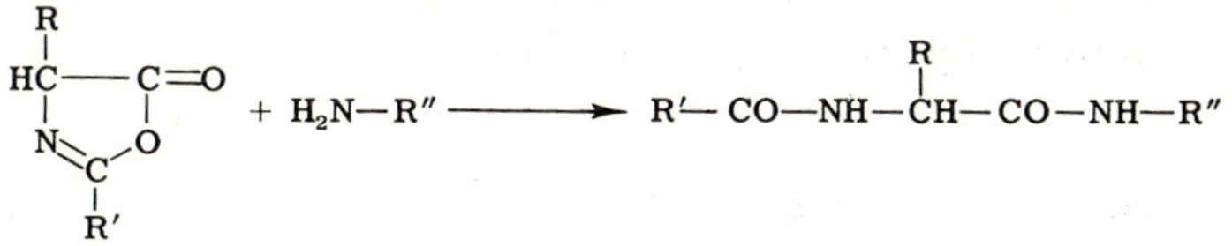

[85]

If the azlactone formation is carried out with acylglycine in the presence of an aldehyde, unsaturated compounds are formed. They can also be cleaved by aminolysis to give rise to peptides with α,β-unsaturated amino acids [86]:

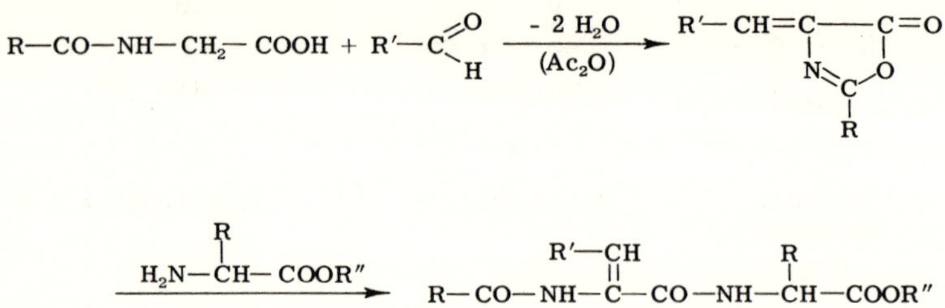

[86]

This coupling method was once of some importance since azlactones are relatively easily accessible. The possible tautomerism [87], however, causes very rapid racemization (*639*):

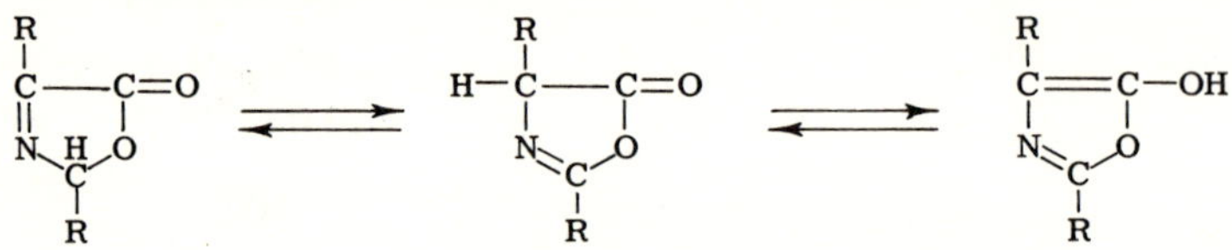

[87]

The oxazolones that are obtained from C^{α}-substituted amino acids, such as α-aminoisobutyric acid or isovaline, are sterically stable. The fact that only acyl-type protecting groups are capable of the azlactone formation is another disadvantage. No oxazolone derivatives are known for amino acids blocked with protecting groups of the urethan type (*175*). Azlactones of *N*-protected dipeptides [88] have been prepared (*1239, 2117*):

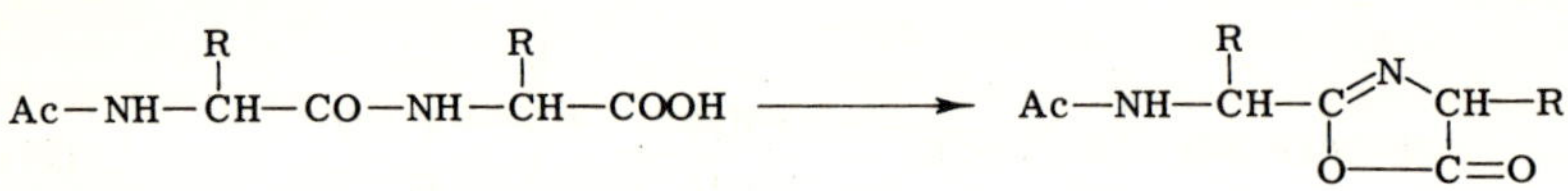

[88]

The importance of the azlactone formation for racemization is discussed elsewhere (see Chapter X,B).

b. Oxazolidinone-5

Ben-Ishai (*148*) has studied the reaction of acylamino or carbobenzoxyamino acids with formaldehyde in the presence of toluenesulfonic acid. It leads to the formation of *N*-acyloxazolidinones [89] when the water is removed by azeotropic distillation:

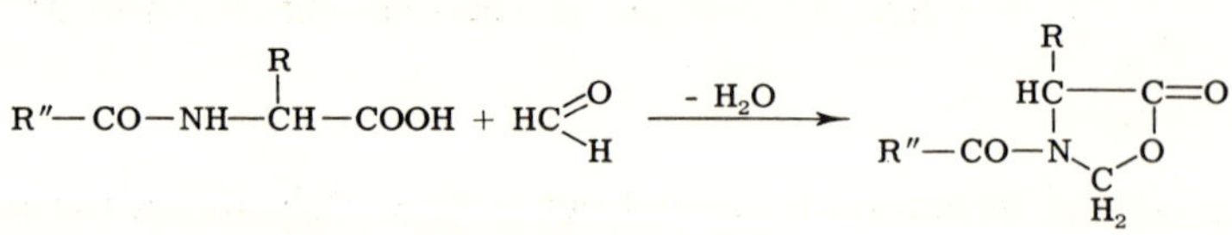

[89]

The formation of a dimeric product was observed in the case of carbobenzoxyglycine [90]:

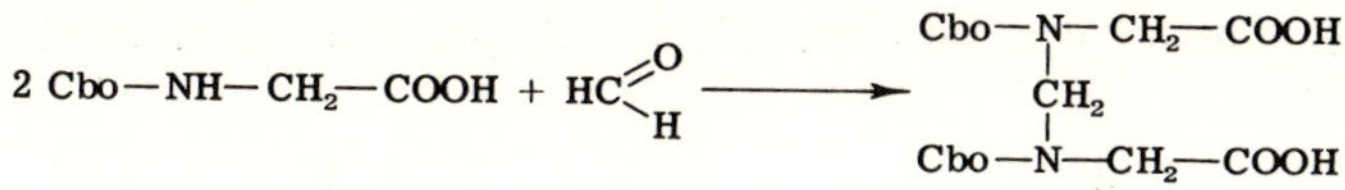

[90]

Micheel and Thomas (*1555*) described at approximately the same time the formation of *N*-tosyloxazolidinones from *N*-tosylamino acids. The reaction was carried out in the presence of acetic anhydride to bind the water released. Sulfuric acid or thionyl chloride was used as catalyst (*1552, 1555*). Acetaldehyde, benzaldehyde, and chloral have also been used for the formation of oxazolidinones. These reagents give rise to 4-substituted compounds (*1552*). Formaldehyde may be used as paraformaldehyde (*148*) or as trioxane (*1548*), acetaldehyde as paraldehyde (*1552*).

The cleavage of oxazolidinones with amines or amino acid esters results in the formation of amide or peptide bonds (*148, 1547, 1548, 1552, 1555*). The coupling with only 1 equivalent of benzylamine did not give the *N*-protected amino acid benzylamide. Ring opening without the elimination of aldehyde led to the formation of an *N*-substituted compound [91] (*148, 1552*):

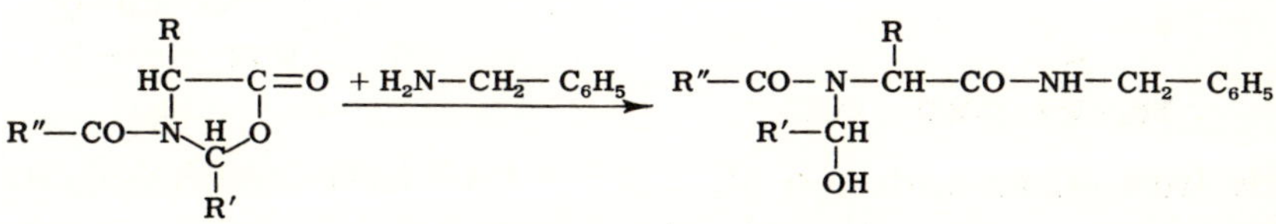

[91]

The substituent at the nitrogen atom can be cleaved under alkaline conditions or with an excess of benzylamine. Therefore an excess of the amino component is required for the synthesis of a peptide (*1552*). The oxazolidinone also is presumably formed via this intermediate (*148, 1552, 1555*). Certain precautions must be observed, especially during the subsequent alkaline hydrolysis, to take care of the aldehyde released during the coupling. Acetaldehyde may cause polycondensations resulting in contaminations which can be removed only with great difficulty. Chloral may decompose to chloroform and formic acid, thus requiring the presence of an extra equivalent of reagent (*1552*). Racemization to a minor extent was observed in the case of cystine and serine, but not during the synthesis of other dipeptides so far reported (*1548*). The oxazolidinone derivatives of *N*-protected dipeptides are not yet known.

The reaction of oxazolidinones with alcohols does not lead to the formation of amino acid esters. Micheel and Haneke (*1548*) have reported ring opening and formation of an *N*-alkoxymethyl compound. It may lose the alkoxymethyl group and form the *N*-protected amino acid

(*1555*), or alcohol will be eliminated and the cyclic derivative obtained again [92]:

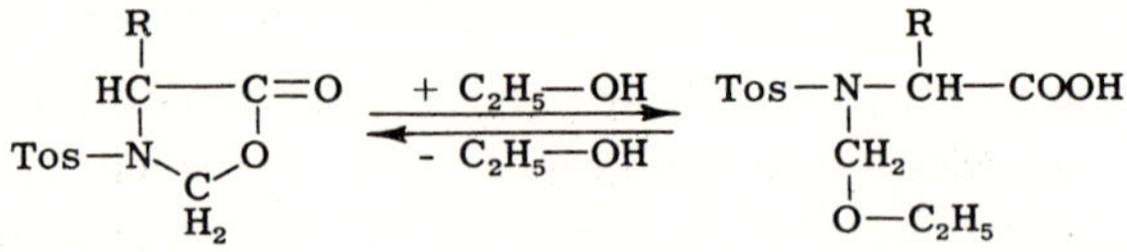

[92]

Participation of the hydroxyl group of serine leads to ring closure in addition to the regular oxazolidinone formation (*1548*).

c. Oxazolidinedione-2,5 (*N*-Carboxylic Anhydrides)

Leuchs *et al.* (cf. *1175*) prepared for the first time *N*-carboxylic anhydrides of amino acids [93]. Their starting material was methoxycarbonylamino acids. The anhydrides are more readily accessible in an analogous reaction from carbobenzoxyamino acid chlorides [94] or

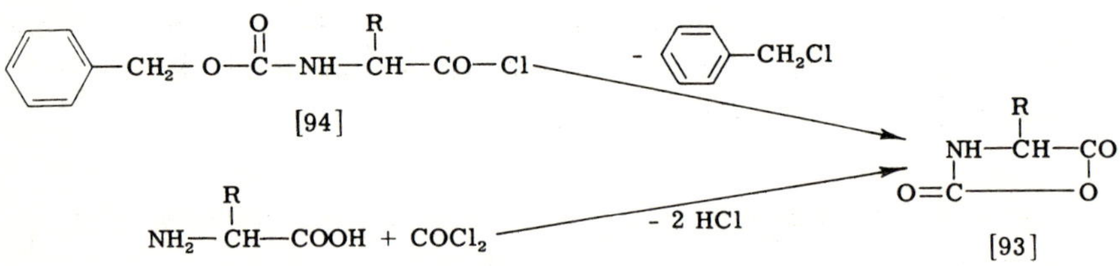

[94] [93]

directly from amino acids and phosgene. *N*-Carboxylic anhydrides can be cleaved by aminolysis, alcoholysis, and hydrolysis [95]:

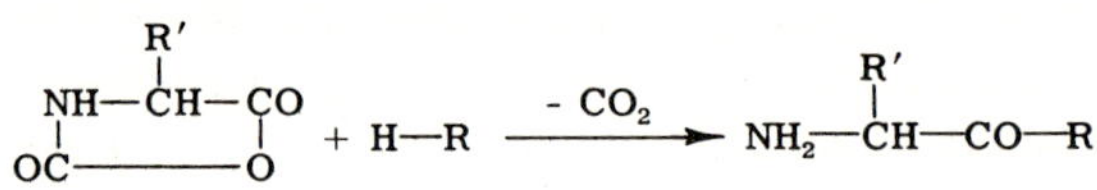

[95]

R = OR″, NH—R″, OH

The amino group released during the reaction is available for a new attack of the *N*-carboxylic anhydride. Thus the method is applied mainly to the synthesis of polyamino acids. The polycondensations of amino acid *N*-carboxylic anhydrides have been the subject of a large number of investigations. This special branch of peptide chemistry was able to make available polyamino acids of a degree of polymerization of 1000 and more, thus making important contributions to the understanding of proteins and protein structure. Most valuable knowledge has emanated from preparative work and from kinetic considerations. Excellent monographs on the topic have been written by Bamford *et al.* (*104*) and by Katchalski (*1175*). The use of the term "polyamino acids" for those peptides resulting from the polycondensation of suitable amino acid derivatives makes confusion less likely than use of the term "polypeptides" that has now become familiar, e.g. in the expression "biologically active polypeptides."

The reaction with water, according to Leuchs, proceeds via the carbamic acid [96]. The acid can be isolated as barium salt:

[96]

Bailey (*95*) found that a corresponding intermediate is also formed upon cleavage with 2 moles of primary amine or amino acid ester [97]:

[97]

It depends upon the basicity of the individual components whether these compounds remain final products of the reaction or whether a polycondensation is initiated. Bamford *et al.* (*104;* cf. *103*) found that the polycondensation is favored by cleavage with a weak base and the formation of a strongly basic amino group. The reaction will not go beyond the cleavage if the reverse is true.

Hunt and du Vigneaud (*1082*) allowed alanine *N*-carboxylic anhydride to react with 2 moles of histidine methyl ester, and were able to isolate the dipeptide via the copper complex. Wessely *et al.* (*2472*) have applied the *N*-carboxylic anhydride method to the synthesis of certain di- and tripeptides of DL-phenylalanine. The yields of purified products, however, were far below 50%. Bailey (*94, 95*) has described conditions allowing a controlled peptide synthesis via the *N*-carboxylic anhydrides. The aminolysis with 2 moles of amino acid ester gives rise to peptide ester *N*-carboxylic acid salts (analogous to [97]). One mole of amino acid ester in the presence of a tertiary base leads to the formation of the carbamate of the tertiary base. It decomposes on heating to 30°–40°C with carbon dioxide evolution and elimination of the tertiary base to yield the peptide ester. A number of peptide esters have been prepared using triethylamine or methyldioctylamine. The cleavage of the anhydride was carried out at —40°C (cf. *1653*). Racemization was not observed. Rudinger and Šorm (*1861*) and Zaoral *et al.* (*2656*) have used *N*-methylpiperidine as tertiary base. The method was modified by foregoing the isolation of the dipeptide ester and thus permitting the synthesis of tripeptides in one continuous operation. When the acid chloride method is used for the second coupling reaction, the base already present will act as acceptor for the released hydrochloric acid. This modification was applied to the synthesis of partial sequences of oxytocin.

The yields, however, were not always satisfactory (*1057, 2652*). Langenbeck and Kresse (*1326*) have reported that a poorly soluble carbamate is formed with tribenzylamine as base. It precipitates immediately and additional condensations with the *N*-carboxylic anhydride are prevented.

Brenner and Hofer (*347*) have studied the acylation of the acid hydrazide group by *N*-carboxylic anhydrides. The reactions were performed in glacial acetic acid to prevent polycondensations. Protonation of the hydrazide group by stronger acids, e.g. trifluoroacetic acid, prevents the acylation. Certain esters, not easily accessible by other methods, have been prepared via the cleavage of *N*-carboxylic anhydrides with alcohols. The synthesis of *O*-glycyl-*N*-carbobenzoxyserine methyl ester was also carried out successfully in this way (*1569*). Birkhofer and Kachel (*245*) and Birkhofer and Modic (*248*) have synthesized *N*-carboxylic anhydrides of β-amino acids.

d. Sulfur and Phosphorus Analogues of Oxazolidinediones

α. *Thiothiazolidone*. Compounds of the 2-thiothiazolidone-5 type [98] represent the dithio analogues of *N*-carboxylic anhydrides. They are prepared from carbon disulfide and α-amino acid amides [99] or α-amino acid nitriles [100] (*520, 521, 976*). The cleavage of the thio-

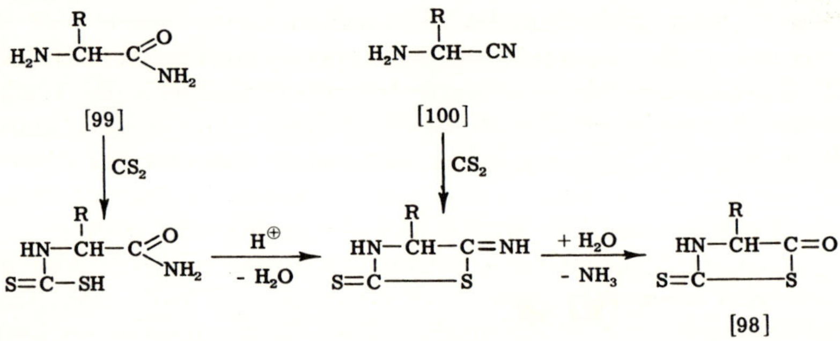

thiazolidones with amino acid esters proceeds via the same intermediate [101] as in the case of the *N*-carboxylic anhydrides:

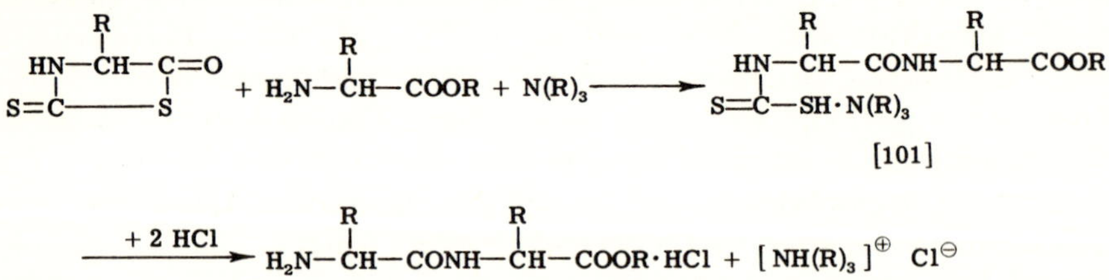

The thiothiazolidones are more stable than the *N*-carboxylic anhydrides and can be coupled with free amino acids in aqueous alkaline solution or in glacial acetic acid (*524, 525*). It is thus possible to obtain free peptides directly. They are, however, contaminated by the unreacted amino acid and by products of higher condensation. Separation by column chromatography is possible. The aminolytic cleavage of the glycine derivative [98] (R = H) gave rather satisfactory yields (*524*). The results with the alanine derivative [98] (R = CH_3), however, were much poorer. Cook and Levy (*525*) assume on the basis of their results from the cleavage with amines (*522, 523*) that the ammonium salt of dithiocarbamic acid is formed in good yield. On acidification, however, the cyclic derivative is formed again. Davis and Levy (*569*) observed a great tendency to racemization that further restricts the application of this method. Hofmann *et al.* (*1024*) arrived at the same conclusion when attempting to prepare amino acid *N′*-carbobenzoxy hydrazides via this method.

Polycondensation of the thiothiazolidones can be brought about only in boiling pyridine (*523*) or methanol (*522*). Even under these conditions, the degree of polymerization remains relatively low with 10 or fewer amino acid residues.

β. Phosphorus Analogues of N-Carboxylic Anhydrides. Keller *et al.* (*1205*) have prepared the intramolecular anhydrides of monophenylphosphoamino acids by reacting the sodium salts of monophenylphosphoamino acids with sulfur trioxide in phosphorus oxychloride. The substances are very sensitive to moisture. They are immediately hydrolyzed in aqueous alkaline solution. It was possible to obtain hydroxamic acids by cleavage with hydroxylamine in aqueous solution. The analogous reactions with amines or amino acids are not yet known.

6. Synthesis of Peptides by Intramolecular Rearrangement

a. Aminoacyl Insertion

Brenner *et al.* (*354–358a, 561*) have discovered and extensively studied an intramolecular rearrangement known as "aminoacyl insertion." During the reaction a derivative of an aminoacyl hydroxy acid is converted to the derivative of a hydroxyacylamino acid [102]. Prerequisite for the reaction is the presence of a hydroxyl group in *β*-position to the carboxyl group. Free carboxyl groups (Y = OH), acid amides (Y = NH_2), substituted acid amides (Y = NH — R), or esters (Y = OR) are all suitable for the insertion reaction:

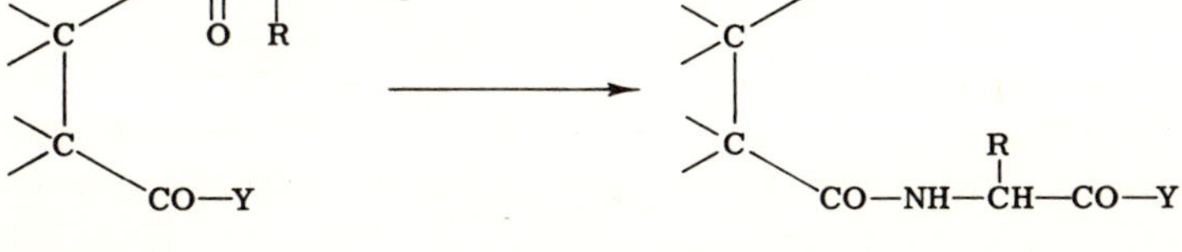

[102]

Since *N*-acylamino acids are also substituted amides (Y = NH — CHR — CO — Y′), it is possible to synthesize peptides by repeating the reaction. Brenner *et al.* (*343, 354, 358*) have discussed the mechanism of the reaction [103].

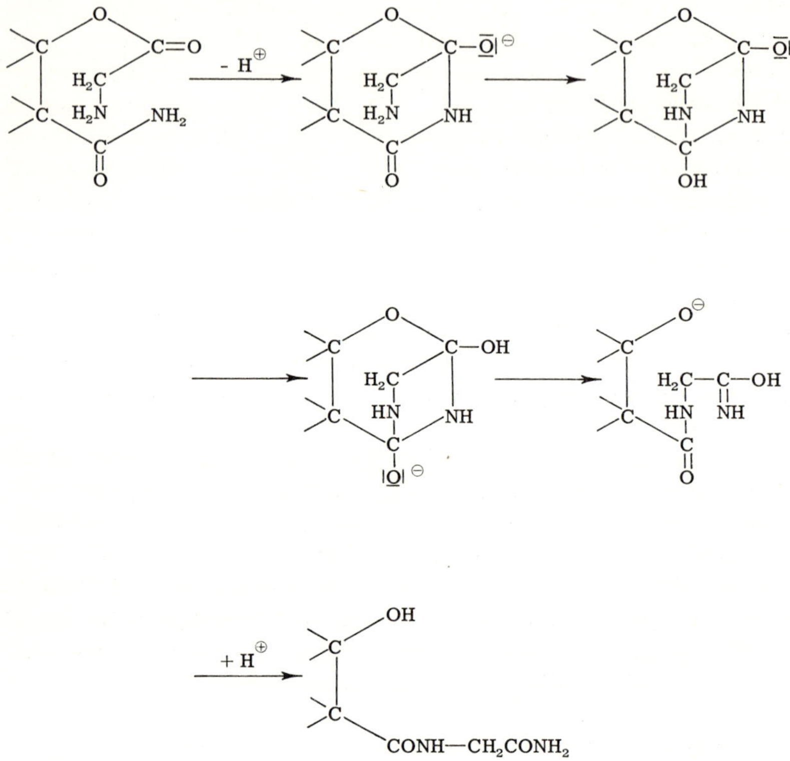

[103]

The rearrangement proceeds most easily with salicylic acid as hydroxy component. It offers, as is evident from the mechanism discussed, a favorable steric arrangement. The insertion is spontaneous after the amino-protecting group has been removed. It is impossible, therefore, to isolate the precursor, namely, *O*-aminoacyl salicylic acid [104], or its amides or esters:

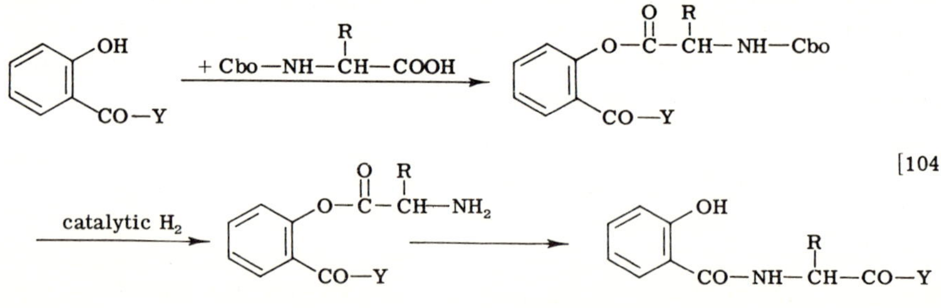

[104]

It is possible to isolate the perchlorate of [104] if the catalytic hydrogenation has been carried out in the presence of perchloric acid. The insertion can then be brought about by the addition of a base. It was demonstrated for *O*-phenylalanylsalicylic acid amide (*358*) that salicyloylphenylalanine amide can be obtained in yields of over 90% either with inorganic bases such as potassium hydrogen carbonate, sodium carbonate, ammonia, and sodium hydroxide in aqueous solutions, or with diethylamine or triethylamine in water, methanol, or chloroform. Hydroxyphenylimidazolones [105] are formed at pH values lower than 8:

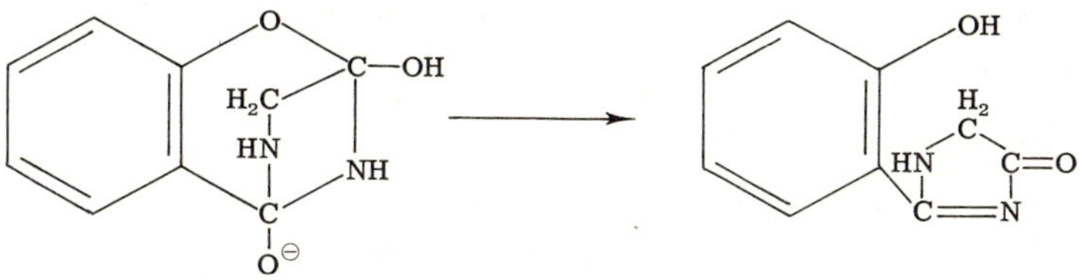

[105]

The same imidazolone formation was observed in organic nonpolar solvents, such as tetrahydrofuran. Aminoacyl salicylic acid esters will be hydrolyzed prior to the rearrangement by an excess of a strong base (*354*). The aminoacyl insertion of salicylic acid derivatives proceeds without racemization.

More vigorous reaction conditions are required for the rearrangement of derivatives of aliphatic β-hydroxy acids. The rearrangement of *O*-glycyl-β-hydroxybutyric acid amide (*358*) or of *O*-glycyl-*N*-benzoylthreonine amide (*132*) could be brought about only by the application of potassium *tert*-butylate. The expected danger of racemization can be reduced by the right choice of reaction conditions.

b. *N,N*-Bisaminoacyl Imides

Diacyl imides [106], like acid anhydrides, can be cleaved aminolytically to form amides:

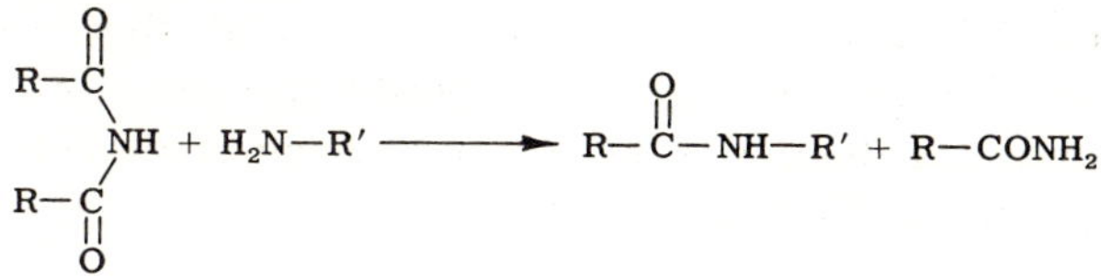

[106]

Mejer *et al.* (*1530*) were able to synthesize the corresponding acetyl or benzoyl amides from diacetimide ($R = CH_3$) or dibenzimide ($R = C_6H_5$) and different amines ($R' = C_6H_5$, $CH_3—C_6H_5$, $C_6H_5—CH_2$, etc). Wieland *et al.* (*2543, 2555*) have described the reactions of bisaminoacyl imides [107]. Intramolecular rearrangement leads to the formation of dipeptide amides. The mixed imides (R′ = R″) produce a mixture of the two possible compounds:

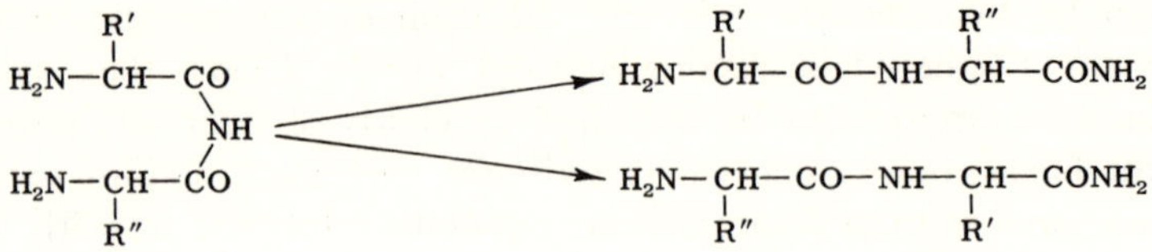

[107]

The glycine residue always displays a greater tendency to acylation when present as one of the two aminoacyl groups. This is certainly so for a sterically hindered second aminoacyl group.

c. *N,N′*-Bisaminoacyl Hydrazines

The *N,N′*-bisaminoacyl hydrazines [108] are structurally related to the bisaminoacyl imides. They have been investigated by Brenner and Hofer (*346, 347*; cf. *2434*). These compounds are stable in aqueous solution and under neutral or basic conditions in organic solvents. They can be rearranged to peptide hydrazides in the presence of organic acids such as glacial acetic acid, propionic acid, or pivalic acid, and in solvents such as, dioxane, tetrahydrofuran, dimethylformamide, etc. The

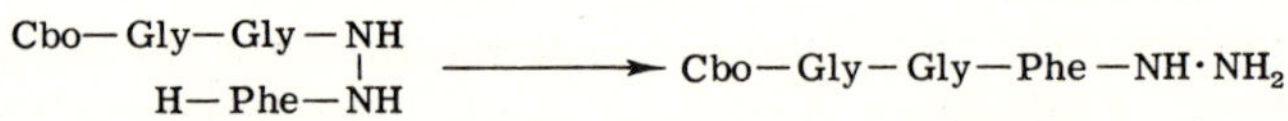

[108]

rearrangement is strongly dependent upon the solvent used and the amount of acid added. The reaction proceeds under ideal conditions with practically quantitative yields. The required starting material can be readily prepared from *N*-protected amino acid or peptide hydrazides and amino acid *N*-carboxylic anhydrides. The reaction is carried out in glacial acetic acid to prevent polycondensation. Efforts to prepare *N*-protected amino acid hydrazides by rearrangement of *N*-carbobenzoxy-(or *tert*-butyloxycarbonyl)-*N′*-aminoacyl hydrazines were not successful.

Kurtz and Niemann (*1318*) have shown that this rearrangement reaction may also proceed in the opposite direction. They obtained *N*-acetyl-*N′*-tyrosylhydrazine from acetyltyrosine hydrazide in 0.05 *N* hydrochloric acid at 95°C.

C. SYNTHESIS OF PEPTIDES BY ACTIVATION OF THE AMINO GROUP

The amino group, in contrast to the carboxyl group, is capable of amide formation by way of its free electron pair without additional activation. Therefore the assumption was made that peptide syntheses by activation of the amino group proceed also via an intermediate activated at the carboxyl group. This intermediate would result from the addition of the activated amino component to the carboxyl group ([118], page 132). The intermediate is a mixed anhydride and presents the same problems with regard to possible application and racemization as the mixed anhydrides. Several cases have been reported where peptide syntheses via activation of the carboxyl group gave poor results, while the amino group could be activated successfully (*296, 493*).

1. Isocyanate Method

Goldschmidt and Wick (*822, 823*) were the first to apply to peptide synthesis the long known reaction of isocyanic acid esters with carboxylic acids which, with elimination of carbon dioxide, leads to the formation of acid amides. The carbonylamino acid esters [109] react with *N*-pro-

[109]

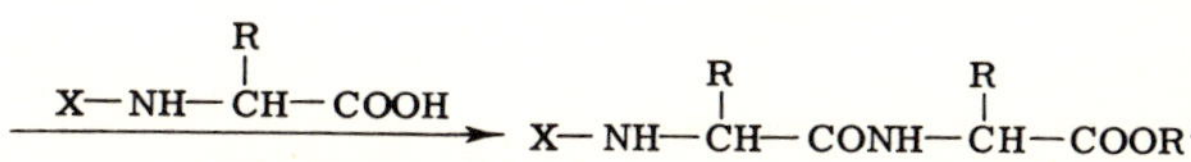

tected amino acids to form peptides in high yield. The carbonylamino acid esters are surprisingly stable compounds that can be distilled without decomposition. They show practically no tendency to polymerization, which is otherwise common to isocyanates. They are synthesized by the reaction of phosgene with amino acid ester hydrochlorides, carried out at 120°C in toluene (*823*). The compounds are very sensitive to moisture. Traces of water lead to the formation of urea derivatives (cf. *496, 1354*), whose removal at the completion of the coupling reaction causes difficulties (*2592*).

To form a peptide, the carbonylamino acid ester and the *N*-protected amino acid are normally heated to 110°C in toluene until the evolution of carbon dioxide ceases (*823*). The reaction starts already at room temperature when carried out in pyridine, and heating to 50°–60°C is sufficient for its completion (*815, 823*). Apparently the reaction proceeds via the mixed anhydride of the carboxylic acid and carbamic acid. These anhydrides of simple isocyanic acid esters and carboxylic acids have been isolated [110]. Racemization was not observed for

$$R{-}COOH + O{=}C{=}N{-}R' \longrightarrow R{-}CO{-}O{-}CO{-}NH{-}R' \longrightarrow R{-}CONH{-}R' + CO_2$$

[110]

reactions with *N*-protected amino acids, and high yields were obtained. Racemization, due to the intermediate formation of a mixed anhydride, may have to be expected for *N*-protected peptide acids; no information is available on this at present. The reaction with *N*-protected hydroxy-amino acids (free hydroxyl group) leads preferentially to the formation of urethans [111] (*2592*):

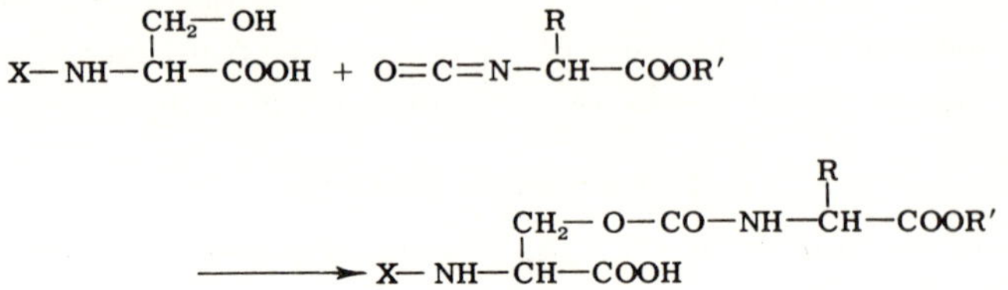

[111]

Acid amides, e.g. *N*-protected glutamine or asparagine, react to form derivatives of urea [112] (*2592*):

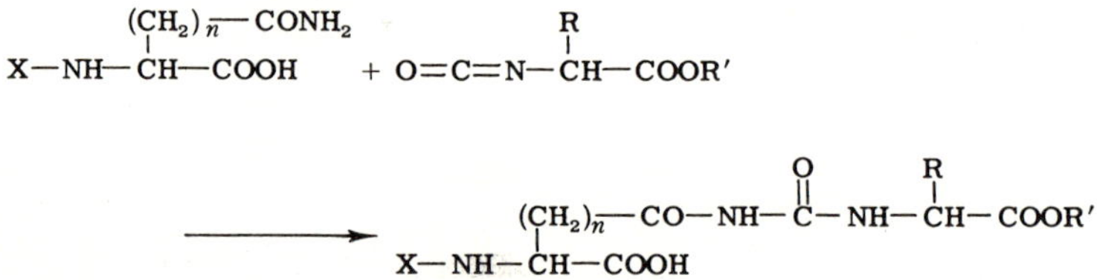

[112]

The reaction of *N*-protected amino acids with glutamic acid or aspartic acid diesters as amino components leads to the formation of carbonyl-bisamino acid esters (*2592*) as the result of a disproportionation of the mixed anhydrides. The isocyanate method cannot be applied to peptide esters. The initially formed carbonyl compounds would undergo cyclization to the hydantoins [113] (*823*):

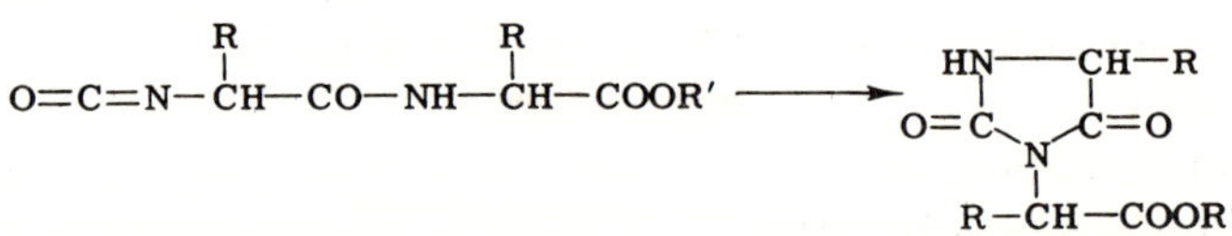

[113]

Losse and Weddige (*1461, 1462*) have used the α-isothiocyanato fatty acid esters instead of the α-isocyanato fatty acid esters to synthesize peptides. They prepared them from amino acid esters and carbon disulfide. The reaction with *N*-protected amino acids is carried out without solvent at 150°–160°C.

2. Phosphazo Method

Phosphazo compounds [114] were first prepared from aniline and phosphorus trichloride. Investigations on their structure have led to the conclusion that they are present in a bimolecular [115] state (*817, 849*). The reaction of the phosphazo compounds of amino acid esters

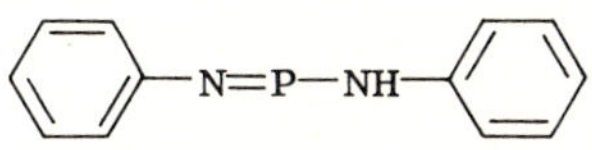

[114]

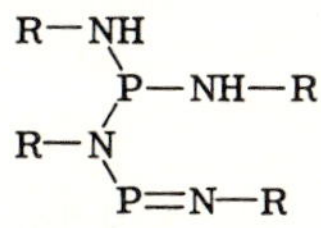

[115]

[116] with 4 moles of an *N*-protected amino acid leads to the formation of the peptide:

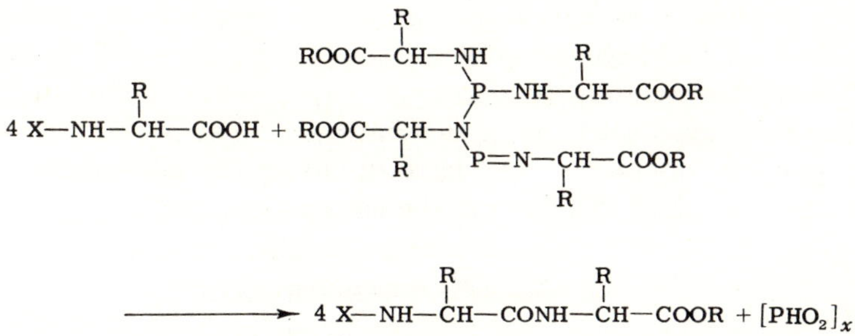

[116]

The reaction is carried out in the following manner. The amino acid ester (or its hydrochloride) is dissolved in pyridine, the phosphazo compound is formed by the addition of phosphorus trichloride, and is then combined with the *N*-protected amino acid. To complete the reaction, the mixture is kept at room temperature (*818, 978*) or, more frequently, heated for several hours to 80°–100°C (*815, 849*). According to Goldschmidt (*814*), the use of pure phosphorus trichloride and the absolute exclusion of moisture are important prerequisites for the success of this coupling method. Wünsch (cf. *814*) has commented in a discussion on the formation of hydrogen phosphides that they may cause difficulties during the subsequent catalytic hydrogenation. Several authors have compared the yields of coupling reactions via phosphazo derivatives with other methods and found them to be lower (*979, 1246*). The steric hindrance of the trityl group remains a disturbing factor in spite of the activation of the amino group (*2203*). Cbo-Val-Val-OEt, however, gave yields of 85–90% although it is also sterically hindered (*2593*). Secondary amino acids react also with phosphorus trichloride. Wünsch (*2592*) assumes, at least for proline, the formation of an intermediate, chlorophosphite diamide [117]. The phosphazo method leads

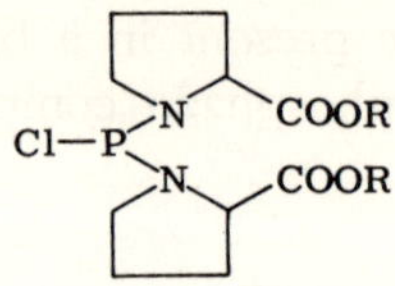

[117]

to the formation of impure products with hydroxy acids if the hydroxyl group is not protected (*2592, 2637*). No side reactions were observed with *O*-protected derivatives (*2596*). The synthesis of tyrosine-containing peptides has been described without mention of complications (*820*). Peptides containing nitroarginine can be prepared by this method but will be obtained in low yield only (*803, 2593*).

The question of racemization has not been clarified unequivocally for this method. Goldschmidt *et al.* (*816, 818, 821*) observed no racemization during the coupling of dipeptide acids with amino acid esters. Grassmann *et al.* (*852*) have demonstrated that racemization must be expected for tripeptides; they have discussed the intermediary formation of a mixed anhydride [118] as the cause. Süs (*2237*) has published a

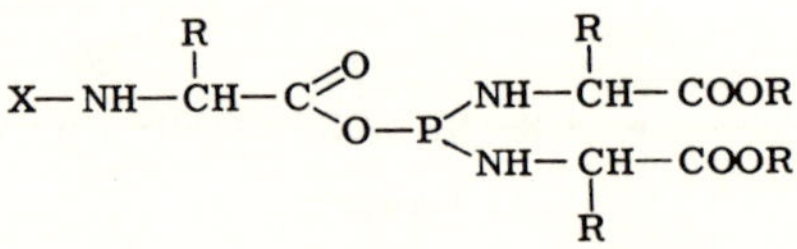

[118]

peptide synthesis in which phosphorus trichloride is added to the mixture of *N*-protected amino acid and amino acid ester in benzene. Apparently this synthesis proceeds also via the phosphazo compound or a mixed anhydride. An acid chloride does not seem to occur as an intermediate.

Goldschmidt and his co-workers and Grassmann and Wünsch and their co-workers have systematically investigated the phosphazo method but otherwise only few applications of this method have become known. They were generally confined to special amino acids, blocking groups, esters, etc. (*435, 978, 979, 2203, 2496*). The phosphazo method has been applied to the synthesis of naturally occurring polypeptides only in the case of glutathione (*815*) and oxytocin (*1854*).

3. Use of Esters of Phosphorous Acid

The methods in which derivatives of phosphorus trichloride are used as coupling reagent may be considered as modifications of the phosphazo method.

a. Ethyl Dichlorophosphite

The peptide synthesis with ethyl dichlorophosphite is very similar to the phosphazo method. Goldschmidt and Obermeier (*819*) assume that alcohol is eliminated from the initially formed ethyl phosphite diamide [119], which is then rearranged to a genuine phospho compound:

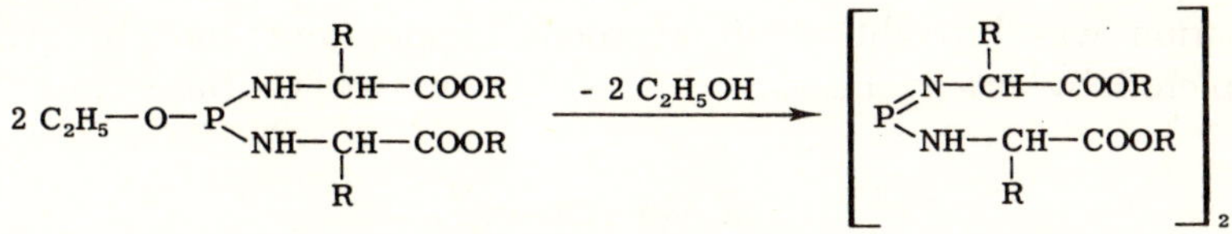

[119]

Young *et al.* (*2611*) have described a reaction for the formation of the diamide [119] whereby the components are allowed to react in a molar ratio of 2:1. They have also described the formation of an imide [120] that requires a molar ratio of 1:1. Normally the yields are relatively low, but compare favorably with those of other methods. Higher yields are obtained when ethyl dichlorophosphite is used as anhydride-forming reagent [121]. Both procedures proceed via well-defined inter-

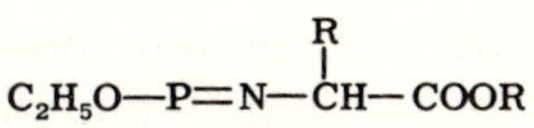

[120]

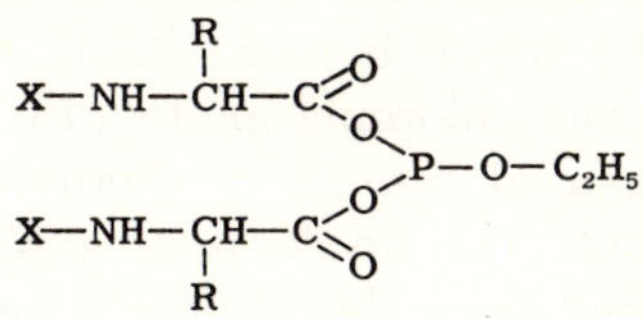

[121]

mediates. An additional third one is known as "standard" procedure: Ethyl dichlorophosphite is added in the presence of triethylamine to the mixture of *N*-protected amino acid and amino acid ester dissolved in diethyl phosphite. This procedure gives the highest yields.

b. Diethyl Chlorophosphite

Anderson *et al.* (*46, 52*) found that diethyl chlorophosphite may be used instead of phosphorus trichloride. Some of the diethyl phosphite amides [122] formed with amino acid esters are distillable compounds.

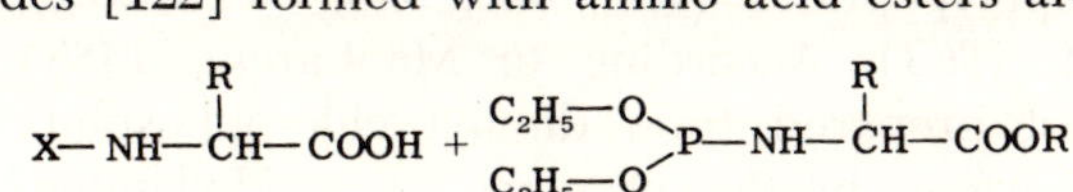

[122]

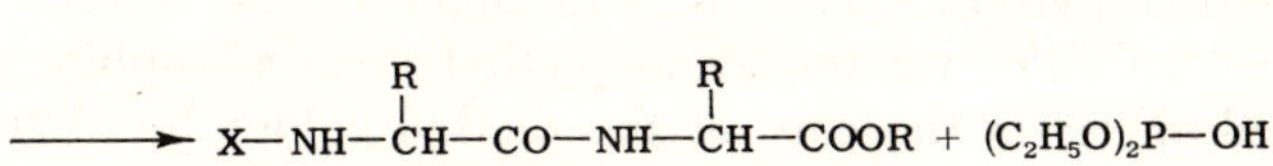

The coupling reaction is carried out by keeping the components at room temperature or by refluxing. The most suitable solvents are toluene, benzene, chloroform, and benzonitrile. Tetrahydrofuran or ethyl acetate is not as good. The phosphite amides of dipeptide esters have also been prepared. Diethyl chlorophosphite may also be used to activate the carboxyl group. A mixed anhydride [123] will be formed (*53*). Benzene and toluene have been recommended as solvents. The anhydride formation was brought about at room temperature and the reaction completed by briefly heating the solution to 100°C. Boissonnas *et al.*

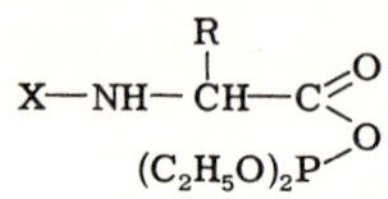

[123]

(*296*) applied this method to the synthesis of dipeptide sequences of oxytocin. They used it either in its "amide" form or as "standard" procedure, i.e. reacting diethyl chlorophosphite in pyridine with the mixture of acylamino acid and amino acid ester.

The use of *o*-phenylene chlorophosphite (*46, 53, 1247, 1354, 1863, 1890*), which is more stable and can be prepared more easily, and of ethylene chlorophosphite (*1863, 2611*), has been suggested. Both have been used instead of diethyl chlorophosphite in diethyl phosphite as solvent. The addition of a tertiary base to triethyl phosphite can be omitted since this solvent is capable of binding hydrogen halides [124]:

$$(C_2H_5O)_3P + HCl \longrightarrow (C_2H_5O)_2P{-}OH + C_2H_5Cl \qquad [124]$$

Roeske *et al.* (*1863*) have pointed out that higher yields will be obtained with the *o*-phenylene derivative.

c. Tetraethyl Pyrophosphite

The tetraethyl pyrophosphite method is the only modification of the phosphazo method that has been used more frequently for the synthesis of naturally occurring polypeptides (*296, 642, 643, 806, 1057, 1190, 1191, 1802, 1807, 1854, 1863*). According to MacLaren (*1488*), the coupling reagent is best prepared from diethyl chlorophosphite and diethyl phosphite in benzene in the presence of triethylamine. The tetraethyl pyrophosphite method proceeds as "amide" or "anhydride" procedure (*45*) via the same diethyl chlorophosphite amide or mixed anhydride as in the case of the diethyl chlorophosphite method [125]:

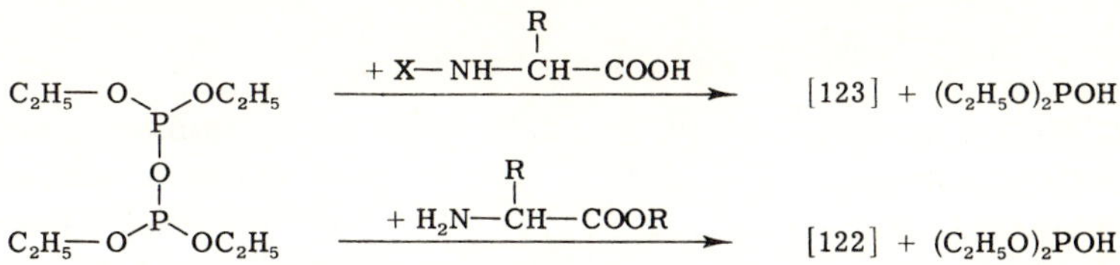

[125]

The "standard" procedure is also applicable (*45, 48*): Tetraethyl pyrophosphite will be added to the mixture of carboxyl and amino components. Racemization was not observed with *N*-protected amino acids as carboxyl components. The "amide" and "standard" procedures show little tendency to cause racemization during the coupling of *N*-protected peptide acids with amino acid esters. There is more danger of racemization when the anhydride method is employed, especially when the hydrochloride and triethylamine are used instead of the free amino acid ester. Diethyl phosphite (*45, 1745*), pyridine (*1489*), or dimethoxyethane (*48*) is used as solvent. The coupling reaction is carried out with a large excess (50–100%) of tetraethyl pyrophosphite (*45, 806, 1190*) and the reaction is completed by heating to 80°–100°C. Protection of the guanido function by salt formation (*42, 806*) is sufficient for the synthesis of arginine-containing peptides. The synthesis of a peptide via the tetraethyl pyrophosphite method proceeds more smoothly in the presence of imidazole (*50*).

Rydon (*1885*) and Crofts *et al.* (*542*) prefer bis-*o*-phenylene pyrophosphite to tetraethyl pyrophosphite. It can be synthesized more easily from pyrocatechol and phosphorus trichloride (*541*). The coupling reactions are carried out in an analogous manner in pyridine, normally following the "standard" procedure. Sterically pure products are obtained only when *N*-protected amino acids are used as carboxyl components. The hydroxyl group of tyrosine must be protected. The use of diethylethylene pyrophosphite has also been reported (*49*).

Kitaoka *et al.* (*1250*) applied tetraethyl pyrophosphite and Furuyama *et al.* (*778*) bis-*o*-phenylene pyrophosphite to the polycondensation of tripeptides. Rothe *et al.* (*1872*), using tetraethyl pyrophosphite or diethyl chlorophosphite, obtained cyclic peptides.

4. Use of Esters of Arsenious Acid

Vaughan (*2361*) has suggested the use of diethyl chloroarsenite for the synthesis of peptides. These syntheses are conducted either as "amide" or as "anhydride" procedures. The greater stability and readier accessibility of the coupling reagent are claimed to be an advantage over the analogous diethyl chlorophosphite method.

5. Phosphorus Pentoxide

The active intermediate of the activation of the amino group by phosphorus pentoxide (*691, 996, 1959*) is supposedly a phosphoric acid monoethyl ester monamide (*1959*). It apparently results from the reaction of phosphorus pentoxide with the diethyl phosphite used as solvent. Metaphosphoric acid ethyl ester is presumably an intermediate. For the synthesis of a peptide, the solution of phosphorus pentoxide in diethyl phosphite can be heated up to 3 hours to 100°C with the mixture of the carboxyl and amino components while a tertiary base is present (*691, 1959*). It is also possible to activate first the amino component only and add the carboxyl component afterwards (*1959*). Racemization must be expected when *N*-protected peptides are used as carboxyl components (*691*).

IV. AMINO ACIDS

A. MONOAMINOMONOCARBOXYLIC ACIDS

1. α-Amino Acids

Peptides containing glycine, alanine, valine, leucine, isoleucine, or phenylalanine as well as peptides composed of only one of these amino acids have been described frequently in the literature. The applicability of new protecting groups and coupling methods, the problems of optical rotation and racemization, and the specificity of various enzymes were studied with these simple peptides. The synthesis of these peptides requires no special preparative methods. Therefore a detailed citation of pertaining references will be omitted and this chapter restricted to a few characteristic approaches.

α-Aminobutyric acid and phenylglycine have been found only in recent years in peptide antibiotics. Very few peptides containing these amino acids are known. [Cf. (*1851a*) for a review on peptides containing "unnatural" amino acids.]

a. Glycine and Alanine

Zahn *et al.* (*1946, 2618, 2626, 2629, 2636, 2637*) and Schnabel (*1938, 1939, 1940*) have synthesized peptides containing glycine and alanine, which they used for X-ray and infrared spectroscopic comparisons in their studies on the chemistry of silk fibroin (cf. *1472, 2176, 2675*). Some of the DL-compounds were subjected to resolution to obtain the optically uniform components. The azide method was used preferentially for these syntheses. The carbobenzoxy group and the benzyl ester [1] served as protecting groups. According to Marsh *et al.* (*1506*), the X-ray data of the crystalline regions of *Bombyx mori* silk fibroin are best explained by the sequence H-Ser-Gly-Ala-Gly-Ala-Gly-OH or X-Gly-X-Gly-X-Gly (X = any amino acid except proline). The X-ray data and infrared spectra of β-poly-L-alanine and tussah fibroin are in good agreement. The isolation of di-, tri-, and tetraalanine from tussah

H-Ser-Gly-Ala-Gly-Ala-Gly-OH	(DL, DL, DL)	(*1938,2636*)
H-Ser-Gly-Ala-Gly-Ala-Gly-OH	(L, L, L)	(*1946*)
H-Gly-Ala-Gly-Tyr-Gly-Ala-OH	(DL, L, DL)	(*1938*)
H-Ser-Ala-Gly-Gly-Ala-Gly-OH	(DL, DL, DL)	(*1939*)
H-Ala-Gly-Gly-Ser-Ala-Gly-OH	(DL, DL, DL)	(*1939*)
H-Ser-Gly-Ala-Gly-Ala-Gly-Tyr-OH	(L, L, L, L)	(*1940*)
H-Gly-Val-Gly-Ala-Gly-Tyr-OH	(L, L, L)	(*2626*)
H-Gly-Arg-Gly-Ala-Gly-OH	(L, L)	(*2618*)

[1]

fibroin (amino acid contents: 45% alanine, 25% glycine) indicates a large accumulation of alanine residues. Zahn and Meissner (*2629*) have synthesized tetra-, penta-, and hexa-L-alanyl-L-alanine as model compounds. Debyeograms of the microcrystalline powders of the peptides showed about the same characteristic reflexes for the pentapeptide as the powder of tussah fibroin. The hexapeptide showed practically quantitative agreement. Savige (*1912*) synthesized Cbo-L-Pro-(L-Ala)$_4$-L-Ala-OBzl as model for X-ray studies of fibrous proteins. Peptides containing alanine have been described in connection with the synthesis of sequences of ribonuclease (*1499a, 1829a*) and of peptides occurring in bacterial cell walls (*2286a*).

Erlanger and Brand (*683*) and Brand *et al.* (*334, 336*) preferred the azide method of synthesizing di-, tri-, tetra-, penta-, and hexapeptides of D- and L-alanine and di- and tripeptides of glycine and D(L)-alanine. These compounds were used to compare specific rotations and to collect data on "residue rotations" (*332*).

b. VALINE

Peptides containing valine are of interest for two reasons: Due to steric hindrance certain reactions, such as the formation of peptide bonds, alkaline hydrolysis, and the formation of hydrazides, are difficult for peptides with carboxyl-terminal valine; the influence of valine-containing peptides on the growth of bacteria has been investigated in systematic studies.

The four possible isomeric dipeptides of valine were prepared by Hinman *et al.* (*1009*) from carbobenzoxyvaline chloride. Later Shankman and Schvo (*2055*) synthesized the eight optically active tripeptides of valine. They used the phthalyl group as protection and *N,N'*-dicyclohexylcarbodiimide as coupling reagent. Phthalylvalyl-*N,N'*-dicyclohexylurea was isolated as a by-product in 20% yield. The acid hydrolysis of Phth-Val-Val-Val-OMe gives only poor yields. Therefore the phthalyl group was removed first by treatment with hydrazine hydrate in methanol for 19 hours at 29°C. The tripeptide methyl ester hydrochloride was then saponified with 0.5 *N* potassium hydroxide (75 minutes, 37°C). Alkaline hydrolyses of peptides with valine or leucine as C-terminal amino acids are very slow (*1009, 2052*). Weygand *et al.* (*2477*, cf. *1777*) selected the synthesis of the sterically hindered TFA-L-Val-L-Val-OMe to study the tendency to racemization of the different coupling methods. The degree of racemization was determined by gas chromatography (see Chapter IX,A,1,c).

The growth-promoting action of L-valine, the isomeric valylvalines, valylvalylvaline, and some of the corresponding phthalyl peptides and peptide ester hydrochlorides was investigated for different strains of bacteria. Di-, tri-, and tetrapeptides of L-leucine were tested in parallel experiments. Several cultures utilized different derivatives for growth. The action of H-L-Val-L-Val-L-Val-OH is approximately equivalent to the content of valine (*2053*). H-L-Val-L-Val-D-Val-OH was found to be a species specific inhibitor (ID_{50} = 20–30 μg/ml) for *Lactobacillus plantarum* and *Pediococcus cerevisiae* (*2050, 2054*). Several peptides are able to reduce this inhibitor action (*2051*). The remaining possible stereoisomeric divalylvalines were not active or only at considerably higher concentrations. H-L-Val-L-Val-D-Leu-OH was the only one of the eight possible tripeptides of valine and/or leucine with L-L-D-configuration that showed a significant inhibitor action (*2052*). Shankman *et al.* (*2050, 2054a*) synthesized other di- and tripeptides of different amino acids and tested them against bacteria to correlate bacteriostatic activity and structure, composition, and configuration of the peptides. [Cf. the papers by Kihara *et al.* (*1229, 1230*) and Florsheim *et al.* (*731*) and the chapter in Volume II on strepogenin-active peptides for the growth-stimulating action of peptides on bacteria.]

c. Leucine and Isoleucine

Leucine-containing peptides have been described by various authors (*131, 1957, 2126*) and were used for studies of the specificity of leucine aminopeptidase (*198, 199, 200, 1748, 2144, 2151, 2153*). The occurrence of D-leucine in etamycin, esperin, and sporidesmolide I is of interest.

Isoleucine occurs less frequently in biologically active polypeptides. Those that do contain it are $Ileu^5$-angiotensin I and II, eledoisin, oxytocin, and bacitracin. The relatively frequent use of tosyl-L-isoleucine for the synthesis of intermediates of oxytocin (*225, 272, 1184, 1854*) must be mentioned. Carbobenzoxy-L-isoleucine (*1133*) and L-isoleucine methyl ester hydrochloride (*296*) are the preferred starting materials.

D-Alloisoleucine was found in actinomycins. Peptides containing D-alloisoleucine have been prepared for the total synthesis of actinomycin (*381, 382, 399, 1319*). D-Alloisoleucine has also been found in mycoside C of *Mycobacterium avium* and mycoside J_2 of *Mycobacterium paratuberculosis* (*1087*). Winitz *et al.* (*2567*) and Izumiya *et al.* (*1124*) have synthesized peptides with carbobenzoxy-L-alloisoleucine and D-allo-isoleucine.

d. PHENYLALANINE

The synthesis of peptides containing phenylalanine presents no special problems (e.g. *774, 775, 1114, 2368*). Besides glycine and alanine, phenylalanine is the standard amino acid used to test new protecting groups and coupling methods and the possibly occurring racemization. Cbo-Gly-L-Phe-Gly-OEt is frequently used for these tests (see Chapter X,A,1,b). Phenylalanine-containing peptides have been synthesized repeatedly to check the specific cleavage of the phenyl-alanyl—aminoacyl bond by chymotrypsin. Phenylalanine occurs very frequently in biologically active polypeptides. Strikingly enough, the D-isomer is present in many peptide antibiotics. Peptides with phenyl-alanine show ultraviolet absorption at $\epsilon = 187$, which facilitates the determination of molecular weights (*2389, 2393*).

e. α-AMINOBUTYRIC ACID

α-Aminobutyric acid has been found in animal and plant tissue and in normal human urine. The presence of this amino acid in proteins has not yet been demonstrated. D-α-Aminobutyric acid, however, is contained in the peptide antibiotics staphylomycin factor S and ostreogrycin. Izumiya *et al.* (*1124–1127*) employed the anhydride method and prepared, from carbobenzoxy-L-α-aminobutyric acid, H-L-Abu-L-Tyr-OEt·HCl and —NH_2·HCl as well as H-L-Abu-L-Lys-OMe·2HCl and —NH_2·2HCl (cf. *1114*). Losse (*1452*) has described the resolution of the racemic mixture of H-Gly-DL-Abu-OBzl via the diastereoisomeric dibenzyl hydrogen D-tartrates. α-Aminobutyric acid-containing analogues of biologically active peptides were synthesized by several authors (*630a, 1029a, 1904c*).

f. *C*-PHENYLGLYCINE

C-Phenylglycine is present in several heteromeric heterodetic cyclic peptide antibiotics, e.g. ostreogrycin and staphylomycin factor S. Cbo-DL-C-Ph·Gly-Gly-OEt and Cbo-Gly-DL-C-Ph·Gly-OMe(-OEt) were synthesized via the isocyanate method (*823*) or by employing phenyl dichlorophosphite (*819*). DL-*C*-Phenylglycyl peptides were also prepared from DL-phenylglycine *N*-carboxylic anhydride according to the method of Bailey as modified by Langenbeck and Kresse (*1326*).

2. C^{α}-Substituted α-Amino Acids

C$^{\alpha}$-Substituted amino acids [2] have not been found as constituents of biologically active polypeptides. α-Hydroxy-α-amino acids [2] (X= OH) are present in peptide-like natural products, i.e. the ergot alkaloids, cephalosporin, lycomarasmin, etc. (cf. *2080*). α-Aminoisobutyric acid [2] (X = CH_3, α-methylalanine) was found in protein hydrolyzates and in an antibiotic (cf. *1367*).

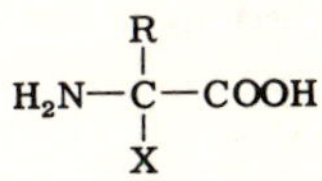

[2]
(X = OH or CH_3)

a. Amino Acids with Functional C^{α}-Substituents

Shemyakin and his school (*59, 2079–2081, 2089*) have pursued most extensively the problems of introducing an additional functional group in the α-position. Oxazolones are accessible from acylamino acids and can be readily substituted at the *C*$^{\alpha}$ atom by halogen [3]:

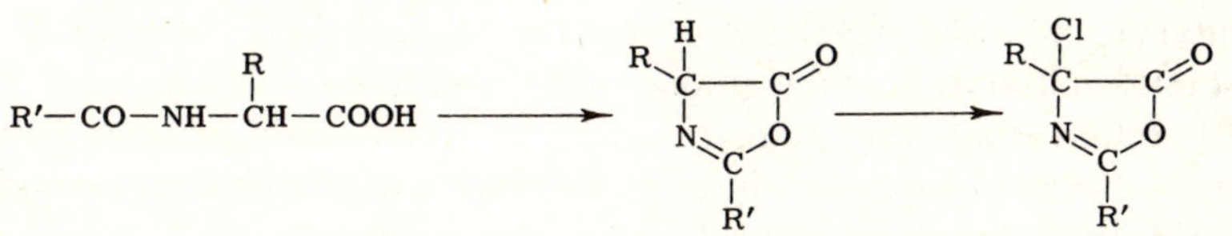

[3]

The reaction with water, alcohols, mercaptans, or amines leads to the exchange of halogen and to cleavage of the oxazolone ring [4]:

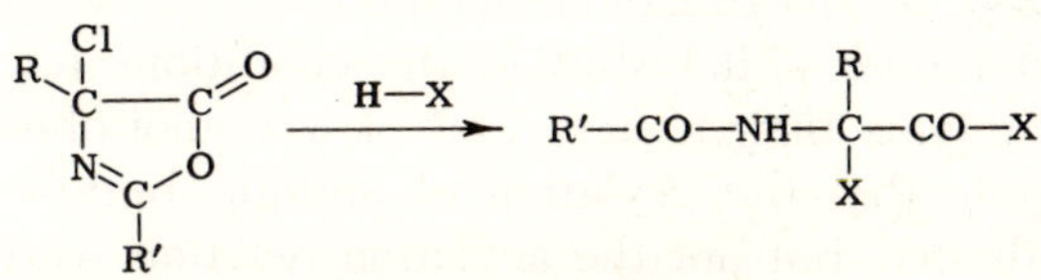

[4]

(X = OH, O—R, S—R, NH—R)

Certain conditions permit the opening of the ring and the introduction of the α-substituent in subsequent steps. *C*$^{\alpha}$-Hydroxy-*N*-acylamino acid esters are obtained by the successive reactions with alcohol and water, *C*$^{\alpha}$-alkoxy-*N*-acylamino acids by the reaction with water followed by treatment with alcohol. Cleavage with amino acids leads directly to the formation of peptides. Lead tetraacetate not only introduces the acetoxy residue in oxazolones, it is also capable of direct substitution at the methylene group of benzoylglycine phenyl ester without azlactone formation. Efforts to carry out the same reaction with phthalyl- or carbobenzoxyglycine were not successful (*2079*).

Peptides have been synthesized via activated esters, which had been obtained by opening the oxazolone ring with the corresponding alcohols (*2080*). The azide (*59*) and carbodiimide (*59*) methods were also employed. Alkaline hydrolysis of the esters and catalytic hydrogenation in the case of the benzyl esters yield the *N*-protected free acids (*59, 2080*). All of these reactions produce only racemic mixtures. The diastereoisomers were separated in a few cases by fractional crystallization (*59*).

b. AMINO ACIDS WITH C^{α}-ALKYL SUBSTITUENTS

α-Aminoisobutyric acid is of some importance as C^{α}-alkylated amino acid. Bergmann *et al.* (*200;* cf. *196*) synthesized the first α-aminoisobutyric acid peptides, using the ethyl ester as starting material. They employed the carbobenzoxy compound for the synthesis of α-aminoisobutyryl peptides. The acid chloride provided the activation in either case.

Leplawy *et al.* (*1367*) have recently showed renewed interest in this strongly hindered amino acid. In contrast to the results of Bergmann *et al.* (*200*), they did not succeed in obtaining carbobenzoxy-α-aminoisobutyryl chloride. Cyclization to the *N*-carboxylic anhydride was constantly observed, even under mild conditions. Faust and Lange (*710*) found that the acid chloride is stable only at temperatures below —30°C. The acid chloride method was successfully employed with the tosyl group as protection for the amino group. The mixed anhydride with pivalic acid proved to be very useful for coupling reactions. Tosyltri-α-aminoisobutyryl-α-aminoisobutyric acid methyl ester (*1367*) was synthesized via the oxazolone method.

Faust and Lange (*710*) studied the conditions for the synthesis of aminoacyl-α-aminoisobutyric acid and of α-aminoisobutyrylamino acids. It is remarkable that the acylation of aminoisobutyric acid is effected by steric hindrance, but not the acylation reaction with aminoisobutyric acid. Satisfactory results were obtained with esters when the phosphorus oxychloride and carbodiimide methods were employed. It was not possible to carry out coupling reactions via the azide method. Attempts to obtain the hydrazine from carbobenzoxy-α-aminoisobutyric acid methyl, ethyl, or benzyl ester were unsuccessful.

Other C^{α}-alkylamino acids have been described by Stapleton and Swan (*2191*), α-methyl-DL-cysteine hydrochloride and the racemic mixture of α,α′-dimethylcystine; by Arnstein (*72*), α-methyl-DL-cystine; by Otani and Winitz (*1668*) and Mejer (*1529*), α-alkylated serines; and by Almond *et al.* (*31*), α-methylphenylalanine and α-methyltyrosine.

3. *N*-Methylamino Acids

N-Methylamino acids have so far not been found as constituents of peptide hormones and proteins. They are, however, of great importance in several peptide antibiotics. No systematic studies have been conducted on the synthesis of peptides containing *N*-methylamino acids. Nevertheless a number of peptides with *N*-methylamino acids, preferably *N*-methylglycine (sarcosine), have been synthesized. This was the result of investigations on the specificity and the reaction mechanism of proteases (e.g. *207, 1114, 2146, 2606*) and on problems concerning secondary structure and its influence on a given biological activity. A number of analogues of biologically active polypeptides with *N*-methylamino acids have become known for the same reasons (*469, 1079, 1160, 1525*).

N-Methylamino acids are normally prepared by the method of Fischer *et al.* (*718, 720*). Tosylamino acids are methylated by methyl iodide or dimethyl sulfate, followed by the removal of the tosyl group (*627, 1109, 1111, 1160, 1367*). This cleavage requires very vigorous reaction conditions. Therefore Quitt *et al.* (*1778a, 1779*) have suggested the methylation of *N*-benzylamino acids with formaldehyde/formic acid and reduction with sodium fluoborate. Catalytic hydrogenation produces the *N*-methylamino acids in good yield. The substitution of the amino hydrogen is sometimes omitted, and the Schiff's base obtained by treatment of amino acid esters with formaldehyde is reduced. The *N,N*-dimethylamino derivative is also formed during the reduction. It must therefore be possible to separate the two methylated amino acids, once the hydrolysis of the esters has been carried out (*1079*). Methods based on the reaction of methylamine with the corresponding α-halogen carboxylic acids are also known (*23, 1110, 1111, 1112, 1118*). They are limited in application since it is difficult to obtain optically active α-halogen carboxylic acids, and Walden inversion occurs during the amination. [Cf. (*1730*) for differences in the chromatographic behavior of amino acids and their methylated derivatives.]

The carbobenzoxy group is used almost exclusively for the protection of *N*-methylamino acids during peptide synthesis (*675, 1079*). Contrary to the corresponding unmethylated derivatives, they show poor crystallization properties (*675*). *p*-Nitrocarbobenzoxy-*N*-methylamino acids have been recently described (*2087*). The application of the standard methods for the synthesis of peptides (*1079, 1114*) and the cleavage of the protecting group by catalytic hydrogenation (*1079, 1114, 1525*) or with hydrogen bromide in glacial acetic acid (*469, 675, 1079*) proceed without complication. Of particular interest is the protection of the secondary amino group by nitrosation, as suggested by Quitt *et al.* (*1779b*). The *N*-protected derivatives are synthesized by treatment of the *N*-methylamino acids with isoamyl nitrite. The protecting group is removed by treatment with hydrogen chloride in benzene at 10°C. An unfavorable influence is exerted by the strong tendency of *N*-nitroso-*N*-methylamino acids to racemize upon application to peptide synthesis.

Methyl and ethyl esters (*1160*), *tert*-butyl esters (*1079*), and benzyl esters (*776*) are known as carboxyl-protected derivatives. Rather vigorous reaction conditions are required for the esterification, especially for *N*-methylamino acids with branched side chains. Amides have also been used. They were synthesized via the mixed anhydride method from carbobenzoxy derivatives and ammonia, followed by the removal of the protecting group (*469, 1525*). The following methods were used for coupling reactions: *p*-nitrophenyl ester (*1131*), mixed anhydrides (*469, 1114, 1525*), carbodiimide (*1079*), azide (*776*), and acid chloride (*183, 1160, 1367, 1840*). The latter method proved to be very useful for coupling reactions with sterically hindered *N*-methylamino acids (*208, 1367*). The acylation of a secondary amino group normally compares unfavorably with that of a primary group (*1131*). Peptides with *N*-methylamino acids show little tendency to crystallize.

4. ω-Amino Acids

a. β-Alanine

β-Alanine is a constituent of carnosine (β-alanyl-L-histidine), anserine (β-alanyl-1-methyl-L-histidine), ophidine (β-alanyl-2-methyl-L-histidine), pantothenic acid (α,γ-dihydroxy-β,β-dimethylbutyryl-β-alanine), and coenzyme A. These compounds and a number of other β-alanine peptides were synthesized via carbobenzoxy-β-alanine azide (*135, 935a, 2119*), phthalyl-β-alanine chloride (*1665, 2339*), and the *p*-nitrophenyl ester (*97*). Baddiley and Mathias (*82*) prepared a number of derivatives of pantothenic acid in order to test them as possible precursors of coenzyme A in microorganisms. They synthesized peptides of pantothenic acid with glycine, DL-alanine, β-alanine, DL-serine, and L-glutamic acid, and corresponding β-alanyl derivatives. Hanson and Smith (*935*) and Adams *et al.* (*20*) used simple β-alanine and β-alanyl peptides for investigations of the specificity of peptidases. The peptides were prepared by the acid chloride or azide method. Ried and Marquard (*1816*) have reported that β-alanyl peptides can be prepared conveniently by the pyrazolide method. Carbobenzoxypenta-β-alanyl-β-alanine methyl ester was prepared by this procedure. β-Alanyl and β-alanine peptides can also be prepared in good yield by the isocyanate method (*1931*). Birkhofer and Hartwig (*244*) have described peptides of β-alanine with other amino acids. Carbobenzoxy-β-alanyl-β-alanine on treatment with base undergoes the same rearrangement as peptides of α-amino acids to form carboxylbis-β-alanine, Aonuma *et al.* (*60*) have synthesized various peptides with β-alanine, using the azide method. Studies by the Japanese authors have demonstrated a uterus-contracting action for Cbo-β-Ala-β-Ala-NH_2, Cbo-β-Ala-β-Ala-β-Ala-NH_2, and Cbo-β-Ala-β-Ala-L-His-NH_2.

b. γ-Aminobutyric Acid

γ-Aminobutyric acid has been found in its free form in the brain and in plant tissue. It does not occur as a constituent of proteins. Evans and Irreverre (*699*) have described the synthesis of H-γ-Abu-γ-Abu-OH via the anhydride method, using carbobenzoxy-γ-aminobutyric acid and γ-aminobutyric acid benzyl ester as starting components. Noguchi *et al.* (*1627*) have used phthalyl-γ-aminobutyric acid for coupling reactions. Phth-γ-Abu-γ-Abu-OEt can be hydrolyzed for 30 minutes at 100°C with 2 *N* sulfuric acid in glacial acetic acid. The phthalyl group can be removed from the substituted dipeptide by treatment with hydrazine hydrate in ethanol. Phthalyl-γ-aminobutyric acid *p*-nitrophenyl ester was used for the synthesis of Phth-γ-Abu-L-His-OH (Phthalylhomocarnosine) (*97*). Hunt and du Vigneaud (*1082*) used carbobenzoxy-γ-aminobutyryl chloride to synthesize free H-γ-Abu-L-His-OH. Poduška and Rudinger (*1745*) converted tosyl-γ-aminobutyryl acid to *N*-tosylpyrrolidone. The reaction was carried out under the conditions of a peptide synthesis with thionyl chloride or *sec*-butyl chlorocarbonate in the presence of a tertiary base. The lactam can be cleaved with ethylglycinate in boiling acetone to form Tos-γ-Abu-Gly-OEt. The tosyl group was removed with sodium in liquid ammonia or with hydrogen bromide in glacial acetic acid (2 hours, 70°C, phenol present) after the ester had been hydrolyzed. Rothe (*1869*) has synthesized *cyclo*-di-(tri-, tetra-, and hexa)-γ-aminobutyryl.

5. Heterocyclic Amino Acids

The group of heterocyclic amino acids includes proline and tryptophan. Both amino acids are present in various biologically active polypeptides. They are entirely different in behavior. Other heterocyclic amino acids, such as histidine and hydroxyproline which is closely related to proline in origin and reactivity, will be described, according to their functional groups, in Section B and E, respectively, of this Chapter.

a. Proline

Numerous biologically active polypeptides contain one or more residues of proline. Bradykinin has a strikingly high proline content. Collagen must be mentioned as a protein composed mainly of proline, hydroxyproline, and glycine residues. Extensive experience in the synthesis of biologically active polypeptides and model substances of collagen has demonstrated there to be no special difficulties with the synthesis of proline or prolyl peptides. In contrast to *N*-methylamino acids, the acylation of the imino groups proceeds smoothly during the synthesis of proline peptides. Proline, due to its imino group, is exceptional in its effect on the secondary structure of polypeptides. It also displays special properties during the enzymatic hydrolysis of biologically active polypeptides and proteins for the purpose of structural elucidation, or when synthetic polypeptides are tested for purity. The aminoacyl—proline bond is stable to enzymes such as trypsin, chymotrypsin, carboxypeptidase, and leucine aminopeptidase.

The review of Fruton (*772*) contains the first reports on attempts to synthesize proline-containing peptides. Following the introduction of the carbobenzoxy group, prolyl and proline peptides were first prepared by salt coupling reactions via the acid chloride method (*5, 7, 208, 1606, 2125, 2148*). Carbobenzoxy-L-proline (*5, 7, 179, 850, 934, 1091, 1863, 1886*) is, also after the development of modern coupling techniques, the most important starting material for prolyl peptides. The carbodiimide and anhydride methods are frequently used for coupling reactions of long chain C-terminal proline peptides. The activation by ethoxyacetylene (*1690*) or *o*-phenylene chlorophosphite (*1863*) is also applicable to the synthesis of prolyl peptides. The carbobenzoxy-L-proline *p*-nitrophenyl ester (*273, 902, 1616*) has been used more frequently in recent years than the *p*-nitrophenyl thiolester (*996, 2533*). In contrast to other optically active amino acids, both esters of *N*-protected peptides can be prepared (*899, 2027, 2029, 2030*). It appears that the anhydride method cannot always be applied with success (*916*).

Beside carbobenzoxy-L-proline, other derivatives of proline protected at the imino group, e.g. trifluoroacetyl- (*2490*), *p*-nitrocarbobenzoxy- (*463*), *p*-methoxycarbobenzoxy- (*2487*), *p*-phenylazocarbobenzoxy- (*2037*), *p*-(*p'*-methoxyphenylazo)-carbobenzoxy- (*2037*), cyclohexyloxycarbonyl- (*575a*), and *tert*-butyloxycarbonyl-L-proline (*49*), have been applied only infrequently. Tosyl-L-proline (*1478, 1770*) was used in the form of its acid chloride (*129, 130*) for coupling reactions. Wünsch (*2595*) and Wünsch and Heidrich (*2597a*) have used with advantage *p*-phenylazocarbobenzoxy-L-proline for the synthesis of substrates for collagenase. The *p*-phenylazocarbobenzoxy group is unstable at pH 13 during alkaline hydrolysis.

L-Proline methyl ester hydrochloride (*692, 890, 996*), the benzyl ester hydrochloride (*1606, 1690, 1756*), and the *tert*-butyl ester (*48, 1173, 2024*) have been used for the synthesis of proline peptides. A very useful derivative, 1-L-prolyl-2-(*tert*-butyloxycarbonyl)-hydrazine, was used by Guttmann (*890*) for the synthesis of Val^5-angiotensin I. The carbodiimide, anhydride, and *p*-nitrophenyl ester methods can be applied equally well to the synthesis of proline and prolyl peptides. The azide (*916, 1616*), ethoxyacetylene (*1690*), and phosphazo (*2592, 2595a*) methods are employed less frequently. Sakakibara and Nagai (*1900*) have pointed out that the azide method is superior to the anhydride method (isovaleryl chloride) with regard to purity and yield (cf. *916*). There is no danger of racemization in the case of proline. Therefore the "breaking point" during the synthesis of biologically active polypeptides is chosen, whenever possible, in such a way that it follows a proline residue.

While the steps leading to the synthesis of proline-containing peptides proceed without complication, the alkaline hydrolysis and catalytic hydrogenation are frequently accompanied by side reactions. Numerous investigations, dealing with the synthesis of model peptides of collagen (*56, 574, 916, 996, 1250, 1755, 1756, 1760, 1761, 1900, 2101, 2111*), have produced a wealth of information. The hydrolysis of Cbo-Gly-L-Pro-OMe with 1 *N* sodium hydroxide can be conducted with practically no cleavage of the peptide bond (*1761, 2101*), but Cbo-L-Pro-Gly-OMe is hydrolyzed to the extent of 70%, yielding Cbo-L-proline and glycine (*2101*). It is also difficult to hydrolyze higher peptides containing the sequence Pro-Gly in satisfactory yield (*573, 1761*). Boissonnas *et al.* (*292*) have reported another example with a sequence of bradykinin, namely, Cbo-Arg(NO_2)-Pro-Pro-Gly-OMe (cf. *2025*). It was possible to hydrolyze Cbo-L-Pro-Gly-OMe without side reaction under very mild conditions (0.1 *N* sodium hydroxide in acetone, 40 minutes, 22°–25°C) (*2101*). Difficulties during the alkaline hydrolysis of proline- and glycine-containing peptides caused Debabov and Shibnev (*575*) to use the *p*-nitrobenzyl ester, which can be cleaved by hydrogenolysis. The *p*-nitrobenzyl ester of proline itself has not yet been described. Shibnev *et al.* (*2101*) have noted cleavage of the peptide bond during normal esterification of H-L-Pro-Gly-OH and H-Gly-L-Pro-OH with methanol/hydrogen chloride. Attempts to prepare the esters by hydrogenolysis of the carbobenzoxy compounds in the presence of hydrogen chloride were also accompanied by partial hydrolysis of the peptide bond. Better results were obtained with glacial acetic acid. The authors were able to prepare analytically and chromatographically pure dipeptide ester hydrochlorides in practically quantitative yield only when they carried out the hydrogenation under neutral conditions with a palladium catalyst activated previously by hydrochloric acid. The hydrogenation is fast enough so that diketopiperazine formation does not occur. Poroshin *et al.* (*1759*) demonstrated by their investigations on the hydrogenolytic stability of glycine-, proline-, and hydroxyproline-containing di- and tripeptides under acidic and alkaline conditions that H-Pro-Gly-OH is less stable than H-Gly-Pro-OH. The glycylamino acid bond in tripeptides is the first to be cleaved during acid hydrolysis. The acid hydrolysis shows a higher degree of specificity than the alkaline hydrolysis. Wünsch (*2595*) has reported on the lability of the Pro-X bond in H-Gly-Pro-X-OH (X = Gly or Ala). Cleavage of the Pro-X bond and diketopiperazine formation occur during crystallization from water/ethanol or water/methanol/ether.

Abderhalden and Nienburg (*7*) observed the great tendency to cyclization of proline-containing dipeptide derivatives when Cbo-L-Pro-L-Pro-OH was hydrogenated in the presence of glacial acetic acid. Smith and Bergmann (*2148*) obtained the diketopiperazine from Cbo-Gly-L-Pro-NH_2 by the same reaction eliminating ammonia. The ready diketopiperazine formation of proline-containing peptides has been reported again and again (*1606, 1900*). Rydon and Smith (*1886*) have studied in detail the autocondensation of di-, tri-, and tetrapeptide ethyl esters of glycine and proline. The ring closure reaction of H-Gly-Pro-OEt is considerably faster than that of H-L-Pro-Gly-OEt. Furthermore, diketopiperazine is not the only reaction product for the latter compound. It is not the basicity of the two compounds (p*K* 7.6–8.4 for the amino compound, p*K* 9.7 for the imino compound) that governs the cyclization, but rather geometrical and stereochemical factors. It can be stated in general that proline-containing peptides in combination with glycine and hydroxyproline residues are rather labile and difficult to prepare.

Investigations on the amino acid sequence of collagen and its degradation by collagenase, and studies on synthetic substrates and the specificity of this enzyme, have been described in publications by Grassmann *et al.* (*844, 845, 846, 1632*), Heyns and Leger (*995, 997*), Nagai *et al.* (*1591, 1592, 1593, 1594*), Poroshin *et al.* (*1755, 1757, 1761*), Wünsch as well as Wünsch and Heidrich (*2595b, 2597a*), and in a review on collagen and gelatine by Harrington and Hippel (*941*).

b. Tryptophan

Tryptophan, in addition to its presence in plant and animal proteins, occurs in several peptide hormones such as ACTH, α- and β-MSH, and glucagon; in peptide antibiotics, e.g. tyrocidin B; and in an interesting structural arrangement in phalloidin. Through the presence of the indole nucleus, tryptophan becomes a highly reactive amino acid and the synthesis of tryptophan-containing peptides is frequently accompanied by undesirable side reactions.

Very few tryptophan peptides (cf. *772*) were prepared by the Fischer method before 1948. Several tryptophyl peptides were then synthesized via carbobenzoxy-L-tryptophan chloride in coupling reactions with salts (*572, 2145*); they were needed for studies of behavior toward carboxypeptidase. Today carbobenzoxytryptophan (*2145*), L-tryptophan methyl ester hydrochloride (*2, 291*), the ethyl ester hydrochloride (*174, 1089*), and the benzyl ester hydrochloride (*2561a*) are used for the synthesis of peptides. Almost all tryptophyl peptides were synthesized via the anhydride method (*337, 572, 2301, 2365, 2531, 2606*) with ethyl or isobutyl chlorocarbonate and phosphorus oxychloride as anhydride components, or via the carbodiimide method (*37, 1035, 1521, 1784, 2703*). Schwyzer and Li (*2021*) and Li *et al.* (*1415*) have used with great success the carbobenzoxy-L-tryptophan cyanomethyl ester (*2009*) and Wilchek and Patchornik (*2561a*) the nitrophenyl ester. The azide method was applied to the synthesis of Try^3-oxytocin and Try^3-Lys^8-vasopressin (*896*). No side reactions were described for the azide formation in acidic medium. Brandenburg (*337*), however, isolated Ac-Gly-N-nitroso-DL-Try-Gly-OEt [5] from the reaction of Ac-Gly-DL-Try-NH_2NH_2 with sodium nitrite (30% excess) and ethylglycinate:

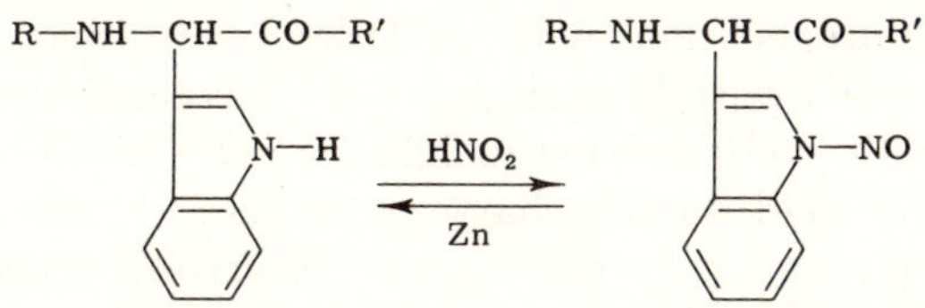

[5]

This side reaction could be prevented when only 1 equivalent of nitrite was used. The *N*-nitroso compound was reduced with zinc dust to Ac-Gly-DL-Try-Gly-OEt. This compound can be converted to the *N*-nitroso derivative by treatment with 1 equivalent of sodium nitrite in acid. There is less danger of nitrosation of the NH group of the indole nucleus in the case of carbobenzoxy derivatives. The carbobenzoxy peptide azides are more strongly hydrophobic and go faster as acetyl derivatives into the organic phase. It is not possible to exclude this side reaction completely.

Other tryptophan derivatives suitable for peptide coupling reactions are: *p*-phenylazocarbobenzoxy- (*2037*), *p*-(*p′*-methoxyphenylazo)-carbobenzoxy- (*2037*), trifluoroacetyl- (*2480, 2498*), and *tert*-butyloxycarbonyl-L-tryptophan (*49*). The latter compound was obtained in only 36%

yield from *tert*-butyloxycarbonyl *p*-nitrophenyl ester. Stelakatos *et al.* (*2203*) prepared trityl-L-tryptophan by direct tritylation of the diethylammonium salt of tryptophan in water/isopropanol. Schwyzer and Kappeler (*2703*) synthesized the compound by tritylation of L-tryptophan methyl ester followed by alkaline hydrolysis. Trityltryptophan was used in coupling reactions with carbodiimide.

Tryptophan peptides have been prepared via the acid chloride (*2145*), anhydride (*37, 337, 1028, 2301, 2531, 2598b*), carbodiimide (*337, 896, 1089, 1784, 2301*), azide (*337*), and *p*-nitrophenyl ester (*337*) methods. Ingle (*1089*) found that the anhydride method (isobutyl chlorocarbonate) led to racemization and the formation in 60% yield of Cbo-Gly-DL-Try-OH when carbobenzoxyglycine was allowed to react with L-tryptophan in 1 equivalent of 1 *N* sodium hydroxide. The decarbobenzoxylation of tryptophan-containing peptides with hydrogen bromide in glacial acetic acid is accompanied by a decomposition of the indole nucleus (appearance of a violet color, changes in the ultraviolet spectrum and the chromatogram) (*291*). The yields are frequently low (*1415*). Theodoropoulos and Fruton (*2301*) have discussed the formation of a β-oxindolylalanine derivative. The catalytic hydrogenation is normally carried out in the presence of palladium black in methanol/glacial acetic acid (*291*), alcohol/hydrochloric acid (*1120, 1521*), glacial acetic acid (*2365*), or dilute acetic acid (*1035*). Keil *et al.* (*1204*) reported reduction to octahydrotryptophan on hydrogenation of tryptophan or tryptophan-containing peptides in the presence of platinum oxide. Difficulties are sometimes encountered during isolation procedures following the hydrogenation with palladium. A violet discoloration appears if air is not carefully excluded (*337*). Esterifications of tryptophyl peptides with methanol/hydrochloric acid (*1521*) or benzyl alcohol/hydrochloric acid (*1035*) have been carried out without complication. Diketopiperazine formation was observed on occasion during the hydrogenation of tryptophan-containing carbobenzoxy dipeptide esters under neutral conditions or in acetic acid (*337, 771*). Davis (*572*) described another undesirable side reaction, the formation of 5-indolylhydantoin-3-acetamide upon amidation of Cbo-L-Try-Gly-OEt with ammonia-saturated methanol.

Patchornik *et al.* (*1696*) and Ramachandran and Witkop (*1789*) have reported the selective cleavage of *C*-tryptophyl bonds in peptides and proteins with *N*-bromosuccinimide; [cf. Cohen and Witkop (*506*)].

B. BASIC AMINO ACIDS

All known biologically active polypeptides contain multifunctional amino acids. Such additional functional groups pose special problems for the synthesis of peptides and make necessary a detailed discussion of the amino acids in question. The group of basic amino acids comprises lysine, ornithine, α,γ-diaminobutyric acid, α,β-diaminopropionic acid, arginine, norarginine, and histidine. With the exception of α,β-diaminopropionic acid and norarginine, these amino acids are constituents of biologically active polypeptides. Citrulline and norcitrulline, though not of basic nature, will also be described in this chapter because of their close relationship to ornithine, arginine, and α,γ-diaminobutyric acid. The main difficulty with the basic amino acids is the proper choice of α- and ω-protecting groups. A compilation (Tables II-VI; cf. *1848*) of amino acid derivatives of preferentially the L-series, suitable for peptide synthesis, provides information on the various possibilities.

1. Lysine

Lysine participates mainly through its α-amino and carboxyl groups in the structure of polypeptides. Lysine-vasopressin, α-and β-MSH, ACTH, glucagon, and eledoisin are examples of this type of lysine incorporation. Sometimes, as in kallidin, lysine occupies the amino-terminal position. The ϵ-amino group of lysine is not often engaged in the formation of peptide bonds. N^{ϵ}-Peptide bonds of lysine were found to occur in bacitracin and in various natural products that, are not peptides in the strict sense, such as biocytin (*1708, 2576*) and streptolin (*2360*). The isolation of N^{ϵ}-(glycyl-L-α-glutamyl)-L-lysine from partial hydrolyzates of collagen (*1516*) furnished proof that the ϵ-function of lysine participates also in the structure of proteins. Thus it appears to be possible that lysine introduces branching in polypeptide chains.

a. Lysyl Peptides

Dicarbobenzoxy-L-lysine hydrazide (*201, 203, 333, 1771*) and dicarbobenzoxy-L-lysine (*204, 291, 1976*) were used for the unequivocal synthesis of lysyl peptides. The latter derivative was coupled by the acid chloride (*204*), phosphazo (*820, 849*), and, preferably, the anhydride method (*1114, 1028, 1381, 1932, 2419, 2420*), or via the dicarbobenzoxy-L-lysine *p*-nitrophenyl ester (*1739*). Di-*tert*-butyloxycarbonyl-L-lysine *p*-nitrophenyl ester has also been described (*1904a*). Amiard and Goffinet (*35*) applied the carbodiimide method to the synthesis of peptides with N^{α},N^{ϵ}-ditrityl-L-lysine. Both trityl groups of an N^{α},N^{ϵ}-ditrityllysyl peptide can be removed by boiling for 5 minutes with 50% acetic acid. The N^{α}-position can be detritylated by heating for 15 minutes with 5 *N* hydrochloric acid in acetone (*35*).

b. α-Lysine Peptides

α. Use of N^{ϵ}-Carbobenzoxylysine. Bergmann *et al.* (*204*) obtained N^{ϵ}-carbobenzoxy-L-lysine by hydrolyzing N^{ϵ}-carbobenzoxy-L-lysine *N*-carboxylic anhydride. The *N*-carboxylic anhydride was prepared from dicarbobenzoxy-L-lysine by treatment with phosphorus pentachloride. Neuberger and Sanger (*1604,* cf. *1317*) were able to synthesize N^{ϵ}-carbobenzoxy-L-lysine by carbobenzoxylation of the copper complex [6] of L-lysine. This method is now used almost exclusively (*1252a,*

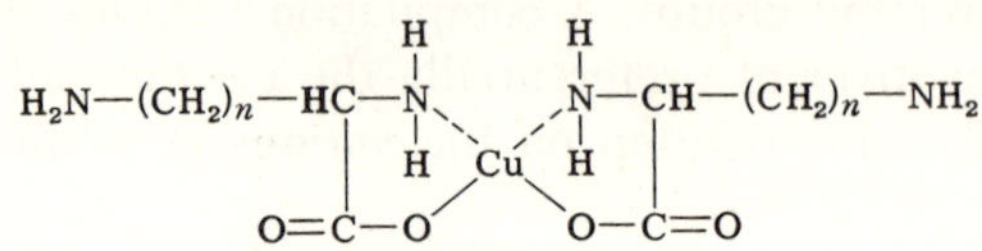

[6]

1932). The copper of the carbobenzoxy-L-lysine copper complex can be precipitated with hydrogen sulfide (*1252a, 1604*) or, more conveniently, with thioacetamide (*748*). It may also be removed as potassium tetracyanatocuprate(I) (*2643*) by treatment with potassium cyanide or by exposure to 2 *N* hydrochloric acid at 60°C (*2033*). Zahn and Falkenburg (*2620*) synthesized N^{ϵ}-carbobenzoxy-L-lysine directly from free lysine by treatment with phenylbenzyl carbonate in aqueous alcohol. The hydrolysis of N^{ϵ}-carbobenzoxy-L-lysine *N*-carboxylic anhydride with 1 *N* methanolic hydrochloric acid (*204*), 1 *N* ethanolic hydrochloric acid, or benzyl alcohol/hydrochloric acid-saturated ether (*683, 684*) yields N^{ϵ}-carbobenzoxy-L-lysine methyl, ethyl, or benzyl ester. Shiba and Kaneko (*2099*) have developed a simple preparative procedure for synthesizing the methyl ester by direct esterification of N^{ϵ}-carbobenzoxy-L-lysine with methanol/thionyl chloride (1.5 hours, 40°C; 20 hours, 25°C; yield, 84%). The benzyl ester was obtained by treatment of N^{ϵ}-carbobenzoxy-L-lysine with benzyl alcohol/thionyl chloride (4 hours, 40°C; 23 hours, 29°C; yield, 24%). Falkenburg (*701*) synthesized the ethyl ester with 1 *N* ethanolic hydrochloric acid (24 hours; room temperature). Peptides with N^{ϵ}-carbobenzoxy-L-lysine esters were synthesized via the azide (*333, 335, 684, 2419*), anhydride (*291, 737, 1122, 1127, 1381, 2342, 2419, 2620*), and phosphazo (*820, 2592*) methods. The peptides L-lysyllysine through tetra-L-lysyl-L-lysine (*333, 684, 820, 2419*) were prepared by these methods. Detailed studies were carried out on the behavior of lysine-containing peptides and peptide derivatives toward proteolytic enzymes, such as papain (*204*) and trypsin (*1122, 1127, 1381, 2420*). Transpeptidations have been observed in the presence of trypsin (*1381, 2420*).

Normally the hydrolysis and the reaction with hydrazine hydrate of carbobenzoxy-substituted lysine-containing peptides proceed without complication, but Waley and Watson (*2419*) observed the formation of urea derivatives during the hydrolysis with 1 *N* sodium hydroxide of Cbo-L-Lys(Cbo)-L-Lys(Cbo)-L-Lys(Cbo)-OMe and of the corresponding tetralysine derivative.

The α-amino-protecting group of N^ϵ-carbobenzoxy-L-lysine-containing peptides must be removable selectively, if in the course of a peptide synthesis it becomes desirable to continue condensation reactions at the α-amino group. Boissonnas *et al.* have used the combination N^α-trityl/N^ϵ-carbobenzoxy for the synthesis of biologically active polypeptides, e.g. for sequences of α-MSH (*291*) and ACTH (*295, 299*) and for eledoisin (*1904*). There are no reports available on the application of other protecting groups that could be cleaved selectively in the presence of the N^ϵ-carbobenzoxy group, e.g., the *tert*-butyloxycarbonyl, the phthalyl, the trifluoroacetyl, and the formyl groups (cf. however, the N^α-formyl-N^ϵ-carbobenzoxy combination of diaminobutyric acid, Volume II, Chapter III,B,1,a,β). Schwyzer and Sieber (*2033*) have applied the N^ϵ-*p*-(*p*′-methoxyphenylazo)-carbobenzoxy protection in combination with the *tert*-butyloxycarbonyl group to the synthesis of Lys^2-Lys^7-gramicidin S.

β. *Use of N^ϵ-Tosyllysine.* The *p*-toluenesulfonyl group is used most frequently as protection for the ϵ-amino group. Roeske *et al.* (*1863*) prepared for the first time N^ϵ-tosyl-L-lysine via the copper complex. It was used in combination with the carbobenzoxy group for the synthesis of Lys^8-vasopressin (*1523, 1863*), kallidin (*1739*), a linear gramicidin S analogue (*686*), an ACTH nonadecapeptide (*1413, 1414*), and substituted α-MSH sequences (*1032, 1043*). The combination with the trityl group was applied to the synthesis of Lys^2-Lys^7-gramicidin S (*2029*). The required detosylation of N^ϵ-tosyl-L-lysine that must be carried out with sodium in liquid ammonia is a disadvantage.

Cipera (*493*) has described by-products occurring during the reaction of N^α-carbobenzoxy-N^ϵ-tosyl-L-lysine with, for example, N^ϵ-tosyl-L-lysine methyl ester whenever the activation was brought about at the carboxyl group (anhydride, carbodiimide, and azide methods). The possibility of lactam formation caused by the fact that the tosyl amide nitrogen can be readily acylated has been discussed. The use was suggested of the tetraethyl pyrophosphite method ("amide" procedure) to prevent these side reactions. According to Nagamatsu *et al.* (*1594a*), N^α-carbobenzoxy-N^ϵ-tosyl-L-lysine can be coupled to N^ϵ-tosyl-L-lysine benzyl ester via the carbodiimide method without complication.

γ. *Use of N^ϵ-Formyllysine.* N^ϵ-Formyl-L-lysine, like the N^ϵ-carbobenzoxy and N^{ϵ}-tosyl compounds, can be prepared via the lysine-copper complex. Ethyl formate was used for its synthesis (*1030*). Free lysine can be formylated selectively at the ε-amino group in aqueous alkaline solution by treatment with *p*-nitrophenyl formate (*1656a*). The synthesis of N^{ϵ}-formyl-L-lysine methyl ester requires esterification of the N^{α}-carbobenzoxy compound with diazomethane and removal of this protecting group by catalytic hydrogenation. Peptides with N^{ϵ}-formyl-L-lysine derivatives were synthesized via the anhydride method (*1025, 1030*) and modified azide method (*1025*) (aqueous tetrahydrofuran/dimethylformamide). The N^{ϵ}-formyl group is labile to strong acids. The exact conditions for the removal of the formyl group from lower peptides with N^{ϵ}-formyllysine are not known. An ACTH tricosapeptide was deformylated by Hofmann *et al.* (*1044*) by exposure to 0.5 *N* hydrochloric acid at 100°C for 60–80 minutes. The N^{ϵ}-formyl group is stable to aqueous acetic or formic acid. Thus it is possible to remove selectively the N^{α}-carbobenzoxy group by hydrogenolysis in these solvents (*1025, 1030*).

δ. *Use of N^{ϵ}-Tritryllysine.* N^{ϵ}-Trityl-L-lysine methyl ester, described by Zervas and Theodoropoulos (*2668*) and by Stelakatos *et al.* (*2203*), is a suitable starting material for the preparation of N^{α}-lysine peptides. The compound can be isolated as dihydrochloride [7] (X = Cl) after selective detritylation of N^{α},N^{ϵ}-ditrityl-L-lysine methyl ester [8] with methanolic hydrochloric acid (*35*). The dihydrobromide [7] (X = Br) was obtained by tritylation of N^{α}-carbobenzoxy-L-lysine methyl ester [9] and decarbobenzoxylation with 10% hydrogen bromide in glacial acetic acid/ether (room temperature, 24 hours) (*236*). Thus the trityl

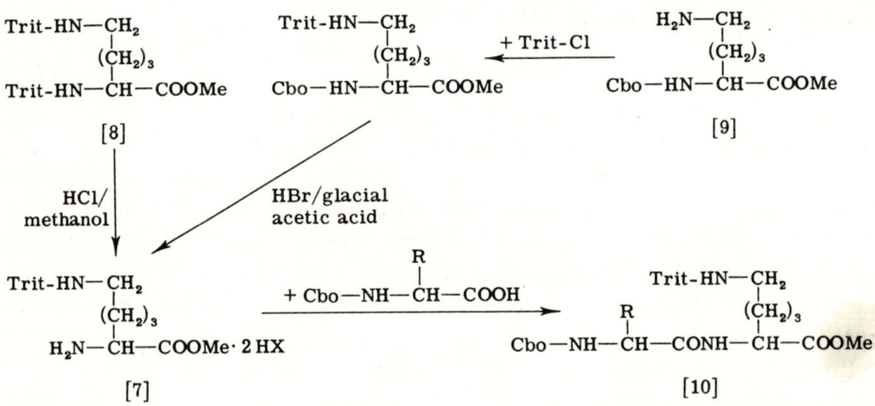

group, normally labile to acids, is stable under these conditions when attached to the ε-amino group of lysine. N^{ϵ}-Trityl-L-lysine methyl ester was used for the synthesis of α-lysine peptides [10] via the anhydride

(*236*) and carbodiimide (*35*) methods. Alkaline hydrolysis yields the free acid, detritylation with acetic acid and hydrogenolysis or hydrogenolysis alone yield the free peptide. The derivatives of N^ϵ-trityl-L-lysine have not been used on a broader basis for the synthesis of higher peptides.

ε. *Use of N^ϵ-Cyclohexyloxycarbonyllysine.* The cyclohexyloxycarbonyl group also offers protection of the ε-amino group with sufficient differentiation from the N^α-carbobenzoxy group (*1341, 1983*). N^ϵ-Cyclohexyloxycarbonyl-L-lysine can be prepared via the L-lysine-copper complex, the corresponding methyl ester via the *N*-carboxy anhydride (*1983*). Treatment with 1.5 moles hydrogen bromide for 2 hours will selectively remove the carbobenzoxy group. The cleavage of the cyclohexyloxycarbonyl group requires heating to 60°C for 1 hour. The protecting group combination was applied to the synthesis of an L-lysine-containing cyclic disulfide.

ζ. *Use of N^ϵ-tert-Butyloxycarbonyllysine.* Schwyzer *et al.* (*2024*) made a most welcome contribution to the synthesis of biologically active polypeptides by the introduction of N^ϵ-*tert*-butyloxycarbonyl-L-lysine. This derivative of lysine is prepared by the reaction of the copper complex with *tert*-butyloxycarbonyl azide in the presence of magnesium oxide (*2022*). The introduction of the N^ϵ-*tert*-butyloxycarbonyl group by the action of *tert*-butyloxycarbonyl *p*-nitrophenyl ester on N^α-carbobenzoxy-L-lysine represents a somewhat more rational approach (*2228*). The methyl ester was prepared by treatment of N^α-carbobenzoxy-N^ϵ-*tert*-butyloxycarbonyl-L-lysine with diazomethane. Hydrogenolysis gave N^ϵ-*tert*-butyloxycarbonyl-L-lysine methyl ester (*2022*). N^ϵ-*tert*-Butyloxycarbonyl-L-lysine can also be prepared from free lysine by treatment with *tert*-butyloxycarbonyl phenyl or *p*-nitrophenyl ester (*2022; cf. 2620*).

Schwyzer and Rittel (*2022*) have presented data on the stability of the N^ϵ-*tert*-butyloxycarbonyl (BOC) group shown in the following tabulation:

Acid	Time	Temp. (°C)	BOC group
Acetic acid			
2 *N*	24 hr	25	No cleavage
Glacial	24 hr	25	No cleavage
50% or 75%	30 min	30	No cleavage
Trifluoroacetic acid			
Anhydrous	1 hr	23	Complete cleavage
Hydrochloric acid			
2 *N*	30 min	25	Gas evolution and complete cleavage
Concentrated	A few sec	25	Foaming and complete cleavage

The stability of the *tert*-butyloxycarbonyl group to dilute acetic acid makes it possible to use it together with the trityl residue. The N^{α}-trityl group can be removed with 75% acetic acid at 30°C within 30 minutes. The *tert*-butyl ester is stable under these conditions. The carbobenzoxy and *p*-phenylazobenzyloxycarbonyl groups have also been used as N^{α}-protecting groups. Schwyzer *et al.* (*2024*) and Kappeler and Schwyzer (*1173*) took advantage of the excellent suitability of N^{ϵ}-*tert*-butyloxycarbonyl-L-lysine for the synthesis of complex peptides in the case of a nonadeca- and a tetracosapeptide of ACTH.

c. ε-Lysine Peptides

α. *Syntheses via the Copper Complex.* N^{ϵ}-Lysine peptides were described for the first time by Theodoropoulos and Craig (*2299*). The masking of the α-amino group in the lysine-copper complex was used with advantage to prepare N^{ϵ}-(tosyl-L-leucyl)-L-lysine by treatment with tosyl-L-leucyl chloride at pH 10. A pure product was isolated after countercurrent distribution. Theodoropoulos (*2292*) prepared other N^{ϵ}-lysine derivatives via the copper complex using, however, the anhydride method and the carbobenzoxy protection. The free N^{ϵ}-lysine peptides can be isolated after catalytic hydrogenation or treatment with hydrogen bromide in glacial acetic acid. Zahn and Pätzold (*2633a*, cf. *1698*) prepared by the same method ε-L-glutamyl-α-L-lysine and ε-L-glutamyl-γ-L-lysine. The α,ε-isomer rearranges via the imide to the γ,ε-isomer on recrystallization from boiling water (see Chapter IV,C,1,a,δ). Theodoropoulos and Craig (*2299*) found that the ε-peptide bond is extremely stable to acid hydrolysis. The ε-peptide bond is also stable to the enzymes pepsin and ficin.

β. *Syntheses with the α-Amino Group Substituted.* The second and more frequently applied method for the synthesis of N^{ϵ}-lysine peptides uses lysine derivatives substituted at the α-amino group.

β_1. *Use of N^{α}-tosyllysine.* Swallow *et al.* (*2244*) synthesized N^{ϵ}-(L-aspartyl-α)-L-lysine and N^{ϵ}-(L-aspartyl-β)-L-lysine via the anhydride method, using carbobenzoxy-L-aspartic acid α- or β-benzyl ester and N^{α}-tosyl-L-lysine benzyl ester hydrochloride as starting materials. The protecting groups were removed by hydrogenolysis, followed by treatment with sodium in liquid ammonia. The first compound is also accessible from carbobenzoxy-L-aspartic acid anhydride and the L-lysine-copper complex. A mixture of the α- and β-isomers was formed by aminolysis of the anhydride by N^{α}-tosyl-L-lysine benzyl ester in aqueous potassium hydrogen carbonate/ethyl acetate.

Mechanic and Levy (*1516*) synthesized N^{ϵ}-(glycyl-L-glutamyl-α)-L-lysine, a sequence they had earlier isolated from a partial hydrolyzate of collagen. The carbodiimide method was used to couple carbobenzoxy-

L-glutamic acid γ-benzyl ester with N^{α}-tosyl-L-lysine benzyl ester. The carbobenzoxy group was removed by treatment for 2 hours with hydrogen bromide in glacial acetic acid. Tosylglycine azide was used in the next coupling reaction and the detosylation carried out with sodium in liquid ammonia.

β_2. *Use of N^{α}-carbobenzoxylysine.* Bezas and Zervas (*236*) opened new synthetic routes with the preparation of N^{α}-carbobenzoxy-L-lysine. Prior to that N^{α}-tosyl-L-lysine or a corresponding ester had to be used as starting material for the synthesis of N^{ϵ}-lysine peptides and the free peptides could be obtained only by the reaction with sodium in liquid ammonia. N^{α}-Carbobenzoxy-L-lysine was prepared by carbobenzoxylation of N^{ϵ}-benzylidene-L-lysine (10 minutes at —5° to —10°C in 1 *N* sodium hydroxide and 10 minutes at room temperature). The benzyl ester of N^{α}-carbobenzoxy-L-lysine was used for the synthesis of peptides via the carbodiimide and anhydride methods. The free ε-lysine peptides were obtained under mild conditions by catalytic hydrogenation. Kornguth *et al.* (*1286a*) have prepared N^{ϵ}-glutamyl-α- and N^{ϵ}-glutamyl-γ-lysine via the same method.

β_3. *Use of N^{α}-trifluoroacetyllysine.* Weygand and Geiger (*2482*) have described N^{α}-trifluoroacetyl-L-lysine methyl ester. So far there are no reports on the application of this substance to the synthesis of N^{ϵ}-lysine peptides.

d. N^{α},N^{ϵ}-LYSINE PEPTIDES

α. *Symmetrical Peptides.* Two types of N^{α},N^{ϵ}-lysine peptides must be distinguished. To the first group belong the compounds where the same peptide chain substitutes the N^{α}- and the N^{ϵ}-amino groups. The synthesis of these derivatives—starting from L-lysine methyl, ethyl, benzyl, or *p*-nitrobenzyl ester by one of the conventional coupling methods—presents no great difficulties (*1976*). Zahn and Pätzold (*2633a*, cf. *1698*) have synthesized interesting N^{α},N^{ϵ}-lysine peptides with glutamic acid, e.g.:

N^{α},N^{ϵ}-bis-(TFA-Glu-α-OBzl)-γ-Lys-OEt
N^{α},N^{ϵ}-bis-[TFA-Glu-α-Lys(Cbo)-OEt]-γ-Lys-OEt

These derivatives were used for the preparation of the free peptides. The autohydrolysis of the compounds, their p*K* values and infrared spectra have been discussed.

β. *Unsymmetrical Peptides.* Theodoropoulos and Craig (*2299*) obtained via the anhydride method N^{α}-[carbethoxy-DL-Ileu]-N^{ϵ}-[Cbo-L-Asp(NH_2)]-L-Lys-OMe from N^{α}-carbethoxy-DL-Ileu-L-Lys-OMe and carbobenzoxy-L-asparagine. They did not describe the conversion to the free peptide.

Mechanic and Levy (*1516*) prepared N^{α}-glycyl-N^{ϵ}-(L-glutamyl-γ)-L-lysine and N^{α}-glycyl-N^{ϵ}-(L-glutamyl-α)-L-lysine. The precursor of the first compound, N^{α}-(Tos-Gly)-N^{ϵ}-(Tos-Glu-γ)-Lys-OBzl was obtained by aminolysis of tosyl-L-pyroglutamic acid with N^{α}-(Tos-Gly)-L-Lys-OBzl.

It is more advantageous to use for the synthesis of unsymmetrical peptides compounds that have been described by Bezas and Zervas (*236*), namely, N^{ϵ}-trityl-L-lysine methyl ester dihydrochloride or, the dihydrobromide (cf. page 154). The application of the carbodiimide or anhydride method leads to the formation of the substituted α-peptide [10]. It can be detritylated with acetic acid [11] and reacted to form the mixed N^{α},N^{ϵ}-peptide derivative [12]. Alkaline hydrolysis and hydrogenolysis yield the free peptide [13]. The reaction analogous to

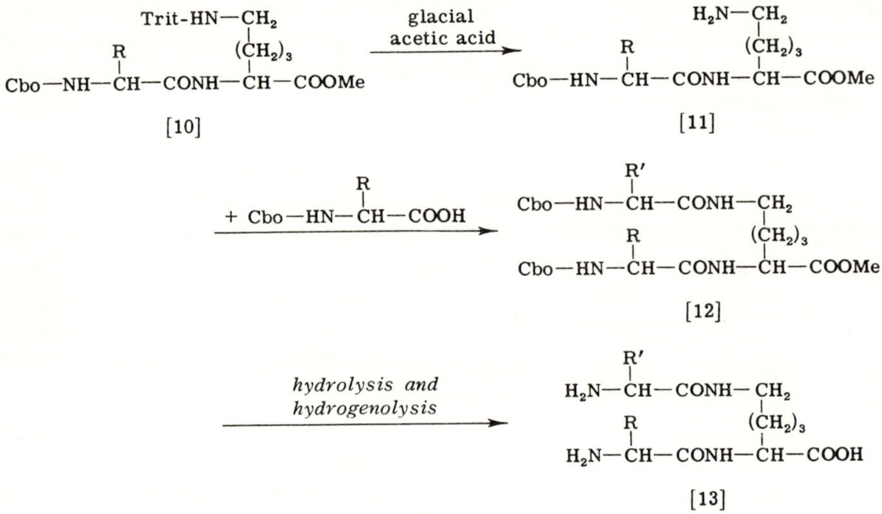

[10] [11] [12] [13]

the carbobenzoxylation, the treatment of N^{ϵ}-benzylidene-L-lysine with carbobenzoxyamino acid chlorides, gives only poor yields. Peptide coupling reactions at the ϵ-amino group of free lysine can be carried out via the *p*-nitrophenyl ester method (*1224a*).

2. Ornithine

The relationship to lysine is so close that the experience obtained with it can be applied to ornithine almost everywhere without complication. A considerable amount of information can be found in publications on the synthesis of ornithine-containing antibiotics. Gramicidin S and tyrocidin A and B contain L-ornithine; bacitracin contains only D-ornithine.

a. Ornithyl and Ornithine Peptides

Synge (*2251*) synthesized the first ornithyl peptides, using dicarbobenzoxy-L-ornithine chloride. Ornithyl peptides were prepared later by the azide (*958*) and phosphazo (*820*) methods. N^δ-Carbobenzoxy-L-ornithine for the synthesis of α-ornithine peptides was prepared from the L-ornithine-copper complex by treatment with carbobenzoxy chloride in sodium hydroxide (*2251*). The carbobenzoxylation of the copper complex proceeds more smoothly in the presence of an excess of magnesium oxide (*111*). N^δ-Carbobenzoxy-L-ornithine methyl ester (prepared by esterification with 1 *N* methanolic hydrochloric acid or from N^δ-carbobenzoxy-L-ornithine *N*-carboxylic anhydride) was used for the synthesis of peptides via the acid chloride (*2251*), azide (*958, 1415*), phosphazo (*820*), anhydride (*1986*), and carbodiimide (*1415*) methods. Erlanger *et al.* (*687, 692*) synthesized the methyl, ethyl, and benzyl ester hydrochlorides of N^δ-tosyl-L-ornithine. The trifluoroacetyl group (*2480*) and the formyl group (*1656a*) were also used as N^δ-protecting group for ornithine.

b. Rearrangement Reactions of Ornithine Derivatives

Barras and Elmore (*112*, cf. *551*) observed that N^α-tosyldiaminocarboxylic acid methyl ester hydrochlorides [14] ($n = 2$: α,γ-diaminobutyric acid; $n = 3$: ornithine; $n = 4$: lysine) will be cyclized upon

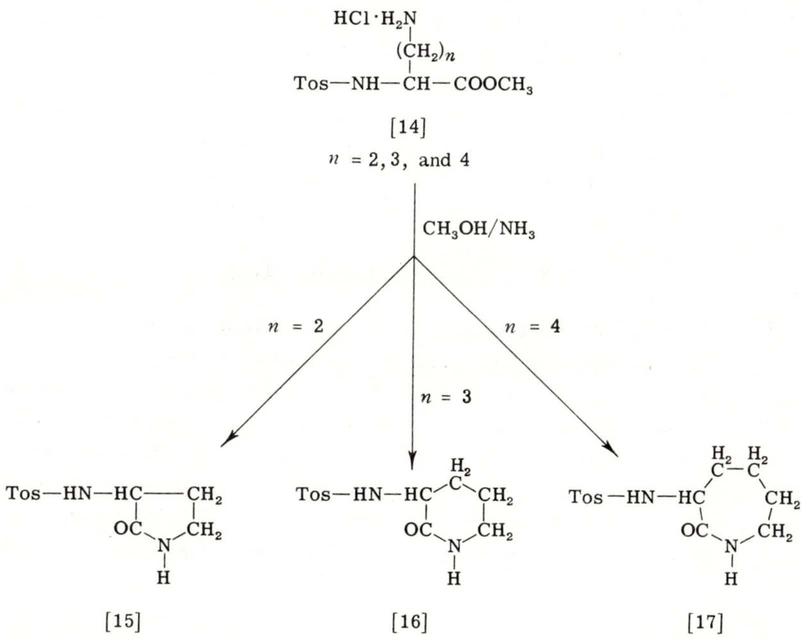

treatment with methanolic ammonia to form 3-tosyl-L-aminopyrrolidone-2 [15], 3-tosyl-L-aminopiperidone-2 [16], and 3-tosyl-L-aminohomopiperidone-2 [17]. The yields of 88, 82, and 49% demonstrate that the formation of the seven membered ring is less likely. 3-Tosyl-L-aminopiperidone-2 [16] will also be formed from [14] ($n = 3$) by treatment with triethylamine in methanol. The three lactams are otherwise accessible from the corresponding N^{α}-tosyl-N^{ω}-carbobenzoxydiaminocarboxylic acid amides by hydrogenolysis in methanol in the presence of acetic acid. The yields were quantitative for [15] and [16] but only 7% in the case of [17]. N^{α}-Tosyl peptides of the constitution represented by [18] lose glycine

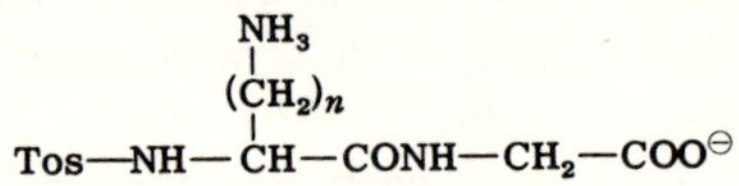

[18]

in the case of α,γ-diaminobutyric acid ($n = 2$) and L-ornithine ($n = 3$) on treatment with methanolic or ethanolic ammonia at room temperature. Tos-L-Lys-Gly-OH is stable even at 60°C.

L-Ornithine can be prepared from tosyl-L-glutamine by dehydration of the carboxamido group and reduction of the resulting cyano group (*2654*). Zaoral and Rudinger (*2653*) applied the reaction to peptides and obtained H-L-Orn-Gly-OH from Tos-L-Glu(NH_2)-Gly-OMe by dehydration, hydrolysis, and reduction with sodium in liquid ammonia. It is also possible to dehydrate phthalyl-DL-glutamine methyl ester and convert it by catalytic hydrogenation to N^{α}-phthalyl-DL-ornithine methyl ester hydrochloride.

Losse and Augustin (*1453*) have reported the resolution of the racemic mixture of synthetic ornithine with dibenzyl-D-tartaric acid. Izumiya *et al.* (*1123*) used leucine aminopeptidase to resolve the racemic mixture of N^{δ}-benzoyl-DL-ornithine amide.

3. α,γ-Diaminobutyric Acid

Homodetic cyclic heteromeric peptide antibiotics, such as the polymyxins, circulin, and colistin, contain α,γ-diaminobutyric acid as a main constituent. The synthesis of these biologically active polypeptides of relatively complex nature required extensive investigations on the reactivity of α,γ-diaminobutyric acid. An important difference between this amino acid and lysine is the ready lactam formation of N^{γ}-acyl-α,γ-diaminobutyric acid. Derivatives of the resulting 3-amino-pyrrolidone-2 are used as starting material for the synthesis of peptides.

a. SYNTHESIS OF α,γ-DIAMINOBUTYRIC ACID

Most of the amino acids present in biologically active polypeptides can be obtained by fractionation of protein hydrolyzates or by enzymatic resolution of racemic mixtures of synthetic DL-amino acids. Some amino acids can be readily converted into others by simple chemical reactions. All this is not true for α,γ-diaminobutyric acid. It is accessible only

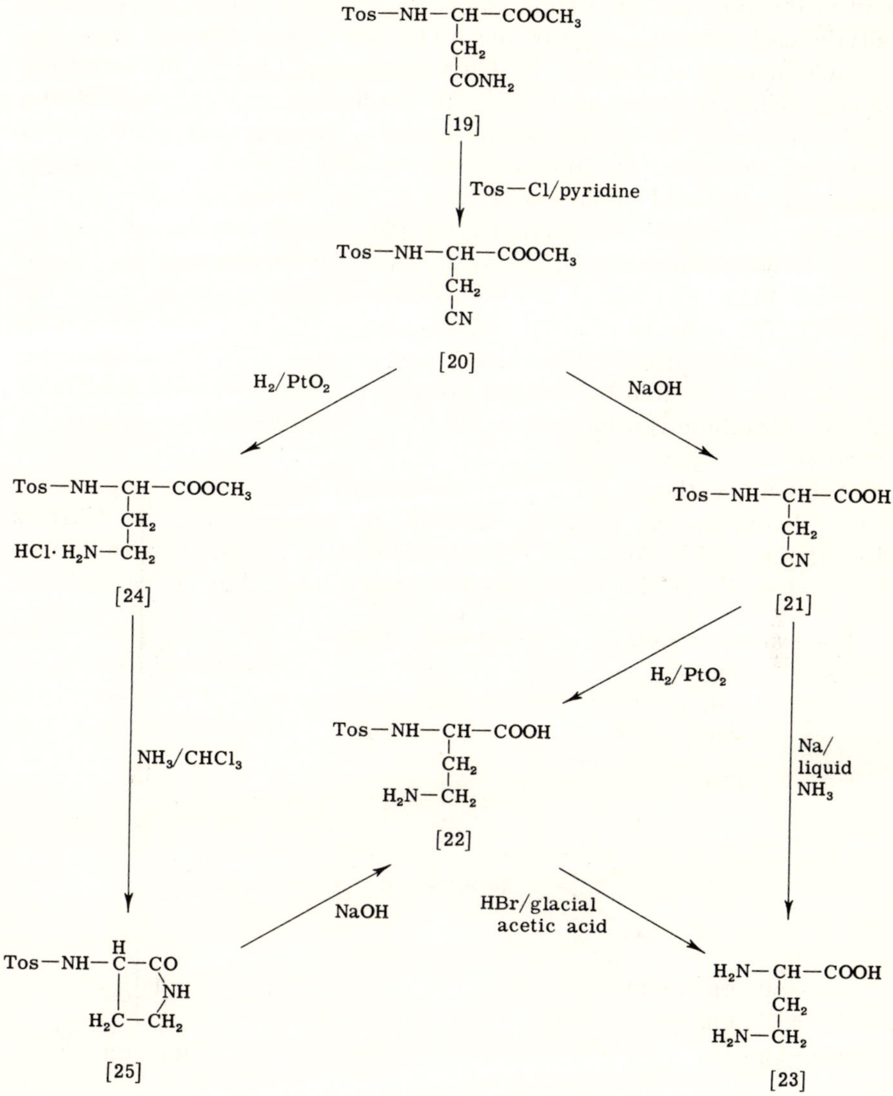

by a series of reactions of many steps. Zaoral and Rudinger (*2654*) have shown that tosyl-L-asparagine methyl ester [19] can be dehydrated with tosyl chloride in pyridine to form β-cyano-α-tosyl-L-aminopropionic acid methyl ester [20]. Optically pure tosyl-L-asparagine can be recovered after treatment with 35% hydrogen bromide under mild conditions, thus assuring that the dehydration proceeds without racemization. The hydrolysis of compound [20] is followed by a hydrogenation in acetic acid in the presence of Adams' catalyst yielding N^{α}-tosyl-L-α,γ-diaminobutyric acid [22]. Cleavage of the protecting group with 35% hydrogen bromide in acetic acid at 70°–80°C under pressure leads to the formation of L-α,γ-diaminobutyric acid [23]. The reduction and cleavage of the tosyl group may be carried out in one step by treatment with sodium in liquid ammonia. In another synthetic approach, β-cyano-α-tosyl-L-aminopropionic acid methyl ester [20] is reduced catalytically to tosyl-L-α,γ-diaminobutyric acid methyl ester [24], which can be cyclized to the pyrrolidone derivative [25] by treatment with ammonia in chloroform (cf. *112*). The lactam can be hydrolyzed by 4 *N* sodium hydroxide at 70°C. The series of reactions [19] → [20] → [21] may also be carried out with carbobenzoxy-L-asparagine methyl ester. L-α,γ-Diaminobutyric acid is then accessible by catalytic hydrogenation under mild conditions (*2653*). Ornithine can be prepared from tosylglutamine in an analogous reaction.

Another synthesis of α,γ-diaminobutyric acid uses tosylglutamic acid γ-hydrazide as starting material. It proceeds via the Curtius degradation and hydrolysis of the initially formed tosylaminopyrrolidone-2 [27] (*2654*). If the rearrangement of the azide [26] is carried out

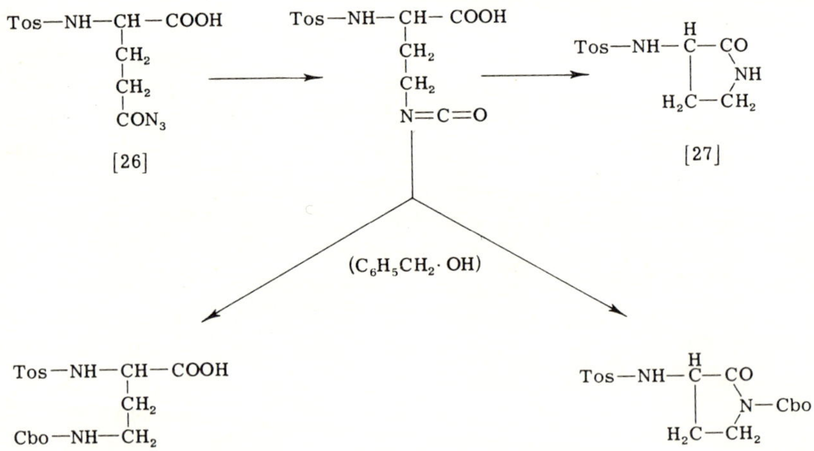

[26] [27] [28] [29]

in the presence of benzyl alcohol, N^{α}-tosyl-N^{γ}-carbobenzoxy-L-α,γ-diaminobutyric acid [28] (*37%*) is formed in addition to the corresponding lactam [29] (*25%*) (*1846*).

The Hofmann degradation of tosyl-L-glutamine developed by Rudinger *et al.* (*1858*) represents the most convenient synthesis. Pure tosyl-L-α,γ-diaminobutyric acid is obtained in 70% yield.

b. α,γ-DIAMINOBUTYRYL AND α,γ-DIAMINOBUTYRIC ACID PEPTIDES

Zaoral *et al.* (*2656*) have synthesized α,γ-diaminobutyryl peptides with amino-terminal α,γ-diaminobutyric acid from N^{α},N^{γ}-dicarbobenzoxy-α,γ-diaminobutyric acid via the anhydride (carbobenzoxy chloride, *N*-methylpiperidine in chloroform) and azide methods. N^{α},N^{γ}-Dicarbobenzoxy-L-α,γ-diaminobutyric acid hydrazide was prepared from the oily N^{α},N^{γ}-dicarbobenzoxy-L-α,γ-diaminobutyric acid ethyl ester, or by hydrazinolysis of 1,3-dicarbobenzoxy-L-aminopyrrolidone-2 in ethanol at room temperature. The carbobenzoxy groups were removed with hydrogen iodide in glacial acetic acid.

L-α,γ-Diaminobutyryl peptides containing α,γ-diaminobutyric acid in the chain were synthesized predominantly with N^{γ}-carbobenzoxy- and N^{γ}-tosyl-L-α,γ-diaminobutyric acid or the corresponding esters. The N^{γ}-carbobenzoxy and N^{γ}-tosyl compounds can be prepared most conveniently, like the corresponding derivatives of lysine and ornithine, by acylation of the L-α,γ-diaminobutyric acid-copper complex (*1746, 2386*). The direct carbobenzoxylation in strongly alkaline solution gave poorer results although the two amino groups are of different basicity (*1746, 2656*). The hydrochlorides of the methyl (*1119, 2386*) and ethyl (*1745, 2656*) esters of the N^{γ}-carbobenzoxy compound and the sulfate of the methyl ester of the N^{γ}-tosyl compound (*2389*) have been described. These derivatives were used for the synthesis of peptides via the anhydride (*2386, 2656*) and carbodiimide (*2386, 2389*) methods or via the *N*-carboxylic anhydride in chloroform at −65°C (*2656*). N^{γ}-Carbobenzoxy- and N^{γ}-tosyl-α,γ-diaminobutyric acid (L and D) were used for the synthesis of unsymmetrically substituted bis derivatives that were coupled to peptides via the carbodiimide (*1745, 2384, 2389*), tetraethyl pyrophosphite (*1745*), and azide (*2227*) methods. The N^{α}-protecting group was removed selectively and other coupling reactions were carried out. The reaction of N^{α}-carbobenzoxy-N^{γ}-tosyl-L-α,γ-diaminobutyric acid gave, in addition to the desired peptide ester, 1-tosyl-3-carbobenzoxy-L-aminopyrrolidone-2. N^{α},N^{γ}-Diacyl-L-α,γ-diaminobutyric acid can be converted to 1,3-diacyl-L-aminopyrrolidone-2 [30] by treatment with phosphorus pentachloride (*2565*), thionyl chloride, or ethyl chlorocarbonate (*1745*):

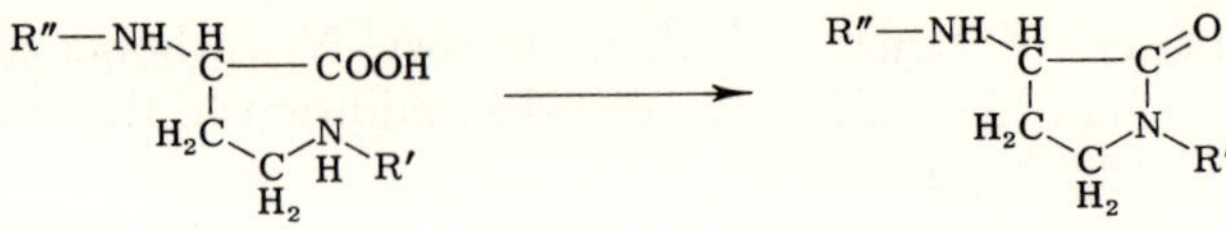

[30]

The lactams [30a–d] are useful products for the synthesis of peptides. They are subject to ready cleavage by ammonia, hydrazine, amino acid esters, or sodium hydroxide to form the corresponding derivatives of N^{α},N^{γ}-diacyl-L-α,γ-diaminobutyric acid [31]. The aminolysis of the

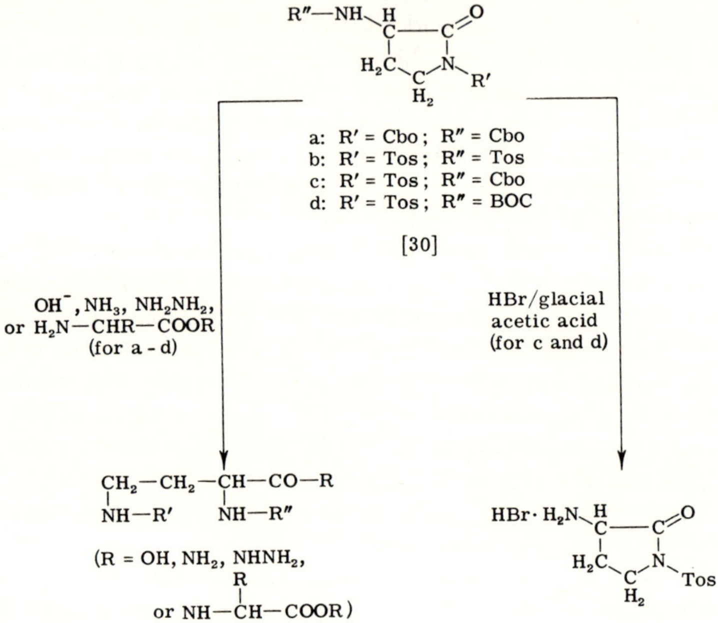

[30] [31] [32]

lactam ring is dependent upon the solvent. Acetonitrile and nitromethane were shown to be suitable, while dioxane, pyridine, diethyl phosphite, and dimethylformamide gave less satisfactory results (*1746*). The derivatives [30c and d] are converted to 1-tosyl-L-3-aminopyrrolidone-2 hydrobromide [32] by treatment with hydrogen bromide in glacial acetic acid.

Treatment by phosgene in dioxane of N^{γ}-carbobenzoxy-L-α,γ-diaminobutyric acid [33] leads to the formation of the N-carboxylic anhydride [34] (*1745*). If the reaction is carried out in boiling toluene, a small amount of 1-carbobenzoxy-3-L-isocyanopyrrolidone-2 [35] is formed (*2656*). This compound can be obtained more readily from 1-carbo-

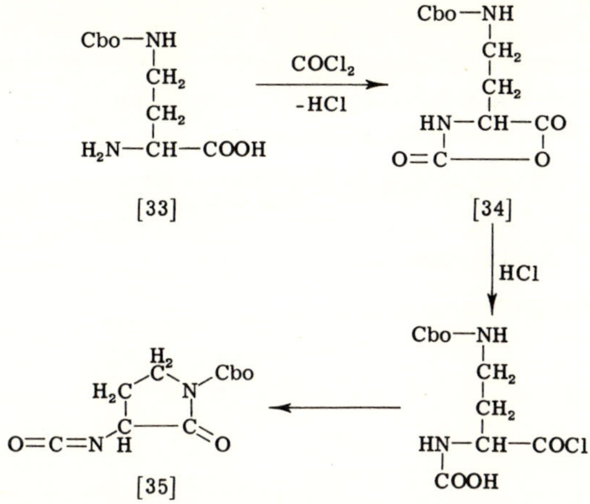

benzoxy-L-3-aminopyrrolidone-2 by treatment with phosgene. The simultaneous activation of the amino and carboxyl groups in one molecule is of interest. It is possible to react this compound first with a carbobenzoxyamino acid at the amino terminal and extend it subsequently at the carboxyl terminal by cleavage of the lactam ring by an amino acid ester [36]. 1-Tosyl-3-L-isocyanopyrrolidone-2 [37] (*1746*) can be pre-

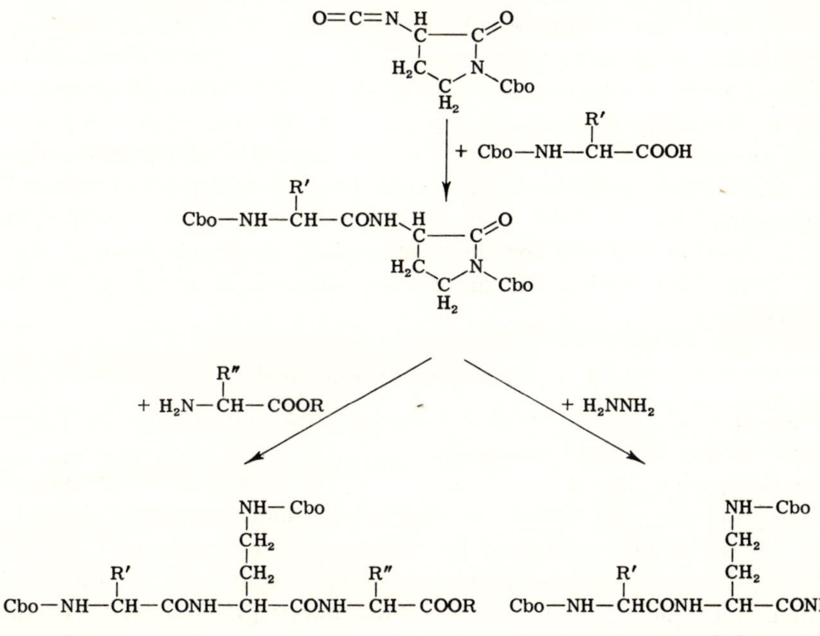

[36]

pared in an analogous reaction from 1-tosyl-L-3-aminopyrrolidone-2. The latter compound is accessible by treatment under pressure of N^{α},N^{γ}-

ditosyl-L-α,γ-diaminobutyric acid with hydrogen bromide in glacial acetic acid in the presence of phenol (3 hours, 70°C). In contrast to the corresponding carbobenzoxy compound, activation at the amino group in the form of the isocyanate [37] did not give satisfactory results when

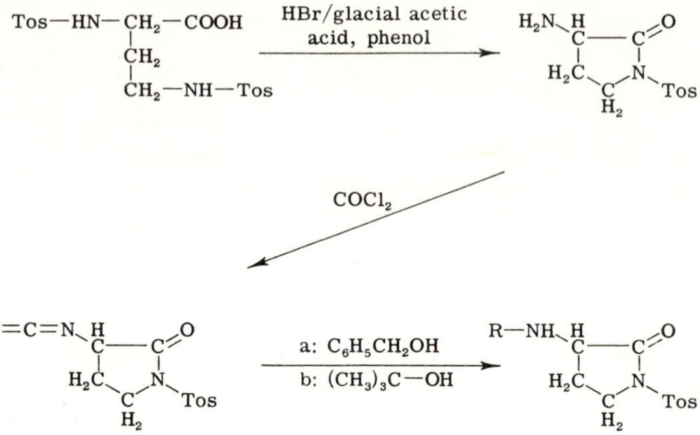

[37] [38]

(a: R = Cbo
b: R = BOC)

reacted with, for example, N^{α}-carbobenzoxy-N^{γ}-tosyl-L-α,γ-diaminobutyric acid. It is more advantageous to react 1-tosyl-L-3-aminopyrrolidone-2 with *N*-substituted amino acids by the tetraethyl pyrophosphite or the azide method. The reaction of the isocyanate [37] with benzyl alcohol or *tert*-butanol leads to the formation of 1-tosyl-3-carbobenzoxy-L-aminopyrrolidone-2 [38a] and 1-tosyl-3-*tert*-butyloxycarbonyl-L-aminopyrrolidone-2 [38b], respectively. γ-Butyrolactone may also be used as starting material for the synthesis of α,γ-diaminobutyryl peptides (*2094, 2256, 2647*). The derivatives obtained, however, are of the DL-series.

4. α,β-Diaminopropionic Acid

Haskell *et al.* (*959*) found α,β-diaminopropionic acid to be a constituent of the antibiotic viomycin.

In the earlier literature Miyanoki (*1565*) described N^{β}-glycyl-α,β-diaminopropionic acid and N^{α},N^{β}-diglycyl-α,β-diaminopropionic acid (L or DL ?), which he synthesized by chloroacetylation. Schneider (*1948*) synthesized N^{α}-glycyl-L-α,β-diaminopropionic acid, N^{α},N^{β}-diglycyl-L-α,β-diaminopropionic acid, and L-α,β-diaminopropionylglycine via the acid chloride method.

Poduška *et al.* (*1747*) have employed other methods of peptide chemistry to synthesize several peptides containing α,β-DL-diaminopropionic acid. N^{α},N^{β}-Dicarbobenzoxy-DL-α,β-diaminopropionylglycine ethyl ester, the corresponding free dipeptide, and the dipeptide amide were synthesized via the anhydride method (*sec*-butyl chlorocarbonate, *N*-ethylpiperidine in chloroform) and the isocyanate method. Dicarbobenzoxy-DL-α,β-diaminopropionic acid had been prepared by Kjaer (*1251*). The reaction of N^{α},N^{β}-dicarbobenzoxy-DL-α,β-diaminopropionic acid with PCl_5 in chloroform led to the formation of the *N*-carboxylic anhydride [DL-4-(carbobenzoxyaminomethyl)-oxazolidinedione-2,5]. It was converted to N^{β}-carbobenzoxy-DL-α,β-diaminopropionic acid methyl and ethyl esters. N^{α}-Glycyl-DL-α,β-diaminopropionic acid was synthesized from these compounds via the azide method. Kjaer (*1251*) had observed that in strongly alkaline solution the β-position of α,β-diaminopropionic acid is preferentially acylated. Thus N^{β}-Tos-Gly-DL-Dap-OH is formed in 30% yield during the reaction of tosylglycine chloride with DL-α,β-diaminopropionic acid, besides the N^{α},N^{β}-ditosyl compound in 10% yield. N^{β}-Glycyl- and N^{α},N^{β}-diglycyl-DL-α,β-diaminopropionic acid were obtained by detosylation with hydrogen bromide in glacial acetic acid (2 hours, 70°C) in the presence of phenol or by treatment with sodium in liquid ammonia. Schneider (*1949*) has described dicarbobenzoxy-L-α,β-diaminopropionic acid and the hydrochloride of β-carbobenzoxy-L-α,β-diaminopropionic acid methyl ester.

N^{α}-Tosyl-L-α,β-diaminopropionic acid can be prepared by Hofmann degradation of tosyl-L-asparagine. Treatment with potassium cyanate leads to the formation of β-ureido-α-tosyl-L-aminopropionic acid, and subsequent detosylation to β-ureido-L-α-aminopropionic acid. The latter amino acid is identical with albizziine (*810, 1252*), which was isolated from mimosa.

5. Arginine

The guanido group of arginine readily gives rise to side reactions. Therefore *N*-guanido(N^{G})-substituted arginine derivatives are used for the synthesis of peptides. On the other hand, the strongly basic guanido group ($pK = 12.5$)—as compared to the α-amino group ($pK = 9.25$) and to α-amino acid esters ($pK = 8$)—can be protonated selectively (see Chapter I,E), and peptides synthesized at pH 9 without side reaction (*805*).

a. Syntheses with N^{G}-Substituted Arginine

α. Use of N^{G}-Nitroarginine. Nitroarginine is the most important N^{G}-substituted derivative of arginine. Bergmann *et al.* (*202;* cf. *1288*) used it for the first time for the synthesis of peptides. The structure of nitro-

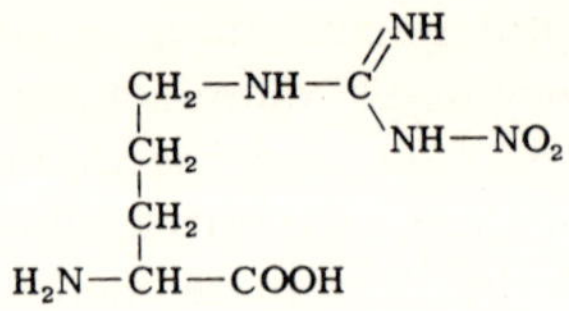

[39]

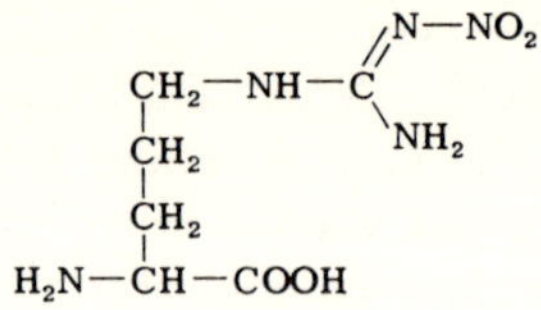

[40]

arginine—the nitroamino [39] or nitrimino [40] form—have been discussed by several authors. Physicochemical data are not in agreement with either of the two forms.

Hofmann *et al.* (*1028, 1029*), van Orden and Smith (*2359*), Zahn and Diehl (*2617;* cf. *602*), and Gibian and Schröder (*803*) have carried out systematic investigations on the application of nitroarginine to the synthesis or arginyl and arginine peptides. Carbobenzoxynitro-L-arginine (*202, 1028, 2359*) was preferred as starting material for the synthesis of arginyl peptides. It was coupled with amino acid esters via the anhydride (*217, 803, 1028, 1117, 2359*) or carbodiimide method (*803, 1117, 2617*). The phosphazo (*803*) and isocyanate (*803*) methods as well as coupling with phosphorus pentoxide in diethyl phosphite (*996*) are also applicable in principle. Arginyl peptides cannot be prepared via the azide method. The high reactivity of the nitroguanido group with hydrazine makes impossible the synthesis of carbobenzoxynitro-L-arginine hydrazide (*772, 1486,* cf. *2116a*). Attempts to apply the acid chloride method were likewise unsuccessful. Berse and Piché (*215*) have coupled *p*-nitrocarbobenzoxynitro-L-arginine via the anhydride method. Phthalyl- (*2359*) and tosylnitro-L-arginine (*2043*) have not yet been used for the synthesis of peptides, while *tert*-butyloxycarbonylnitro-L-arginine (*1020b*) and tritylnitro-L-arginine (*2024*) were applied successfully. According to Haas (*904*), it is not possible to hydrolyze tritylnitro-L-arginine methyl ester satisfactorily.

Several arginine-containing peptides had been prepared earlier via the azlactone (*197*) and acid chloride (*202*) methods. The hydrochlorides of nitro-L-arginine methyl (*1028*), ethyl (*2617, 2618*), and benzyl (*602, 2322, 2617*) esters enjoyed first choice as derivatives for the synthesis of arginine peptides via the anhydride (*803, 1028*), azide (*2617*), and carbodiimide (*803, 2617*) methods. Carbethoxynitro-L-arginine methyl ester was isolated after application of the anhydride method as the result of cleavage of the mixed anhydride at the wrong bond. Formation of acylurea during application of the carbodiimide method was observed several times. The application of the phosphazo method appeared to be less promising (*803*). Wünsch (*2593*) pointed out the formation of by-products caused by the reductive action of polymetaphosphorous acid on the nitroguanido group. Boissonnas *et al.*

(*292*) have used the hydrobromide of nitro-L-arginine *p*-nitrobenzyl ester for the synthesis of peptides. Substituted peptides containing nitroarginine can be saponified under normal conditions (*803, 1028, 2617*). More vigorous reaction conditions, however, are sometimes required (*217*).

Bergmann *et al.* (*202*) have already reported that the nitro group in nitroarginine can be readily reduced in the presence of catalysts with regeneration of the guanido function. Apparently the nitroguanido group [41] is first converted to the aminoguanido group [42], which is then cleaved hydrogenolytically to the guanido group and ammonia. If the reaction were to follow this course, 4 moles of hydrogen should be consumed, but so far there are only reports on the consumption of 3 moles of hydrogen (*202, 803*). In general, the reduction of a nitroarginine residue located in a peptide chain frequently requires long hydrogenation times. There are also cases where even carbobenzoxy-nitro-L-arginylamino acids could not be hydrogenated under acidic, neutral, or alkaline conditions (*217*). Clubb *et al.* (*505*) found that nitro-L-arginine can be converted quantitatively to L-arginine by electrolytic reduction at a mercury cathode. Gros *et al.* (*866*), however, reported that the electrolytic and catalytic reduction frequently stops at the level of aminoguanidine [42]. The nitroguanido group cannot be reduced

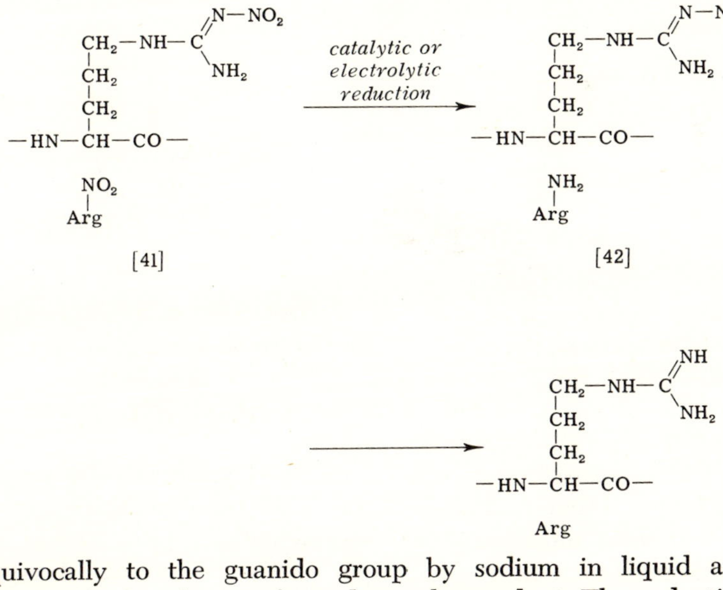

[41] [42]

unequivocally to the guanido group by sodium in liquid ammonia (*2000, 2322*). Ornithine is formed as a by-product. The reduction with zinc/hydrochloric acid also did not proceed satisfactorily.

Though nitro-L-arginine and carbobenzoxynitro-L-arginine are used very frequently for the synthesis of even arginine-containing higher peptides, the protection of the guanido group by introduction of the nitro group does not completely exclude the possible occurrence of side reactions. Carbobenzoxynitro-L-arginine lactam [43] (1-nitroguanyl-3-carbobenzoxyaminopiperidone-2) was the main product (*280*) when carbobenzoxynitro-L-arginine *p*-nitrophenyl ester was to be prepared by the carbodiimide method. In contrast to the reactivity of other amino

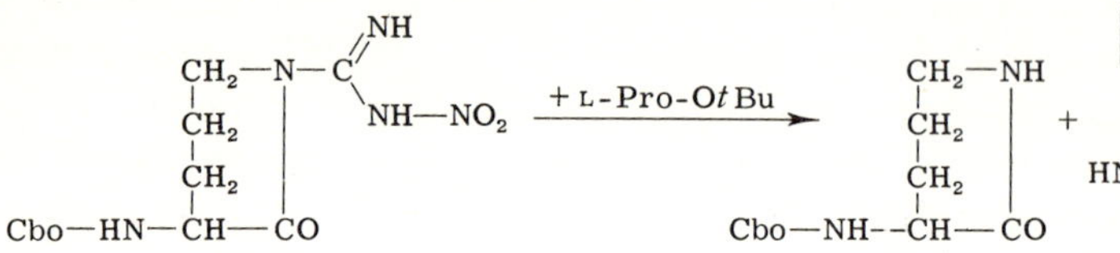

[43] [44]

acid lactams, the lactam of arginine does not react with amino acid esters to form the corresponding dipeptide derivatives. The nitroguanido group is transferred to the amino component with formation of carbobenzoxy-L-ornithine lactam [44] (3-carbobenzoxy-L-aminopiperidone-2) (*1702*). Paul *et al.* (*1702*) isolated 2-nitrimino-4-carboxy-1,3-diazocycloheptane [45] when they attempted to prepare *tert*-butyloxycarbonyl-nitro-L-arginine by heating nitro-L-arginine with *tert*-butyloxycarbonyl *p*-nitrophenyl ester and *tert*-butanol in aqueous sodium hydrogen carbonate solution:

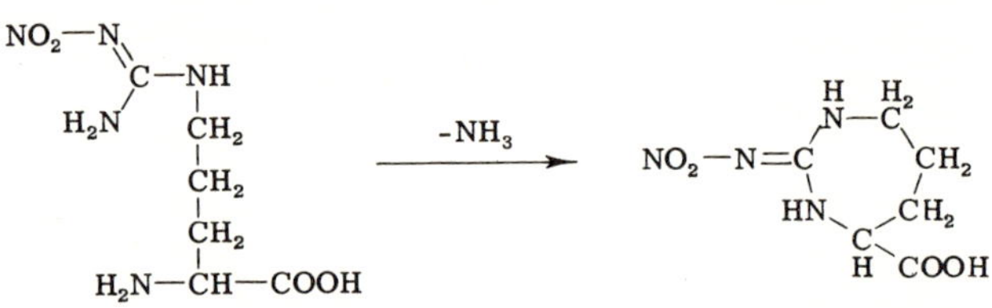

[45]

Similar ring closure reactions are known from the earlier literature. N^{α}-Acetylanhydronitro-DL-arginine (N^{α}-acetylnitro-DL-arginine lactam) (*252*) was formed on boiling nitro-L-arginine in acetic anhydride/acetic acid. L-Arginine was racemized and converted to triacetyl-DL-arginine lactam by boiling it in an excess of acetic anhydride (*186*).

The carbobenzoxy group can be removed selectively by treatment with hydrogen bromide/glacial acetic acid from peptides (or their esters) containing nitro-L-arginine. Several authors (*1023, 1971*) found that the bromide content of the hydrobromides was too high. Frequently it was impossible to lower it by renewed precipitation or recrystallization.

β. Use of N^G-Tosylarginine. The preparation of N^α-carbobenzoxy-N^G-tosyl-L-arginine [46] by tosylation of N^α-carbobenzoxy-L-arginine at pH 11 to 12 offers another reversible protection of the guanido group (*1785, 1943, 2704, 2705*). Tosylation at pH 8.2 leads to the formation

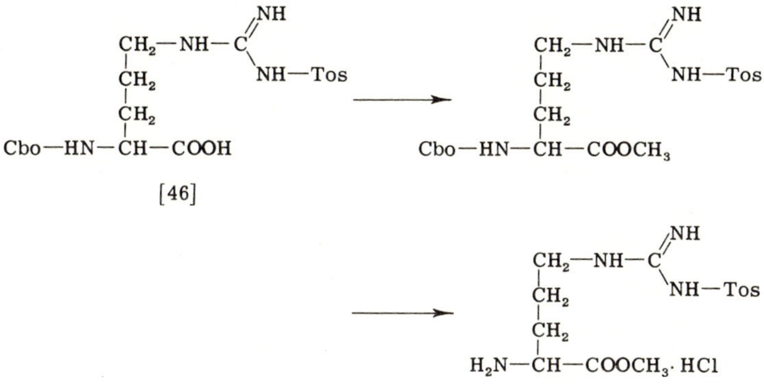

[46] [47]

of N^α-carbobenzoxy-N^G-tosyl-L-arginine lactam (*1943*). Attempts to prepare N^G-tosyl-L-arginine, like N^α-tosyl-L-lysine, via the copper complex have not been so far successful. Here, as in the case of nitro-L-arginine, it has not been elucidated whether the tosylimino or tosylamino compound is present. Other derivatives suitable for the synthesis of peptides are N^α-[*p*-(*p′*-methoxyphenylazo)-carbobenzoxy]-N^G-tosyl-L-arginine (*2021, 2037*) and N^α-*tert*-butyloxycarbonyl-N^G-tosyl-L-arginine (*1785*). The hydrochloride of N^G-tosyl-L-arginine methyl ester [47], which is required for the synthesis or arginine peptides, can be prepared by esterification with diazomethane of N^α-carbobenzoxy-N^G-tosyl-L-arginine or by treatment with thionyl chloride in methanol and subsequent catalytic hydrogenation (*902*). The corresponding benzyl ester may be obtained by reacting N^G-tosyl-L-arginine with benzyl alcohol/hydrogen chloride (*1785*). N^G-Tosyl-L-arginine is accessible by catalytic hydrogenation of the N^α-carbobenzoxy compound (*902, 1785*). It was used as triethylammonium salt in a salt coupling reaction via the *p*-nitrophenyl ester method (*902*). Peptide syntheses with derivatives of N^G-tosylated arginine have been carried out preferentially via the carbodiimide method (*1414, 1943, 2021*). The method of activated esters has been made applicable to the synthesis of arginyl peptides (*902*) with the preparation of the cyanomethyl and *p*-nitrophenyl esters of N^α-carbobenzoxy-N^G-tosyl-L-arginine. Lactam formation occurs only in the case of the latter compound to a minor extent. The N^G-tosyl group can be removed only with sodium in liquid ammonia.

The N^G-tosyl protection was used extensively for the synthesis of peptide hormones (*902, 1411, 1414, 2696*).

γ. *Use of N^G-Mono- and N^G,N^G-Dicarbobenzoxyarginine.* The carbobenzoxylation of L-arginine with 1 equivalent of carbobenzoxy chloride in a solution of sodium hydrogen carbonate or in the equivalent amount of sodium hydroxide results in the formation of N^{α}-carbobenzoxy-L-arginine (*193, 291*). The sodium salts of N^{α},N^G-dicarbobenzoxy- and

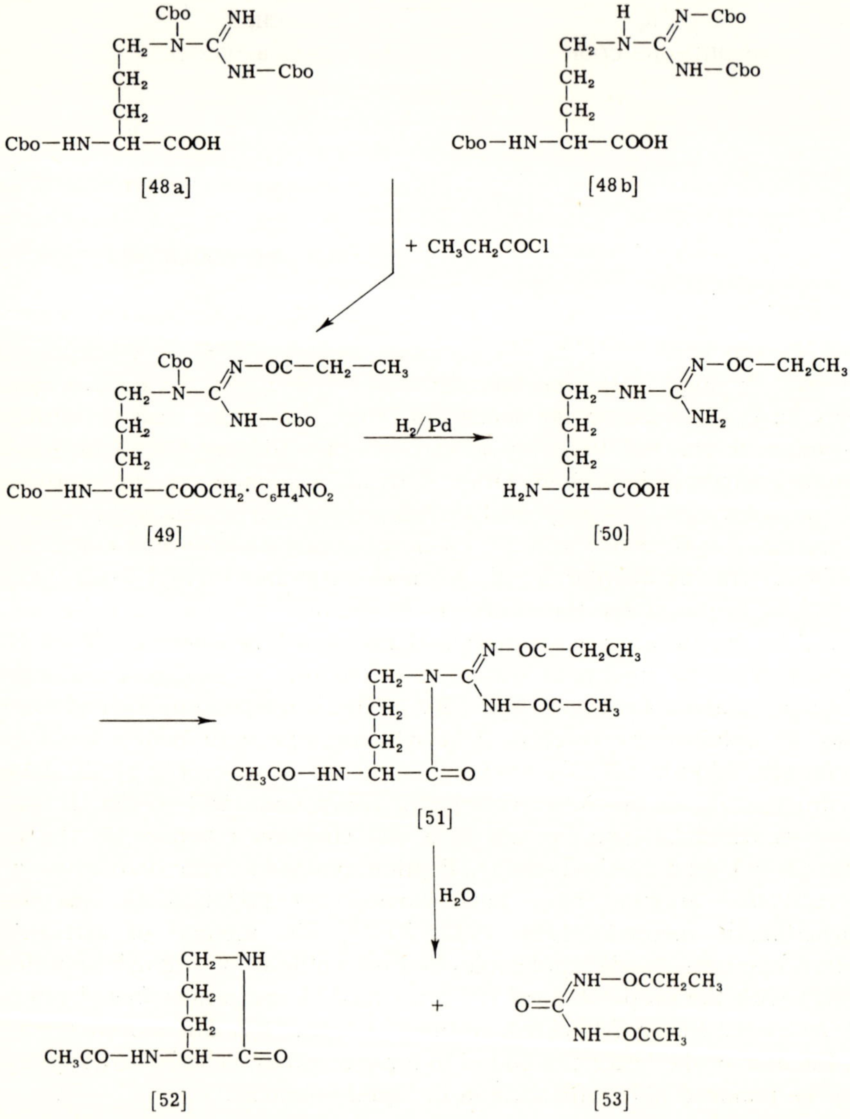

N^{α},N^{G}, N^{G}-tricarbobenzoxy-L-arginine are formed on treatment with 4 equivalents of carbobenzoxy chloride in strong alkali. These salts can be separated by crystallization from ethanol and converted to the free acids (*2669, 2670*). N^{α},N^{G}-Dicarbobenzoxy-L-arginine, in contrast to the N^{α}-mono derivative, can be carbobenzoxylated to the tricarbobenzoxy compound. One of the N^{G}-carbobenzoxy groups of tricarbobenzoxy-L-arginine is cleaved preferentially by 1 equivalent of sodium hydroxide in cold ethanol.

According to Zervas *et al.* (*2671*), [48a] and [48b] are possible structures of N^{α},N^{G},N^{G}-tricarbobenzoxy-L-arginine. The trisubstituted derivative was converted to the *p*-nitrobenzyl ester and acylated with propionic acid anhydride [49] to determine which sturcture really exists. Catalytic hydrogenation yielded N^{G}-propionyl-L-arginine [50], and boiling with acetic anhydride N^{α},N^{G}-diacetyl-N^{G}-propionylarginine lactam [51]. The latter compound was hydrolyzed by water to form DL-α-acetylaminopiperidone [52] and *N*-acetyl-*N′*-propionylurea [53]. N^{α},N^{G}-Dicarbobenzoxy-L-arginine can be converted to N^{α},N^{G}-dicarbobenzoxy-L-arginine lactam [54] by treatment with phosphorus pentachloride or *N*,*N′*-dicyclohexylcarbodiimide. Acetylation and cleavage by water of N^{α},N^{G}-dicarbobenzoxy-N^{G}-acetyl-L-arginine lactam [55] to form DL-α-acetylaminopiperidone [52] and *N*-acetyl-*N′*-carbobenzoxyurea [56] furnished the structural proof.

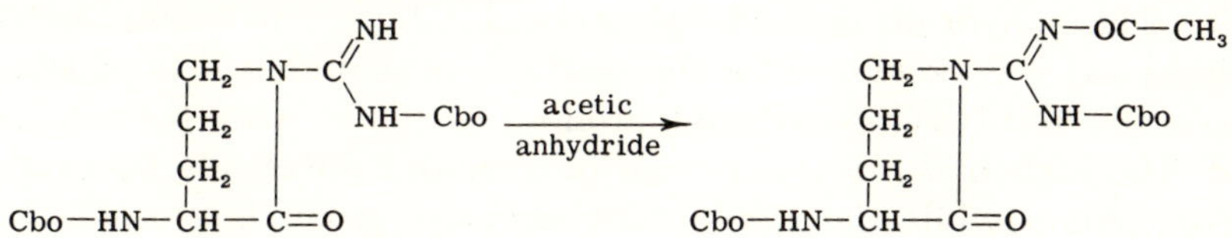

[54] [55]

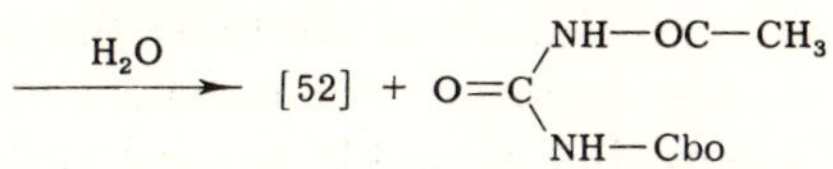

[56]

Zervas *et al.* (*2663*) used N^{α},N^{G},N^{G}-tricarbobenzoxy-L-arginine to synthesize several tricarbobenzoxyarginyl peptide benzyl esters via the anhydride and carbodiimide methods. The protecting groups can be removed by catalytic hydrogenation without difficulty. The use of the tricarbobenzoxy compound is more advantageous, since coupling of N^{α},N^{G}-dicarbobenzoxy-L-arginine with amino acid methyl esters via the acid chloride or carbodiimide method is accompanied by formation of N^{α},N^{G}-dicarbobenzoxy-L-arginine lactam. N^{α},N^{G}-Dicarbobenzoxy-L-arginine can be readily converted to N^{G}-carbobenzoxy-L-arginine methyl or benzyl ester via the *N*-carboxylic anhydride (*2670*). L-Arginyl-L-arginine (as diflavianate and tripicrolonate) was prepared for the first time from N^{α},N^{G},N^{G}-tricarbobenzoxy-L-arginine and N^{G}-carbobenzoxy-L-arginine benzyl ester via the anhydride method (*2662, 2663*).

The somewhat cumbersome synthesis of the carbobenzoxy derivatives described has so far limited a broader application of these compounds. N^{α},N^{G},N^{G}-Tricarbobenzoxy-L-arginine *p*-nitrophenyl ester (*1619*) is the only derivative that has been used for the synthesis of higher peptides by Nicolaides and De Wald (*1616*). Due to the lability to alkali of one N^{G}-carbobenzoxy group, the alkaline hydrolysis of tricarbobenzoxyarginyl peptide methyl esters leads to the formation of N^{α},N^{G}-dicarbobenzoxyarginyl peptide acids. The formation of N^{α},N^{G}-dicarbobenzoxy peptide methyl ester was observed during peptide synthesis (*1616*).

An interesting derivative of arginine was obtained with the synthesis of N^{α}-*tert*-butyloxycarbonyl-N^{G},N^{G}-dicarbobenzoxy-L-arginine (*866*). N^{G},N^{G}-Dicarbobenzoxy-L-arginine (*2670*) was prepared from N^{α},N^{G},N^{G}-tricarbobenzoxy-L-arginine via the *N*-carboxylic anhydride and used as starting material for this synthesis. The *tert*-butyloxycarbonyl group was introduced with the aid of *tert*-butyloxycarbonyl azide. The anhydride method (pivalyl chloride, triethylamine in tetrahydrofuran) was employed for the synthesis of peptides. It is possible to cleave the *tert*-butyloxycarbonyl group selectively with trifluoroacetic acid. The carbobenzoxy groups can be removed selectively by catalytic hydrogenation while the *tert*-butyloxycarbonyl group is present.

Gish and Carpenter (*804, 805*) have prepared N^{α},N^{G}-di-*p*-nitrocarbobenzoxy- and N^{α}-*p*-nitrocarbobenzoxy-L-arginine. The dicarbobenzoxy compound can be converted to N^{G}-*p*-nitrocarbobenzoxy-L-arginine via the *N*-carboxylic anhydride by treatment with thionyl chloride.

According to Guttmann (*901a,b*) the N^{G}-*p*-nitrocarbobenzoxy residue is more readily removed by catalytic hydrogenation than the N^{G}-carbobenzoxy group, but is stable toward trifluoroacetic acid and hydrogen bromide/acetic acid.

δ. *Use of* N^{G}*-2-(Isopropyloxycarbonyl)-3,4,5,6-tetrachlorobenzoylarginine.* Recently Guttmann (*901a,b*) reported on the N^{G}-2-(isopropyloxycarbonyl)-3,4,5,6-tetrachlorobenzoyl group, which is stable against hydrogenation as well as trifluoroacetic acid and hydrogen bromide/acetic acid. Therefore the carbobenzoxy and *tert*-butyloxycarbonyl groups are selectively removable. The new protecting groups are removed quantitatively in boiling acetic acid.

b. Syntheses with N^G-Unsubstituted Arginine

The investigations of several authors have demonstrated that arginine-containing peptides can be synthesized without masking the guanido function by an actual protecting group. The guanido group is far more basic than the amino group of an amino acid ester, thus allowing a protection during the condensation reaction by simply blocking the functional group by the addition of protons. The direct substitution of the guanido group is omitted more and more frequently, especially for the synthesis of higher peptides (*295, 1023, 1025, 1032, 1037, 1043, 1170, 1172, 1173, 2019, 2024*). The acid chloride (*805*) or anhydride method (*217*), and preferably the carbodiimide and azide methods, have been employed for the synthesis of peptides with arginine protected by proton addition. Arginine was present in these cases either in the carboxyl (*1170, 2020, 2024*) or in the amino component (*291, 295, 1025, 1032, 1037, 1043, 1170, 1172, 2019, 2020*). The guanido group of arginine can be used to liberate the amino group of the amino component (*291, 295*). Anderson (*42*) has carried out a peptide synthesis with tetraethyl pyrophosphite in diethyl phosphite, using the hydrobromide of N^{α}-carbobenzoxy-L-arginine. Chlorophosphite in diethyl phosphite was also used for coupling reactions (*2611*).

The acid chloride of N^{α}-*p*-nitrocarbobenzoxy-L-arginine can be used for the synthesis of peptides. The compound is not very stable and yields amount to only about 50% (*805, 806*). Other derivatives that are only N^{α}-protected are trityl-L-arginine (*291*), tosyl-L-arginine (*185, 2043, 2654*), and trifluoroacetyl-L-arginine (*1923, 2480*).

The dihydrochloride or the dihydrobromide of arginine methyl ester is used for the synthesis of peptide derivatives of arginine where the amino acid is not N^G-substituted (*291*). The carbodiimide method was applied after the hydrogen chloride had been removed from the α-amino group with, for example, triethylamine or tri-*n*-butylamine. Zahn and Diehl (*2616*) synthesized the hydrochloride of L-arginine benzyl ester and coupled it via the azide method in ethanol or triethylamine/ethyl acetate, or via the carbodiimide method in acetonitrile.

The conversion of ornithine-containing peptides to peptides of arginine by reacting them with S-methylisothiuronium iodide is not of great interest. N^{α}-Tos-L-Arg-Gly-OH was prepared via this method (*111*). Nitro-L-arginine can be prepared from ornithine and 2-methyl-1-nitroisourea (*993;* cf. *772*). Bodanszky *et al.* (*277b*) synthesized Arg^8-vasopressin, Arg^8-vasotocin, and H-His-Phe-Arg-Try-Gly-OH from the corresponding Orn-derivatives by guanidation with 1-guanyl-3,5-dimethylpyrazole.

6. Citrulline

Citrulline (δ-carbamylornithine) is closely related to arginine. This amino acid has not yet been found in biologically active polypeptides. Izumiya and Shito (*1120*) have prepared carbobenzoxy-L-citrulline and coupled it via the carbodiimide method with a series of L-amino acid benzyl esters. The free citrullyl peptides were obtained by catalytic hydrogenation. Bodanszky *et al.* (*270, 278*) and Ondetti (*1663*) used carbobenzoxy-L-citrulline *p*-nitrophenyl ester for the synthesis of citrulline containing analogues of peptide hormones (oxytocin, vasopressin, bradykinin, and partial sequences of α-MSH). The carbobenzoxy group was removed without complication by treatment with hydrogen bromide in glacial acetic acid.

7. Norarginine and Norcitrulline

Rudinger *et al.* (*1858*) have recently synthesized the lower homologues of arginine and citrulline, norarginine [57] (γ-guanido-α-aminobutyric acid) and norcitrulline [58] (γ-ureido-α-aminobutyric acid). Hofmann degradation of tosyl-L-glutamine yields N^{α}-tosyl-L-α,γ-diaminobutyric acid (see Chapter IV,B,3,a), which can be reacted with *O*-methylisourea under alkaline conditions to form γ-guanido-N^{α}-tosyl-L-α-aminobutyric acid, while the reaction with potassium cyanate in aqueous solution at 50°–60°C leads to the formation of γ-ureido-N^{α}-tosyl-L-α-aminobutyric acid. Cleavage of the N^{α}-tosyl group and conversion

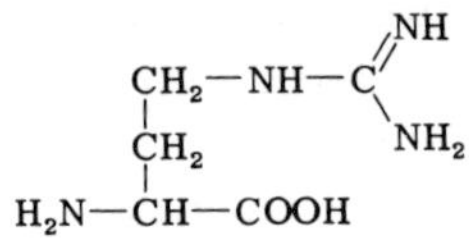

[57]

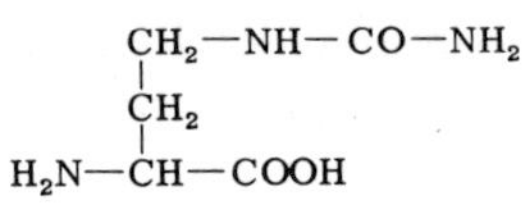

[58]

to norarginine or norcitrulline can be accomplished by treatment with 35% hydrogen bromide in glacial acetic acid in the presence of phenol (2.5 hours at 75°C or 36 hours at 40°C). Norarginine can be obtained by benzoylation of N^{γ}-carbobenzoxy-L-α,γ-diaminobutyric acid, decarbobenzoxylation with hydrogen bromide in glacial acetic acid, reaction of the resulting N^{α}-benzoyl-L-α,γ-diaminobutyric acid with S-methylisothiourea, and subsequent acid hydrolysis of the γ-guanido-N^{α}-benzoyl-L-α-aminobutyric acid.

8. Histidine

Histidine is an important constituent of biologically active polypeptides and part of the active center of many hydrolytic enzymes.

Photooxidation of the histidine residue in peptide hormones (e.g. glucagon) or enzymes (e.g. ribonuclease, trypsin, or lysozyme) results in the loss of hormonal or enzymatic activity. The weakly basic imidazole ring in the side chain of histidine poses a number of problems for the synthesis of histidine-containing peptides. This may well have been the reason that until 1953 only a few DL-histidyl peptides (*191, 722*) and histidine peptides (*201, 636, 719, 1082, 1083, 2119*; cf. *772*) were known.

a. SYNTHESES WITH N^{im}-UNSUBSTITUTED HISTIDINE

Carbobenzoxy-L-histidine hydrazide (*1052*) is the derivative of choice (*1022, 1950, 1951, 2441*) for the synthesis of histidyl peptides. Due to the basic nature of the imidazole ring, carbobenzoxy-L-histidine azide can be extracted with organic solvents only after the addition of 50% potassium carbonate (pH 9) to its solution in hydrochloric acid. Potassium hydrogen carbonate (pH 7) is not sufficient for the precipitation of the azide (*724*). Chloroform/triethylamine can also be used for the extraction (*572*). Yields of coupling reactions vary widely (25–82%).

Carbobenzoxy-L-histidine had already been prepared by Bergmann and Zervas (*193*), although presumably racemized. A somewhat modified procedure for its preparation was recently reported by Patchornik *et al.* (*1695*). This substance has not been used for the synthesis of peptides. The phthalyl and tosyl residues have also been used as protection of the amino group (*978*). Phthalylhistidine, however, can be synthesized only over a number of steps from phthalic anhydride and L-histidine methyl ester. The resulting *o*-carboxybenzoyl-L-histidine methyl ester is hydrolyzed under alkaline conditions and *o*-carboxybenzoyl-L-histidine cyclized by drying over phosphorus pentoxide at 110°C (cf. page 10). Ring closure with thionyl chloride yields only a racemized product. Tosyl-L-histidine is accessible by alkaline hydrolysis of N^{α},N^{im}-ditosyl-L-histidine (*978*). The phosphazo and carbodiimide methods were applied to the synthesis of peptides. Sheehan *et al.* (*2067*) described N^{α}-*p*-nitrocarbobenzoxy-L-histidine. Treatment with *N*,*N'*-diisopropylcarbodiimide led to the formation of the lactam with the imidazole ring, N^{β}-(*p*-nitrobenzyloxycarbonyl)-β, *N'*-carbonylhistamine [59]. The lactam

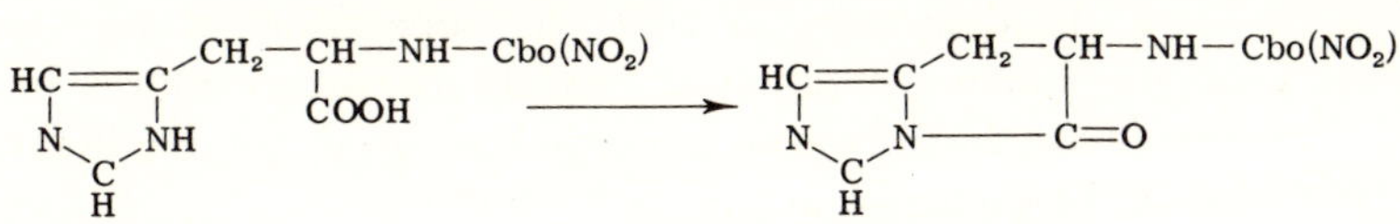

[59]

was opened by aminolytic cleavage (see Chapter III,B,4,a). Peptide syntheses with N^{α}-formyl-L-histidine (*719*) and N^{α}-trifluoroacetyl-L-histidine (*2498*) have not been reported.

L-Histidine methyl ester dihydrochloride is used almost exclusively as starting material for the synthesis of histidine peptides. The compound can be prepared by esterification in methanol/hydrochloric acid (*714, 719*) or methanol/sulfuric acid (*572*), best, however, by treatment of histidine with methanol/thionyl chloride (*348, 349*). The formation of the free ester from the hydrochloride, in contrast to other amino acid esters, is not complete upon treatment with triethylamine in tetrahydrofuran or dimethylformamide. Chloroform/triethylamine, chloroform/ammonia (*95, 1007*), or methanol/sodium methylate (*719*) is more suitable for this purpose. The azide (*201, 724, 1538, 1950, 1951, 2119*), acid chloride (*636, 1082, 1083*), and carbodiimide (*1459, 1535, 1655, 1950*) methods were used for coupling reactions. The possible formation of carbobenzoxyacylureas was stressed (cf. *978*). Mixed anhydrides (*1950, 1951, 2298*) and activated esters [*p*-nitrophenyl ester (*1951*) and cyanomethyl ester (*1950*)] have been employed less frequently for the synthesis of peptides. The alkaline hydrolysis or the conversion to hydrazides of histidyl and histidine peptide esters with the unsubstituted imidazole ring proceeds without complication (*724, 1050, 1051, 1052, 2441*). The carbobenzoxy group was always removed by catalytic hydrogenation.

The phthalyl group together with the azide method (*724*) and the acid chloride method in combination with amino acid salts (*1459, 2339*) have been applied to the synthesis of histidine peptides in addition to the carbobenzoxy group. Akabori *et al.* (*28*) have synthesized L-histidine benzyl ester di-*p*-toluenesulfonate and coupled it with several carbobenzoxyamino acids to form carbobenzoxy dipeptide benzyl esters, applying the azide, carbodiimide, and anhydride methods (anhydride formation with isovaleryl chloride). L-Histidine *p*-nitrobenzyl ester dibenzenesulfonate was described by Shields *et al.* (*2103*), but has not yet been used for the synthesis of peptides. Sometimes the formation of a by-product can be observed during the synthesis of peptides via the azide method. This by-product is the result of an intramolecular reaction with the imidazole ring and no longer gives the Pauly reaction (*1538, 1950, 1951*). The application of unsubstituted histidine frequently proves to be a disadvantage, in that the separation of the reaction product from the starting components by standard procedures becomes more difficult. Free histidine is nevertheless used predominantly for the synthesis of biologically active polypeptides.

b. SYNTHESES WITH N^{im}-SUBSTITUTED HISTIDINE

du Vigneaud and Behrens (*627*) had already synthesized *im*-benzyl-L-histidine [60] [1(or 3)-benzyl-L-histidine] by reacting L-histidine with

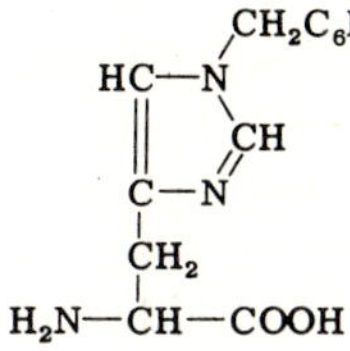

[60]

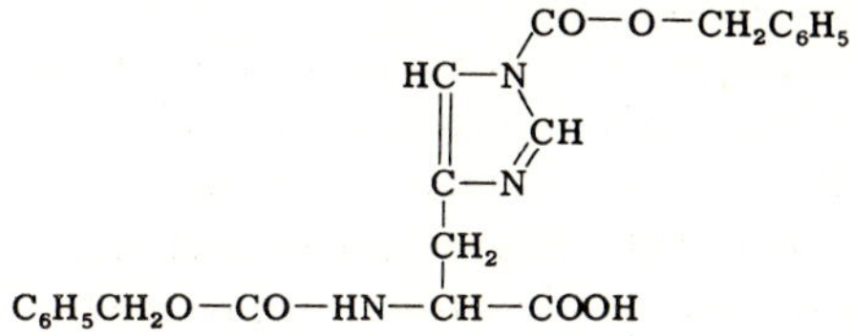

[61]

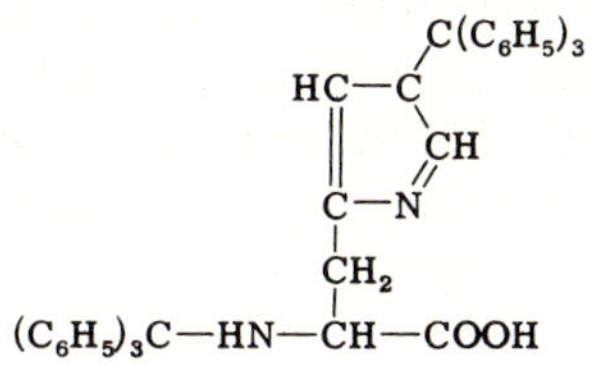

[62]

benzyl chloride in the presence of sodium in liquid ammonia. This derivative of histidine was used to synthesize carbobenzoxy-*im*-benzyl-L-histidine (*275, 1672, 2300, 2570*), *im*-benzyl-L-histidine methyl ester dihydrochloride (*1539, 2291, 2300*), and the di-*p*-toluenesulfonate (*2293*) and dibenzenesulfonate (*2300*) of the benzyl ester.

Histidyl peptides were synthesized via the carbodiimide (*1459, 2300, 2302*) or *o*-phenylene chlorophosphite method (*360*). The application of the phosphazo method (*360*) was less successful. The anhydride (*2300*) and carbodiimide (*360, 2300, 2302*) methods were employed to synthesize histidine peptides. Carbobenzoxy-*im*-benzyl-L-histidine is poorly soluble in organic solvents. Therefore the reaction with *N,N'*-dicyclohexylcarbodiimide is carried out in a large volume of dimethylformamide (*2300*), methylene chloride (*2293*), or dimethylformamide/diethyl phosphite (*1972*) at room temperature, or in a large volume of pyridine at −10°C (*1459*). Carbobenzoxy dipeptide methyl esters with amino- or carboxyl-terminal *im*-benzylhistidine can be readily converted to hydrazides (*1539, 1972, 2300*). The carbobenzoxy group can be removed selectively with hydrogen bromide in glacial acetic acid (*360, 2302*), or by catalytic hydrogenation in anhydrous ethanol (*2300*). The carbobenzoxy group and the *im*-benzyl residues will be removed simultaneously if the time for the hydrogenation in 50% ethanol is extended, or if the hydrogenation is carried out at 50°–60°C in 80% acetic acid or glacial acetic acid (*360, 1459, 2293, 2300*), or if sodium in liquid ammonia is used for the reductive cleavage (*627, 2119, 2300*). The trityl group served as additional protection (*1344, 2300*) in *im*-benzyl-L-histidine peptides where the anhydride and carbodiimide methods afforded the coupling. Tosyl- (*627*) and *p*-(*p'*-methoxyphenylazo)-carbobenzoxy-*im*-benzyl-L-histidine (*2037*) have been synthesized.

im-Benzylhistidine has been used for the synthesis of several angiotensin (*360, 2302, 2593*) and ACTH sequences (*1170, 1413, 1415, 1943*).

Patchornik *et al.* (*1695*) synthesized N^{α},N^{im}-dicarbobenzoxy-L-histidine [61] by carbobenzoxylation of L-histidine in sodium hydroxide at pH 9 to 10. It is better to prepare the compound in sodium hydrogen carbonate solution (*26, 1903*). The derivative can be applied to the synthesis of peptides via the anhydride or *p*-nitrophenyl ester method, best, however, via the carbodiimide method (*1093, 1902, 2048a*). The carbobenzoxy group attached to the imidazole ring behaves like an activated acyl group (see Chapter III,B,4,a) and is cleaved at higher temperatures by aniline, ammonia, or sodium methylate to form the corresponding benzylurethans (*1695*). N^{α},N^{im}-Dicarbobenzoxy-L-histidine is unstable if stored over long periods of time at room temperature (*1902*). The N^{im}-carbobenzoxy group is unstable to alkali. Thus, the peptide acids obtained on hydrolysis of histidine-containing N^{α},N^{im}-dicarbobenzoxy peptide esters are protected only at the amino group. The reaction consumes 3 equivalents of alkali (*1902*). The N^{α}-carbobenzoxy group can be cleaved with hydrogen bromide in glacial acetic acid or dioxane under standard conditions (*671, 1092, 1093*). The dihydrobromide of N^{im}-carbobenzoxy-L-histidine methyl ester becomes accessible by this procedure after dicarbobenzoxy-L-histidine has been esterified with diazomethane (*1695*). The monocarbobenzoxy derivative can be coupled via, for example, the carbodiimide method (*1092*). The N^{im}-carbobenzoxy group and the N^{α}-substituent may be removed simultaneously by catalytic hydrogenation (*1092*) or by heating to 37°C for 50 minutes in concentrated hydrochloric acid (*1903*). Several hours of reaction time are required for the cleavage with hydrogen bromide in glacial acetic acid (*671*). It is surprising that this protecting group can be removed selectively by brief heating with glacial acetic acid (*671*).

A third possibility for the synthesis of histidyl peptides is offered by the application of N^{α},N^{im}-ditrityl-L-histidine [62]. Amiard *et al.* (*37*) prepared the compound by tritylation of L-histidine methyl ester dihydrochloride and subsequent hydrolysis. Later, Stelakatos *et al.* (*2203*) were able to isolate a small amount of N^{α}-trityl-L-histidine as by-product of the tritylation of histidine in water/isopropanol/diethylamine. N^{im}-Trityl-L-histidine methyl ester hydrochloride is accessible by selective detritylation (*2203*). Only N^{α},N^{im}-ditrityl-L-histidine has been used for the synthesis of peptides. Several peptides were prepared via the carbodiimide method in methylene chloride (*37*), pyridine/acetonitrile (*291*), or dichloromethane (*1459*) at temperatures of 0 to —10°C. [Cf. (*2048a*) on the cleavage of histidyl peptide bonds with *N*-bromosuccinimide.]

C. ACIDIC AMINO ACIDS

1. Glutamic Acid

Glutamic acid is very widely distributed in animals, plants, and bacteria. It is incorporated in proteins and peptides most commonly via an α-peptide bond, while the γ-carboxyl group is free (glutamyl, e.g. α-MSH, β-MSH, and gastrin) or is present as amide (glutaminyl, e.g. oxytocin, vasopressin, glucagon, etc.). Naturally occurring glutamyl γ-peptides are also known: glutathione, ophthalmic acid, norophthalmic acid, hypoglycine B (*1147, 1148*), and pteroylglutamic acid (conjugates of folic acid, cf. *57, 303, 304, 1582, 1583, 2041* for syntheses). Other glutamyl γ-peptides have been isolated from vegetable sources (*1330, 1576, 1819, 2379, 2381*). The capsular substance of *Bacillus anthracis* (anthrax polypeptides) and polypeptides isolated from culture media of various strains of *Bacillus subtilis* (subtilis polypeptides) are of special interest. These polypeptides have been identified as a poly-γ-D-glutamic acid and poly-γ-glutamic acid of mixed configuration (up to 14% of the L-form) (cf. *420, 1292*). Eisenin, pelvetin, eledoisin, physalaemin, and gastrin are naturally occurring polypeptides that contain pyroglutamic acid.

a. Glutamyl Peptides

Glutamyl α- or γ-peptides can be prepared from intramolecular cyclic derivatives of glutamic acid, and from derivatives of *N*-protected glutamic acid monosubstituted at the α- or γ-carboxyl group. Derivatives of *N*-protected glutamic acid are capable of forming intramolecular anhydrides [63] or intramolecular diacyl imides [64]. They may also cyclize with formation of γ-lactams [65], i.e. derivatives of pyroglutamic acid

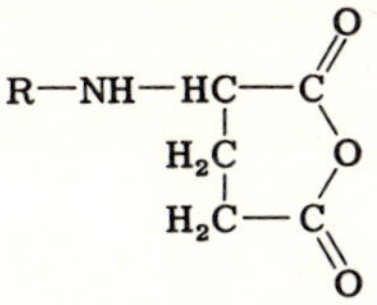

[63]

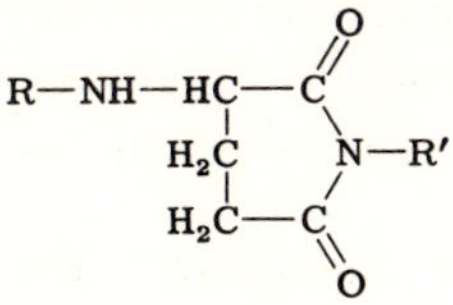

[64]

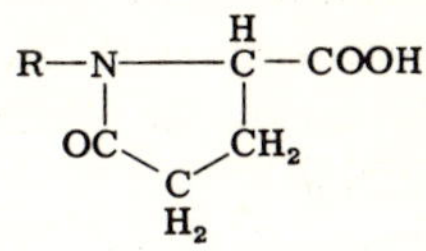

[65]

(pyrrolid-5-one-2-carboxylic acid). Glutamyl α- or γ-peptides and derivatives selectively substituted at the α- or γ-carboxyl group are directly accessible from intramolecular anhydrides and from pyroglutamic acid.

α. *N-Acylglutamic Acid Anhydrides.* *N*-Acylglutamic acid anhydrides are cleaved by aminolysis, ammonolysis, hydrazinolysis, and alcoholysis. These reactions lead to the formation of α- and γ-derivatives which cannot be separated satisfactorily in each case (cf. *2236*).

α_1. *Carbobenzoxy and other urethan-protected glutamic acid anhydrides.* Bergmann and Zervas (*193*) first described carbobenzoxy-L-glutamic acid anhydride, which was prepared by heating a mixture of carbobenzoxy-L-glutamic acid and acetic anhydride. Le Quesne and Young (*1369*) recommend that the reaction be carried out at room temperature to reduce the danger of racemization. *N,N'*-Dicyclohexylcarbodiimide or ethoxyacetylene may also be used for the ring closure reaction (*799*). Aminolysis will open the anhydride with formation of the α-derivative (*193, 206*). Carfully conducted investigations demonstrated the simultaneous formation of the γ-compound (*199, 1533*). According to Le Quesne and Young (*1369*), the opening of the anhydride ring is governed by the type of cleaving agent and the solvent used. The γ-isomers formed as by-products during the reaction with ammonia or amino acid esters have different p*K* values and can be removed by fractional extraction with sodium hydrogen carbonate solution. Although many glutamyl α-peptides were prepared by this procedure (*198, 638, 847, 938, 1424, 1844, 2041, 2095, 2100, 2620*), its application is strongly restricted by the low yields, the elaborate preparative efforts, and the uncertainty about the ratio in which the two isomers may form. In addition to the carbobenzoxy derivative, *p*-chlorocarbobenzoxy- (*1246*), *p*-methoxycarbobenzoxy- (*2488*), *p*-nitrocarbobenzoxy- (*478*), and *tert*-butyloxycarbonyl-L-glutamic acid (*489a, 1975*) have been converted to the corresponding intramolecular anhydrides by treatment with acetic anhydride or *N,N'*-dicyclohexylcarbodiimide. These derivatives are used for the preparation of selectively substituted *N*-protected derivatives of glutamic acid.

Upon catalytic hydrogenation, carbobenzoxyglutamic acid anhydride will be rearranged via a carbamic acid intermediate to the *N*-carboxylic acid anhydride. This product is decomposed by water to form pyroglutamic acid. Decarbobenzoxylation of the intramolecular anhydride by treatment with hydrogen bromide in glacial acetic acid leads to the formation of glutamic acid, pyroglutamic acid, and acetic anhydride (*1293b*).

α_2. *Phthalylglutamic acid anhydride.* King and Kidd (*1241;* cf. *242, 1227, 2314*) found that only the DL-compound is formed from phthalic anhydride and L-glutamic acid. The optically active derivative may be prepared from phthalic anhydride and L-glutamic acid diethyl ester, ring closure of the *o*-carboxybenzoyl-L-glutamic acid diethyl ester, and subsequent hydrolysis (cf. *308*). The anhydride formation with acetic anhydride proceeds with preservation of the optical activity. Phthalyl-L-glutamic acid can be prepared by a very simple procedure that uses carbethoxyphthalimide (*1600*). Experiments by King *et al.* (*1234, 1237, 1238, 1240, 1241*) have shown that the aminolysis and alcoholysis of phthalyl-L-glutamic acid anhydride lead to the formation of γ-derivatives with preservation of the optical activity. These results were confirmed by several authors (*496, 933, 2563*). Only the reaction with sodium methylate in methanol yields a mixture of α- and γ-esters (*2059, 2314*).

α_3. *Trifluoroacetylglutamic acid anhydride.* Weygand and Reiher (*2496*) found that *N*-trifluoroacetyl-L-glutamic acid anhydride will be formed directly upon treatment of L-glutamic acid with trifluoroacetic anhydride. The aminolytic cleavage was studied with aniline and the formation of a mixture of the α- and γ-derivatives demonstrated. Ring opening with ethanol yielded the α- and γ-ethyl esters in a ratio of 2:1. The esters were separated via the dicyclohexylammonium salts (*2481*).

α_4. *Phenylthiocarbonylglutamic acid anhydride.* Kollonitsch *et al.* (*1273*) cleaved phenylthiocarbonyl-L-glutamic acid anhydride with amino acid esters in chloroform. The α-derivative were obtained in 50% yield. No statement was made on the extent of isomer formation.

β. Glutamic Acid γ-Esters and Glutamyl α-Peptides. The γ-ethyl, γ-methyl, γ-benzyl, and γ-*tert*-butyl esters are important for the synthesis of glutamyl α-peptides. Abderhalden and Nienburg (*6*) had already prepared L-glutamic acid γ-ethyl ester by partial esterification with ethanolic hydrochloric acid. This preparative technique was later modified in various ways (*195, 510, 931, 972, 1560, 1769, 1843;* cf. *2322*). L-Glutamic acid γ-methyl ester can be prepared by the analogous reaction with methanolic hydrochloric acid (*510, 1293, 2415*) or with thionyl chloride (*296*) or acetyl chloride (*834, 931*) in methanol. The γ-carboxyl group may also be esterified by transesterification with methyl acetate in the presence of perchloric acid (*241, 2273*). The synthesis of glutamic acid γ-benzyl ester by partial esterification with hydroiodic acid (*931*), hydrogen bromide (*259*), *p*-toluenesulfonic acid (*498*), or benzenesulfonic acid (*418, 428*) proceeds with yields of only 30–45%. Guttmann and Boissonnas (*894*) obtained good results with 98% sufuric acid (cf. *967*). It is also possible to synthesize the γ-monoester by partial alkaline hydrolysis of carbobenzoxy-L-glutamic acid dibenzyl ester (*931*). L-Glutamic acid γ-*tert*-butyl ester was prepared by transesterification with *tert*-butyl acetate (*2266*). Esterifications of carbobenzoxy-L-glutamic acid α-benzyl (*2007*), α-ethyl (*2007*), and α-methyl (*1260*) esters with isobutylene/sulfuric acid have also been reported. Various derivatives, as they are required, can be obtained by catalytic hydrogenation, alkaline hydrolysis, or reaction with hydrazine hydrate (*1260, 2019*). The carbobenzoxy-L-glutamic acid γ-ethyl (*6, 931, 972, 2415*), γ-methyl (*296, 931*), and γ-benzyl (*931, 996*) esters are accessible without difficulty by the reaction of carbobenzoxy chloride with L-glutamic acid γ-monoesters. Losse *et al.* (*1455a, 1456, 1457*) prepared derivatives activated at the α-carboxyl group, e.g. α-phenyl thiolester and α-*p*-nitrophenyl ester from carbobenzoxy-L-glutamic acid γ-methyl ester and γ-benzyl ester. Beside the carbobenzoxylation, the introduction of the *p*-phenylazocarbobenzoxy (*2700*), *p*-(*p'*-methoxy-phenylazo)-carbobenzoxy (*2037*), *tert*-butyloxycarbonyl (*1975*), formyl (*306*), and trityl (*490*) groups have been described.

A basically different method for the preparation of *N*-substituted glutamic acid γ-esters is offered by the alcoholysis of the corresponding intramolecular anhydrides. As in the case of the aminolysis, the way in which the ring is opened is dependent upon the type of protecting group used. The reaction of carbobenzoxy-L-glutamic acid anhydride with benzyl alcohol yields only 20–40% of the γ-ester. Losse *et al.* (*1456*) separated it from the α-ester by fractional extraction with sodium hydrogen carbonate, Weygand and Hunger (*2488*) by fractional crystallization of the dicyclohexylammonium salts. The isolation of the γ-phenyl ester, resulting from the reaction with phenol, can also be achieved via the dicyclohexylammonium salt (*1260*). The alcoholysis of the carbobenzoxy-L-glutamic acid anhydride, however, is applied to the synthesis of α-esters. The phthalyl-L-glutamic acid γ-methyl and γ-benzyl esters (*1238*), the tosyl-L-glutamic acid γ-methyl ester (*1846*), and the trifluoroacetyl-L-glutamic acid γ-ethyl ester (*2481*) have found little preparative application. The trityl-L-glutamic acid γ-methyl ester was prepared by alkaline hydrolysis of trityl-L-glutamic acid dibenzyl ester with sodium methylate in methanol. Transesterification occurs simultaneously and afterward the α-benzyl ester group is cleaved selectively by hydrogenolysis (*38*).

N-Substituted glutamic acid γ-esters were used for the synthesis of glutamyl α-peptides. Various combinations of amino- and carboxyl-protecting groups were employed and the acid chloride (*638, 1560*), carbodiimide (*2322*), anhydride (*590, 834, 979, 1286a, 1425, 1521, 1605, 1888, 1930, 1950, 2322*), phosphazo (*979*), and *p*-nitrophenyl ester (*830, 834, 996, 1414*) methods applied. Carbobenzoxy-L-glutamic acid γ-esters were used most frequently. The γ-benzyl ester was preferred since the γ-methyl and γ-ethyl esters cannot be hydrolyzed under alkaline conditions without the danger of $\alpha \rightarrow \gamma$ transpeptidation (*1425, 1930*). The γ-benzyl ester group was introduced with benzyl alcohol/ hydrochloric acid subsequent to the synthesis of a peptide with carbobenzoxy-L-glutamic acid anhydride (*2620*). The carbobenzoxy group has been removed by treatment with hydrogen bromide in glacial acetic acid. The introduction of the carbobenzoxy-L-glutamic acid γ-*tert*-butyl ester (*1260, 2019*) brought a considerable improvement. The synthesis of long chains of naturally occurring peptides with free γ-carboxyl groups is now possible under very mild preparative conditions.

The azide method found application in the form of carbobenzoxy-L-glutamic acid α-hydrazide while the γ-carboxyl group was unprotected. The α-hydrazide is accessible by hydrazinolysis of the α-methyl or α-ethyl ester (*1370, 2173*). The direct cleavage of the intramolecular anhydride with hydrazine leads to the formation of a mixture of the α- and γ-isomers that cannot be separated (*1369*). The α-hydrazides were used to prepare various glutamyl α-peptides (*1370, 2173*). The isomeric γ-compounds are formed as by-products. Glutamyl α-peptides can be synthesized via the oxazolidone method while the γ-carboxyl group is unprotected (*1547*).

Goodman *et al.* (*830–835*) have used glutamyl α-peptides of the type Cbo-Glu(OMe)-[Glu(OMe)]$_n$-Glu(OEt)-OEt ($n = 0$–9) for investigations on the tertiary structure of polypeptides in relation to the chain length (cf. *615*).

γ. *Glutamic Acid α-Ester and Glutamyl γ-Peptides.* The safest approach to the synthesis of unequivocal glutamyl γ-peptides is the application of *N*-substituted glutamic acid α-esters. The alcoholysis of carbobenzoxy-L-glutamic acid anhydride yields predominantly α-esters. Methyl (*639a, 937*), ethyl (*815, 1370, 1425, 2486*), benzyl (*205, 1370, 1425, 1456, 2488*), and *p*-nitrobenzyl (*1260*) esters were used without special purification for the synthesis of peptides at one time. Craig distribution (*1845*), fractional extraction (*815, 1370, 1456*), or, better, fractional crystallization of the dicyclohexylammonium salts (*1260, 2486, 2488*) allows separation of the γ-esters that are always present. Activated esters (*1259, 1260, 1893, 2559*) can be prepared by cleavage with phenol, thiophenol, or mercaptoethanol. Wieland and Weidenmüller (*2559*) have shown that the reaction of carbobenzoxy-L-glutamic acid anhydride with thiophenol in a weakly polar solvent, such as tetrahydrofuran in the presence of a weak base, yields as much or more of the α-compound as there is γ-compound formed. The application of triethylamine causes here, and especially in the case of phthalyl-L-glutamic acid anhydride, the formation of predominantly the γ-ester without racemization. The reaction in strongly polar solvents in the presence of triethylamine is accompanied by racemization and leads almost exclusively to the formation of γ-ester. The opening of the ring of carbobenzoxy-L-glutamic acid anhydride with mercaptoethanol in the presence of dicyclohexylamine leads to the formation of the dicyclohexylammonium salt of carbobenzoxy-L-glutamic acid α-ethyl thiolester in more than 70% yield. The ethyl thiolester acts as protecting group and provides activation (*1259*). In a next step the substitution at the γ-carboxyl group may be carried out. Hydrazinolysis will yield the α-hydrazide and open the way for the formation of an α,γ-bispeptide. Isoglutamyl peptides are accessible by ammonolysis. Carbobenzoxy-L-glutamic acid α-benzyl ester γ-*p*-nitrophenyl ester or γ-phenyl thiolester are also suitable derivatives (*1456, 1457*). Various α-esters of *p*-chlorocarbobenzoxy- (*1246*), *p*-methoxycarbobenzoxy- (*1975, 2488*), and *tert*-butyloxycarbonyl-L-glutamic acid (*1975*) were obtained by alcoholysis of the intramolecular anhydride.

Sachs and Brand (*1889*) have reported another synthesis of *N*-substituted α-benzyl esters. Glutamic acid dibenzyl ester can be converted to the mono-α-ester by treatment with hydroiodic acid in glacial acetic acid. The cleavage reaction is followed by *N*-substitution. Carbobenzoxy-L-glutamic acid α-*tert*-butyl ester was synthesized by esterification of carbobenzoxy-L-glutamic acid γ-methyl ester with *tert*-butyl acetate, followed by alkaline hydrolysis (*2281, 2283*). Phthalylglutamic acid α-esters are accessible by esterification of phthalylglutamic acid γ-benzyl ester and catalytic hydrogenation (*1238*) of the γ-benzyl ester group, or by opening the ring of phthalyl-L-glutamic acid anhydride with methanol/sodium methylate (*597*). In contrast to the carbobenzoxy compound, the γ-esters is formed preferentially if the cleavage is carried out with methanol alone. Tritylglutamic acid α-ethyl ester (*38, 2668*), α-methyl ester (*38*), and α-benzyl ester (*531, 2716*) are easily accessible by alkaline hydrolysis of the corresponding diesters. The α-ester group is protected by steric hindrance from attack by the base. Trifluoroacetyl-L-glutamic acid α-ethyl ester is formed via the dicyclohexylammonium salt in addition to the γ-ester upon alcoholysis of the intramolecular anhydride (*2481, 2496*). Jöhl and Stoll (*1148*) used this compound for the synthesis of hypoglycine B (L-glutamyl-γ-hypoglycine A). Hypoglycine B was isolated from unripe fruits of *Blighia sapida* (cf. *1147*). Tosyl-L-glutamic acid α-*tert*-butyl ester can be prepared by alkaline hydrolysis of tosyl-L-pyroglutamic acid *tert*-butyl ester (*2281, 2283*). Selective esterification of the α-carboxyl group (free of the corresponding γ-esters) is possible by reaction of carbobenzoxy-, tosyl-, and phthalyl-L-glutamic acid with alkyl halides or dialkyl sulfates in the presence of an equimolecular amount of triethylamine or dicyclohexylamine (*1598a*).

Derivatives of glutamic acid carrying substituents at the α-carboxyl group were used for the synthesis of glutathione and of folic acid conjugates. Various coupling methods were employed to synthesize glutamyl γ-peptides. Derivatives with the carbobenzoxy group were coupled via the acid chloride (*303, 639a, 937, 938, 1861, 2173*), carbodiimide (*2488*), phosphazo (*815, 816, 1425*), anhydride (*450, 1425, 1888, 1890, 2283, 2488*), and pyrazolide (*1815*) methods. *p*-Methoxycarbobenzoxy-L-glutamic acid α-benzyl ester was coupled via the anhydride and carbodiimide methods (*2488*), and *p*-chlorocarbobenzoxy-L-glutamic acid α-methyl ester via the carbodiimide method (*1246*). Trifluoroacetyl- and phthalylglutamic acid α-esters were coupled via the γ-chlorides or via the anhydride method (*1148, 1238, 1930, 2481*). Phthalyl-L-glutamic acid γ-phenyl thiolester is easily accessible and also applicable (*2559*). Tritylglutamic acid α-esters were activated with *N,N′*-dicyclohexylcarbodiimide for the synthesis of peptides (*38*). Attempts to couple tosyl-L-glutamic acid α-*tert*-butyl ester with amino acid esters via the anhydride or carbodiimide method did not lead to the formation of the desired peptide derivatives. Another reaction, ring closure to tosyl-L-pyroglutamic acid *tert*-butyl ester, took place (*2281, 2283*). The sulfonamide nitrogen is apparently activated by the *tert*-butyl ester group, thus favoring the formation of the lactam. The ring closure reaction was also observed with the phosphazo and azide methods, but here the expected γ-peptide was also formed. This is remarkable, since tosylglutamic acid γ-hydrazide (*2559*) can be used for the synthesis of peptides while the α-carboxyl group is not protected (*786, 1930, 2347*). In the latter case, however, the formation of structural isomers was also observed (*786*).

While certain activation methods require the protection of the second carboxyl function, a large number of glutamyl γ-peptides were prepared via the azide method and the α-carboxyl group remained unprotected. Carbobenzoxy-L-glutamic acid γ-hydrazide (*734, 972, 1370, 2173*) was applied to the synthesis of peptides by a number of authors (*427, 734, 972, 1368, 1370, 1843, 1844, 2095, 2173, 2415*). Sachs and Brand (*1888, 1891*) have shown that the coupling reaction does not always proceed unequivocally; a mixture of the α- and γ-peptides may be formed. The formation of a cyclic intermediate [66] was discussed; it can be opened in both ways by aminolysis. These findings were confirmed by Rowland

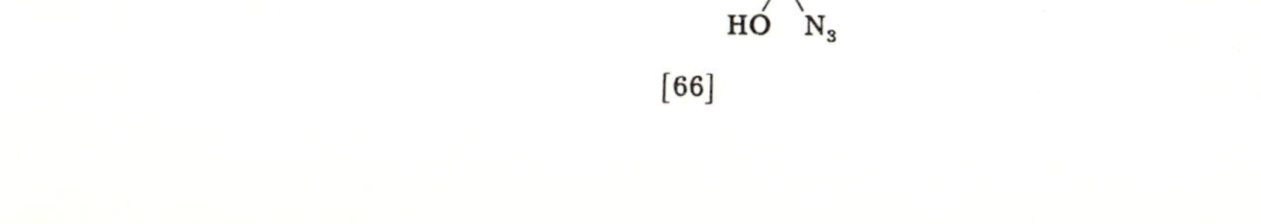

[66]

and Young (*1844*) who pointed out that it is possible to separate the γ-isomer if the reaction products are crystalline; thus, the application of the method remains entirely feasible. The formation of structural isomers was also observed by Bruckner *et al.* (*417, 428, 431;* cf. *418, 420*). The formation of isomers can be prevented by esterification with diazomethane (*2173;* cf. *1424*) or ethanolic hydrochloric acid (*651*) of the α-carboxyl group of the γ-hydrazide. Taschner *et al.* (*2281, 2283*), found it more advantageous to esterify with *tert*-butyl acetate to form carbobenzoxy-L-glutamic acid α-*tert*-butyl ester γ-hydrazide. In addition to carbobenzoxy-L-glutamic acid γ-hydrazide, the corresponding tosyl compound (*786, 1930, 2347*) and the carbobenzoxy- and phthalyl-DL-glutamic acid γ-phenyl thiolesters (*2509*) have been used for the synthesis of peptides while the α-carboxyl group was not protected. Amiard *et al.* (*39*) have found that the α-carboxyl group in trityl-L-glutamic acid is so strongly hindered that, upon reaction with a peptide ester and *N,N'*-dicyclohexylcarbodiimide, only the γ-derivative is formed. Morris (*1577*) found, however, that the α-carboxyl group is not sufficiently hindered and that the α-isomer is formed in addition to the desired γ-peptide.

δ. *α → γ Transpeptidation.* Certain reactions during the synthesis of glutamyl α- and γ-peptides are accompanied by side reactions. Alkaline

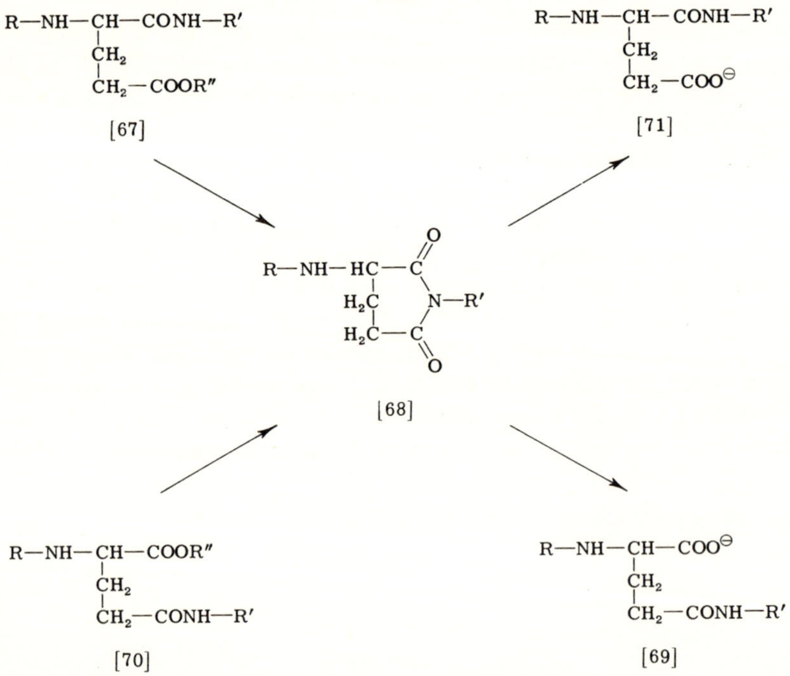

hydrolysis must be mentioned first. Intermediate imide formation is responsible for the appearance of the α- and γ-isomers (cf. *2162*). Kovács *et al.* (*1296*) have called this rearrangement $\alpha \rightarrow \gamma$ transpeptidation. The reaction was studied systematically by Battersby and Robinson (*120*). Treatment of Ac-Gly-Gly-L-Glu(OEt)-Gly-*n*-hexyl amides [67] with 0.1 *N* sodium hydroxide at room temperature, or with 0.28 *N* sodium carbonate solution, leads to the formation of an optically inactive mixture of the structurally isomeric tetrapeptides in which the γ-compound predominates. As a consequence of the preferential cleavage of the glutarimide intermediate [68] with formation of the γ-derivative [69], only a small amount of the α-isomer [71] is formed upon alkaline treatment of glutamyl γ-peptides [70]. Clayton *et al.* (*498*) have demonstrated the presence of the pyrrolidone-5 derivative [73], beside the glutarimide [72], after treatment of glutamyl α-peptides with thionyl chloride/pyridine. The hydrolysis of the glutarimide leads preferentially to the formation of the γ-isomer [74], whereas from the pyrrolidone the starting material is recovered together with a small amount of decompo-

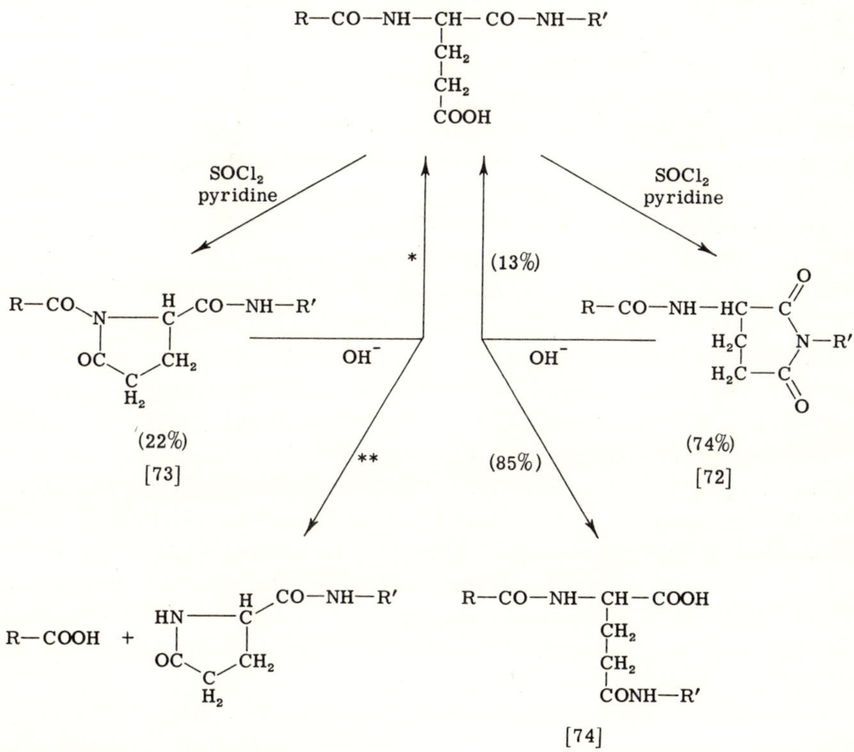

* Predominantly; ** to a small extent.

sition products. Battersby and Robinson (*121*) obtained from another model substance, H-DL-Glu-Gly-*n*-hexyl amide, the pyrrolidone derivative [73] with thionyl chloride, while treatment with ethyl chlorocarbonate/triethylamine yielded the glutarimide derivative [72]. According to Kovács *et al.* (*1295, 1296*), brief heating with acetic anhydride of Cbo-L-Glu-α-L-Glu-OH or γ-Glu-OH causes intramolecular transpeptidation via the glutarimide. The equilibrium is strongly in favor of the γ-glutamyl component. Medzihradszky and Bruckner (*1520*) observed the intramolecular transpeptidation also in structural isomers of polyglutamic acids and were able to isolate polyglutamic acids of the mixed type (cf. *426, 1517*). Investigations by Battersby and Reynolds (*119*) have demonstrated that the individual cleavage and rearrangement reactions depend upon steric effects caused by amino acids in the neighborhood of glutamyl residues. Bruckner *et al.* (*419*) succeeded in almost completely suppressing the partial $\alpha \rightarrow \gamma$ transpeptidation during alkaline hydrolysis by adding copper(II) hydroxide, i.e. by forming a biuret complex. So far it has not been possible to establish to what extent transpeptidation still occurs. The extent of racemization observed frequently under alkaline conditions was reduced by the presence of copper(II) hydroxide. This method made possible an improved synthesis of optically pure α-polyglutamic acid of the L- and D-series (*421, 425;* cf. *422, 424*). Liefländer (*1425*) reported that it was not possible to prepare pure glutamyl α-peptides via this method. Bruckner *et al.* (*429;* cf. *1518*) used L-glutamyl-γ-L-glutamic acid α,α'- or α,γ'-dimethyl ester [75a] ($R = OCH_3$, $R' = OH$) or [75b] ($R = OH$, $R' = OCH_3$) to demonstrate that the carboxymethyl groups can be reduced selectively with lithium borohydride in dimethylformamide at 20°C without cleavage of the peptide bonds (cf. page 191). This is of importance for the determination of the type of bond present in glutamyl peptides. The amino alcohols were isolated via the dinitrophenyl (DNP) derivatives after acid hydrolysis. Sachs and Brand (*1892*) identified glutamyl γ-peptides by their reaction with nitrous acid. In contrast to the glutamyl α-peptides, the nitrogen is completely liberated as amino nitrogen and the γ-peptide bond cleaved (cf. *450*). A long established fact is noteworthy in this connection: Glutamyl γ-bonds are sensitive to hydrolytic reagents, even to hot water (*815, 937*). Le Quesne and Young (*1370, 1372*) studied the autohydrolysis and found that glutamyl γ-peptides as well as glutamyl α-peptides will be hydrolyzed (24 hours, 100°C) (cf. *2100*).

ϵ. *Glutaminyl and Isoglutaminyl Peptides.* The first synthesis of glutaminyl peptides used glutamyl α-peptides as starting material. The amide group was introduced via the acid chloride (*938, 1533*) or by

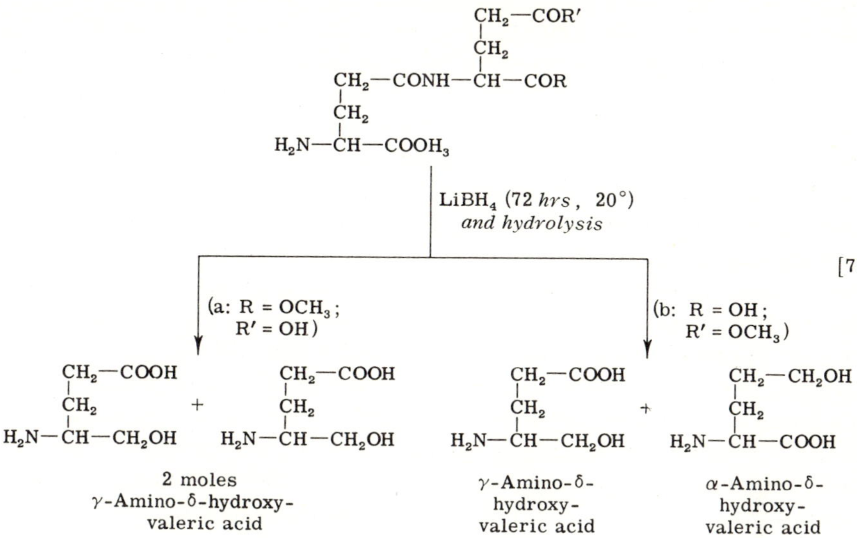

[75]

ammonolysis of the γ-ethyl ester (*773, 1560*). Today carbobenzoxy-L-glutamine is used primarily as starting material. The compound is accessible from L-glutamine (*193, 1809*) or from L-glutamic acid γ-methyl (*296*) or γ-ethyl ester (*1560*) and subsequent amidation. A number of procedures are known for the synthesis of glutamine (*1560, 1859, 2715*). Carbobenzoxy-L-glutamine was coupled with amino acid or peptide esters via the anhydride (*296, 297, 1191, 1860, 1972, 2098*), carbodiimide (*1170, 1191*), carbonyldiimidazole (*1170*), tetraethyl pyrophosphite (*1191*), phosphazo (*1424*), and phosphorus oxychloride (*1491*) methods. The latter two methods frequently gave rather low yields. Katsoyannis *et al.* (*1191*) were able to isolate the γ-cyano compound formed as a by-product by dehydration of the γ-carboxylamido group when they applied the tetraethyl pyrophosphite method. Ressler and Ratzkin (*1809*) isolated γ-cyano-L-aminobutyric acid from the reaction of carbobenzoxy-L-glutamine with *N,N'*-dicyclohexylcarbodiimide, followed by reduction with sodium in liquid ammonia. Although the majority of authors do not describe it, the possibility of dehydration of carbobenzoxy-L-glutamine during coupling reactions must not be overlooked. Carbobenzoxy-L-glutamine α-*p*-nitrophenyl ester (*273*) proves to be a very suitable derivative for the coupling reaction since no side reactions occur in this case. The α-cyanomethyl ester (*1950*) has not been often used. Sondheimer and Holley (*2161*) obtained carbobenzoxy-L-glutamine α-hydrazide by selective hydrazinolysis at 0°C of the α-methyl esters and used it successfully for coupling reactions (*1033, 1043, 1170, 1424, 2161*). Hydrazinolysis of the γ-carboxyl amide group was not observed during this reaction, but was reported to occur in *n*-butanol at 40°C (*1491;* cf. *651*). Other derivatives applicable to the synthesis of glutaminyl peptides are *p*-chlorocarbobenzoxy- (*1247*), *p*-(*p*′-methoxyphenylazo)-carbobenzoxy- (*1170*), *p*-methoxycarbobenzoxy- (*1975*), formyl- (*306*), trityl- (*36, 2203*), and tosyl-L-glutamine (*1846, 2247*) as well as *tert*-butyloxycarbonyl-L-glutamine (*1020b*) and *tert*-butyloxycarbonyl-L-glutamine hydrazide (*1967*). Phthalyl-L-glutamine may be obtained from phthalyl-L-glutamic acid anhydride—accessible by a synthesis of several steps—by treatment with ammonia in ether or dioxane (*1241*), or simply by reacting glutamine with carbethoxyphthalimide (*1975*). A very convenient approach to the synthesis of glutaminyl peptides is offered by the cleavage of the pyrrolidone ring of tosylpyroglutamyl peptides with ammonium hydroxide solution; since the tosyl protection requires reduction with sodium in liquid ammonia and other protecting groups have been suggested for pyroglutamic acid, it would appear to be more feasible to use such derivatives of pyroglutamic acid for this purpose.

The carbobenzoxy compound was also used for the synthesis of isoglutaminyl peptides. The carbobenzoxy derivative can be readily prepared from carbobenzoxy-L-glutamic acid anhydride by treatment with ammonia (*193, 1807*). A simpler synthesis of carbobenzoxyisoglutamine was reported by Ressler (*1805*); it is based on the stronger acidity of the γ-carboxyl group. During the reaction of carbobenzoxy-L-glutamic acid with 1 mole of isobutyl chlorocarbonate/triethylamine only the α-carboxyl group will be activated. It can be readily converted to the amide. Swan (*2246*) synthesized several isoglutamyl peptides by aminolysis of tosyl-L-pyroglutamic acid amide [76] with amino acid esters.

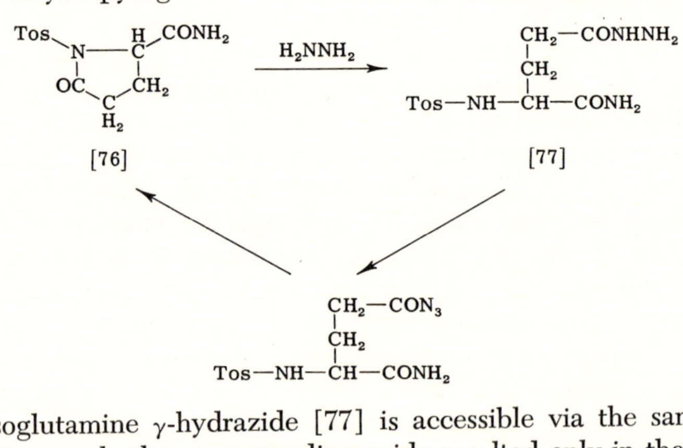

Tosyl-L-isoglutamine γ-hydrazide [77] is accessible via the same route. Attempts to couple the corresponding azide resulted only in the isolation of tosyl-L-pyroglutamic acid amide.

With the exception of the dehydration in the course of certain coupling reactions, no side reactions have been reported to occur during the synthesis of glutaminyl peptides. Alkaline hydrolysis proceeds without complication. Formation of pyroglutamyl derivatives was observed during the cleavage of the phthalyl group by hydrazinolysis (*1967*). Transesterification may occur during the cleavage of the formyl group with hydrogen chloride in alcohol. Melville (*1533*) has reported on the acid lability of ω-amide groups (cf. *489*). Hofmann *et al.* (*1026, 1044*) took advantage of this fact during the synthesis of ACTH peptides. The acidic cleavage (0.5 *N* HCl, 60–80 minutes, 100°C) of N-terminal acetyl and ε-formyl groups of lysine leads to the concurrent liberation of the γ-carboxyl group.

ζ. Pyroglutamic Acid. Harrington and Moggridge (*939*) have shown that the reaction of tosyl-L-glutamic acid with acetic anhydride or with acetyl chloride does not lead to the formation of the intramolecular anhydride. Lactamization occurs instead and 1-tosyl-L-pyrrolid-5-one-2-carboxylic acid (tosyl-L-pyroglutamic acid) is formed. This is in contrast to the behavior of the acetyl, carbobenzoxy, phthalyl, trifluoroacetyl, and other derivatives of glutamic acid. Rudinger (*1846*) found thionyl chloride at room temperature and tosyl chloride in pyridine or phosphorus trichloride more suitable for the ring closure reaction. Tosyl-L-pyroglutamyl chloride is formed on heating tosyl-L-glutamic acid in thionyl chloride. The action of phosphorus pentachloride produces tosyl-L-glutamic acid dichloride, which Swan and du Vigneaud (*2247*) originally thought to be pyroglutamyl chloride. Stedman (*2198*) found that this compound can be obtained only from the unstable dichloride in boiling benzene.

Gibian and Klieger (*798, 799*) have reported that carbobenzoxy-L-pyroglutamic acid can be readily obtained as dicyclohexylammonium salt by rearrangement of the intramolecular anhydride with dicyclohexylamine in anhydrous tetrahydrofuran/ether. Carbobenzoxy-L-glutamic acid α-benzyl ester can be cyclized with phosphorus pentachloride to form carbobenzoxy-L-pyroglutamyl benzyl ester (*173*). *p*-Methoxycarbobenzoxy-, *tert*-butyloxycarbonyl-, and formyl-L-glutamic acid can also be rearranged via the intramolecular anhydrides to form the corresponding L-pyroglutamyl compounds (*799, 1975,* cf. *489a*). It is not clear if the optical activity was preserved in the case of the formyl compound. Treatment of trifluoroacetyl-L-glutamic acid anhydride with dicyclohexylamine (DCHA) and water leads to the elimination of the trifluoroacetyl group and with ring closure to the formation of racemic pyroglutamic acid (*2496*). If water is excluded the rearrangement also proceeds with elimination of the protecting group; however, optically active L-pyroglutamic acid can be isolated (*799*).

The γ-carboxyl group in *N*-protected L-pyroglutamic acids [78] is moderately activated. Under certain reaction conditions it is possible to open the lactam ring with formation of glutamyl γ-peptides [79] (*1847*). The application of the acid chloride (*590, 1847, 1852, 1854, 2247*), anhydride, dicyclohexylcarbodiimide (DCCI), or *p*-nitrophenyl ester method (*799, 1258*) led to the formation of pyroglutamyl peptides [80], from which glutamyl α-peptides [81] or glutaminyl peptides [82] are accessible by hydrolysis or ammonolysis. The carboxyl group can be liberated selectively if the benzyl or *tert*-butyl ester is applied, while the lactam ring [83] remains unaffected. Methanolysis, hydrogenolysis, or ammonolysis (*799, 1258, 1846*) produces *N*-protected glutamic acid γ-methyl esters, γ-hydrazides, and γ-amides [84] (cf. page 195). Tosyl-L-pyroglutamic acid was used for the synthesis of L-proline (*1770*) and, via ammonolysis of the lactam ring and detosylation with hydrogen bromide in glacial acetic acid, for the synthesis of L-glutamine (*1859*). The action of hydrogen bromide in glacial acetic acid on tosyl-L-isoglutamine causes ring closure and formation of tosyl-L-pyroglutamic acid amide (*1859*). A shortcoming of the synthesis of glutaminyl and glutamyl peptides from tosyl-L-pyroglutamic acid is the necessity to detosylate with sodium in liquid ammonia. With the frequently observed complications, this reaction cannot be considered ideal although it has been repeatedly described (*1577, 1847, 1860, 2247*) and applied to the synthesis of eisenin and pelvetin and certain sequences of oxytocin. Carbobenzoxy-L-pyroglutamic acid was used successfully for the synthesis of eledoisin (*1904*).

L-Pyroglutamic acid can be obtained by cyclization of L-glutamic acid γ-methyl or γ-ethyl ester with ammonia-saturated methanol (*127, 510;* cf. *126*). Angier *et al.* (*57*) have described the synthesis of peptides with unsubstituted pyroglutamic acid azide; the yields, however, were moderate. The ring of pyroglutamic acid can be opened by ethanolic hydrochloric acid with simultaneous esterification of the γ-carboxyl group. Pyroglutamyl peptide azides (*1904*) can be coupled in good yield. Pyroglutamyl peptides may also be obtained from glutaminyl peptides by boiling with water (*489a, 1860*) or from glutamyl γ-methyl ester peptides by treatment with methanol saturated with ammonia (*2098*).

η. *Glutamyl α,γ-Bispeptides.* Goldschmidt and Wick (*823*) succeeded in simultaneously coupling the α- and γ-carboxyl group of carbobenzoxy-L-glutamic acid via the phosphazo method. Sachs and Brand (*1890*) achieved this via the *o*-phenylene chlorophosphite method. Several authors (*1698, 1976, 2099*) were less successful with regard to yield and homogeneity in the application of the mixed anhydride method, using esters of chlorocarbonic acid or isovaleryl chloride. Anhydrides

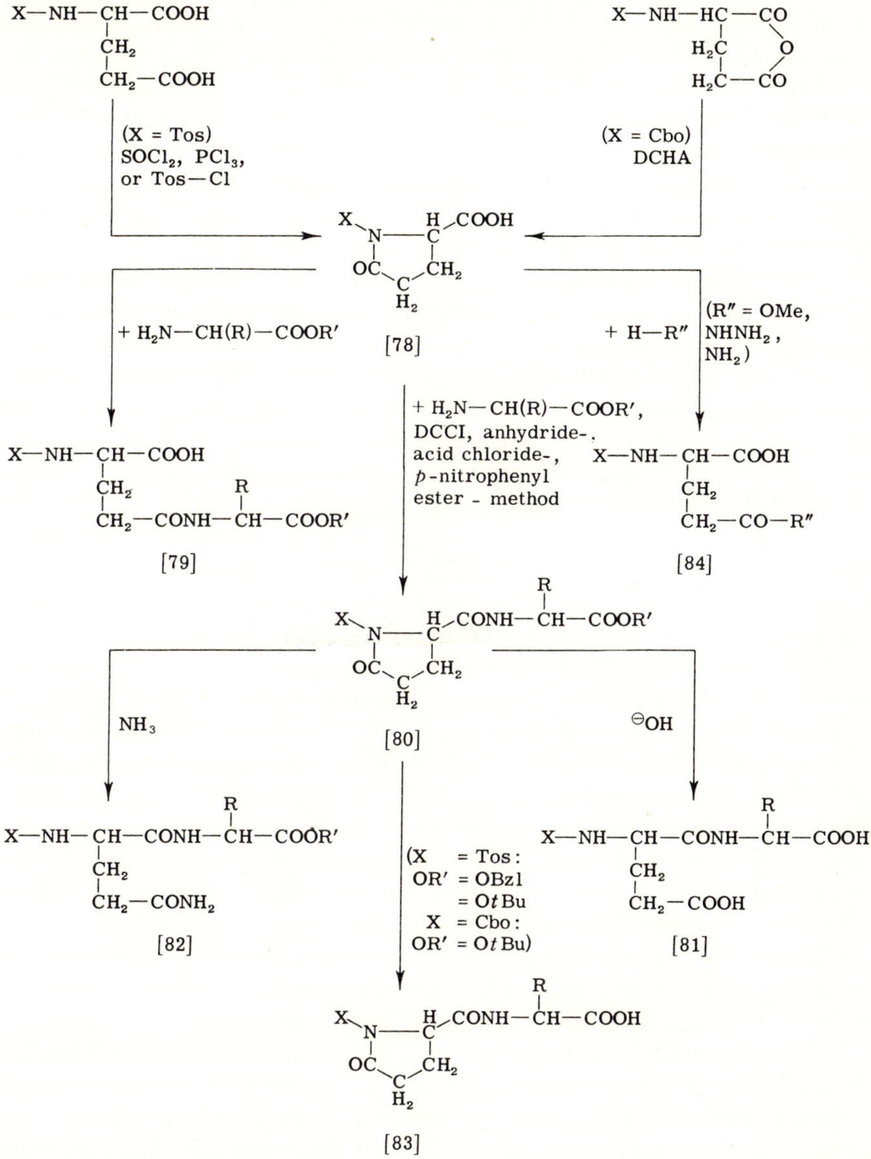

with phosphorus oxychloride seem to be better. Carbobenzoxy-L-glutamic acid diazide (*2099*), in analogy to the already described coupling with *p*-nitrobenzoyl-L-glutamic acid diazide (*1582*), also seems to be more advantageous. Shiba and Kaneko (*2099*) used α-peptides as starting material and introduced the γ-substituent via the anhydride and carbodiimide methods (cf. *1890*), while Mowat *et al.* (*1583*) made use of the acid chloride method. Yields were not always satisfactory. The simultaneous substitution of the α- and γ-carboxyl groups via activation with *N,N'*-dicyclohexylcarbodiimide (*1698, 1976, 2099*) is superior to the methods already mentioned. Zahn and Pätzold (*2633a;* cf. *1698*) obtained good yields from the reactions of *N*-carbobenzoxy-L-glutamic acid and *N*-trifluoroacetyl-L-glutamic acid via the carbodiimide method, and *N*-trifluoroacetyl-L-glutamic acid bis-*p*-nitrophenyl esters with N^{ϵ}-carbobenzoxy-L-lysine ethyl ester (cf. *2099*).

The alkaline hydrolysis of carbobenzoxy-L-glutamyl bispeptides, hydrazinolysis of esters, and decarbobenzoxylations rarely follow the regular course. However, under special reaction conditions, it has been possible to obtain the desired derivatives (*1698, 1976, 2099*). Investigations on the hydrolysis of glutamyl bispeptides have demonstrated that, for example, L-glutamylbis-L-glutamic acid is decomposed in water at 100°C to form pyrrolidonecarboxylic acid, glutamic acid, and pyroglutamylglutamic acid (*2099*).

b. Glutamic Acid and Glutamine Peptides

A variety of esters is available for the synthesis of peptides with glutamic acid as the carboxyl-terminal amino acid: glutamic acid diethyl ester hydrochloride (*834, 1950*), the dimethyl ester hydrochloride (*979*), the dibenzyl ester toluenesulfonate (*1116, 2278, 2670*), the dibenzyl ester benzenesulfonate (*431, 979, 2103*), and the dibenzyl ester hydrochloride (*431, 734, 979, 1889, 2103*). Di-*tert*-butyl ester (*48, 2265*) proved to be very useful. All of these compounds were applied to the synthesis of glutamic acid peptides via various coupling methods (*201, 979, 1249, 1424, 1535, 1538, 1771, 1844, 1950, 2322, 2583*). The alkaline hydrolysis normally poses no problems (cf. *1538*). Acid hydrolysis of alkyl esters has also been described (*1538*). In any event, the application of the benzyl ester is more advantageous since it can be removed by hydrogenolysis. [Cf. (*1021*) for coupling reactions with free glutamic acid.]

Glutamic acid α- or γ-monoesters were employed in salt coupling reactions whenever it was desired to add amino acids to the carboxyl-terminal glutamic acid residues (*894, 1228, 1890*). A number of α,γ-disubstituted derivatives of glutamic acid have recently been reported that make it possible to carry out additional coupling reactions at the carboxyl end after the one or the other carboxyl-protecting group has been removed selectively. Derivatives containing one activated carboxyl group while the other bears a readily removable protection fulfill the same purpose (cf. Table VII).

The decarbobenzoxylation of peptides with carboxyl-terminal glutamic acid esters by treatment with hydrogen bromide in glacial acetic acid can be dangerous. Bayer *et al.* (*125;* cf. *1248*) found that exposure of glutamic acid α-benzyl ester or γ-ethyl ester to a 30% solution of hydrogen bromide in glacial acetic acid for 15 minutes causes the respective formation of 18% and 14% of free glutamic acid. The γ-benzyl ester was hydrolyzed to the free acid in 73% yield. The diethyl ester yielded 36% of α-monoester.

Peptides with carboxyl-terminal glutamine were synthesized repeatedly via salt coupling reactions (*217, 1033, 1354*). Other derivatives, however, such as glutamine methyl ester hydrochloride (*2161*), the *p*-nitrobenzyl ester hydrobromide (*2163*), and the *tert*-butyl ester (*48*), are more suitable for the synthesis of peptides. Carbobenzoxy-L-glutamine and tosyl-L-glutamine can be esterified with alcohol/thionyl chloride (*1108*) in spite of their lability to acid (*2654*). It is better, however, to use diazoalkanes (*1033, 2160, 2654*). Taschner and Wasielewski (*2279*) were successful with acid-catalyzed transesterifications employing suitable mildly acting catalysts (*2277*). The reaction of formyl-L-glutamine with phenyl diazomethane yielded the benzyl ester. Efforts to deformylate, however, have not so far been successful. Carbobenzoxy-L-glutamine benzyl ester can be decarbobenzoxylated with hydrogen bromide in glacial acetic acid (*2163*). Care must be exercised during the liberation of the free ester from L-glutamine methyl ester hydrochloride for coupling reactions. The formation of derivatives of isoglutamine is encountered during the alkaline hydrolysis of fully protected peptides. Investigations by Sondheimer and Holley (*2160*) have shown that carbobenzoxy-L-glutamine and carbobenzoxy-L-isoglutamine are formed via the labile glutarimide upon alkaline treatment of carbobenzoxy-L-glutamine methyl ester (see Chapter IV,C,1,a,δ). 1-L-Glutamine-2-*tert*-butyloxycarbonylhydrazine (*2038;* cf. *2236*) is a very suitable derivative for the synthesis of peptides with glutamine inside the chain.

Akabori *et al.* (*27*) and Shimonishi *et al.* (*2104*) used xanthydrol in glacial acetic acid (for 2 days at room temperature) to convert carbobenzoxy-L-glutamine to carbobenzoxy-L-glutamic acid γ-*N*-xanthylamide in order to prevent side reactions at the carboxylamido group. This amide-protecting group is stable to alkali. It can be removed together with the carbobenzoxy group by treatment with 20% hydrogen bromide in glacial acetic acid (2 hours, room temperature). The carbobenzoxy group can be cleaved selectively by catalytic hydrogenation. Esterification of the carbobenzoxy compound can be brought about with diazomethane.

[Cf. (*216, 969, 1976*) for symmetrical and unsymmetrical peptides of simple dicarboxylic acids.]

2. Aspartic Acid

Aspartic acid is present in proteins and biologically active polypeptides as aspartyl or asparaginyl residue. β-MSH, ACTH, glucagon, the angiotensins, and eledoisin contain one or several residues of aspartic acid; oxytocin, vasopressin, insulin, and tyrocidin A and B contain asparaginyl residues. In bacitracin an aspartyl residue is linked through its α- and β-carboxyl groups. The molecule contains an additional residue of D-asparagine. Asparagine has been used for the synthesis of many naturally occurring polypeptides, since it offers certain advantages over preparative procedure required for peptides that contain the aspartyl residue.

a. Aspartyl Peptides

N-Protected aspartic anhydrides and *N*-protected derivatives of aspartic acid that are selectively substituted at the α- or β-carboxyl group are used as starting material for the synthesis of aspartyl peptides.

α. N-Acylaspartic Anhydrides. α_1. *Carbobenzoxyaspartic anhydride.* Bergmann and Zervas (*193*) prepared carbobenzoxyaspartic anhydride by heating carbobenzoxy-L-aspartic acid with acetic anhydride. The reaction was later carried out at room temperature to prevent racemization (*1371, 1474, 1559, 2100*). Differences in the reported melting points could be explained by dimorphism (*1149*). Cleavage of the anhydride with amino acid esters leads to the formation of aspartyl peptides (*193, 206, 847*). Investigations of Le Quesne and Young (*1371*) have demonstrated that a mixture of the α- and β-isomers is formed during the aminolysis of the anhydride with amino acid esters in ethyl acetate/sodium hydrogen carbonate. The isomers can be separated by fractional extraction with sodium carbonate. Generally the α-compound is formed predominantly. This approach was applied to the synthesis of aspartyl α-peptides (*725, 860, 1149, 1354, 2100*); the yields, however, were not very satisfactory. Ammonolysis permits the preparation of carbobenzoxy-L-isoasparagine (*193, 1474;* cf. *478*). Carbobenzoxy-L-aspartic acid anhydride can be decarbobenzoxylated without complication by treatment with hydrogen bromide in glacial acetic acid (*1293b*).

α_2. *Phthalylaspartic anhydride.* King and Kidd (*1242*) found that phthalyl-L-aspartic acid, like the corresponding derivative of glutamic acid, can be prepared only via phthalyl-L-aspartic acid diethyl ester. It is more advantageous to use *o*-carbethoxythiobenzoic acid under milder conditions instead of phthalic anhydride (*101*). The direct introduction of the phthalyl group is possible by the method of Nefkens *et al.* (*1600*). The anhydride will be cleaved by aminolysis to form aspartyl β-peptides (*1242*). Tanenbaum (*2262*) was able to demonstrate that the ammonolysis is strongly dependent upon the solvent. Under anhydrous conditions the β-isomer is formed, in aqueous ethanol the α-isomer predominates.

α_3. *Trifluoroacetylaspartic anhydride.* L-Aspartic acid reacts with trifluoroacetic anhydride to form trifluoroacetyl-L-aspartic anhydride without racemization (*2490*). The anhydride will be cleaved at room temperature in tetrahydrofuran by ammonia, aniline, or amino acid esters to form the α-derivatives (*2478, 2490*). The resulting compounds are first isolated as the crystalline dicyclohexylammonium salts. They are subsequently liberated by passage over acidic ion-exchange resins. The yields of the reactions with amino acid esters are only 10–20%. Attempts to prepare the β-acid amide from trifluoroacetyl-L-aspartyl-α-glycine ethyl ester via the β-acid chloride and reaction with liquid ammonia failed. Trifluoroacetylamino-*N*-carbethoxymethyl succinimide [85] was

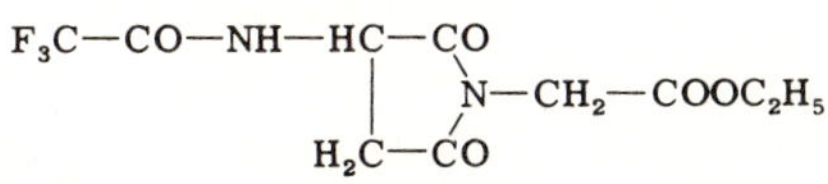

[85]

obtained. The reaction of ethanol with trifluoroacetyl-L-glutamic anhydride leads to the formation of a mixture of the α- and γ-esters, while the α-ester is formed almost entirely by the same treatment of trifluoroacetyl-L-aspartic anhydride. The ester can be converted to the stable *N*-trifluoroacetyl-γ-oxo-L-norvaline ethyl ester (*2490;* cf. *1435*) via the acid chloride and treatment with diazomethane. Glutamyl γ-peptides thus become available by Wolff's rearrangement (*2490*).

α_4. *Tosylaspartic anhydride.* Tosyl-L-aspartic acid can be converted to the intramolecular anhydride by treatment with acetyl chloride (*939*) or acetic anhydride (*330*). Alcoholysis leads preferentially to the formation of the α-ester (*330, 939*). Ammonolysis produces tosyl-L-isoasparagine (*1805*).

α_5. *Benzylaspartic anhydride.* Treatment with acetic anhydride of *N*-benzyl-DL-aspartic acid results in the acetylation of the secondary amino group and with ring closure the formation of *N*-acetyl-*N*-benzyl-DL-aspartic anhydride (*1442*). *N*-Benzyl-DL-aspartic anhydride is formed by treatment with cold acetyl chloride/acetic acid in a ratio of 1:1. A ratio of 3:1 of acetyl chloride/acetic acid produces only the hydrochloride of *N*-benzyl-DL-aspartic acid (*2676*). Aminolysis of the anhydride with amino acid esters in water or acetone leads preferentially to the formation of the β-isomer. The reaction with ammonia yields the α-isomer. *N*-Benzylated amino acids are not of great value for the synthesis of peptides. Therefore the interesting investigations of Liwschitz *et al.* (*763, 1434, 1436–1441, 1444*) will not be discussed in detail. The authors used maleic anhydride or maleic acid as starting material for the synthesis of DL-aspartyl α- or β-peptides.

β. *Aspartic Acid β-Ester and Aspartyl α-Peptides.* Carbobenzoxy-L-aspartic acid β-esters are used almost exclusively for the synthesis of aspartyl α-peptides. The hydrochloride of L-aspartic acid β-methyl ester can be obtained by selective esterification with anhydrous methanol/hydrogen chloride (*510*) or methanol/thionyl chloride (*1993*). The β-methyl and β-ethyl esters were prepared from aspartic acid by treatment with methyl or ethyl acetate/perchloric acid. Unesterified material was removed with Amberlite IR-4B (*241*). Carbobenzoxylation in sodium hydroxide/sodium hydrogen carbonate or, better, in the presence of magnesium oxide leads to the formation of carbobenzoxy-L-aspartic acid β-methyl ester (*1993*). *p*-Phenylazocarbobenzoxy- (*2236*) and *p*-(*p′*-methoxyphenylazo)-carbobenzoxy-L-aspartic acid β-methyl esters (*2037*) have also been used at starting material for the synthesis of peptides (*2032, 2236*). L-Aspartic acid β-benzyl ester can be prepared by selective esterification with benzyl alcohol/concentrated hydrochloric acid (*1124*) or benzyl alcohol/sulfuric acid (*150, 967*). As in the case of the β-methyl ester, carbobenzoxy-L-aspartic acid β-benzyl ester (*150, 1124*) thus becomes accessible. This compound has been prepared more frequently via a different method. Carbobenzoxy-L-aspartic acid dibenzyl ester was hydrolyzed selectively with potassium hydroxide in benzyl alcohol (*759*), sodium hydroxide in aqueous dioxane (*178*), or lithium hydroxide in aqueous acetone (*436*). Skeggs *et al.* (*2134*) added an additional purification step—fractionation between 0.2 *M* phosphate buffer (pH 5.8) and ethyl acetate. The required dibenzyl ester can be prepared from the disilver salt of carbobenzoxy-L-aspartic acid by treatment with benzyl iodide (*759*). It is more advantageous, however, to esterify with benzyl alcohol/*p*-toluenesulfonic acid (*178*) or to carbobenzoxylate the dibenzyl ester *p*-toluenesulfonate (*1515*). Goodman *et al.* (*828*) were able to synthesize L-aspartic acid β-*p*-nitrobenzyl ester by treatment of the corresponding β-benzyl ester with nitronium fluoborate in acetic acid [86]:

$$\begin{array}{l} \quad CH_2-COOCH_2C_6H_5 \\ \quad | \\ H_2N-CH-COOH \end{array} + NO_2^{\oplus}\,BF_4^{\ominus} \longrightarrow \begin{array}{l} \quad CH_2COOCH_2C_6H_4NO_2 \\ \quad | \\ H_2N-CH-COOH \end{array} \qquad [86]$$

The exclusive nitration in the *p*-position was established by hydrolysis with 10% hydrochloric acid and isolation of *p*-nitrobenzyl alcohol. Carbobenzoxy-L-aspartic acid β-*tert*-butyl ester is especially useful for the synthesis of peptides. The compound can be synthesized by esterification with isobutylene/sulfuric acid of carbobenzoxy-L-aspartic acid α-benzyl or α-ethyl ester and subsequent selective alkaline hydrolysis (*2007, 2598*). Other suitable derivatives are accessible by introducing via the azides the *tert*-butyloxycarbonyl and methoxycarbobenzoxy groups in L-aspartic acid α-benzyl ester β-*tert*-butyl ester (*1974,* cf. *2598a*).

Carbobenzoxy-L-aspartic acid β-esters were used for the synthesis of substituted aspartyl α-peptides via the anhydride (*63, 210, 436, 827b, 1124, 1558, 1972, 1993, 2134, 2295*), carbodiimide (*152, 436, 752a, 890, 1104, 1124, 1515, 1904, 2218, 2236, 2295*), and *p*-nitrophenyl ester (*827a, 890, 1972*) methods.

Esters of aspartyl α- and β-peptides, like those of glutamyl peptides, readily rearrange to a mixture of the isomeric compounds upon exposure to alkaline hydrolysis conditions. Battersby and Robinson (*120*) demonstrated for benzoyl-DL-aspartyl-α-glycine *n*-hexylamide β-ethyl ester that the β-isomer is formed predominantly (cf. *1294*). The intermediary succinimide is more stable than the corresponding derivatives of glutarimide and can be isolated. Bernhard *et al.* (*210*) selected *N*-alkyl amides of carbobenzoxy-L-aspartyl-α-serine β-benzyl ester to study the alkaline hydrolysis, succinimide formation, subsequent hydrolysis by potentiometric titration, and optical rotation during the reaction. They also discussed a mechanism for the reaction. Hanson and Rydon (*935b*) also described the ready rearrangement upon amidation of Cbo-L-Asp(OBzl)-L-Ser-OMe to the β-peptide diamide. Schneider (*1951a*) has reported on the high lability of the aspartic acid β-ester bond in peptides containing serine and threonine. Aspartyl α-peptides show in boiling water a tendency to rearrange to the β-compound (*1149*). Iselin and Schwyzer (*1104*) reported another example. They attempted to hydrolyze selectively with 0.2 *N* potassium hydrogen carbonate (pH 9.8–10.2) the β-benzyl ester of carbobenzoxy-L-aspartyl glycine *p*-nitrophenyl ester but isolated carbobenzoxy-L-aminosuccinimidoglycine *p*-nitrobenzyl ester. ε-(L-Aspartyl-α)-L-lysine and ε-(aspartyl-β)-lysine were cyclized to ε-(L-aminosuccinyl)-L-lysine by treatment with 11 *N* hydrochloric acid at 80°C (cf. *2243*). As a consequence of these observations, ring closure with formation of derivatives of succinimide cannot be completely excluded during the removal of amino-protecting groups by treatment with hydrogen bromide in glacial acetic acid (*436*).

γ. *Aspartic Acid α-Esters and β-Aspartyl Peptides.* Bergmann *et al.* (*205*) synthesized carbobenzoxy-L-aspartic acid α-benzyl ester (cf. *1559*) by reacting carbobenzoxy-L-aspartic anhydride with benzyl alcohol in a sealed tube at 100°C. The selective hydrolysis at the β-carboxyl group of L-aspartic acid dibenzyl ester *p*-toluenesulfonate by hydroiodic acid represents an alternative synthesis. Carbobenzoxylation furnishes a very pure isomer (*436*). The symmetrical anhydride of *N*-carbobenzoxy-L-aspartic acid α-benzyl ester was isolated as a by-product of the carbobenzoxylation. This compound is also formed upon treatment of carbobenzoxy-L-aspartic acid α-benzyl ester with acetic anhydride. Carbobenzoxy-L-aspartic acid α-ethyl ester can be prepared by alcoholysis of the intramolecular anhydride and fractionation of the mixture of the esters (*1371*). The compound can be readily purified via the dicyclohexylammonium salt (*2598*). *tert*-Butyloxycarbonyl- and *p*-methoxycarbobenzoxy-protected α-esters are accessible via the same method (*1974*). Alcoholysis of the intramolecular anhydride of aspartic acid with the unprotected amino group leads almost exclusively to the formation of the α-ester (*1293b*). The acid chloride method has been applied several times to carbobenzoxy-L-aspartic acid α-benzyl ester to synthesize aspartyl β-peptides (*206, 636, 847, 938, 1559*). Swallow *et al.* (*2244*) synthesized ε-L-aspartyl-β-L-lysine via the anhydride method. Grassmann and Schneider (*847*) used carbobenzoxy-L-aspartic acid dichloride for the synthesis of L-aspartyl-bisglycine. Losse *et al.* (*1455a*) prepared some activated β-esters of carbobenzoxy-L-aspartic acid-α-benzyl esters.

δ. *Asparaginyl and Isoasparaginyl Peptides.* Extensive studies of problems in the synthesis of oxytocins and angiotensins have demonstrated that the synthesis of peptides containing asparagine is confronted with certain difficulties despite the fact that there are many advantages over the synthesis of aspartyl peptides. The most important starting material for the synthesis of asparaginyl peptides is carbobenzoxy-L-asparagine. The compound is synthesized by carbobenzoxylation under mild conditions, i.e. in the presence of magnesium oxide (*193*), in sodium hydrogen carbonate solution (*296*), or in sodium hydroxide while a pH value of about 8 (*1809*) is maintained. Carbobenzoxy-D-asparagine was obtained from the DL-compound by resolving the racemic mixture via the cinchonine salt (*209*). Practically all methods for the synthesis of peptides were tested for applicability—the anhydride method [with ethyl chlorocarbonate: 0–22% yield (*725, 1354*); with *sec*-butyl chlorocarbonate: 52% yield (*1854*); with isovaleryl chloride: 55% yield (*1191*); pivalyl chloride, however, gave a yield of 80% (*2648*)] and the carbonyldiimidazole (54%) (*1701*), carbodiimide (39–45%) (*1104, 1191*), phosphazo (55%) (*1854*), isocyanate (30%) (*1354*), tetraethyl pyrophosphite (35–40%) (*296, 642, 1191*), diethyl chlorophosphite (24–50%) (*296, 1354, 1558, 1560, 1823, 1827*), and phenylene chlorophosphite (41%) (*1354*) methods. The application of carbobenzoxy-L-asparagine hydrazide, admittedly in a salt coupling reaction, did not give satisfactory results (*1987*). Today the nitrophenyl ester is the ideal starting material for the synthesis of asparaginyl peptides, despite the fact that the synthesis of this compound with *N,N'*-dicyclohexylcarbodiimide leads to the formation of dehydration products and the yields are only 40–50%. The ester was first introduced by Bodanszky and du Vigneaud (*273*) and applied by several authors (*230, 1192, 1972*), producing yields of up to 90%.

Dehydration of ω-acid amide groups was observed during the synthesis of oxytocin sequences (*807, 1190, 1803*). Detailed investigations revealed that this dehydration is caused by different reagents and is dependent upon the type of amino-protecting group, the α-substituent, and the reaction conditions. Anhydro compounds were isolated after activation of the carboxyl group by the carbodiimide (*273, 1191, 1809*), phosphorus oxychloride (*1419*), and tetraethyl pyrophosphite (*1191*) methods. They were found even after exposure to mild conditions, i.e. in aqueous solutions with chloroformates, *p*-toluenesulfonyl chloride, or benzenesulfonyl chloride (*1420*) as anhydride-forming reagents. They were not encountered when isovaleryl chloride was used for the anhydride formation. Carbobenzoxy-L-asparagine can be converted to carbobenzoxy-β-cyano-L-alanine very readily (*273, 1419, 1420, 1809*). Carbobenzoxy-L-asparagine esters or carbobenzoxy-L-asparaginyl peptides are stable in the presence of *N,N'*-dicyclohexylcarbodiimide (*273, 1420*). The elimination of water by *N,N'*-dicyclohexylcarbodiimide from carbobenzoxy-L-glutamine occurs to a much smaller extent (*1809*). The dehydration of carbobenzoxy- and tosylasparagine esters and of tosyl- and phthalylglutamine esters with *p*-toluenesulfonyl chloride in pyridine has been described (*2654*). The carbobenzoxy group of carbobenzoxy-β-cyano-L-alanine can be removed with sodium in anhydrous liquid ammonia without affecting the nitrile group. α,γ-Diaminobutyric acid will be formed in the presence of methanol (*1809;* see Chapter IV,B,3,a). Tosyl-β-cyano-L-alanine can be reconverted to L-asparagine by treatment for 10 minutes with 35% hydrogen bromide in glacial acetic acid (*2654*). The hydrolysis of the nitrile group to the acid amide group was carried out successfully with 30% hydrogen peroxide in acetone/1 *N* sodium carbonate solution (24 hours at 25°C). It was also possible to hydrolyze Cbo-L-Ala(CN)-Gly-O*t*Bu but not the corresponding methyl ester (*1419*). Liberek (*1420, 1421;* cf. *1422*) prepared a number of carbobenzoxy-β-cyano-L-alanyl peptides via the *p*-nitrophenyl ester, phosphazo, phosphorus oxychloride, and carbodiimide methods.

The formation of ring structures during coupling reactions for the synthesis of asparaginyl peptides will lead to the appearance of additional by-products. Surbeck-Wegmann (*2236*) isolated *p*-phenylazo-carbobenzoxy-α-aminosuccinimide in 40% yield from the reaction of *p*-phenylazocarbobenzoxy-L-asparagine with L-glutamine-1-(*tert*-butyloxy-carbonyl)-hydrazine in the presence of *N,N'*-dicyclohexylcarbodiimide. The coupling reaction had to be carried out with *p*-phenylazocarbo-benzoxy-L-aspartic acid methyl ester. The β-amide group was introduced subsequently to obtain the desired peptide. Carbobenzoxyasparaginyl peptides are apparently more stable to alkaline hydrolysis than peptides with C-terminal asparagine. Benzoyl-DL-asparaginyl-α-glycine *n*-hexyl-amide will be rearranged to a very minor extent to the benzoyl-DL-aspartyl β-peptide by treatment with 0.1 *N* sodium hydroxide for 6 hours at room temperature (*120*). Riniker *et al.* (*1820, 1824*) have recently shown that asparaginyl peptides undergo changes upon exposure to tempera-tures of 50°–100°C in water. Via intermediary imide structures aspartyl α-peptides are formed, and via $\alpha \rightarrow \beta$ transpeptidation aspartyl β-pep-tides. The N-terminal asparaginyl residue may be split off as aspartic acid and ammonia, and the compounds are subject to extensive destruction.

Schwyzer *et al.* (*2016, 2017*) and Rittel *et al.* (*1827*) obtained asparaginyl peptides for the synthesis of angiotensins by alkaline and acidic hydrolysis of the corresponding asparaginyl peptide esters. They isolated only very little aspartyl peptide besides the asparaginyl peptide from the alkaline hydrolysis (1 *N* K_2CO_3 solution, methanol, PH 10.5–11, 75 minutes, 30°C; or 0.1 *N* sodium hydroxide, methanol, pH 10.5–11, 30 minutes). The acid hydrolysis (concentrated HCl, 40°C, 90–100 min-utes), however, produced the aspartyl peptide predominantly. Leach and Lindley (*1352, 1353, 1354*) have reported on the hydrolysis of amide groups of free and peptide-bound asparagine. Decarbobenzoxylations of carbobenzoxy-L-asparaginyl peptides by catalytic hydrogenation or by reaction with hydrogen bromide follow the normal course.

Other important derivatives for the synthesis of asparaginyl peptides are *p*-methoxycarbobenzoxy- and *tert*-butyloxycarbonyl-L-asparagine and the corresponding *p*-nitrophenyl ester (*1974*). These protecting groups, which can be readily cleaved under acidic conditions, offer an advantage in the case of peptides containing methionine or cysteine as well as peptides with ω-protecting groups where catalytic hydrogenation is not desirable. Phthalyl-L-asparagine prepared from asparagine and car-bethoxyphthalimide serves a similar purpose (*1600*). Again, the *p*-nitrophenyl ester is the most suitable derivative for coupling reactions (*1974*). The phthalyl group can be removed selectively by treatment at room temperature for 2 days with hydrazine hydrate in dioxane (*230*), or by boiling in methanol for 1 hour (*1967*). The reactions proceed with-out attack of the acid amide group. Zervas and Theodoropoulos (*2668*) and Stelakatos *et al.* (*2203*) prepared trityl-L-asparagine in 40–50% yield by direct tritylation of L-asparagine in aqueous isopropanol in the presence of diethylamine. Amiard and Heymès (*36*) obtained trityl-L-asparagine trityl ester in 85% yield by tritylation in water/triethylamine. The ester can be hydrolyzed in good yield in aqueous acetone/pyridine. The synthesis of trityl-L-asparagine by tritylation of L-aspartic acid dibenzyl ester is somewhat more elaborate (*36*). The β-carboxyl group must be liberated selectively and amidated with ammonia/*N,N'*-dicyclohexylcarbodiimide. The α-carboxyl group is liberated selectively by catalytic hydrogenation. Peptide syntheses via the carbodiimide method have been described as giving moderate to good yields (*36, 490*). Side reactions caused by dehydration were not reported.

Carbobenzoxyisoasparagine (*193, 1474*) was used for the synthesis of isoasparaginyl peptides via the anhydride method with isobutyl chlorocarbonate as anhydride-forming reagent (*1474*).

b. Aspartic Acid and Asparagine Peptides

Peptides with carboxyl-terminal aspartic acid pose no problems (*820, 1369, 1425, 1558, 1810, 2283, 2583*). L-Aspartic acid dimethyl ester (*849*), the diethyl ester (*820, 1369, 2583*), the di-*tert*-butyl ester (*2283*), and the dibenzyl ester (*1425, 1558*) were employed for their synthesis. They were coupled via the anhydride (*1425, 1558, 2283*), azide (*2583*), and phosphazo (*820*) methods. The lability of the ester groups to 30% hydrogen bromide in glacial acetic acid must be taken into consideration if it is desired to continue with coupling reactions at the amino end (*125*). As in the case of glutamic acid, 18% of α-monoester and 6% of acid are formed from aspartic acid dibenzyl ester, 16 and 10% of acid from the β-benzyl and α-benzyl esters, respectively. The β-ethyl ester is more stable; it is hydrolyzed to the extent of only 4%. By contrast, one frequently finds decarbobenzoxylations described without mention of side reactions (*149, 2163, 2217*). Coupling reactions with salts of aspartic acid monoesters are applicable where additional condensations at the carboxyl end are intended (*1249, 2350a*). L-Aspartic acid α-*tert*-butyl ester β-benzyl ester (*1862, 2265*) or the α-benzyl ester β-*tert*-butyl ester (*1974*) appears to be promising for this purpose.

Peptides with carboxyl-terminal asparagine have been prepared predominantly via salt-coupling reactions (*217, 1354, 1750, 1807*). The anhydride method was applied most frequently. During the acid-catalyzed esterification of, for example, tosyl-L-asparagine with methanol/hydrochloric acid, tosyl-L-aspartic acid dimethyl ester (9%) is formed in addition to the expected tosyl-L-asparagine methyl ester (*2654;* cf. *1849*). This is in remarkable contrast to tosyl-L-glutamine where, during the same reaction, the diester is formed in a yield of 24%. It was surprising that alcoholysis of the amide group upon exposure to ethanol/hydrochloric acid ocurs to a small extent. Tosyl-L-glutamic acid diethyl ester was isolated in only 9% yield. Tosyl- (*2654*) and carbobenzoxy-L-asparagine methyl esters (*193, 1987, 2654*) are accessible without difficulty by esterification with diazomethane. Sulfuryl chloride was used as catalyst for the esterification of both compounds. Transesterification of *N*-acyl derivatives with methyl formate in the presence of 96% sulfuric acid or toluenesulfonic acid (*2279*) leads to the formation of the same products. Alcoholysis of the amide group does not occur. L-Asparagine α-benzyl ester can be prepared from the carbobenzoxy compound by treatment with hydrogen bromide/glacial acetic acid (*2163, 2217*). Several methods are known for its synthesis: introduction of the β-amide group in carbobenzoxy-L-aspartic acid benzyl ester via the acid chloride (*205*), reaction of carbobenzoxy-L-asparagine with phenyl diazomethane (*2163*), and, finally, reaction of the silver salt of carbobenzoxy-L-asparagine with benzyl bromide (*2217*). Carbobenzoxy-L-asparagine can also be esterified by treatment with *p*-nitrobenzyl chloride (*2163*) in dimethylformamide/triethylamine. L-Asparagine *p*-nitrobenzyl ester was used for the synthesis of an insulin sequence (*1192*).

Alkaline hydrolyses during the synthesis of asparagine peptides, as in the case of glutaminyl derivatives, can lead to complications. Sondheimer and Holley (*2160*) did not obtain the corresponding acids, but isolated carbobenzoxy-L-aminosuccinimide as main product upon alkaline treatment of carbobenzoxy-L-asparagine methyl ester and -isoasparagine methyl ester.

3. α-Aminoadipic Acid

α-Aminoadipic acid was found in its free form in pea sprouts (*962*) and bound as δ-peptide in an extract of *Penicillium chrysogenum* (*74, 75*). Morris and Rudinger (cf. *1850*) reported the synthesis of this peptide without experimental details. Its structure (H-Aad-δ-Cys-Val-OH) [87] shows great similarity to cephalosporin [88] (cf. *10, 1612*). The tosyl group was used as *N*-terminal protection. The α-carboxyl group was blocked by the oxazolidone ring. The acid chloride method was applied to the coupling reaction with H-Cys(Bzl)-Val-OMe. The oxazolidone ring and the methyl ester were cleaved by alkaline hydrolysis. The tosyl group was removed with sodium in liquid ammonia.

Gut and Rudinger (*886a*) have compared the reactivity of α-aminoadipic acid and glutamic acid. It is more difficult to prepare the derivative corresponding to tosyl-L-pyroglutamic acid, namely, 1-tosyl-6-oxo-L-piperidine-2-carboxylic acid. The compound was obtained only upon

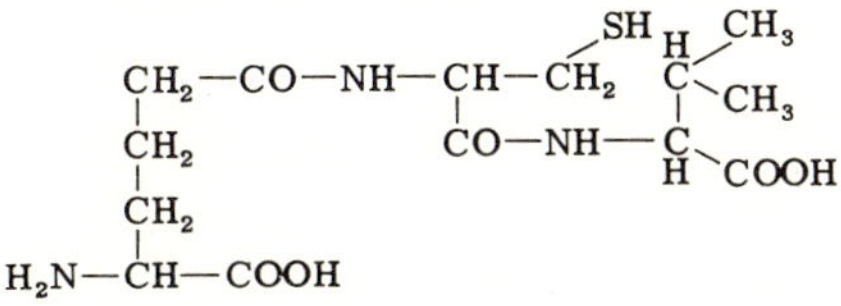

[87]

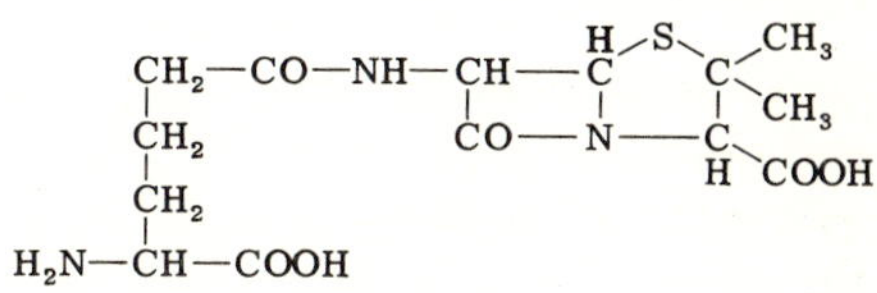

[88]

treatment with toluenesulfonyl chloride in pyridine. The reactions of this "pyroaminoadipic acid" proceed in the same way as those of pyroglutamic acid. Besides the α- and δ-amides, α-aminoadipyl-α- and -δ-glycine were synthesized. L-α-Aminoadipic acid can be synthesized by Arndt-Eistert synthesis from tosyl-L-glutamic acid (*1852a*).

D. DIAMINODICARBOXYLIC ACIDS

1. α,ε-Diaminopimelic Acid

The isolation from bacterial cell walls of peptides containing α,ε-diaminopimelic acid (*1151, 2588*) created a demand for the synthesis of corresponding model peptides. Work *et al.* (*2589*) synthesized LL-, DD-, and *meso*-α,ε-diaminopimelic acid. Bricas *et al.* (*363*) and Nicot and Bricas (*1621*) synthesized peptides containing *meso*-α,ε-diaminopimelic acid via the anhydride and carbodiimide methods. Dicarbobenzoxy-*meso*-α,ε-diaminopimelic acid (*2404*) and *meso*-α,ε-diaminopimelic acid dimethyl ester hydrochloride were used as starting material. These peptides were tested for behavior toward leucine aminopeptidase and carboxypeptidase (*361, 362, 1621*).

E. AMINO ACIDS WITH THE ALCOHOLIC HYDROXYL GROUP

1. Serine

Serine is frequently present in biologically active polypeptides, e.g. ACTH, MSH, glucagon, insulin, eledoisin, and bradykinin. With its reactive hydroxyl group serine plays an important role in the active center of many enzymes. The important class of phosphopeptides is derived from linking a residue of phosphoric acid to serine. The reactivity of the hydroxyl group as well as the lability of seryl peptide bonds (*589, 948*) and the tendency to β-elimination reactions and N $\rightarrow$ O acyl shifts make it difficult to synthesize serine-containing peptides.

a. Syntheses with the Unprotected Hydroxyl Group

α. Seryl Peptides. Seryl peptides are synthesized preferentially from carbobenzoxy-L-serine hydrazide (*22, 430, 589c, 724, 725, 740, 769, 894, 949, 1021, 1538, 1951, 2617, 2636, 2637*). A rearrangement of the azide is the cause of widely differing yields. 4-Carbobenzoxyaminooxazolidinone-2 [89] (*86, 769*) is formed via the isocyanate:

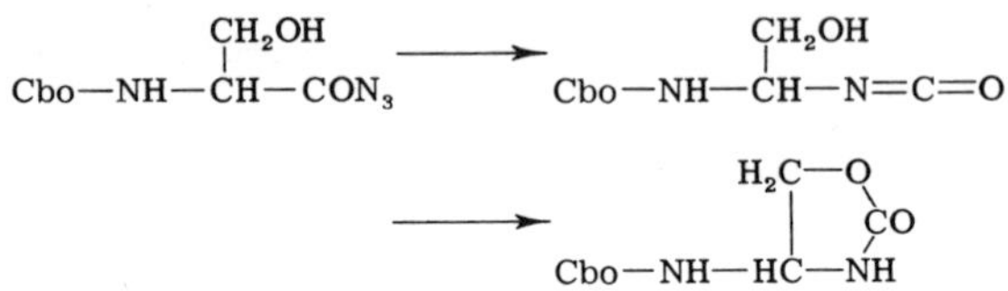

[89]

According to Schnabel and Zahn (*1946*), the hydrazinolysis of carbobenzoxy-L-serine methylester is often accompanied by racemization, even if an insufficient amount of hydrazine hydrate is used (cf. *2005*). An optically pure product, however, can be obtained by recrystallization. During the carbobenzoxylation of serine, *N,O*-dicarbobenzoxyserine (*762, 1379*) is formed in addition to the *N*-protected product. Therefore *N*-carbobenzoxy-L-serine is prepared by carrying out the acylation at pH 9.8–10 and keeping the reaction mixture for 30 minutes at pH 10 to hydrolyze the *N,O*-bis derivative (*1569*). It is more advantageous to carry out the carbobenzoxylation in the presence of magnesium oxide (*85*) or sodium hydrogen carbonate (*894*). Carbobenzoxy-L-serine was applied successfully to the synthesis of peptides via the carbodiimide method (*87, 589c, 740, 741, 894, 896, 1102, 1538, 2020, 2617*). The formation of acylureas was observed on occasion (*1538, 2617*). The anhydride method has been employed less frequently (*345, 740*). The method of activated esters was described for carbobenzoxy-L-serine 2,4-dinitrophenyl ester (*1499a, 1829a*). The application of the acid chloride or phosphazo method to the synthesis of seryl peptides requires protection of the alcoholic hydroxyl group (*851*).

Modified carbobenzoxy protecting groups such as the *p*-nitrocarbobenzoxy- (*804*), the phenylazocarbobenzoxy- (*2037*), and the *p*-(*p'*-methoxyphenylazo)-carbobenzoxy (*2037*) groups have been used to block the amino group of serine. These derivatives have not yet been applied to the synthesis of peptides. Anderson and McGregor (*49*) obtained only modest yields when they tried to synthesize *tert*-butyloxycarbonyl-DL-serine. By contrast, *tert*-butyloxycarbonyl-L-serine hydrazide can be prepared very readily via the methyl ester, which in turn is accessible by treatment with *tert*-butyloxycarbonyl azide in pyridine (*1103*). Di- and tripeptide derivatives of serine were synthesized by using tosyl-DL- and -L-serine hydrazide (*1021, 2219, 2583*) as well as tosyl-L-serine in combination with the oxazolidinone method (*932, 1548*), or by coupling trifluoroacetyl-DL- and -L-serine via the carbodiimide method (*2497*). The *N*-trifluoroacetyl group is labile to acid when attached to a β-hydroxyamino acid. Weygand and Rinno (*2497*) attribute this to an N → O acyl shift. Due to the fast hydrolysis of *O*-trifluoroacetyl compounds this shift is not reversible. Other ninhydrin-positive products are formed during the cleavage of the tosyl group of tosylserine by sodium in liquid ammonia (*345*). By contrast, Cbo-Ser-Met-OH can be decarbobenzoxylated by treatment with sodium in liquid ammonia without complication (*345, 1021*). Phthaloylation of L-serine with phthalic anhydride has not yet been described (*2064*). The method of Nefkens (*1600*) was applied only to the synthesis of phthalyl-DL-serine. Trityl-L-serine (*891;* cf. *37*) was converted to the β-lactone in 15% yield (*2067*). Aminolytic cleavage of the lactone leads to the formation of peptide derivatives whereby the optical activity is preserved. *N,N'*-Dicyclohexylcarbodiimide was used to activate trityl-L-serine for the synthesis of peptides (*891, 1904*). *N*-Benzyl- (*2370*) and *N,N'*-dibenzylserine (*2371*) are known only in the form of their DL-derivatives. They are not of interest for the synthesis of peptides.

The lability to alkali of the seryl peptide bond is responsible for the low yields of hydrolysis (*894, 949, 2637*). Schnabel and Zahn (*1946*) caution on the partial racemization under alkaline conditions. Thus, serine in Cbo-D-Ser-Gly-L-Ala-OEt will be racemized in aqueous methanolic solution by the addition of triethylamine (*1941*). Guttmann and Boissonnas (*894*) found that, under the normal conditions of decarbobenzoxylation with hydrogen bromide in glacial acetic acid, the alcoholic hydroxyl group was acetylated completely while the phenolic hydroxyl group was not affected (cf. *1616a, 1951*). Addition of 30% water to the hydrogen bromide/glacial acetic acid solution prevents the *O*-acetylation. It is more advantageous to carry out the reaction in trifluoroacetic acid or diethyl phosphite (*894, 895*). Dioxane was also used as solvent.

β. *Serine Peptides.* The synthesis of serine peptides with the unprotected hydroxyl group poses fewer problems than that of seryl peptides. However, caution must also be exercised during alkaline hydrolysis and the reaction with hydrogen bromide in glacial acetic acid. The methyl ester (*894, 949*) is used preferentially as well as the ethyl ester (*1793*) and benzyl ester (*741, 2617*) as ester component. *O*-Benzyl-DL-serine benzyl ester was isolated as hydrochloride from the esterification of DL-serine with benzyl alcohol in the presence of benzenesulfonic acid (*2617*). Peptides were synthesized via the azide (*430, 949, 1021, 1235*), anhydride (*82, 210, 740, 896, 2296*), carbodiimide (*740, 935b, 1904a, 2020, 2300, 2617*), acid chloride (*4, 323, 1235, 2370*), and phosphazo (*2637*) methods. The latter method gave low yields (cf. tyrosine). *N,O*-Di-(dicarbobenzoxy-L-tyrosyl)-DL-serine was isolated from the condensation of *N,O*-dicarbobenzoxy-L-tyrosine with the sodium salt of DL-serine via the acid chloride method (*4*). Schneider (*1951a*) has reported on the relatively easy hydrolysis of the aspartyl—serine bond, which was explained by an interaction between the β-carboxyl group and the hydroxyl group.

b. Syntheses with the Protected Hydroxyl Group

Efforts to find a reversible protection for the hydroxyl group of serine have been made by the various schools of peptide synthesis. Such a protection would safely exclude the occurrence of side reactions. Protection of the alcoholic hydroxyl function by acylation enhances the chances of β-elimination. Therefore etherification is more suitable for serine as well as for threonine.

α. *O-Acetylserine.* *O*-Acetyl-L-serine can be prepared by treatment of L-serine with an excess of glacial acetic acid saturated with hydrogen chloride (*709, 2064*). Kienhuis *et al.* (*1228*) succeeded in the *O*-acetylation of H-L-Ser-Gly-OEt·HCl by applying the same procedure. The resulting product was used in additional coupling reactions via the anhydride method (*1228, 2350a*). Theodoropoulos *et al.* (*2297*) describe the *O*-acetylation of a fully protected L-serine-containing tripeptide with acetic anhydride in pyridine. The *O*-acetyl group is stable under the conditions of hydrogenolysis in methanol. *O*-Acetylseryl peptides undergo an O → N acyl shift at pH values > 7 (*894, 2297*). Phthalyl-*O*-acetyl-L-serine was used in coupling reactions via the carbodiimide method (*2064*). Frankel and Halmann (*762*) synthesized *N*-carbobenzoxy-*O*-acetyl-DL-serine from *N*-carbobenzoxy-DL-serine and acetyl chloride. The reaction was carried out in benzene. The compound was used for the synthesis of poly-*O*-acetyl-DL-serine after conversion to *O*-acetyl-DL-serine *N*-carboxylic anhydride.

β. *O-Benzoylserine.* Bergmann and Miekeley (*189*) and Elliot (*658*) have described *O*-benzoyl-DL-serine. Baganz and Dransch (*91*) prepared *O*-benzoyl-L-serine by hydrolysis of L-α-benzamido-β-propionic acid methyl ester with potassium hydrogen carbonate solution. The *O*-benzoyl compound has not yet been used for the synthesis of peptides.

γ. *O-Trifluoroacetylserine.* Weygand and Rinno (*2497*) have reported that *N,O*-bistrifluoroacetyl-L-(and DL)-serine are not suitable for the synthesis of peptides. They demonstrated for the reaction with tyrosine methyl ester, using *N,N'*-dicyclohexylcarbodiimide as condensing agent, that the *O*-trifluoroacetyl group will be transferred with formation of *N*-trifluoroacetyltyrosine methyl ester. The *O*-trifluoroacetyl group is rapidly hydrolyzed in aqueous solution.

δ. *O-Carbobenzoxyserine.* Levene and Schormüller (*1379*) had already demonstrated that the carbobenzoxylation of DL-serine leads to the formation of *N,O*-dicarbobenzoxy-DL-serine in addition to *N*-carbobenzoxy-DL-serine. Frankel and Halmann (*762*) separated the two compounds successfully by fractional crystallization. *O*-Carbobenzoxy-DL-serine was prepared via *O*-carbobenzoxy-DL-serine *N*-carboxylic anhydride

ε. *O-Benzylserine.* The *O*-benzyl residue is a very suitable protecting group for the hydroxyl function. Okawa (*1652*) used sodium benzylate and bromoacetate for the synthesis of *O*-benzyl-DL-serine. The resulting *O*-benzylglycol aldehyde was subjected to cyanohydrin synthesis. Grassmann *et al.* (*851*) employed the reaction of α,β-dibromopropionic acid ethyl ester with sodium benzylate in benzyl alcohol and subsequent amination with concentrated ammonia. Okawa (*1653*) was able to prepare *O*-benzyl-L-serine by enzymatic cleavage of *N*-acetyl-*O*-benzyl-DL-serine with Taka-diastase. Wünsch and Fürst (*2597*) obtained the compound from *N*-formyl-*O*-benzyl-DL-serine by resolution of the racemic mixture via the brucine and quinine salts. *N*-Carbobenzoxy-*O*-benzyl-L-(DL)-serine (*851, 1654*) and *O*-benzyl-L-(DL)-serine ethyl ester hydrochloride (*851, 1655*) and methyl ester hydrochloride (*1090, 1655, 2082a*) were used as starting material for the synthesis of peptides. The phosphazo (*851*), acid chloride (*1653*), anhydride (*1090*), *p*-nitrophenyl ester (*278*), and carbodiimide (*1654, 1655, 1903*) methods as well as opening of the *N*-carboxylic anhydride of *O*-benzyl-DL-serine were applied (*1653*). It was possible to use *N*-phthalyl-*O*-benzyl-L-serine (*2082a*), *N*-formyl-*O*-benzyl-L-serine (*752a*), and *N*-trifluoroacetyl-*O*-benzyl-DL-serine (*2497*) for the synthesis of peptides via the carbodiimide method. The benzyl ether bond is stable under the conditions of alkaline hydrolysis (*851, 1653, 1655*) and hydrazide formation (*1655*). Cleavage of the seryl peptide bond, which is frequently encountered when the hydroxyl group of serine is not protected, was not observed. Accordingly, the yields are considerably higher (*1090*). The benzyl group is stable under the conditions of esterification. However, *O*-acetyl-L-serine was isolated when selective decarbobenzoxylation of *N*-carbobenzoxy-*O*-benzyl-L-serine by treatment with hydrogen bromide in glacial acetic acid (50 minutes) was attempted (*1655*). *N*-Deformylation with 1 *N* hydrochloric acid in methanol by allowing the solution to stand overnight can be carried out without adverse effects on the ether bond (*1090*). The benzyl group can be removed by catalytic hydrogenation in aqueous dioxane (*851*), in methanol/hydrochloric acid (*752a*), or in methanol/acetic acid (*1653*), or by treatment for 1 hour at room temperature (*278*) with 25% hydrogen bromide solution in dioxane.

The benzyl group was applied for the first time to the synthesis of a biologically active polypeptide during the synthesis of the *N*-terminal sequence of ACTH, namely, Cbo-Ser(Bzl)-Tyr-Ser(Bzl)-Met-OH (*1090*). De Haas and Van Deenen (*579*) prepared *N*-phthalyl-L-serine *tert*-butyl ester by catalytic hydrogenation of *N*-phthalyl-*O*-benzyl-L-serine *tert*-butyl ester. The compound is applicable to the synthesis of *O*-peptides and also to the preparation of L-serine *tert*-butyl ester.

ζ. O-Tetrahydropyranylserine. Iselin and Schwyzer (*1101*) were able to prepare *N*-carbobenzoxy-*O*-tetrahydropyranyl-DL-serine methyl ester from carbobenzoxy-DL-serine methyl ester and 2,3-dihydropyrane in the presence of catalytic amounts of hydrochloric acid. The tetrahydropyranyl group is stable to alkali, as was demonstrated by hydrolysis and conversion to *N*-carbobenzoxy-*O*-tetrahydropyranyl-DL-serine cyanomethyl ester. The carbobenzoxy group can be removed selectively by hydrogenation in methanol in the presence of an equivalent amount of acetic acid. The tetrahydropyranyl group can be readily cleaved by boiling for 5 minutes with 2 *N* hydrochloric acid in ethanol (*1101*), or by boiling for 10 minutes in 50% acetic acid (*1154*). Jones *et al.* (*1154*) used *N*-carbobenzoxy-*O*-tetrahydropyranyl-L-serine for preparation of *N*-(L-seryl)-D-glucosamine hydrochloride, coupling it to 1,3,4,6-tetra-*O*-acetyl-*β*-D-glucosamine via the carbodiimide method. Trifluoroacetyl-*O*-tetrahydropyranyl-DL-serine methyl ester was prepared by Weygand and Rinno (*2497*) but not used for the synthesis of peptides.

η. O-tert-Butylserine. Beyermann and Bontekoe (*228, 229*) have reported that the acid-catalyzed esterification of carbobenzoxyhydroxyamino acids with isobutylene leads to the formation of *tert*-butyl ester *tert*-butyl ethers. Subsequent hydrogenolysis yielded the hydrochloride of *O*-*tert*-butyl-L-serine *tert*-butyl ester. This compound represents the ideal starting material for the synthesis of peptides with C-terminal serine. Both protecting groups can be removed with trifluoroacetic acid in 16 hours at room temperature (*229*), with hydrogen bromide in glacial acetic acid in 30 minutes (*457*), or with hydrogen chloride in trifluoroacetic acid (*1962*). Formation of *O*-acetylserine was observed during the reaction with hydrogen chloride in glacial acetic acid. Other suitable derivatives of *O*-*tert*-butyl-L-serine such as the *N*-carbobenzoxy compound, the corresponding hydrazide, as well as *O*-*tert*-butylserine and *O-tert*-butyl-L-serine methyl ester hydrochloride, have been described by Callahan *et al.* (*457*), by Schröder (*1962*), and Wünsch and Jentsch (*2597c*).

c. N → O Acyl Shift

The reversible shift of an acyl residuc between hydroxyl and amino groups (acyl migration, peptidyl migration, acyl shift) (cf. *182, 189*) proceeds via the hydroxyoxazolidine [90] as intermediate (*96, 506, 660*).

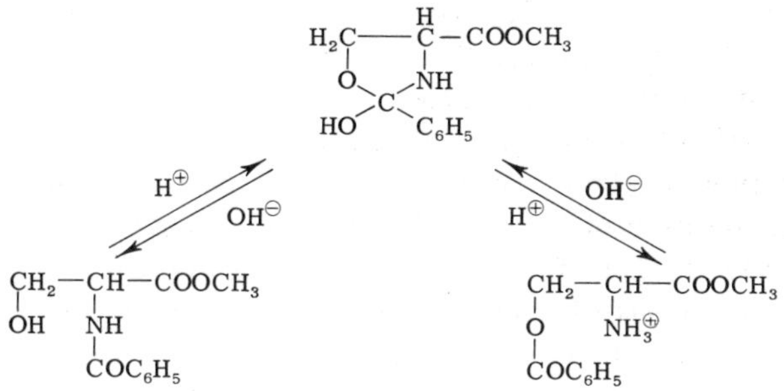

[90]

The same reaction mechanism was discussed for the S → N acyl shift (see Chapter VI,C,2). The migration of the acyl residue is dependent upon the pH value, the nature of the acyl group (*506*), and the hydroxy-amino acid involved (*777*).

Efforts have been made to devise a chemical method for the specific cleavage of peptide chains at the residues of serine and threonine, utilizing the acyl migration. A technique developed in the case of silk fibroin uses concentrated sulfuric acid. The amino groups released were determined as DNP derivatives after treatment with 2,4-dinitrofluorobenzene (*660;* cf. *540a*). The method was also applied to other proteins (*1787, 2573*). It was possible to convert a poly-DL-serine by treatment with concentrated sulfuric acid almost completely to the polyester (*708*). However, low yields, nonspecific cleavage, and partial destruction of other amino acid residues have been pointed out (*506*). The peptidyl migration at serine residues of silk fibroin can also be brought about by anhydrous phosphoric acid (*1471, 2636*). Anhydrous hydrofluoric acid appears to be a very suitable agent for this purpose. The rearrangement proceeds with high yields and without nonspecific cleavage of peptide bonds (*2105*). The migration of acyl residues was also observed in the presence of the following reagents: alcoholic hydrochloric acid (*96*), hydrogen chloride in nitromethane (*540a*), formic acid/boron trifluoride (*96*), hydrogen chloride in ether (*1380*), and acetic anhydride in methanol (*1064*) under the conditions of esterifications. In this connection, kinetic studies carried out by Harris *et al.* (*948*) in the case of the acid hydrolysis of dipeptides with 2 *N* hydrochloric acid at 99°C are of interest. According to these authors, H-Gly-Ser-OH is cleaved 5 times as fast as H-Ser-Gly-OH or H-Gly-Ala-OH. A reversible inactivation of lysozyme and ribonuclease, originally ascribed to acyl migration induced by anhydrous formic acid (*1158*), was later recognized as formylation of hydroxyl groups (*747, 1596, 1781*). The reverse reaction of the N → O acyl shift, the O → N migration, usually occurs in water at pH values > 7 (*152, 1157c, 1380*). Triethylamine in water also causes an O → N acyl shift (*1380*). Kinetic studies were conducted with *O*-acetylserine and -serylleucine (*1158*), with *O*-acetylserine and -threonine (*777*), and with *O*-diazoacetyl-, *O*-glycolyl-, and *O*-acetylserine (*779*). Zahn and Siepmann (*2639*) studied an alkyl migration, analogous to the acyl shift, in the case of *O*-dinitrobenzylserine and threonine.

The N → O acyl shift was used for the synthesis of *O*-peptides (*151a*). In a synthesis of azaserine, use was made of the rearrangement of *N*-glycyl-DL-serine to *O*-glycyl-DL-serine (*1569*). Guttmann and Boissonnas (*897*) were able to rearrange an *O*-acetyl derivative to the *N*-acetyl compound by keeping it at pH 7.9 for 1 hour. The *O*-acyl compound had been obtained from the decarbobenzoxylation of a substituted N^{α}-carbobenzoxy α-MSH sequence by treatment with hydrogen bromide in glacial acetic acid/diethyl phosphite/methyl ethyl sulfide = 4:1:1.

2. Threonine

Compared with other amino acids, the number of threonine-containing peptides is surprisingly low. Sequences of higher peptides with threonine were synthesized only when efforts to synthesize polymyxin B, circulin A, insulin, and glucagon were being made.

a. Syntheses with the Unprotected Hydroxyl Function

Carbobenzoxy-L-threonine (*84, 1535, 1903, 2638*), carbobenzoxy-L-threonine hydrazide (*583, 1814, 1972*), and L-threonine methyl (*1972, 2386*) and ethyl (*1746*) esters are the derivatives of choice. They are used almost exclusively for the synthesis of peptides. The carbobenzoxylation of threonine is carried out in the presence of sodium hydroxide (*1903, 2638*) or magnesium oxide (*84, 1535*). Bergel and Wade (*176*) isolated 5-methyl-2-oxooxazolidine-4-carboxylic acid from the reaction of the DL-compound in sodium hydroxide (pH 8–10, 5°C). The same compound was formed upon direct treatment of carbobenzoxy- or *p*-nitrocarbobenzoxy-DL-threonine with 2 moles of alkali at room temperature. The free esters of threonine are crystalline compounds.

Guttmann and Chang (*888*) described L-threonine benzyl ester *p*-toluenesulfonate. The esterification with benzyl alcohol by the method of Miller and Waelsch (*1561*) requires a reaction time of 15 hours, otherwise L-threonine *p*-toluene sulfonate will be isolated. Racemization was not observed under these more vigorous reaction conditions. Threonyl and threonine peptides were synthesized preferentially by the carbodiimide (*1193, 1200, 1535, 1903, 2386, 2567, 2638*) and azide (*1200, 1746, 1972, 2386*) methods. These methods were applied to the synthesis of sequences of the insulin B-chain (*1193, 1200, 1308, 2638*), of glucagon (*230, 1972*), of polymyxin (*2227, 2384, 2386, 2392*), and of circulin A (*1850*). The 2,4-dinitrophenyl ester of carbobenzoxy-L-threonine was also used for the synthesis of peptides (*1829a*). *tert*-Butyloxycarbonyl-L-threonine hydrazide prepared from L-threonine methyl ester and *tert*-butylcarbonyl azide proved to be very useful (*1972*). Phthalyl-L-threonine was prepared by the direct reaction of phthalic anhydride with L-threonine in dioxane suspension. The compound was used for the synthesis of simple substituted peptide derivatives via the carbodiimide method (*2064*). Phthalyl-L-threonyl-*N,N'*-dicyclohexylurea was isolated as by-product. Weygand and Rinno (*2497*) used *N*-trifluoroacetyl-L-threonine hydrazide for the synthesis of threonyl peptides. *N*-Trifluoroacetyl-L-threonine hydrazide and *N*-tosyl-L-threonine hydrazide (*353*) alike have not been used on a broader basis for the synthesis of peptides. Side reactions are associated with the cleavage of the tosyl group by sodium in liquid ammonia (*345*). *N,N'*-Dibenzyl-DL-threonine (*2371*) was used in the form of the *threo*-1-*p*-nitrophenyl-L-aminopropanediol-(1) salt for the resolution of the racemic mixture. Peptides of allothreonine have also been described (*75b, 325*). Kaneko and Inui (*1168*) and Elliott (*658, 659*) have studied the inversion of *erythro*- to *threo*-threonine via oxazolidone esters.

b. Syntheses with the Protected Hydroxyl Function

α. *O-Acetylthreonine.* Sakami and Toennies (*1901*) have already described *O*-acetyl-DL-threonine. Fujiwara *et al.* (*777*) synthesized *O*-acetyl-DL- (and D-) threonine by reacting DL- or D-threonine with acetic acid saturated with hydrogen chloride. The O → N shift takes place within 3 hours upon exposure to 0.5 *N* ammonium hydroxide.

β. *O-Benzoylthreonine.* Elliot (*658*) obtained *O*-benzoyl-DL-threonine by acidic cleavage of the corresponding oxazoline. Alkali very readily induces the O → N acyl migration. Treatment with ethanolic hydrochloric acid converts *N*-benzoyl-DL-allothreonine ethyl ester to the *O*-benzoyl-DL-allo compound. *N*-Formyl-*O*-benzoyl-DL-threonine (*659*) is also known.

The *O*-benzoyl group can be removed by treatment with alkali for 30 minutes. Avaeva and Botvinik (*78*) have prepared several peptides of *N*-(benzoylaminacyl)-*O*-benzoyl-DL-threonine.

γ. *O-Benzylthreonine.* Murase *et al.* (*1588*) prepared *O*-benzyl-L-threonine from *N*-acetyl-DL-threonine by allowing it to react with benzyl bromide in sodium/liquid ammonia. The racemic mixture was resolved by enzymatic digestion with Taka-diastase. There are no reports at hand on the stability or the application to peptide synthesis.

δ. *O-tert-Butylthreonine.* Like tyrosine and serine, carbobenzoxy-L-threonine can be converted to *N*-carbobenzoxy-*O*-*tert*-butyl-L-threonine *tert*-butyl ester by the reaction with isobutylene. The subsequent catalytic hydrogenation leads to the formation of *O*-*tert*-butyl-L-threonine *tert*-butyl ester. Both protecting groups can be removed by allowing the compound to stand for 16 hours in anhydrous trifluoroacetic acid (*228, 229*). The protecting groups were also found to have been removed after the compound had been kept in trifluoroacetic acid for 1 hour at 0°C and 1 hour at room temperature (*230*). In combination with the carbobenzoxy and phthalyl groups, *O*-*tert*-butyl-L-threonine *tert*-butyl ester was very useful for the synthesis of the C-terminal tetrapeptide sequence of glucagon. The phthalyl group was removed selectively by treatment with hydrazine hydrate in dioxane for 2 days at room temperature (*230*), or by boiling with hydrazine hydrate for 1 hour in methanol (*1968*). Other derivatives of *O*-*tert*-butyl-L-threonine were prepared according to procedures described for those of serine (cf. *1962, 2597c*).

ε. *O-Tetrahydropyranylthreonine.* Jones *et al.* (*1153*) have described the synthesis of *N*-carbobenzoxy-*O*-tetrahydropyranyl-L-threonine and its application to the synthesis of *N*-L-threonyl-D-glucosamine. The *O*-tetrahydropyranyl group was removed by boiling for 10 minutes with 50% acetic acid.

3. Hydroxyproline

Hydroxyproline-containing peptides have been synthesized by several authors in connection with studies on the acidic, basic, and enzymatic hydrolysis of collagen to determine the sequence. The specificity of collagenase was studied on hydroxyproline-, glycine-, and proline-containing peptides.

a. Syntheses with the Unprotected Hydroxyl Group

The hydroxyl group was unprotected in all cases where hydroxyproline was used for the synthesis of peptides. The frequently employed carbobenzoxy group protected the amino group (*88, 850, 1694*), and the methyl (*996, 2148*) or benzyl ester (*1116, 1756, 2148*) the carboxyl group.

The azide (*1606, 2148*), carbodiimide (*916, 2064*), and anhydride (*576, 916, 1124, 1756, 1761, 2531, 2532*) methods were applied to coupling reactions for hydroxyprolyl peptides. Hydroxyproline peptides were synthesized via the acid chloride (*1606, 2148*), azide (*916, 2148*), *p*-nitrophenyl ester (*996*), and anhydride (*916, 1761*) methods. Side reactions were not reported for these coupling reactions. Unsatisfactory results were obtained when carbobenzoxy-L-hydroxyproline was to be coupled via the phosphazo method (*2592*). Peptides of glycine, proline, and hydroxyproline and their derivatives do not crystallize well. As in the case of proline, side reactions must be expected during the alkaline hydrolysis (*1758, 1761*), amidation (*996*), and catalytic hydrogenation (*2148*) of peptides containing hydroxyproline and glycine.

In the allo series, Patchett and Witkop (*1694*) synthesized *N*-carbobenzoxy-L(D)-allohydroxyproline lactone from the corresponding acid (cf. *224*), using *N,N'*-dicyclohexylcarbodiimide in methylene chloride or *p*-toluenesulfonyl chloride in pyridine. A dipeptide derivative was obtained by cleavage with ethyl glycinate. Derivatives of L-allohydroxyproline have been described by Adams *et al.* (*21*).

Of the large number of possible *N*-protections beside the carbobenzoxy group, only the tosyl (*1478*) and *p*-nitrocarbobenzoxy (*463, 595*) groups were used to synthesize the derivatives of hydroxyproline.

b. *O*-Substituted Hydroxyproline

α. *O-Tosylhydroxyproline*. Patchett and Witkop (*1694*) synthesized *N*-acetyl-*O*-*p*-tosyl-L-hydroxyproline methyl ester by allowing *N*-acetyl-L-hydroxyproline methyl ester to react with *p*-toluenesulfonyl chloride in pyridine. The corresponding acid was obtained by hydrolysis. Treatment with potassium carbonate and heating in methyl ethyl ketone resulted in the cleavage of the *O*-tosyl group, inversion at the γ-carbon atom, and formation of the lactone of *N*-acetyl-L-allohydroxyproline. *N*-Carbobenzoxy-L- and *N*-carbobenzoxy-L-allohydroxyproline methyl ester can also be converted to the *O*-tosyl compounds by treatment with *p*-toluenesulfonyl chloride in pyridine.

β. *O-tert-Butylhydroxyproline*. The hydroxyl group of hydroxyproline, as in the case of serine and threonine, can be substituted by the *O-tert*-butyl residue by acid-catalyzed addition to isobutylene. The carboxyl group will be esterified simultaneously (*229*). The carbobenzoxy group was used as amino protection during this reaction. *O-tert*-Butyl-L-hydroxyproline *tert*-butyl ester has not been applied to the synthesis of peptides.

4. Hydroxylysine

Hydroxylysine [91] can be synthesized from acetaminomalonic ester and acrolein (*2640, 2644, 2682*):

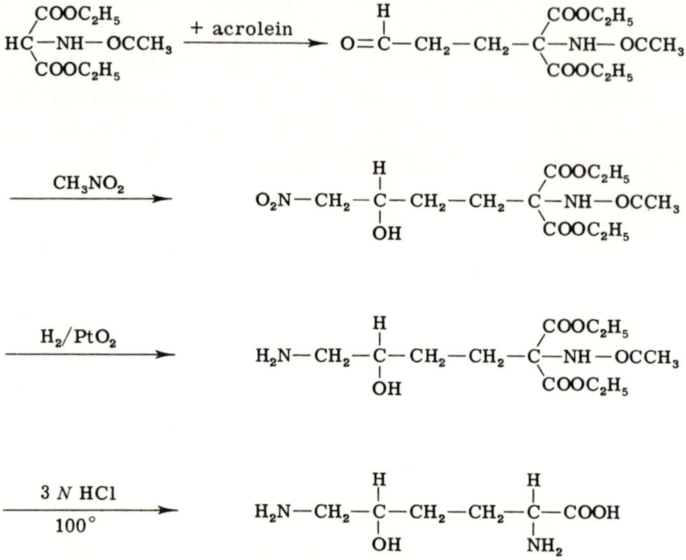

[91]

The diastereoisomers of synthetic hydroxylysine can be readily separated by fractional crystallization from ethyl acetate of the readily accessible dicarbobenzoxyhydroxylysine [92], the larger part of which is present as lactone [93]. Compound [91] was identified as the DL- and com-

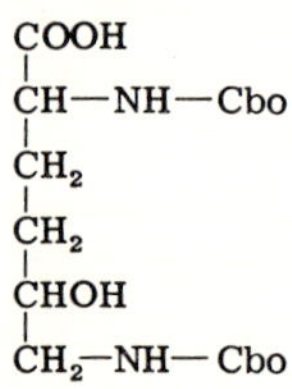

[92]

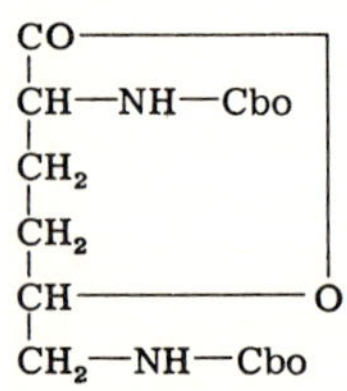

[93]

pound [94] as the DL-allo form (*2635, 2644*). The term hydroxylysine is used arbitrarily for the naturally occurring form of the amino acid. By this no statement is intended to be made on the configuration of the C^{δ} atom with regard to the C^{α} atom. According to Witkop (*2574*), the application of Hudson's lactone rule assigns the *erythro* configuration to the naturally occurring hydroxylysine and the *threo* configuration to allohydroxylysine [91] and [94]. Diphthalyl- and ditosylhydroxylysine

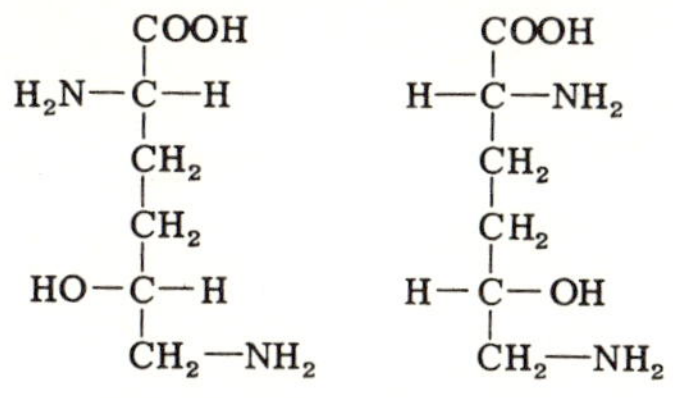

DL-Hydroxylysine

[91]

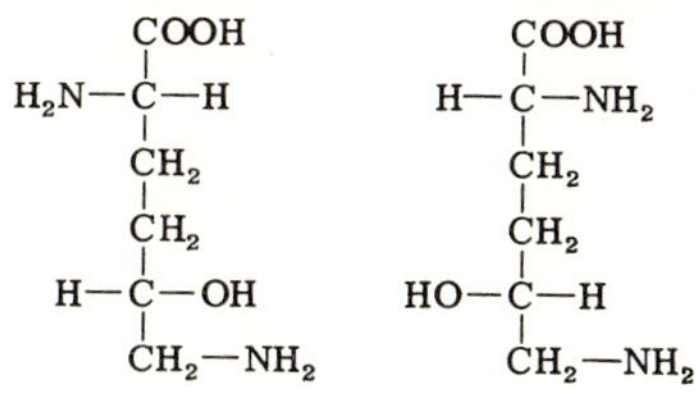

DL-Allohydroxylysine

[94]

are the only known *N*-substituted derivatives of value for the synthesis of peptides (*2635*).

The hydrolysis of DL-hydroxylysine δ-lactone [95] with formation of DL-hydroxylysine is much slower than that of the corresponding allo-form. The first compound is initially converted by acyl migration to DL-hydroxylysine ε-lactam [96], which can be isolated as monohydrate of the hydrochloride. DL-Allohydroxylysine ε-lactam hydrochloride mono-

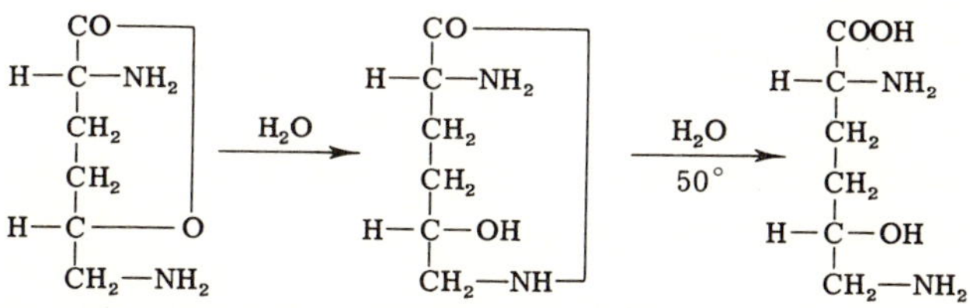

[95] [96] [91]

hydrate (*2635*) is more stable than the analogous DL-hydroxylysine compound. The ammonolysis and hydrazinolysis of dicarbobenzoxy-DL-hydroxylysine δ-lactone or dicarbobenzoxy-DL-allohydroxylysine δ-lactone proceed without epimerization (*2645*). The application of hydrazides to the synthesis of hydroxylysyl peptides did not meet with success. The azides are readily converted to the lactones. The aminolysis of the lactone by amino acids or amino acid methyl esters, ethyl esters, or benzyl esters led only to the formation of the diketopiperazines. However, an attempt to open the lactone by boiling with amino acid amides for 1 hour in dioxane was successful. Hydroxylysyl peptides were thus synthesized for the first time (*1942a, 2645*). Efforts aiming at the partial hydrolysis of DL-allohydroxylysylamino acid amides [97] to obtain DL-allohydroxy-lysylamino acids resulted in the formation of DL-allohydroxylysine and amino acid amide after exposure to 1 *N* HCl at 80°C for only 20 minutes. The lability of the peptide bond can be explained by an acid-catalyzed alcoholysis proceeding via an intermediary δ-lactone ring that is free of tensions. A similar situation is encountered during the N→O acyl migration of aminoacyl compounds of serine, which proceeds via a

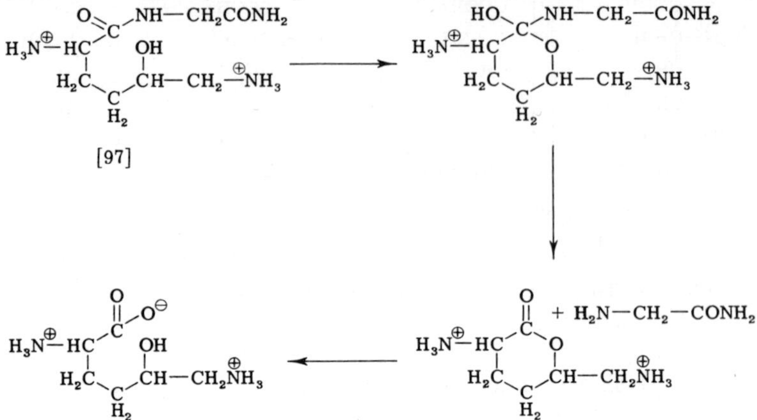

sterically favored oxazolidine ring (*660*). A great lability to acid of hydroxyproline peptides must be expected on the basis of the *cis* arrangement of hydroxyl and carboxyl groups of allohydroxyproline and the possible lactone formation (*2645*).

Wieland (*2522*) has reviewed the occurrence, chemical reactions, and synthesis of other aminohydroxy acids such as β-hydroxyvaline, β-hydroxyleucine, β-hydroxyaspartic acid, γ-hydroxyglutamic acid, γ,δ-dihydroxyleucine, and γ-hydroxyarginine.

F. AMINO ACIDS WITH THE AROMATIC HYDROXYL GROUP

1. Tyrosine

a. Syntheses with the Unprotected Hydroxyl Group

α. Tyrosyl Peptides. The carbobenzoxy group is used preferentially as amino protection for the synthesis of tyrosyl peptides whenever the hydroxyl group remains unprotected. Bergmann and Zervas (*193*) found that carbobenzoxy-L-tyrosine can be prepared only by the reaction of carbobenzoxy chloride with L-tyrosine methyl or ethyl ester and alkaline hydrolysis. It is more advantageous to convert the carbobenzoxy-L-tyrosine ester to the hydrazide (*184*). The direct carbobenzoxylation of L-tyrosine in aqueous alkali requires hydrolysis of the *O*-carbobenzoxy group of the initially formed *N,O*-bis derivative (*593*). *p*-Phenylazo-carbobenzoxy-L-tyrosine is the only known derivative with a modified carbobenzoxy group; it was mentioned in a patent without details of the preparation (*2700*). The synthesis of *tert*-butyloxycarbonyl-L-tyrosine with *tert*-butyloxycarbonyl *p*-nitrophenyl ester gives a yield of only 29%

(*49*). The synthesis of *tert*-butyloxycarbonyl-L-tyrosine hydrazide proceeds more smoothly (*1962*). Formyl-L-tyrosine (*717, 2420*) and trifluoroacetyl-L-tyrosine (*2480*) have not been extensively applied to the synthesis of peptides. They were used only for the preparation of several dipeptide derivatives. The same is true for *N*-tosyl-L-tyrosine which, like the carbobenzoxy compound, can be prepared by tosylation of L-tyrosine ethyl ester and subsequent alkaline hydrolysis (*214, 720*). The tosylation of L-tyrosine in aqueous alkali leads to the formation of *N,O*-ditosyl-L-tyrosine (*718, 1478*).

The azide method is applied preferentially to the synthesis of peptides (*63, 184, 590, 685, 724, 940, 1022, 1032, 1249, 1828, 1993, 2134, 2214, 2628, 2630, 2642*). Investigations by Schnabel and Zahn (*1945*) showed that during the azide formation of Cbo-L-Tyr-Gly-$NHNH_2$ the phenol nucleus is nitrated by excess sodium nitrite, presumably in *o*-position to the hydroxyl group. The reaction of the nitrated azide with DL-alanine benzyl ester yielded carbobenzoxy-L-nitrotryrosyl-glycyl-DL-alanine benzyl ester. Hofmann *et al.* (*1033*) observed the formation of a derivative of urea upon coupling of Cbo-L-Ser-L-Tyr-azide to the triethylammonium salt of L-Met-L-Glu(NH_2)-OH. The product is the result of the rearrangement to the isocyanate. For the synthesis of biologically active polypeptides, *N,N'*-dicyclohexylcarbodiimide (*232, 272, 1200, 1308, 1824, 1827, 2016, 2017, 2344, 2422, 2604, 2696*) and 1-cyclohexyl-3-(2-morpholinoethyl)-carbodiimide (*1827*) were frequently used to couple *N*-protected peptides with C-terminal tyrosine. Racemization was reported for the condensation of Cbo-L-Val-L-Tyr-OH with a tetrapeptide ester (*1821*). Schwarz and Bumpus (*1992*) have pointed out the great tendency to racemization of the tyrosine residue in Cbo-L-Val-L-Tyr-OH. The reaction with *N,N'*-carbonyldiimidazole offers certain advantages (*1701*). The application of the anhydride method (*225, 226, 233, 626, 1185, 1192*) has also been described. The phosphazo (*820*), tetraethyl pyrophosphite (*45*), diethyl chlorophosphite (*2611*), and pyrazolide (*1814*) methods were used less frequently. The acid chloride and isocyanate methods cannot be employed since side reactions occur at the phenolic hydroxyl group. Iselin and Schwyzer (*1102*) synthesized *N*-carbobenzoxy-L-tyrosine *p*-nitrophenyl ester via the carbodiimide and dinitrophenyl sulfite methods.

β. Tyrosine Peptides. The methyl ester (*296, 1993*), ethyl ester (*933*), benzyl ester [as hydrochloride, (*690*), toluenesulfonate (*1117, 2670*), and bezenesulfonate (*803*)] have been applied to the synthesis of tyrosine peptides. L-Tyrosine *tert*-butyl ester (*48;* cf. *1862*) was obtained by acid-catalyzed esterification of *N*-carbobenzoxy-L-tyrosine with isobutylene. The same method of esterification yielded also *O-tert*-butyl-L-tyrosine *tert*-butyl ester (*228, 229*). 1-L-Tyrosine-2-*tert*-butyloxycarbonylhydrazine was used for the synthesis of Val^5-angiotensin I (*890*).

As in the case of tyrosyl peptides, the azide (*1021, 1103, 1938, 1940, 2626, 2627*), carbodiimide (*626, 1124, 1249, 2020, 2569*), and anhydride (*296, 1121, 1125, 1126, 2603, 2605, 2606*) methods were preferred for coupling reactions. The tetraethyl pyrophosphite and chlorodiethyl phosphite methods (*296*) were applied to the synthesis of shorter peptides. While Goldschmidt and Rosculet (*820*) were able to synthesize tyrosine peptides via the phosphazo method, Wünsch *et al.* (*2596*) have reported that the phenolic hydroxyl group must be protected. Iselin (*1098*) observed benzylation of the aromatic nucleus during the acid hydrolysis of carbobenzoxy groups of tyrosine-containing peptides. It was noted more recently that the application of the *p*-nitrophenyl ester method, especially in presence of triethylamine, and of the imidazolide method lead to the formation of *N,O*-bis-(aminoacyl)-tyrosine esters (*1699, 1786*). The products of highest purity were obtained via the anhydride method.

b. Syntheses with the Protected Hydroxyl Group

α. O-Acetyltyrosine. Bergmann (*206*) introduced the acetyl group as protection for the hydroxyl group and applied it preferably in combination with the *N*-carbobenzoxy group. *N*-Carbobenzoxy-*O*-acetyl-L-tyrosine can be prepared by alkaline hydrolysis of *N*-carbobenzoxy-L-tyrosine ethyl ester and subsequent reaction with acetic anhydride. The acid chloride method is applicable to the synthesis of peptides, for example, the coupling reaction with L-tyrosine methyl ester (*184, 206, 774*). Barkdoll and Ross (*108*) adhered to this approach during the stepwise synthesis from the carboxyl end of L-tyrosyl-L-tyrosine and di- and tri-L-tyrosyl-L-tyrosine. *N*-Carbobenzoxy-*O*-acetyl-L-tyrosine was also applied to the synthesis of dipeptide derivatives via the mixed anhydride method (*257, 2420, 2604*). Subsequent catalytic hydrogenation in 0.2 *N* methanolic (*257*) or ethanolic hydrochloric acid (*108*) results in the cleavage of the *N*-carbobenzoxy and *O*-acetyl groups. The *O*-acetyl group will not be attacked when the hydrogenolysis is carried out in 80% acetic acid in the presence of 1.5 equivalents of hydrochloric acid (*255*) or in glacial acetic acid (*108*). It is also stable to treatment (1 hour at 40°C) with 2 *N* hydrogen bromide in glacial acetic acid (*2032*). It is sometimes necessary to acetylate again with acetic anhydride in pyridine after *O*-acetyltyrosine-containing peptide esters have been used in condensation reactions (*2032*). Treatment of *N*-carbobenzoxy-*O*-acetyl-L-tyrosylamino acid esters with aqueous alkali, methanolic ammonia, or hydrazine hydrate causes cleavage of the *O*-acetyl group and the respective formation of the free acid (*108, 184, 206*), the amide (*184, 774*), and the hydrazide (*108*). *O*-Acetyl-L-tyrosine *N*-carboxylic anhydride was prepared by treatment of *N*-carbobenzoxy-*O*-acetyl-L-tyrosine with phosphorus pentachloride (*95*) or by phosgenation of *O*-acetyl-L-tyrosine (*1933*). The latter compound was obtained from the reaction of acetic anhydride with the copper complex of L-tyrosine. The *N*-carboxylic anhydride was applied to the synthesis of peptides.

Besides the carbobenzoxy derivative, *N*-formyl-*O*-acetyl-L-tyrosine (*2420*) was applied to the synthesis of peptides via the anhydride (*1992, 2420*) or carbodiimide method (*1992, 2604*). The *N*-formyl and *O*-acetyl groups can be removed in 40–48 hours at room temperature by treatment with acetyl chloride in benzyl alcohol (*2420*). The application of *N*-acetyl- or *N*-benzoyl-*O*-acetyltyrosine (*2604*) is not of practical value for the synthesis of peptides. *O*-Acetyl-L-tyrosine methyl ester hydrochloride (*108*) and *O*-acetyl-L-tyrosine *p*-nitrobenzyl ester hydrobromide (*2032*) were applied to the synthesis of tyrosine peptides.

β. O-Benzoyltyrosine. Harington and Pitt-Rivers (*940*) synthesized *N*-carbobenzoxy-*O*-benzoyl-L-tyrosine ethyl ester by allowing *N*-carbobenzoxy-L-tyrosine ethyl ester to react with benzoyl chloride in pyridine. Subsequent catalytic hydrogenation in methanol/hydrochloric acid yielded the hydrochloride of *O*-benzoyl-L-tyrosine ethyl ester. In contrast to the *O*-acetyl compound, the *O*-benzoyl group is not attacked. The *O*-benzoyl derivative was allowed to react further via the azide method. The *O*-benzoyl group can be cleaved under the conditions of alkaline hydrolysis (*940, 1099*). The application of *N,O*-dibenzoyl-L-tyrosine (*2604*) is not of importance for the synthesis of peptides.

γ. O-Tosyltyrosine. *O*-Tosyl-L-tyrosine can be prepared by treatment of *N,O*-ditosyl-L-tyrosine with phosphonium iodide/hydroiodic acid at 100°C (*718*), or by tosylation of *N*-acetyl-L-tyrosine methyl ester and hydrolysis for 2 hours with hydrochloric acid in boiling glacial acetic acid (*1128*). In a more feasible approach, the copper complex of L-tyrosine is allowed to react with tosyl chloride (*1190*). *N*-Carbobenzoxy-*O*-tosyl-L-tyrosine was applied to the synthesis of Arg^8-vasopressin and coupled via the anhydride method. The *N*-carbobenzoxy group can be removed selectively by treatment with 2 *N* hydrogen bromide in glacial acetic acid (1 hour, 25°C). Treatment with sodium in liquid ammonia leads also to the cleavage of the tosyl group. *N,O*-Ditosyl-L-tyrosine (*718, 720, 1478*) has not been applied to the synthesis of peptides.

δ. *O-Carbobenzoxytyrosine.* Overell and Petrow (*1672*) synthesized *O*-carbobenzoxy-L-tyrosine by carbobenzoxylation of the copper complex of L-tyrosine. The *N*-carboxylic anhydride of *O*-carbobenzoxy-L-tyrosine was used for the synthesis of poly-*O*-carbobenzoxy-L-tyrosine. The *O*-carbobenzoxy residues were removed by treatment with alkali. Esters of *O*-carbobenzoxy-L-tyrosine are not known. However, *N,O*-dicarbobenzoxy-L-tyrosine (*3, 1176*) was applied to the synthesis of tyrosyl peptides in coupling reactions with amino acid and peptide esters via the anhydride (*590, 593, 1185, 1605, 2214*), carbodiimide (*1515*), azide (*2420*), ethoxyacetylene (*1690*), and acid chloride (*1185*) methods. Schwyzer *et al.* (*2009*) have described the carbethoxymethyl ester and cyanomethyl ester of *N,O*-dicarbobenzoxy-L-tyrosine as activated derivatives, Iselin and Schwyzer (*1102*) the *p*-nitrophenyl ester. *N,O*-Di-*p*-nitrocarbobenzoxy-DL-tyrosine is the only known derivative with a modified carbobenzoxy group (*804*).

ε. *O-Benzyltyrosine.* The protection of the hydroxyl function by benzyl ether formation was suggested by Wünsch *et al.* (*2596*) and independently by Chillemi *et al.* (*490*) and Noguchi *et al.* (*1628*). As in the case of the *O*-carbobenzoxy compound, the benzyl group in *O*-benzyl-L-tyrosine can be introduced by benzylation of the copper complex of L-tyrosine. The availability of *N*-carbobenzoxy-*O*-benzyl-L-tyrosine (*2596*), of *O*-benzyl-L-tyrosine methyl ester hydrochloride (*490, 2596*), and of *N*-carbobenzoxy-*O*-benzyl-L-tyrosine *p*-nitrophenyl ester (*273*) makes possible the synthesis of *O*-substituted tyrosyl as well as tyrosine peptides. Schnabel (*1940*) isolated *O*-benzyl-L-tyrosine benzyl ester as by-product of the esterification of L-tyrosine with benzyl alcohol/benzenesulfonic acid. The compound was separated from L-tyrosine benzyl ester by fractional crystallization.

Peptides containing derivatives of *O*-benzyl-L-tyrosine were synthesized via the phosphazo (*2596*), acid chloride (*490*), azide (*1940*), and *p*-nitrophenyl· ester (*273, 277, 1193, 1232*) methods. The latter method was applied predominantly to the synthesis of biologically active polypeptides. The *O*-benzyl group is stable to treatment with aqueous alkali. This protecting group and the carbobenzoxy group can be removed simultaneously by catalytic hydrogenation in the presence of palladium in aqueous dioxane/glacial acetic acid (*2596*) or methanol/1 *N* hydrochloric acid (*1193*), and by treatment with sodium in liquid ammonia. Schnabel (*1940*) found that the benzyl group is not removed completely from Cbo-Gly-Ala-Gly-Try(Bzl)-OBzl upon treatment with hydrogen bromide/glacial acetic acid. Bodanszky and du Vigneaud (*273*) and Kimbrough and du Vigneaud (*1232*) were able to remove the benzyl group in a smoothly proceeding reaction by treatment for 1–2 hours with hydrogen bromide in glacial acetic acid.

Besides the carbobenzoxy group, the *N-tert*-butyloxycarbonyl residue was used for the protection of *O*-benzyl-L-tyrosine by reacting it with *tert*-butyloxycarbonyl azide in dioxane/aqueous alkali (*2036*). Velluz *et al.* (*2371*) synthesized *N,N,O*-tribenzyl-L-tyrosine, a derivative of no practical value for the synthesis of peptides.

ζ. O-Trityltyrosine. Stelakatos *et al.* (*2203*) prepared *N,O*-ditrityl-L-tyrosine benzyl ester hydrochloride by tritylation of L-tyrosine benzyl ester hydrochloride in chloroform/triethylamine. The compound can be detritylated selectively at the amino group by brief boiling in anhydrous methanol to yield *O*-trityl-L-tyrosine benzyl ester hydrochloride.

η. O-Tetrahydropyranyltyrosine. The investigations of Iselin and Schwyzer (*1101*) have shown that carbobenzoxy-L-tyrosine ethyl ester can be converted to *N*-carbobenzoxy-*O*-tetrahydropyranyl-L-tyrosine ethyl ester by treatment with 2,3-dihydropyrane in the presence of hydrochloric acid. The ester group can be saponified by exposure to a very small excess of alkali. The free acid was converted to the cyanomethyl ester. Heating for 5 minutes to 100°C in 2 *N* hydrochloric acid was sufficient to remove the *O*-tetrahydropyranyl residue (*1099*). The protecting group is stable to catalytic hydrogenation in methanol in the presence of glacial acetic acid.

θ. O-Methyltyrosine. *O*-Methyltyrosine was prepared by methylation of *N*-acyl-L-tyrosine with dimethyl sulfate (*1118*) or diazomethane (*1933, 2494*) and subsequent acidic hydrolysis or acidic and basic hydrolysis. Law and du Vigneaud (*1349*) and Jošt and Rudinger (*1159*) have applied *O*-methyl-L-tyrosine methyl ester hydrochloride and the ethyl ester hydrochloride to the synthesis of *O*-methyloxytocin (*p*-methoxyphenylalanine^{2}-oxytocin). The methyl group in *O*-methyltyrosine represents an irreversible protection of the hydroxyl group. It can be removed only by boiling for 1 hour with phosphonium iodide (*1079*). Prolonged treatment of carbobenzoxy-*p*-methoxyphenylalanine or of fully protected *O*-methyloxytocin with hydrogen bromide in glacial acetic acid will also lead to the cleavage of the *O*-methyl group (*1349*).

ι. O-tert-Butyltyrosine. As in the case of serine and threonine, tyrosine can be converted to *N*-carbobenzoxy-*O-tert*-butyl-L-tyrosine *tert*-butyl ester by reacting carbobenzoxy-L-tyrosine with isobutylene in methylene chloride/sulfuric acid. Subsequent catalytic hydrogenation produces the hydrochloride of *O-tert*-butyl-L-tyrosine *tert*-butyl ester (*229, 457*). The same method yields *N*-carbobenzoxy-*O-tert*-butyl-L-tyrosine ethyl ester, when carbobenzoxy-L-tyrosine ethyl ester is used as starting material. The ester derivative was converted to the corresponding free acid, to *O-tert*-butyl-L-tyrosine ethyl ester hydrochloride, and to *tert*-butyloxycarbonyl-*O-tert*-butyl-L-tyrosine hydrazide (*1962*, cf. *2597c*).

2. 3,4-Dihydroxyphenylalanine

3,4-Dihydrophenylalanine has not so far been found as a constituent of proteins. The compound is of physiological importance as intermediate in the synthesis of adrenaline and the formation of melanin from tyrosine. Several peptides containing DL-3,4-dihydroxyphenylalanine as C-terminal amino acid were synthesized via the phosphorus oxychloride (*1454*) and acid chloride (*1601*) methods. The hydrolysis of the carbobenzoxy dipeptide methyl ester with sodium hydroxide in methanol under nitrogen and the subsequent catalytic hydrogenation proceeded smoothly (*1454*). Losse *et al.* (*1453a*) have synthesized a series of dipeptides, using phthalyl-DL-3,4-dihydroxyphenylalanine as starting material. [Cf. (*2383*) for the resolution of DL-3,4-dihydroxyphenylalanine.]

G. SULFUR AMINO ACIDS

Peptides and proteins containing sulfur can be classified into the following types of compounds:

(1) Compounds containing cysteine with the free mercapto group (e.g. glutathione)

(2) Compounds with a disulfide bridge resulting from the dehydrogenation of two residues of cysteine and formation of cystine; the residues of cysteine may belong to the same peptide chain (e.g. oxytocin, vasopressin, ribonuclease), or they may connect two different peptide chains (e.g. insulin)

(3) Sulfur may be present in the form of thioethers, either as alkylated mercaptoamino acids (methionine) or linking two different peptide chains (lanthionine).

1. Cysteine

In spite of the great importance of cysteine, the number of systematic investigations dealing with this amino acid is small. Few detailed studies have been carried out on suitable and sufficiently differentiated protecting groups for the mercapto function. Dehydrogenation to cystine derivatives occurs very readily during the synthesis of peptides with the unprotected mercapto function, e.g. upon exposure to air, and causes complications. Cysteine is therefore applied almost exclusively in its S-protected form to the synthesis of peptides.

a. Use of S-Benzylcysteine

Sifferd and du Vigneaud (*2119*) and Wood and du Vigneaud (*2579*) introduced the S-benzyl group as protection for the mercapto group on the basis of its ready cleavability with sodium in liquid ammonia. The benzyl group remains the most commonly applied protection for the mercapto group. S-Benzylcysteine is prepared by reduction of cystine to cysteine with sodium in liquid ammonia and immediate benzylation in this solvent (*625, 2579*). Cystine can also be reduced with zinc dust and hydrochloric acid, and cysteine benzylated with benzyl bromide in aqueous alkaline solution (*938*). The introduction of an *N*-protecting group or an esterification is carried out according to standard procedures.

The synthesis of peptides with S-benzylcysteine proceeds without difficulty as long as catalytic hydrogenation or reduction with sodium in liquid ammonia is not required. As a result, almost all *N*-protected derivatives and most of the esters of S-benzylcysteine are known. Also, almost all methods for the coupling of peptides have been applied, regardless of whether cysteinyl or cysteine peptides were to be synthesized.

α. S-Benzylcysteinyl Peptides. The carbobenzoxy and the tosyl groups are preferred for the protection of the amino group. Both groups and the S-benzyl group can be removed simultaneously after completion of the synthesis (*938, 1329*). The following coupling methods have been applied: acid chloride (*515, 634, 806, 938, 1057, 1159, 1160*), activated esters, such as the cyanomethyl ester (*1099, 1246*), the *p*-nitrophenyl ester (*265, 272, 273, 469, 891*), or the phenyl ester (*2274*), mixed anhydride (*80, 265, 296, 1059, 1113, 1491*), azide (*80, 296, 515, 940, 1059, 1113*), tetraethyl pyrophosphite (*296, 634, 803, 1802*), carbodiimide (*214, 231, 274, 463, 626, 1079, 1132, 1160, 1950*), phosphazo and isocyanate (*815*), phosphorus oxychloride (*1491*), and pyrazolide (*1815*) methods. The basicity of hydrazine may lead to the elimination of the S-benzyl group as benzyl mercaptan if the hydrazinolysis of S-benzylcysteine-containing peptide esters is carried out at too high a temperature. This can result in the formation of dibenzyl disulfide [98] or cause racemization [99] if the mercaptan recombines with the acrylic acid derivative (*1491*). This side reaction cannot be completely prevented by maintaining temperatures below 50°C. The hydrazides can be readily obtained with 80% hydrazine hydrate in *n*-butanol (18–24 hours, 40°C), or with 60% hydrazine hydrate in 2-methoxyethanol (*1491*). The same side reaction was observed during the alkaline hydrolysis of esters (*626, 1489, 1491*). Many authors, however, have described the hy-

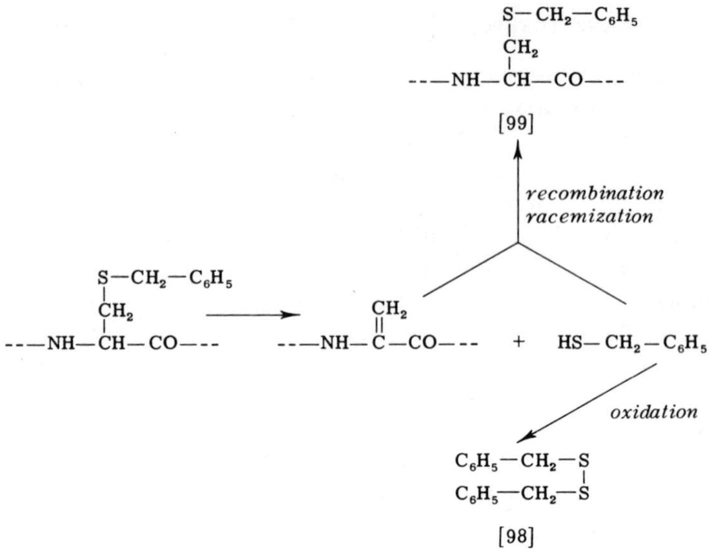

drazinolysis (*296, 940, 972*) and hydrolysis (*296, 940, 1159, 1636, 1950*) following the routine procedures. The carbobenzoxy (*149, 296, 1059, 1113, 1491*) or *p*-chlorocarbobenzoxy group (*1246*) is cleaved in the normal way by treatment with hydrogen bromide in glacial acetic acid. It is also possible to remove these protecting groups with phosphonium iodide in glacial acetic acid (*937*) or with hydrogen chloride in methanol (*815, 1815*). Cleavage of the S-benzyl group occurs at 40°C only upon exposure of the peptide to a 35% solution of hydrogen bromide in glacial acetic acid (*1159*). The mercapto group may be benzylated again (*231, 1159, 1191, 1806*) after the *N*-carbobenzoxy and S-benzyl groups or the *N*-tosyl and S-benzyl groups have been cleaved simultaneously with sodium in liquid ammonia. The reaction is best carried out in the presence of ammonium chloride (*1474*). Benzyl bromide in chloroform in the presence of diethylamine may also be used for the renewed benzylation (*938*). After the reaction with sodium in liquid ammonia the peptides with the free mercapto group are isolated as mercury or copper mercaptides. The mercapto peptides are liberated with hydrogen sulfide (*92, 639a, 938, 940, 1246*) or by electrolytic reduction (*143*). Esters of S-benzylcysteinyl peptides with the unprotected amino group were synthesized originally by cleaving the disulfide bonds of symmetrical *N,N'*-dicarbobenzoxy or *N,N'*-ditosylcystinyl bispeptide esters with sodium in liquid ammonia and subsequent benzylation (*1806*).

β. S-Benzylcysteine Peptides. The benzyl ester is frequently used to protect the carboxyl group during the synthesis of peptides with C-terminal cysteine. This protection and the S-benzyl group can be removed simultaneously by treatment with sodium in liquid ammonia (*1059, 1329*). The methyl ester (*296, 505*) and the ethyl ester (*1113, 2691*) have also been used for this purpose. These protecting groups, however, must be hydrolyzed prior to the cleavage of the S-benzyl group. The *tert*-butyl ester is another known carboxyl derivative (*49, 2265*).

The anhydride (*80, 296, 1059, 1113, 1474, 1854*), phosphazo (*818, 1854*), tetraethyl pyrophosphite (*296, 642, 1181, 1802, 1807*), azide (*900, 940*), acid chloride (*515*), *N,N'*dicyclohexylcarbodiimide (*231, 1860, 2369*), and *p*-nitrophenyl ester (*273, 633, 891*) methods have been applied to the synthesis of S-benzylcysteine peptides.

b. Use of S-*p*-Nitrobenzylcysteine

S-*p*-Nitrobenzylcysteine can be prepared by the reaction of cysteine hydrochloride with *p*-nitrobenzyl chloride in 1 *N* sodium hydroxide at 0°C. According to Berse *et al.* (*212*), the protecting group can be removed again by catalytic hydrogenation in the presence of palladium on charcoal in ethanol containing approximately 10 equivalents of 1 *N* hydrochloric acid.

This protection of the mercapto function was applied to the synthesis of partial sequences of insulin. The carboxyl groups were activated via the anhydride (*1183*), carbodiimide (*1183*), and *p*-nitrophenyl ester (*1192, 1664*) methods. Since the S-*p*-nitrobenzyl group is stable to hydrogen bromide in glacial acetic acid, it was possible to extend the chain after decarbobenzoxylation via the anhydride (*1183, 1192*) and carbodiimide (*1192*) methods and via the *p*-nitrophenyl ester (*1192*). The removal of the S-*p*-nitrobenzyl group from peptide derivatives has not yet been described. Ondetti and Bodanszky (*1664*) have reported the reduction of the nitro group to the amino group in the presence of these catalysts: palladium-charcoal, palladium-barium sulfate, or ruthenium-charcoal.

c. Use of S-Trityl- and S-Diphenylmethylcysteine

N,S-Ditritylcysteine can be obtained by tritylation of free cysteine in a mixture of water and ether in the presence of diethylamine and with exclusion of oxygen (*39*). Zervas and Theodoropoulos (*2668*) have described the *N,S*-ditritylation of cysteine methyl ester but not the hydrolysis to the free acid. The *N*-trityl group can be removed selectively by boiling in aqueous acetic acid/ether (*39*), by exposure to acetone/hydrochloric acid at room temperature or 35°C (*39, 2369*), or by treatment with toluenesulfonic acid in methanol (*2664*). The S-protecting group can be removed selectively from *N,S*-ditritylcysteine methyl ester by treatment with silver nitrate or mercuric chloride (*2664, 2665*).

Only the carbodiimide method has been applied so far to the synthesis of peptides containing S-tritylated cysteine (*39, 2369*). The amino and mercapto protections can be removed simultaneously by briefly passing hydrogen chloride through a solution in chloroform (*39*) or by treatment for 15 minutes with concentrated hydrochloric acid in methylene chloride (*2369*). The trityl group served as protection of the amino as well as the mercapto group in syntheses of glutathione (*39*) and of oxytocin (*2369*).

Zervas and Photaki (*2664, 2665, 2666*) have suggested the use of the diphenylmethyl group as mercapto protection. This protecting group shows a behavior similar to that of the trityl group, but is somewhat more resistant to acids. The group is stable to 0.5 *N* hydrochloric acid and to 0.2 *N* hydrogen bromide in glacial acetic acid/ether at 40°C. A solution of 2 *N* hydrogen bromide in glacial acetic acid (90 minutes, 50°–55°C) causes cleavage to the extent of 45–50%. The S-diphenylmethyl group can be removed by treatment with sodium in liquid ammonia, or by refluxing in trifluoroacetic acid for 20 minutes in the presence of phenol (*2665*).

d. Use of S-Benzylthiomethylcysteine

Pimlott and Young (*1725*) and Young (*2608, 2609*) were the first to apply the benzylthiomethyl group to the protection of the mercapto function. Treatment with benzylthiomethyl chloride in methanol of a suspension of cysteine hydrochloride on the steam bath leads to the formation of a mixture of the S-alkylated amino acid and the S-alkylated amino acid ester. This mixture can be converted by hydrolysis to the acid (*1183, 1725*), or to the methyl ester by completing the esterification by treatment with thionyl chloride in methanol (*1725*). The free acid was also obtained from cystine by reduction with sodium in liquid ammonia and subsequent alkylation with benzylthiomethyl chloride (*1725, 2608*) in the same solvent. S-Benzylthiomethylcysteine can be carbobenzoxylated (*1725*) via the standard method, or converted to the *N*-formyl derivative by treatment with formic acid and acetic anhydride (*1183*).

The cleavage of this S-protecting group was carried out with mercuric chloride in warm 1 *N* hydrochloric acid or with mercuric acetate in 80% formic acid. This treatment leads to the formation of the mercaptide, which is decomposed with hydrogen sulfide or by electrolytic reduction to yield the mercaptan (*416a, 1725, 2608*). Peptides with N-terminal S-benzylthiomethylcysteine, carrying the carbobenzoxy or formyl group as amino protection, have been synthesized via the carbodiimide method (*1183, 2608*). Hydrolysis and hydrazide formation proceeded normally (*2608*). Treatment for 24 hours with 2 *N* sodium hydroxide at room temperature does not affect S-benzylthiomethyl-L-cysteine. Peptides containing S-benzylthiomethylcysteine methyl ester were also synthesized via the carbodiimide method. In this case, the formyl group was used for the protection of the amino group. This protecting group can be removed with 1 *N* hydrochloric acid in boiling methanol (*1183*). The S-benzylthiomethyl group will be attacked upon exposure to 2 *N* hydrochloric acid at 60°C for 5 hours (*1725*). Pimlott and Young have reported that this S-protecting group is stable to 2 *N* hydrogen bromide in glacial acetic acid even after 15 hours of exposure time (*1725*). Katsoyannis (*1183*, cf. *416a*), however, has observed side reactions that could be prevented almost completely by the addition of methyl ethyl sulfide and diethyl phosphite. It is more advantageous to use amino-protecting groups that can be readily removed by acid hydrolysis, e.g. the formyl group. So far, racemization has not been observed during the application of S-benzylthiomethyl-L-cysteine.

S-2,4-Dichlorobenzylthiomethyl- and S-phenylthiomethylcysteine resemble the S-benzylthiomethyl compound very closely in behavior and offer no advantages. Isobutoxymethyl- and *p*-chlorophenoxymethyl-L-cysteine are decomposed immediately upon contact with concentrated acids, including hydrogen bromide/glacial acetic acid. It was found that the isobutoxymethyl derivative, however, is stable to 2 *N* hydrochloric acid at room temperature for a period of 11 hours (*1183*).

e. Use of S-Carbobenzoxycysteine

Berger *et al.* (*180*) prepared *N*,*S*-dicarbobenzoxycysteine by carbobenzoxylation of cysteine in 2 *N* sodium hydroxide. The derivative was converted to the *N*-carboxylic anhydride by treatment with phosphorus pentachloride and used for the synthesis of polycysteine. The carbobenzoxylation in bicarbonate solution leads to the formation of only the S-substituted derivative.

Katsoyannis (*1183*) applied the carbodiimide method to the activation of *N*,*S*-dicarbobenzoxycysteine for coupling reactions with ethyl esters of di- and tripeptides. The esters were hydrolyzed subsequently by treatment with 1 *N* hydrochloric acid in dioxane for 1.5 hours at 90°C. Berger *et al.* (*180*) found that the S-carbobenzoxy bond is not unconditionally stable to dilute aqueous alkali, and that the group will be removed by aqueous ammonia with formation of benzyl urethan. In poly-S-carbobenzoxy-L-cysteine, however, the S-protecting group is stable to aqueous ammonia. The S-carbobenzoxy group can be cleaved selectively by treatment with sodium methylate while the *N*-carbobenzoxy group is present. Upon exposure to hydrogen bromide in glacial acetic acid only the *N*-protecting group will be removed (*2156b*). Under these conditions, Berger *et al.* (*180*) have reported the simultaneous cleavage of both groups. The amino group of S-carbobenzoxycysteine was formylated via the standard method—treatment with formic acid and acetic anhydride—and coupled via the carbodiimide method (*1183*). Sokolovsky *et al.* (*2156b*) have described the synthesis of glutathione and the C-terminal tetrapeptide sequence of oxytocin, using the carbobenzoxy group as *N*- and S-protection. Under alkaline conditions, peptides with amino-terminal S-carbobenzoxycysteine undergo an S → N shift to form the *N*-substituted derivative.

f. Use of S-Ethylcarbamoylcysteine

Guttmann (cf. *1522a*) has reported the application of S-ethylcarbamoylcysteine. The compound can be prepared conveniently from ethyl isocyanate and cysteine hydrochloride (cf. *1841a*). The advantage of this blocking group lies in its stability to acids. Removal can be brought about by treatment with sodium hydroxide, hydrazine, liquid ammonia, or sodium methylate.

g. Use of S-Tetrahydropyranylcysteine

Boiling for 1–2 hours of cysteine methyl ester hydrochloride and dihydropyran in methylene chloride leads to the formation of S-tetrahydropyranylcysteine methyl ester. The ester can be hydrolyzed under standard conditions and acylated with carbobenzoxy chloride. Hydrolysis of the acylated derivative yields the *N*-protected free acid (*1047*). It was demonstrated for Cbo-Cys(Bzl)-Cys-OH that the reaction with 1 equivalent of dihydropyran leads to the formation of only the ester [100], while the reaction with 2 equivalents of the reagent produces the thioether and the ester [101]. Peptides containing S-tetrahydropyranylcysteine methyl ester were synthesized via the azide method. The re-

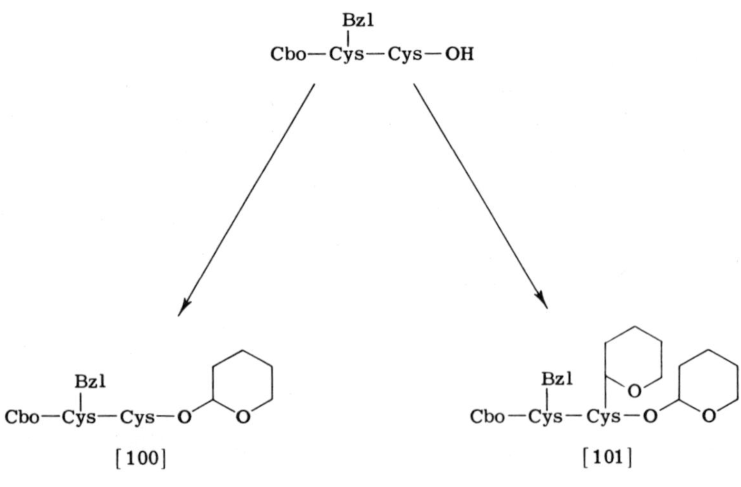

[100] [101]

moval of the protecting group that can be brought about by the reaction with silver ions has so far been described only for simple derivatives but not for peptides. The S-tetrahydropyranyl group is stable to treatment with sodium in liquid ammonia.

h. PROTECTION OF THE MERCAPTO GROUP BY INCORPORATION IN A HETEROCYCLIC SYSTEM

α. Thiazolidine Derivative of Cysteine. Boiling for 6 hours of cysteine hydrochloride in acetone leads to the formation of a thiazolidine ring [102]. This derivative provides complete protection of the mercapto

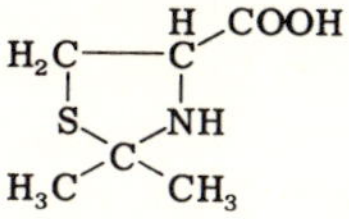

[102]

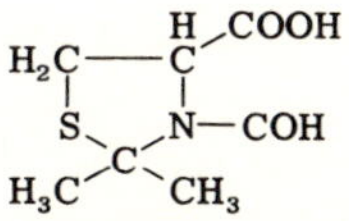

[103]

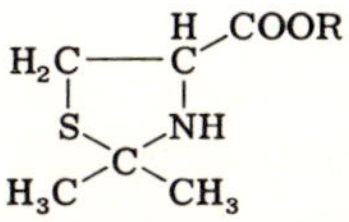

[104]

group but not of the amino group. This function can be protected, e.g. by formylation, and the 3-formyl-2,2-dimethylthiazolidine-4-carboxylic acid [103] coupled via the anhydride or carbodiimide method (*2074*). The thiazolidine ring as well as the formyl group will be cleaved by acid hydrolysis (0.5 *N* methanolic hydrochloric acid, 60 hours at room temperature). It is also possible to hydrolyze in dioxane with 1 *N* sodium hydroxide. The thiazolidine derivative with the unprotected amino function may be coupled to *N*-protected amino acids, e.g. via the anhydride method, to form cysteine peptides. Such coupling reactions may be carried out either directly or after esterification in methanol/hydrogen chloride [104]. The phthalyl group may be removed with hydrazine hydrate while the ring structure is preserved. The thiazolidine ring can be opened by treatment with mercuric chloride.

β. Cysteine Thiolactone. Thiolactone formation of an aminomercapto acid provides simultaneous protection of the carboxyl and the mercapto functions and activation of the carboxyl group at the same time. The thiolactone can be prepared from S-benzylcysteine chloride by treatment with 2 moles of aluminum halide, whereby debenzylation occurs, or from *N*-protected cysteine in dioxane by activation with carbodiimide (*558*). The phthalimido- (*728*), succinimido- (*729*), tosylamido- (*558*,

729), and carbobenzoxyamido-β-propiothiolactones (*558*) have so far been prepared. Cleavage of the thiolactone with acids leads to the formation of the cysteine derivative, which was oxidized and isolated as the cystine compound (*728, 729*). Treatment with bicarbonate results in polymerization (*729*). The lactone is opened by alcohols to form esters; peptides are obtained upon treatment with amino acid derivatives (*558, 728, 729*).

i. Syntheses with the Unprotected Mercapto Group

Peptides containing cysteine with the unprotected mercapto function have been synthesized very rarely (*937*), because it is difficult to separate the resulting compound from by-products formed by oxidation. Purification can be achieved only via mercaptides of heavy metals. *N*-Protected cysteine is accessible by reduction of *N,N'*-bis-protected cystine. Dadić *et al.* (*558*) obtained the tosyl derivative and Foye and Verderame (*753*) the carbobenzoxy derivative by reduction with zinc dust and 2 *N* sulfuric acid in methanol at 65°C. The formyl derivative was obtained by reduction with zinc dust 0.5 *N* hydrochloric acid at 0°C.

Unprotected cysteine can be esterified under standard conditions as long as oxygen is excluded.

[Cf. (*1696a, 2156a*) for a nonenzymatic method for the cleavage by β-elimination and oxidation of the peptide bond preceding cysteine residues.]

k. Subsequent Introduction of the Mercapto Group

Derivatives of cysteine and peptides of cysteine can be synthesized, according to Zervas and Photaki [*2664;* cf. Fischer and Raske (*721*)], by subsequent introduction of the mercapto function. The reaction is

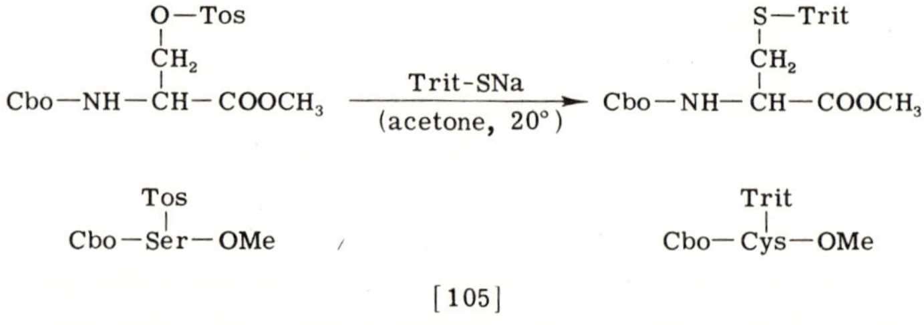

[105]

accompanied by racemization (*1716a*). Therefore the reaction scheme [105] is intended only to indicate the possible synthetic approaches.

2. Cystine

The synthesis of unsymmetrically substituted peptides of cystine still poses great difficulties. The synthesis of symmetrical peptides of cystine presents no problems.

a. Symmetrical Peptides of Cystine

Symmetrical peptides of cystine can be synthesized from suitable derivatives of cystine by allowing them to react with 2 equivalents of an amino acid or peptide ester or by coupling cystine to 2 equivalents of an *N*-protected amino acid or *N*-protected peptide. In an alternative approach, a peptide containing cysteine may be synthesized first and dehydrated to the cystine derivative.

α. Synthesis of Cystine Peptides from Cystine Derivatives. *N,N'*-bis-protected derivatives are available with practically all commonly employed amino-protecting groups. Also, all esters of importance are known. The decarbobenzoxylation of cystine frequently leads to the formation of impure products that cannot be purified by recrystallization (*815, 1053*). Peptides containing *N,N'*-bis-protected cystine have been synthesized via the acid chloride method (*856, 858, 937, 1053, 1243, 1270, 2481, 2508*), *p*-nitrophenyl esters (*2192*), *N*-carboxylic anhydrides (*94, 95*), mixed anhydrides (*497*), and the oxazolidone method (*1548*). Peptides have been coupled only infrequently to esters of cystine, e.g. via the anhydride method (*2668*), or the salts of free cystine via the acid chloride method (*859*).

D-Cystine cannot be prepared by racemization and resolution from the easily accessible L-cystine since in the course of the racemization *meso*-cystine is also formed (*1507, 2579*).

β. Synthesis of Cystine Peptides by Oxidation of Cysteine Peptides. Autooxidation of free mercapto groups leads exclusively to the formation of disulfides. The reaction is very slow; it is strongly catalyzed by ions of heavy metals. Since it is difficult to exclude such ions under normal preparative conditions, the contact with the oxygen of the air is sufficient to isolate as cystine derivative a peptide that originally contained cysteine with the free mercapto group after such operations as evaporation, recrystallization, column chromatography, etc. (*2192*). Cystine bis-peptides have been synthesized on occasion by taking advantage of this fact (*558, 729, 2192, 2322*). In most cases, however, air is passed through the solutions of cysteine peptides for the purpose of oxidation (*73, 940, 1539*). Hydrogen peroxide is not often used as oxidizing agent (*856,*

938). Ferric chloride (*73, 2579*), ferric sulfate (*940*), ferric oxide (*858, 859*), or cupric sulfate (*938*) is frequently added as catalyst. The reaction mechanism of this thiooxidation has been the subject of many investigations. Many of them, however, were not conducted on cysteine but on other mercapto compounds. Cecil and McPhee (*473*) have published an extensive review of these studies. On the basis of kinetic studies of the metal-catalyzed autooxidation, Lamfrom and Nielsen (*1320*) have suggested a mechanism proceeding via a metal complex and the formation of thiol radicals. For the chemical oxidation, however, a mechanism proceeding also via metal complexes but being purely ionic was postulated.

The cleavage of amino- and carboxyl-protecting groups of cystine peptides proceeds normally. In general, catalytic hydrogenation is not applied to the decarbobenzoxylation. The behavior of bis-*p*-nitrocarbobenzoxycystine is of interest. According to Berse *et al.* (*212*), the sodium salt can be reduced to the free cystinyl compound in aqueous solution in the presence of palladium on charcoal within 6–8 hours at room temperature. Bis-*p*-nitrocarbobenzoxycystinyl bispeptides behave similarly in 95% alcohol. Longer hydrogenation times (up to 12 hours) lead to the reduction of the disulfide bond and formation of the cysteine derivative. Bergmann and Michaelis (*188*) investigated this reaction in the case of cystine and dialanylcystine, using palladium black in 1 *N* hydrochloric acid. Decarbobenzoxylation or detosylation with sodium in liquid ammonia leads also to the cleavage of the disulfide bond (*625*). Nevertheless this type of removal of protecting groups is applied on occasion, since dehydration of the cystine derivatives occurs during the isolation through exposure to air (*2192*).

The carbobenzoxy group can be removed while the disulfide bond is preserved by treatment with either hydrogen bromide in glacial acetic acid, concentrated hydrochloric acid (e.g. for 30 minutes at 70°C) (*2508*), or trifluoroacetic acid (refluxing 1 hour) (*2192*). Alkaline hydrolysis of ester groups has been described with 2 *N* ethanolic potassium hydroxide solution in dioxane (*73*), sodium hydroxide in methanol or acetone (*1548*), or 0.37 *N* barium hydroxide for 2 minutes at 10°C (*95*). The formation of the hydrazide proceeds normally (*2668*).

b. Unsymmetrical Peptides of Cystine

α. Disulfide Exchange. The main difficulty during the synthesis of an unsymmetrical disulfide lies in a reaction known as disulfide exchange (*1906*). The reaction [106] itself represents an equilibrium:

$$2\ R{-}S{-}S{-}R' \rightleftharpoons R{-}S{-}S{-}R + R'{-}S{-}S{-}R' \qquad [106]$$

Under alkaline and neutral conditions the disulfide exchange is initiated by mercaptide ions [107]:

$$R{-}S{-}S{-}R' + R{-}S^{\ominus} \rightleftharpoons R{-}S{-}S{-}R + R'{-}S^{\ominus} \qquad [107]$$

Therefore the reaction is enhanced by the addition of thiols and slowed down by compounds that block thiols (e.g. *N*-ethylmaleimide or *p*-chloromercuric benzoate) (*1887*). Under acidic conditions the first step of the reaction is the protonization of the disulfide bond and the formation of a sulfenium ion [108]:

$$R{-}S{-}S{-}R' \xrightarrow{H^{\oplus}} \left[\begin{array}{c} R{-}\underset{\displaystyle H}{\underset{|}{S}}{-}S{-}R' \end{array}\right]^{\oplus} \rightleftharpoons R{-}SH + R'{-}S^{\oplus} \qquad [108]$$

In contrast to the reaction under alkaline conditions, here the reaction is slowed down by thiols (*144*).

β. Synthesis of Unsymmetrical Cystine Peptides. Swan (*2245*) prepared monocarbobenzoxycystine by selective carbobenzoxylation of cystine. Marshall *et al.* (*1507*) applied a somewhat modified procedure for the same purpose. Zervas *et al.* (*2659*) made use of this compound for the synthesis of other derivatives of cystine. Esterification of monocarbobenzoxycystine under standard conditions led to the formation of the dimethyl or dibenzyl ester. The monoesters are accessible under milder conditions. Under such conditions only the carboxyl group of the carbobenzoxylated half cystine will be esterified. Attempts to hydrolyze selectively *N*-carbobenzoxy-*N′*-tritylcystine dimethyl ester or to convert the compound selectively to the hydrazide resulted in a disulfide exchange. As in the case of monocarbobenzoxycystine, it was also possible to obtain monocarbobenzoxyglycylcystine via the acid chloride method in a mixture of 4 *N* sodium hydroxide and dioxane at 0°C. The compound was decarbobenzoxylated within 3 hours at room temperature by treatment with 22% solution of hydrogen bromide in glacial acetic acid. At pH 7.5 disulfide exchange took place in the case of monocarbobenzoxycystine as well as monoglycylcystine. At pH 6.5, monocarbobenzoxycystine proved to be stable over a period of 6 weeks (*2659*).

Unsymmetrical disulfides can also be synthesized by allowing two different thiol derivatives to react. The simplest case is represented by the common oxidation of two different thiols and the isolation of the unsymmetrical component from the reaction mixture (*1433, 1887, 2254*). However, this kind of disulfide formation proceeds in a more uniform fashion if the formation of the unsymmetrical compound is facilitated by the use of properly substituted thiols [109]:

$$R{-}S{-}A + B{-}S{-}R' \longrightarrow R{-}S{-}S{-}R' + AB \qquad [109]$$

Here, the kind of reaction suggested by Fortner and Smiles (cf. *2633*) is worthy of mention: an S-sulfonic acid reacts with a mercaptide ion with elimination of a sulfite ion [110]:

$$R—S—SO_3^{\ominus} + R'—S^{\ominus} \longrightarrow R—S—S—R' + SO_3^{2\ominus} \qquad [110]$$

Thus, Stapleton and Swan (*2193*) were able to obtain monocarbobenzoxycystine by allowing carbobenzoxycystine-S-sulfonic acid and cysteine to react. Zahn and Otten (*2633*) allowed cysteine-S-sulfonic acid [111] and Ac-Gly-Cys-Gly-NH-CH$_3$ or glutathione to react at pH 5 in oxygen-free water. These authors were able to isolate the unsymmetrical cystine peptide [112] and the glutathione derivative [113], as crystalline compounds after separation by carrier-free electrophoresis in a pyridine-acetic acid buffer of pH 3.5. The same compound was obtained

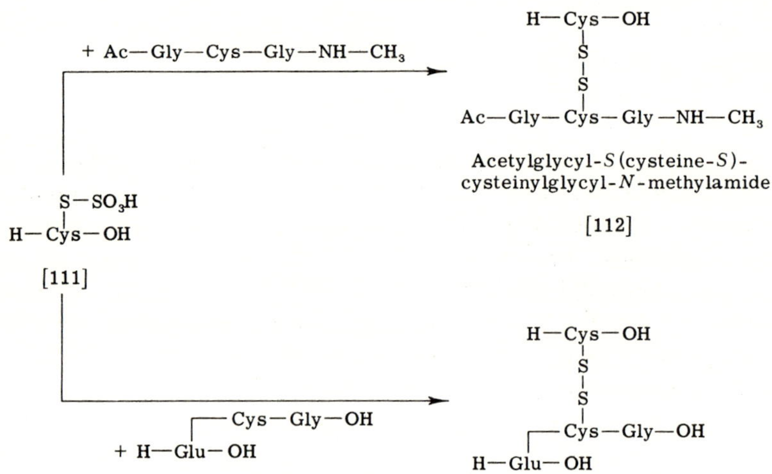

Acetylglycyl-*S*(cysteine-*S*)-cysteinylglycyl-*N*-methylamide

[112]

[111]

γ-Glutamyl-*S*(cysteine-*S*)-cysteinylglycine

[113]

from cystine monosulfoxide [114] and glutathione via a reaction first suggested by Cavallito (cf. *2633*):

$$R—S—SO—R + 2\ R'—SH \longrightarrow 2\ R—S—S—R' + H_2O \qquad [114]$$

Hiskey and Tucker (*1012, 1013*) obtained unsymmetrical derivatives of cystine via the reaction of S-trityl- or S-tetrahydropyranylcysteine with thiocyanogen and subsequent treatment with a second mercapto-protected derivative of cysteine [115]:

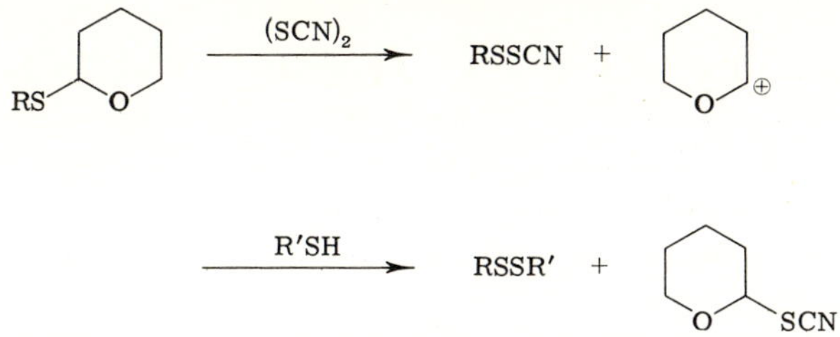

[115]

Weygand and Zumach (*2505*) took advantage of the very fast reaction of cysteine compounds with vicinal dihalides to synthesize unsymmetrical peptides.

Zervas (*2658*) has suggested a basically different approach to the synthesis of unsymmetrical cystine peptides. Two S-protected residues of cysteine are esterified differently. After coupling with a nitrobenzylphosphoryl residue [116] and removal of the S-protecting groups, the two SH groups are oxidized [116] to form an unsymmetrical derivative of cystine [117]. Thus the amino acid is participating in a phosphor-containing ring system that makes the disulfide exchange impossible. Since the carboxyl-protecting groups can be removed selectively the way is open for the synthesis of unsymmetrical peptides.

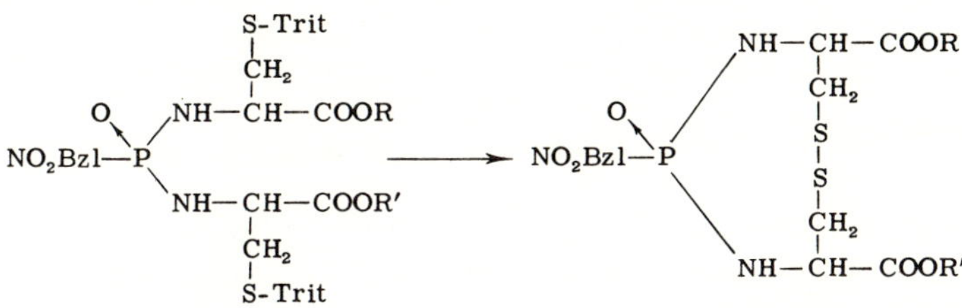

[116] [117]

3. Oxidation Products of Cysteine and Cystine

Complete oxidation of cysteine and cystine leads to the formation of cysteic acid [118]:

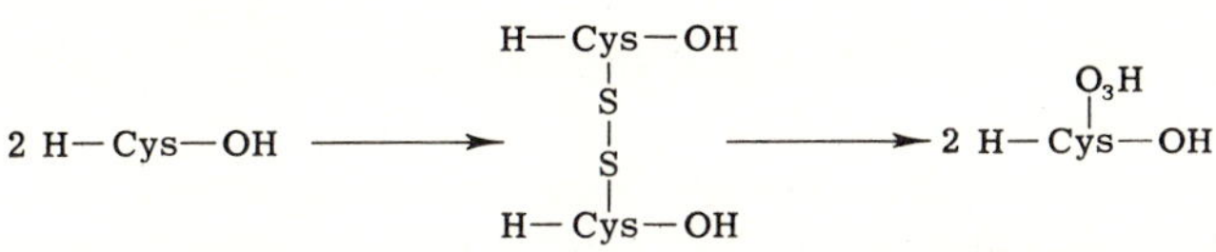

[118]

The monosulfone and cysteine sulfinic acid are the only intermediates of this reaction that have been isolated and well characterized (*1490*).

a. Cystine Monosulfone

Toennies and Lavine (*2317*) obtained the monosulfone of cystine by oxidation of cystine with perbenzoic acid. These authors reported the two alternative structures of the disulfoxide (R—SO—SO—R) and monosulfone (R—S—SO_2—R). Sweetman (*2248*) demonstrated by infrared spectroscopy that the sulfonic acid thiolester is present. Vinkler *et al.* (*2377*) were able to prove the structure by synthesis of the compound from the sulfenyl chloride of cysteine and cysteine sulfinic acid [119]:

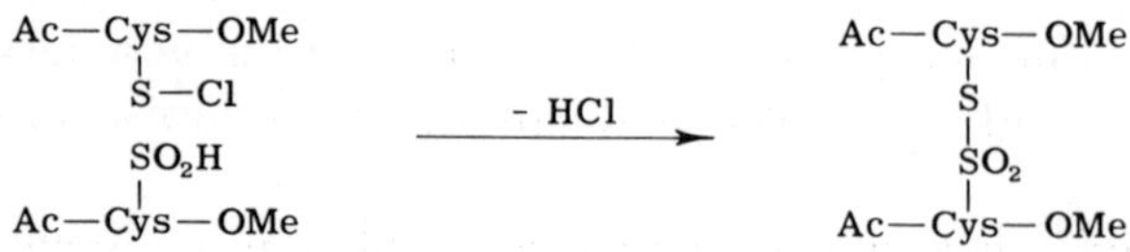

[119]

The monosulfone of cystine can be oxidized to cysteic acid by treatment with an excess of iodine. Reduction with hydroiodic acid stops at the disulfide level.

b. Cysteine Sulfinic Acid

Cysteine sulfinic acid was obtained by decomposing the monosulfone of cystine with alkali [120] or by the reaction of the same compound with thiols [121] (*1348*):

$$3\,R{-}S{-}SO_2R + 4\,NaOH \longrightarrow R{-}S{-}S{-}R + 4\,R{-}SO_2Na + 2\,H_2O \qquad [120]$$

$$R{-}S{-}SO_2R + R{-}SH \longrightarrow R{-}S{-}S{-}R + R{-}SO_2H \qquad [121]$$

The second reaction can be applied to the synthesis of unsymmetrical disulfides. Reduction of sulfinic acid with hydroiodic acid leads to the formation of cysteine.

c. Cysteic Acid

Cysteic acid has not so far been found as a constituent of biologically active polypeptides. Nevertheless peptides containing cysteic acid are of importance for the structural elucidation of biologically active polypeptides and proteins that originally contained cystine (*516*). The compounds that Consden and Gordon (*515*) isolated from a partial acid hydrolyzate of wool are typical examples. In a number of cases the identity of these compounds was established by comparison with synthetic peptides of cysteic acid. Peptides containing cysteic acid are accessible by oxidation of cystine bispeptides with bromine (*2507*). In another approach, *N*-carbobenzoxy-protected *S*-benzylcysteine-containing peptides are synthesized first. The protecting groups are removed by treatment with sodium in liquid ammonia, and the oxidation is carried out either with an aqueous solution of bromine in methylene chloride (*515*) or with hydrogen peroxide in formic acid (*1829*).

d. Taurine

The product of decarboxylation of cysteic acid, taurine, can no longer be considered to be an amino acid. Aminoacyl derivatives of taurine are obtained via the reaction of α-halogen carboxylic acid chlorides with taurine and subsequent amidation (*2507*). Spears and Tieckelmann (*2175*) and Frankel and Moses (*763a*) have used the acid chlorides of phthalyl- and carbobenzoxytaurine to synthesize *N*-protected taurylamino acids and tauryl peptides. In spite of the presence of the phthalyl group, it was possible to hydrolyze the esters under alkaline conditions. Attempts to remove the phthalyl group by treatment with hydrazine, however, failed; they resulted in the cleavage of the sulfonamide bond.

4. Methionine

The sulfur in methionine, present in the form of a thioether, in general is not readily available for chemical reactions. For this reason, no difficulties have been reported for the synthesis of peptides containing methionine or of biologically active polypeptides, e.g. MSH (*894, 895, 2020*), ACTH (*299, 1021, 1103*), or eledoisin (*1904*). Almost all coupling methods were applied to the synthesis of peptides containing methionine, e.g. the azide method (*583, 894, 1021, 1103, 1904*), mixed anhydride method (*894, 1021*), carbodiimide method (*299, 2020*), pyrazolide method (*1814*), and *o*-phenylene pyrophosphite (*542*) or carbonyldiimidazole methods (*1172*). Surprisingly enough, there are cases where even catalytic hydrogenations proceeded smoothly, though extended periods of hydrogenation and the renewal of the catalyst were required (*583, 1021, 1414, 1521*). On the other hand, partial desulfurization of methionine has also been observed (*894*). The hydrolysis and hydrazine formation of methionine-containing peptides pose no problems.

The *S*-methyl group may be cleaved during the removal of protecting groups by treatment with sodium in liquid ammonia. The methyl group can be reintroduced by alkylation with methyl iodide (*350, 583*). In a number of cases this remethylation was not described for higher peptides, occasionally also not for di- and tripeptides (*1021, 1033*). Brenner and Pfister (*350*) observed a complete loss of optical rotation during the decarbobenzoxylation of Cbo-L-Met-L-Met-L-Met-OH by treatment with sodium in liquid ammonia. Free H-L-Met-L-Met-L-Met-OH maintained its full optical activity under the same conditions. The acid-catalyzed decarbobenzoxylation with hydrogen bromide in glacial acetic acid (*786, 894*) or in nitromethane (*30*), as well as with concentrated hydrochloric acid (*581*) or with ethanolic hydrochloric acid

(*785, 786*), may lead to the cleavage of the methyl thioether and formation of S-benzylhomocysteine by benzylation with the benzyl halide resulting from the decomposition of the carbobenzoxy group [cf. (*30, 785*) for a possible reaction mechanism]. Gawron and Draus (*785*) isolated the intermediate, DL-methionine ethyl ester benzylsulfonium chloride. Guttmann and Boissonnas (*894, 895*) have suggested the addition of diethyl phosphite and methyl ethyl sulfide to suppress this reaction and to trap the benzyl halide. According to Iselin (*1097*), side reactions cannot be completely excluded even under these conditions. Methionine sulfoxide—Iselin suggests the use of

$$\begin{array}{c} \text{O} \\ \uparrow \\ \text{—Met—} \end{array}$$

as abbreviation—represents an S-protected derivative that is applicable to the synthesis of peptides.

Oxidation with a small excess of hydrogen peroxide in 1 *N* hydrochloric acid or glacial acetic acid leads only to the formation of sulfoxide. The oxidation is faster than in the case of cystine or cysteine (*2315*). Considerable amounts of sulfone are formed only after longer reaction times (especially in glacial acetic acid) or with a larger excess of peroxide. The addition of ammonium molybdate is required to carry the oxidation to completion (*2316*). Methionine sulfoxide contains an additional center of asymmetry at the sulfur atom. Several derivatives of methionine, e.g. L-methionine methyl ester, are oxidized to the sulfoxide stereospecifically with formation of 80% *d*-sulfoxide in addition to only 20% of the l-form, presumably caused by the fact that an asymmetric atom is already present. This observation, however, has no bearing on the synthesis of peptides. Compounds required for the synthesis of peptides, such as carbobenzoxymethionine sulfoxide, methionine sulfoxide methyl ester, and the hydrazide of carbobenzoxymethionine sulfoxide, can be prepared either from the corresponding derivatives of methionine by oxidation with hydrogen peroxide in methanol, or directly from methionine sulfoxide via standard methods, i.e. esterification or *N*-acylation (*1097*).

The sulfoxide can be readily reduced to the thioether even if it is located in a peptide chain. A suitable reducing agent is thioglycolic acid, used earlier for the reduction of proteins containing oxidized methionine (*577, 955*). The reduction is complete after treatment of the peptide with a 2.5- to fivefold excess of thioglycolic acid in aqueous solution for 8 hours at 50°C. The reduction with thioglycol takes a less favorable course. The reaction is slower in the presence of methanol. Decarbobenzoxylation is best carried out by treatment with concentrated hydrochloric acid for 1 hour at 40°C. Hydrogen bromide in glacial acetic acid also in the case of the sulfoxide causes the formation of S-benzylhomocysteine (10–15%). The main product of this reaction is the S-dibromine derivative that reacts with water to reconvert to the sulfoxide, while the reaction with acetone produces methionine and bromoacetone. This reaction thus represents a simple approach to the reduction of the sulfoxide to methionine. It was demonstrated in the case of the tetrapeptide, H-Pro-Try-Lys(Tos)-Met-OH, that the oxidation can also be conducted with the carbobenzoxy peptide methyl ester and that, after decarbobenzoxylation with concentrated hydrochloric acid, the methionine peptide can be recovered subsequent to treatment with thioglycolic acid (*1097*). The cleavage of the *tert*-butyloxycarbonyl (*1171, 1904, 2020*), trityl (*1904*), or phthalyl group (*299, 894*) of methionine-containing peptides proceeds without complication.

5. Homocysteine

Homocysteine is the next higher homologue of cysteine or desmethyl methionine. The two compounds can be readily converted to each other in reversible reactions. This conversion does not occur only as a side reaction of the various treatments of methionine; it also finds application for preparative purposes. Homocysteine can be prepared from methionine by treatment with 18 *N* sulfuric acid at 130°C (*452*). S-Benzylhomocysteine was obtained from the reaction of carbobenzoxymethionine with hydrogen bromide in glacial acetic acid or with hydrochloric acid. Cleavage of methionine with hydrogen iodide does not lead to the formation of homocysteine but rather to that of homocysteine thiolactone (*90*), which can also be prepared directly from homocysteine by extensive boiling in an aqueous hydrochloric acid solution (*1818*). This conversion of methionine to homocysteine can also be carried out while the amino acid is located within a peptide chain, thus offering a route to the synthesis of homocysteine-containing peptides (*30, 785*). du Vigneaud *et al.* (*632*) have described the synthesis of methionine from homocysteine. The authors subjected the amino acid to reduction with sodium in liquid ammonia and subsequently methylated it. Additional structural proof was provided by the total synthesis of S-benzylhomocysteine from 1-S-benzylmercapto-2-chloroethane and the sodium salt of malonic ester (*640, 1697*).

Herrick and Todd (*2709*) synthesized DL-Homocys^2-glutathione, DL-Homocys^2-isoglutathione, and DL-Homocys^2-asparthione, and Jarvis *et al.* (*1138*) Homocys^1-oxytocin. The key substance for this synthesis was Cbo-Homocys(SBzl)-Tyr-OH, prepared via the carbodiimide method followed by alkaline hydrolysis.

du Vigneaud *et al.* (*641*) have described the cleavage of homocysteine thiolactone with alkali. The reaction was carried out successfully by employing a large excess of alkali. The resulting homocysteine was oxidized to homocystine and isolated as such. Under less basic conditions, aminolysis of the thiolactone ring takes place and the diketopiperazine is formed. Upon oxidation this compound forms an insoluble poly-disulfide. Benesch and Benesch (*141*) took advantage of this ready aminolysis and applied it to the synthesis of peptides. The authors used *N*-acetyl- or *N*-benzoylhomocysteine thiolactone and carried out the coupling reaction with amino acids or peptides in aqueous buffer solutions of pH values from 8 to 10. The peptides were isolated as phenylmercuric salts. The competing hydrolysis can be suppressed quite successfully by maintaining low temperatures. Besides the reaction with amino acids or peptides, the analogous reaction with proteins, e.g. gelatine, has been described. In the latter case the aminolysis of the thiolactone ring was catalyzed by silver ions (*1, 142*). Frankel and Gertner (*761*) synthesized polyhomocysteine from S-carbobenzoxyhomocysteine via the *N*-carboxylic anhydride. The carbobenzoxy groups were removed by treatment with sodium in liquid ammonia.

6. Lanthionine

Lanthionine was isolated by Horn *et al.* (*1065, 1067*) from the acid hydrolyzates of keratins such as wool, human hair, chicken feathers, etc. The keratins had been treated with alkali prior to the acid hydrolysis. After alkaline treatment, lanthionine was also isolated from lactalbumin, the shells of eggs, and insulin (*629, 1066*). Thus lanthionine is not a constituent of these proteins, but rather is formed from the cystine residues contained in the proteins. Lanthionine was found in the hydrolyzates of the peptide antibiotics subtilin (*1391*), Nisin (*211, 1609*), and cinnamycin (*647*). The structures of these antibiotics have not yet been established. Nor is the mechanism of this formation of lanthionine known [cf. Zahn (*2614*) on the various hypotheses].

du Vigneaud and Brown (*628*) obtained lanthionine under alkaline conditions from cysteine and α-amino-β-chloropropionic acid methyl ester [122]:

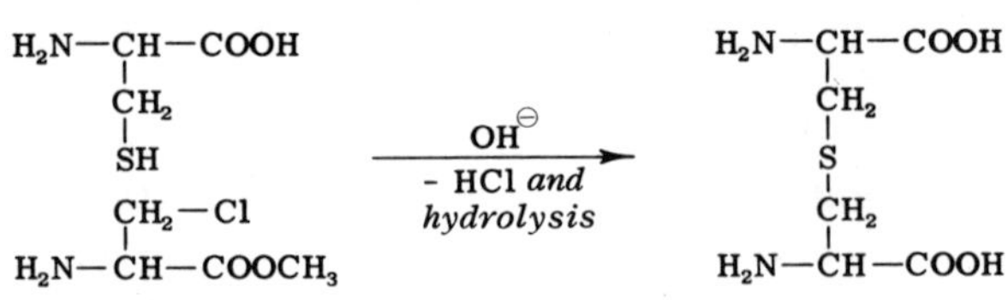

[122]

Since the compound contains two centers of asymmetry, the *meso* form exists in addition to the DL-, D-, and L-forms (cf. *412, 1485, 1795*). The *meso* form was separated by recrystallization from the product mixture of the reaction of L-cysteine with DL-α-amino-β-chloropropionic acid methyl ester. The *meso* form is identical with lanthionine isolated from natural sources (*628*).

In spite of the many publications on the problem of lanthionine formation, the number of synthetic derivatives suitable for the synthesis of peptides is limited. Dicarbobenzoxy-*meso*-lanthionine has long been known (*628*). Zahn and Osterloh (*2631, 2632*) and Zahn and Kessler (*2621*) have described the methyl and ethyl esters of *meso*-lanthionine. Frankel and Gertner (*760*) esterified dicarbobenzoxy-*meso*-lanthionine with benzyl alcohol, following the routine procedure. They also prepared a polymeric compound from the dichloride of dicarbobenzoxy-*meso*-lanthionine via the *N*-carboxylic anhydride; upon hydrolysis, this compound yields lanthionine. The structure of the polylanthionine was not determined in detail. Frankel and Gertner (*760*) also synthesized unsymmetrically substituted derivatives of lanthionine, using cysteine and substituted α-aminoacrylic acid as starting materials. α-Monocarbobenzoxylanthionine α-monobenzyl ester was converted to α'-*N*-carboxylic anhydride by treatment with phosgene and polymerized to polylanthionine. The only known peptides of lanthionine were synthesized by Zahn and Kessler (*2621*). The authors prepared *meso*-lanthionyl-α,α'-diglycine, using dicarbobenzoxy-*meso*-lanthionine dihydrazide as one starting component. In another instance, they condensed carbobenzoxyglycine and lanthionine in a salt coupling reaction to finally obtain α,α'-diglycyl-*meso*-lanthionine. α,α'-Diglycyl-*meso*-lanthionyl-α,α'-diglycine was also synthesized by the same authors.

TABLE II

Substituted Lysine

(a) N^{α}-Substituted lysine

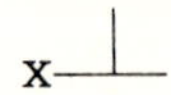

Substituent (X)	References
Cbo	Bezas and Zervas (*236*)
Form	Hofmann *et al.* (*1030*); Wolf *et al.* (*2576*)
Tos	Barras and Elmore (*112*); Kjaer and Larsen (*1252a*)

(b) N^{α}-Substituted lysine esters

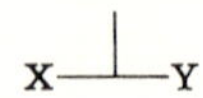

N^{α}-Substituent (X)	Carboxyl substituent (Y)	Salt	References
Cbo	OBzl	–	Bezas and Zervas (*236*)
Tos	OBzl	HBr	Irving and Gutmann (*1095*); Mechanic and Levy (*1516*)
Tos	OBzl	HCl	Swallow *et al.* (*2244*)
Tos	OEt	HCl	Swallow *et al.* (*2244*)
Tos	OMe	HCl	Irving and Gutmann (*1095*); Kjaer and Larsen (*1252a*); Barras and Elmore (*112*)
TFA	OMe	–	Weygand and Geiger (*2482*)

(c) N^{ϵ}-Substituted lysine

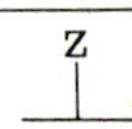

N^{ϵ}-Substituent (Z)	References
Benzylidene	Bezas and Zervas (*236*)
BOC	Schwyzer and Rittel (*2022*)
Cbo	Neuberger and Sanger (*1604*); Fölsch and Serck-Hanssen (*748*); Kjaer and Larsen (*1252a*); Zahn and Falkenburg (*2620*); Schlögl and Fabitschowitz (*1932*); Barras and Elmore (*111*); Bergmann *et al.* (*204*)
Form	Hofmann *et al.* (*1030*)
HOC	Lautsch *et al.* (*1341*)
MZ	Schwyzer and Sieber (*2033*)
NO_2Cbo	Shields *et al.* (*2103*)
Phth	Nefkens *et al.* (*1600*)
Tos	Roeske *et al.* (*1863*); Erlanger *et al.* (*686*); Morozova and Zhenodarova (*1575*)
TFA (DL)	Schallenberg and Calvin (*1923*)
Trit	Amiard and Goffinet (*35*)

(d) N^{ϵ}-Substituted lysine esters

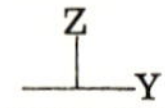

N^{ϵ}-Substituent (Z)	Carboxyl substituent (Y)	Salt	References
BOC	OMe	–	Schwyzer and Rittel (*2022*)
BOC	OMe	HCl	Schwyzer and Rittel (*2022*)
Cbo	OBzl	HCl	Shiba and Kaneko (*2099*); Erlanger and Brand (*684*)

TABLE II (Continued)

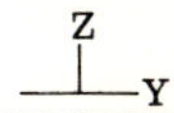

(d) N^{ϵ}-Substituted lysine esters (Continued)

N^{ϵ}-Substituent (Z)	Carboxyl substituent (Y)	Salt	References
Cbo (D)	OBzl	HCl	Erlanger and Brand (*684*)
Cbo	O*t* Bu	HCl	Taschner *et al.* (*2265*)
Cbo	OEt	HCl	Waley and Watson (*2420*)
Cbo	OEt	HCl	Falkenburg (*701*)
Cbo	OMe	HCl	Bergmann *et al.* (*204*); Boissonnas *et al.* (*291*); Shiba and Kaneko (*2099*)
Form	OMe	HCl	Hofmann *et al.* (*1030*)
HOC	OMe	HCl	Schulz (*1983*)
MZ	OMe	HCl	Schwyzer and Sieber (*2033*)
NO_2Cbo	OBzlNO_2	$PhSO_3H$	Shields *et al.* (*2103*)
Tos	OBzl	HCl	Erlanger *et al.* (*686*); Li *et al.* (*1415c*)
Tos	O*t* Bu	HCl	Roeske (*1862*)
Tos	OEt	HCl	Erlanger *et al.* (*686*); Roeske *et al.* (*1863*)
Tos	OMe	HCl	Schwyzer and Sieber (*2029*); Schröder and Gibian (*1971*)
Tos	OBzlNO_2	$PhSO_3H$	Shields *et al.* (*2103*)
Trit	OMe	2 HBr	Bezas and Zervas (*236*)
Trit	OMe	2 HCl	Bezas and Zervas (*236*)

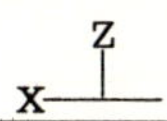

(e) N^{α}, N^{ϵ}-Substituted lysine

N^{α}-Substituent (X)	N^{ϵ}-Substituent (Z)	Salt	References
BOC	BOC	–	Hofmann *et al.* (*1029c*)
BOC	BOC	DCHA	Bernardi *et al.* (*209c*)
BOC	Cbo	–	Anderson and McGregor (*49*)
BOC	Form	–	Hofmann *et al.* (*1029c*)
Cbo	BOC	–	Schwyzer and Rittel (*2022*)
Cbo	Cbo	–	Bergmann *et al.* (*204*); Boissonnas *et al.* (*291*); Schröder *et al.* (*1976*)
Cbo	Form	–	Hofmann *et al.* (*1033*)
Cbo	HOC	–	Schulz (*1983*); Lautsch *et al.* (*1341*)
Cbo	Tos	–	Hofmann *et al.* (*1032*); Roeske *et al.* (*1863*)
Form	Cbo	–	Hofmann *et al.* (*1033*); Klieger *et al.* (*1262*)
Form	Tos	–	Klieger *et al.* (*1262*)
HOC	HOC	–	Schulz (*1983*)
NO_2Cbo	NO_2Cbo	–	Gish and Carpenter (*804*)
POC	POC	–	McKay and Albertson (*1487*)
PZ	BOC	–	Schwyzer and Rittel (*2022*)
Tos	Cbo	–	Irving and Gutmann (*1095*); Kjaer and Larsen (*1252a*); Theodoropoulos *et al.* (*2296*); Barras and Elmore (*111*)
TFA	Tos	–	Weygand and Steglich (*2500*)
Trit	Trit	–	Amiard and Goffinet (*35*)

TABLE II (Continued)

(f) N^{α}, N^{ϵ}- and carboxyl-substituted lysine

Z
X—┴—Y

N^{α}-Substituent (X)	N^{ϵ}-Substituent (Z)	Carboxyl Substituent (Y)	References
BOC	BOC	$NHNH_2$	Hofmann *et al.* (*1029c*)
BOC	BOC	$OPhNO_2$	Hofmann *et al.* (*1029c*)
BOC	Form	$NHNH_2$	Hofmann *et al.* (*1029c*)
BOC	Form	ONSu	Hofmann *et al.* (*1029c*)
BOC	Tos	$OPhNO_2$	Li *et al.* (*1415c*)
Cbo	BOC	OMe	Schwyzer and Rittel (*2022*)
Cbo	BOC	$OPhNO_2$	Schwyzer and Rittel (*2022*)
Cbo	Cbo	$NHNH_2$	Bergmann *et al.* (*201*); Schlögl and Fabitschowitz (*1932*)
Cbo	Cbo	OMe	Bergmann *et al.* (*201*)
Cbo	Cbo	$OPhNO_2$	Pless *et al.* (*1739*)
Cbo	HOC	OCH_2CN	Lautsch *et al.* (*1341*)
Cbo	Tos	$NHNH_2$	Hofmann *et al.* (*1032*)
Cbo	Tos	$OPhNO_2$	Bodanszky *et al.* (*277*)
PZ	BOC	OMe	Schwyzer and Rittel (*2022*)
Tos	Cbo	OBzl	Irving and Gutmann (*1095*); Mechanic and Levy (*1516*)
Tos	Cbo	OMe	Barras and Elmore (*112*); Irving and Gutmann (*1095*)
TFA	Tos	$NHNH_2$-Trit	Weygand and Steglich (*2500*)
TFA	TFA	OCH_2CN	Weygand and Swodenk (*2503*)

TABLE III

Substituted Ornithine

(a) N^{α}-Substituted ornithine

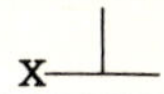

N^{α}-Substituent (X)	Salt	References
Cbo	HCl	Zervas *et al.* (*2663*)
Tos	–	Zaoral and Rudinger (*2654*); Barras and Elmore (*112*)
Tos	HCl	Zaoral and Rudinger (*2654*)
TFA	–	Weygand and Geiger (*2480*)

(b) N^{α}-Substituted ornithine esters

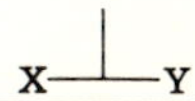

N^{α}-Substituent (X)	Carboxyl substituent (Y)	Salt	References
Tos	OMe	HCl	Zaoral and Rudinger (*2654*); Barras and Elmore (*112*)

(c) N^{δ}-Substituted ornithine

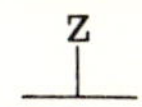

N^{δ}-Substituent (Z)	References
Cbo	Barras and Elmore (*111*); Izumiya *et al.* (*1123*); Harris and Work (*958*); Synge (*2251*); Noguchi *et al.* (*1626*)
Phth	Bodanszky *et al.* (*277b*)
Tos	Noda (*1625*); Erlanger *et al.* (*692*)
Tos (D)	Noda (*1625*)
TFA	Weygand and Geiger (*2480*)

(d) N^{δ}-Substituted ornithine esters

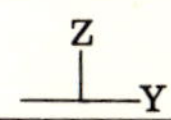

N^{δ}-Substituent (Z)	Carboxyl substituent (Y)	Salt	References
Cbo	OMe	HCl	Harris and Work (*958*); Synge (*2251*)
Tos	OBzl	HCl	Erlanger *et al.* (*687*)
Tos	OBzl	H_3PO_4	Erlanger *et al.* (*687*)
Tos	OEt	HCl	Erlanger *et al.* (*687*)
Tos	OMe	HCl	Noda (*1625*); Erlanger *et al.* (*692*)
Tos (D)	OMe	HCl	Noda (*1625*)

(e) N^{α}, N^{δ}-Substituted ornithine

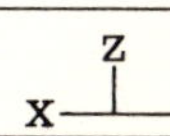

N^{α}-Substituent (X)	N^{δ}-Substituent (Z)	References
Cbo	Cbo	Synge (*2251*)
Cbo	Phth	Bodanszky *et al.* (*277b*)
Cbo	Tos	Li *et al.* (*1415*); Huguenin and Boissonnas (*1080a*)
Form	Cbo	Klieger *et al.* (*1262*)
Phth	Phth	van Tamelen and Smissman (*2360*)
Tos	Cbo	Barras and Elmore (*111*); Zaoral and Rudinger (*2654*)

TABLE IV

Substituted Diaminobutyric Acid

(a) N^{α}-Substituted α,γ-diaminobutyric acid 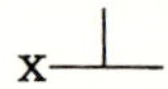

N^{α}-Substituent (X)	Salt	References
Tos	HCl	Zaoral and Rudinger (*2654*)
Tos	–	Zaoral and Rudinger (*2654*); Barras and Elmore (*112*); Poduška and Rudinger (*1746*); Rudinger *et al.* (*1858*)

(b) N^{α}-Substituted α,γ-diaminobutyric acid esters 

N^{α}-Substituent (X)	Carboxyl substituent (Y)	Salt	References
Tos	OMe	2 AcOH	Zaoral and Rudinger (*2654*)
Tos	OMe	HCl	Zaoral and Rudinger (*2654*); Barras and Elmore (*112*)

(c) N^{γ}-Substituted α,γ-diaminobutyric acid 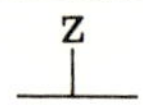

N^{γ}-Substituent (Z)	References
Cbo	Zaoral *et al.* (*2656*); Vogler and Lanz (*2386*); Barras and Elmore (*111*)
Cbo (D)	Vogler and Chopard-dit-Jean (*2384*)
Tos	Poduška and Rudinger (*1746*); Christensen and Riggs (*491*)
Tos (D)	Studer *et al.* (*2227*)

(d) N^{γ}-Substituted α,γ-diaminobutyric acid esters 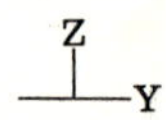

N^{γ}-Substituent (Z)	Carboxyl substituent (Y)	Salt	References
Cbo	OEt	HCl	Zaoral *et al.* (*2656*); Poduška and Rudinger (*1745*)
Cbo	OMe	HCl	Vogler and Lanz (*2386*); Izumiya *et al.* (*1119*);
Tos	OMe	0.5 H_2SO_4	Vogler *et al.* (*2389*)

(e) N^{α},N^{γ}-Substituted α,γ-diaminobutyric acid 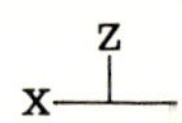

N^{α}-Substituent (X)	N^{γ}-Substituent (Z)	References
BOC	Tos	Poduška and Rudinger (*1746*)
Cbo	Cbo	Zaoral *et al.* (*2656*)
Cbo	Tos	Poduška and Rudinger (*1746*)
Cbo (D)	Tos	Studer *et al.* (*2227*)
Form (D)	Cbo	Vogler and Chopard-dit-Jean (*2384*)
Tos	Cbo	Barras and Elmore (*111*); Poduška and Rudinger (*1746*); Rudinger (*1846*)
Tos	Tos	Poduška and Rudinger (*1746*); Christensen and Riggs (*491*)

TABLE V

Substituted Arginine

(a) N^{α}-Substituted arginine

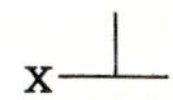

N^{α}-Substituent (X)	References
Cbo	Bergmann and Zervas (*193*); Boissonnas *et al.* (*291*)
MZ	Schwyzer *et al.* (*2037*)
NO_2Cbo	Gish and Carpenter (*805*)
Tos	Serebryanyi *et al.* (*2043*); Barras and Elmore (*111*); Zaoral and Rudinger (*2654*)
TFA	Weygand and Geiger (*2480*); Schallenberg and Calvin (*1923*)
Trit	Boissonnas *et al.* (*291*)

(b) N^{α},N^{G}-Substituted arginine

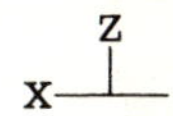

N^{α}-Substituent (X)	N^{G}-Substituent (Z)	References
BOC	$(Cbo)_2$	Gros *et al.* (*866*)
BOC	NO_2	Hofmann *et al.* (*1020b*)
Cbo	Cbo	Zervas *et al.* (*2670*)
Cbo	$(Cbo)_2$	Zervas *et al.* (*2670*)
Cbo	NO_2	Bergmann *et al.* (*202*); Hofmann *et al.* (*1028*); van Orden and Smith (*2359*)
Cbo (D)	NO_2	Gibian and Schröder (*803*)
Cbo	Tos	Schnabel and Li (*1943*)
MZ	Tos	Schwyzer and Li (*2021*)
NO_2Cbo	NO_2	Berse and Piché (*215*)
NO_2Cbo	NO_2Cbo	Gish and Carpenter (*804*)
Trit	NO_2	Schwyzer *et al.* (*2024*)

(c) N^{G}-Substituted arginine

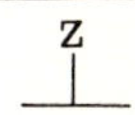

N^{G}-Substituent (Z)	References
Cbo	Zervas *et al.* (*2670*)
$(Cbo)_2$	Zervas *et al.* (*2670*)
NO_2	Bergmann *et al.* (*202*); Hofmann *et al.* (*1028*)
NO_2 (D)	Gibian and Schröder (*803*)
NO_2Cbo	Gish and Carpenter (*805*)
Tos	Guttmann *et al.* (*902*)

(d) N^{G}-Substituted arginine esters

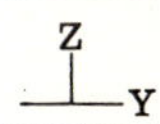

N^{G}-Substituent (Z)	Carboxyl substituent (Y)	Salt	References
Cbo	OBzl	–	Zervas *et al.* (*2670*); Zervas *et al.* (*2663*)
Cbo	OMe	2 HCl	Zervas *et al.* (*2670*)

TABLE V (Continued)

(d) N^G-Substituted arginine esters (Continued)

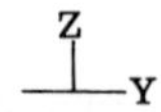

N^G-Substituent (Z)	Carboxyl substituent (Y)	Salt	References
NO_2	OBzl	HCl	Zahn and Diehl (*2617*); Tritsch and Woolley (*2322*)
NO_2	$OBzlNO_2$	HBr	Boissonnas *et al.* (*292*)
NO_2	OEt	HCl	Zahn and Diehl (*2617*)
NO_2	OMe	HCl	van Orden and Smith (*2359*); Hofmann *et al.* (*1028*)
NO_2 (D)	OMe	HCl	Gibian and Schröder (*803*)

(e) N^{α},N^G-Substituted arginine esters

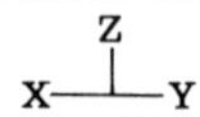

N^{α}-Substituent (X)	N^G-Substituent (Z)	Carboxyl substituent (Y)	References
Cbo	NO_2	$OBzlNO_2$	Boissonnas *et al.* (*292*)
Cbo	$(Cbo)_2$	$OPhNO_2$	Nicolaides and DeWald (*1619*)
Cbo	$(Cbo)_2$	$OBzlNO_2$	Zervas *et al.* (*2671*)
Cbo	NO_2	OMe	Bergmann *el al.* (*202*)
Cbo	Tos	OCH_2CN	Guttmann *et al.* (*902*)
Cbo	Tos	$OPhNO_2$	Guttmann *et al.* (*902*)
Tos	NO_2	OMe·HCl	Serebryanyi *et al.* (*2043*)
Trit	NO_2	OEt	Haas (*904*)

TABLE VI

Substituted Histidine

(a) N^{α}-Substituted histidine

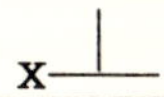

N^{α}-Substituent (X)	References
Cbo	Sakiyama *et al.* (*1903*); Patchornik *et al.* (*1695*); Bergmann and Zervas (*193*)
NO_2Cbo	Sheehan *et al.* (*2067*)
Phth	Helferich and Böshagen (*978*); Fischer and Whetstone (*724*)
Tos	McChesney and Swann (*1478*); Helferich and Böshagen (*978*)
TFA	Weygand and Röpsch (*2498*)
Trit	Stelakatos *et al.* (*2203*)

(b) N^{α}-Substituted histidine derivatives

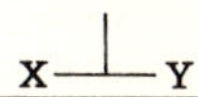

N^{α}-Substituent (X)	Carboxyl substituent (Y)	References
BOC	$NHNH_2$	Schröder and Gibian (*1972*)
Cbo	$NHNH_2$	Holly and Sondheimer (*1052*); Schneider (*1950*); Fischer and Whetstone (*724*)

(c) N^{α},N^{im}-Substituted histidine

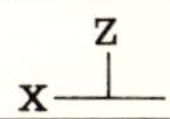

N^{α}-Substituent (X)	N^{im}-Substituent (Z)	References
Cbo	Bzl	Theodoropoulos and Fölsch (*2300*); Bricas and Nicot-Gutton (*360*); Winterstein *et al.* (*2570*)
Cbo	Cbo	Patchornik *et al.* (*1695*); Sakiyama *et al.* (*1903*)
MZ	Bzl	Schwyzer *et al.* (*2037*)
Tos	Bzl	du Vigneaud and Behrens (*627*)
Trit	Trit	Amiard *et al.* (*37*); Stelakatos *et al.* (*2203*)

(d) N^{im}-Substituted histidine derivatives

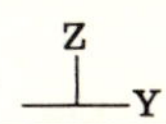

N^{im}-Substituent (Z)	Carboxyl substituent (Y)	Salt	References
Bzl	OBzl	2 $PhSO_3H$	Wiemer (*2561*); Theodoropoulos (*2293*); Theodoropoulos and Fölsch (*2300*)
Bzl	OEt	2 HCl	Stoffel (*2217*)
Bzl	OMe	2 HCl	Theodoropoulos and Fölsch (*2300*)
Cbo	OMe	2 HBr	Inouye and Otsuka (*1092*)

TABLE VII

Protected Glutamic Acid

(a) *N*-Protected glutamic acid X

Amino substituent (X)	Salt	References
BOC	–	Schröder and Klieger (*1975*); Bernardi *et al.* (*489a*)
Cbo	–	Fruton and Bergmann (*773*); Le Quesne and Young (*1369*); Goldschmidt and Jutz (*815*); Bergmann and Zervas (*193*)
Cbo (D)	–	Stoffel (*2217*); Klieger and Gibian (*1258*)
ClCbo	–	Kisfaludy (*1246*)
Form	–	Borek and Waelsch (*306*)
MeOCbo	–	Weygand and Hunger (*2487*)
NO_2Cbo	–	Berse *et al.* (*213*); Gish and Carpenter (*804*)
Phth	–	Hanson and Illhard (*933*); King *et al.* (*1238*); Billman and Harting (*242*); Nefkens *et al.* (*1600*); Clark-Lewis and Fruton (*496*)
Tos (D)	–	McChesney and Swann (*1478*); Rudinger and Czurbová (*1852*)
Tos	–	Rudinger *et al.* (*1859*); Harington and Moggridge (*939*); Rudinger (*1846*)
TFA	–	Weygand and Leising (*2493*)
Trit	$(C_2H_5)_3N$	Amiard *et al.* (*39*)
Trit	$(C_2H_5)_2NH$	Stelakatos *et al.* (*2203*)

(b) *N*-Protected glutamic acid α-derivatives X Y

Amino substituent (X)	α-Carboxyl substituent (Y)	Salt	References
BOC	OBzl	DCHA[a]	Schröder and Klieger (*1975*)
BOC	OMe	DCHA	Schröder and Klieger (*1975*)
Cbo	NH_2	–	Weygand and Hunger (*2486*); Ressler (*1805*); Bergmann and Zervas (*193*); Harington and Moggridge (*939*); Le Quesne and Young (*1369*); Gibian and Klieger (*799*)
Cbo	OBzl	–	Losse *et al.* (*1456*); Bergmann *et al.* (*205*); Le Quesne and Young (*1369*); Sachs and Brand (*1889*); Weygand and Hunger (*2488*); Liefländer (*1425*)
Cbo	OBzl	DCHA	Weygand and Hunger (*2488*); Klieger *et al.* (*1262*)
Cbo	O*t*Bu	–	Taschner *et al.* (*2283*)
Cbo	O*t*Bu	DCHA	Klieger and Gibian (*1260*)
Cbo	OEt	–	Liefländer (*1425*); Le Quesne and Young (*1370*); Goldschmidt and Jutz (*815*); Weygand and Hunger (*2486*)
Cbo	OEt	DCHA	Klieger *et al.* (*1262*); Weygand and Hunger (*2486*)
Cbo (D)	$NHNH_2$	–	Bruckner *et al.* (*418*)
Cbo	$NHNH_2$	–	Šorm and Rudinger (*2173*); Klieger and Gibian (*1259*); Le Quesne and Young (*1370*)

TABLE VII (Continued)

(b) *N*-Protected glutamic acid α-derivatives (Continued)

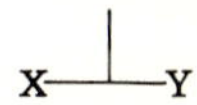

Amino substituent (X)	α-Carboxyl substituent (Y)	Salt	References
Cbo	NHNH-Cbo	–	Hofmann *et al.* (*1024*)
Cbo	OMe	–	du Vigneaud and Miller (*639a*); Harington and Mead (*937*)
Cbo	OMe	DCHA	Klieger *et al.* (*1262*)
Cbo	$OBzlNO_2$	–	Klieger and Gibian (*1260*)
Cbo	OPh	–	Klieger and Gibian (*1260*)
Cbo	SEt	–	Klieger and Gibian (*1259*)
Cbo	SPh	–	Sachs and Waelsch (*1893*); Wieland and Weidenmüller (*2559*)
ClCbo	OMe	–	Kisfaludy (*1246*)
Form	NH_2	–	Borek and Waelsch (*306*)
MeOCbo	OBzl	DCHA	Weygand and Hunger (*2488*)
NO_2Cbo	NH_2	–	Chambers and Carpenter (*478*)
Phth	OEt	–	King and Kidd (*1241*)
Phth	OMe	–	DeWald and Moore (*597*); King *et al.* (*1238*)
Phth (D)	OMe	–	DeWald and Moore (*597*)
Tos	NH_2	–	Harington and Moggridge (*939*)
Tos	O*t*Bu	–	Taschner *et al.* (*2283*)
Tos	$NHNH_2$	–	Rudinger (*1846*)
TFA	OEt	–	Weygand and Geiger (*2481*); Jöhl and Stoll (*1148*)
Trit	OEt	–	Zervas and Theodoropoulos (*2668*); Amiard *et al.* (*38*)
Trit	NH_2	–	Amiard and Heymes (*36*)
Trit	OBzl	–	UCLAF Patent (*2716*); Coutsogeorgopoulos and Zervas (*531*)
Trit	OMe	–	Amiard *et al.* (*38*)

(c) *N*-Protected glutamic acid γ-derivatives

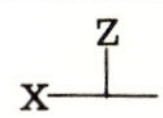

Amino substituent (X)	γ-Carboxyl substituent (Z)	Salt	References
BOC	O*t*Bu	DCHA	Schröder and Klieger (*1975*)
Cbo	OBzl	–	Kraml and Bouthillier (*1301*); Weygand and Hunger (*2488*); Losse *et al.* (*1456*); Hanby *et al.* (*931*); Heyns and Legeler (*996*)
Cbo (D)	OBzl	–	Hanby *et al.* (*931*); Stoffel (*2217*)
Cbo	OBzl	DCHA	Weygand and Hunger (*2488*); Klieger *et al.* (*1262*)
Cbo	O*t*Bu	–	Schwyzer and Kappeler (*2019*)
Cbo	OEt	–	Abderhalden and Nienburg (*6*); Hanby *et al.* (*931*); Hegedüs (*972*)
Cbo	OEt	DCHA	Klieger *et al.* (*1262*)
Cbo	$NHNH_2$	–	Gibian and Klieger (*799*); Fodor *et al.* (*734*); Le Quesne and Young (*1370*); Waley (*2415*); Hegedüs (*972*); Bruckner *et al.* (*428*)
Cbo (D)	$NHNH_2$	–	Bruckner and Kajtar (*417*)
Cbo (D)	OMe	–	Goodman *et al.* (*834*)

TABLE VII (Continued)

(c) *N*-Protected glutamic acid γ-derivatives (Continued)

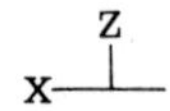

Amino substituent (X)	γ-Carboxyl substituent (Z)	Salt	References
Cbo	OMe	–	Shiba and Imai (*2096*); Hanby *et al.* (*931*); Goodman *et al.* (*834*)
Cbo	OMe	DCHA	Klieger *et al.* (*1262*)
Cbo	OPh	–	Klieger and Gibian (*1260*)
ClCbo	OCH_2CN	–	Kisfaludy (*1246*)
ClCbo	$NHNH_2$	–	Kisfaludy (*1246*)
ClCbo	OMe	–	Kisfaludy (*1246*)
Form	OBzl	–	Borek and Waelsch (*306*)
MeOCbo	OBzl	DCHA	Weygand and Hunger (*2488*)
MeOCbo	O*t* Bu	DCHA	Schröder and Klieger (*1975*)
MZ	OMe	–	Schwyzer *et al.* (*2037*)
PZ	OMe	–	CIBA AG Patent (*2700*)
POC	OEt	–	McKay and Albertson (*1487*)
Phth	OBzl	–	King *et al.* (*1238*)
Phth	OMe	–	King *et al.* (*1238*)
Phth	O*t* Bu	–	Schröder and Klieger (*1975*)
Tos	$NHNH_2$	–	Rudinger (*1846*); Swan (*2246*); Gawron and Draus (*786*)
Tos	OMe	–	Rudinger (*1846*)
TFA	OEt	DCHA	Weygand and Geiger (*2481*)
Trit	OBzl	–	Stelakatos *et al.* (*2203*); Chillemi *et al.* (*490*)
Trit	OMe	–	Amiard *et al.* (*38*)

(d) *N*-Protected glutamine

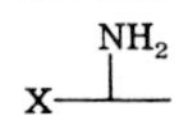

Amino substituent (X)	γ-Carboxyl substituent (NH_2)	References
BOC	NH_2	Hofmann *et al.* (*1020b*)
Cbo	NH_2	Katsoyannis *et al.* (*1191*); Fruton (*770*); Miller and Waelsch (*1560*); Boissonnas *et al.* (*296*)
Cbo	NH-Xanthyl	Akabori *et al.* (*27*)
ClCbo	NH_2	Kisfaludy and Dualszky (*1247*)
Form	NH_2	Borek and Waelsch (*306*)
MeOCbo	NH_2	Schröder and Klieger (*1975*)
MZ	NH_2	Kappeler (*1170*)
Phth	NH_2	Kind and Kidd (*1241*); Schröder and Klieger (*1975*)
Tos	NH_2	Rudinger (*1846*); Swan and du Vigneaud (*2247*)
Tos (D)	NH_2	Rudinger and Czurbová (*1852*)
Trit	NH_2	Amiard and Heymès (*36*); Stelakatos *et al.* (*2203*)

(e) *N*-Protected glutamic acid α,γ-derivatives

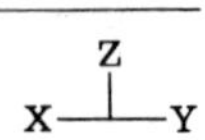

Amino substituent (X)	γ-Carboxyl substituent (Y)	α-Carboxyl substituent (Z)	References
Cbo	OBzl	NH_2	Kraml and Bouthillier (*1301*)
Cbo	OBzl	$OPhNO_2$	Losse *et al.* (*1456*); Fahnenstich (*700*);

TABLE VII (Continued)

(e) *N*-Protected glutamic acid α,γ-derivatives (Continued)

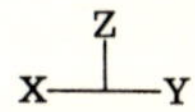

Amino substituent (X)	γ-Carboxyl substituent (Y)	α-Carboxyl substituent (Z)	References
Cbo	OBzl	OPhNO$_2$	Goodman and Stueben (*837*); Li *et al.* (*1413*); Heyns and Legler (*996*)
Cbo	OBzl	SPh	Losse *et al.* (*1456*)
Cbo	O*t* Bu	OBzl	Schwyzer and Kappeler (*2019*)
Cbo	O*t* Bu	OEt	Schwyzer and Kappeler (*2019*)
Cbo	O*t* Bu	NHNH$_2$	Klieger and Gibian (*1260*)
Cbo	O*t* Bu	OPhNO$_2$	Hofmann *et al.* (*1029c*)
Cbo	Cl	OEt	Rudinger and Šorm (*1861*)
Cbo	Cl	OMe	du Vigneaud and Miller (*639a*)
Cbo	OEt	SPh	Sachs and Waelsch (*1893*)
Cbo	NHNH$_2$	NH$_2$	Ebata (*651*)
Cbo	NHNH$_2$	O*t* Bu	Taschner *et al.* (*2283*)
Cbo	NHNH$_2$	NHNH$_2$	Klieger and Gibian (*1260*); Shiba and Kaneko (*2099*)
Cbo	NHNH$_2$·HCl	OEt	Ebata (*651*)
Cbo	NHNH$_2$	OMe	Liefländer (*1424*); Šorm and Rudinger (*2173*)
Cbo	NHNH$_2$·HCl	SEt	Klieger and Gibian (*1259*)
Cbo	NHNH-Trit	SEt	Klieger and Gibian (*1259*)
Cbo	OMe	NH$_2$	Klieger and Gibian (*1259*); Sondheimer and Holley (*2160*)
Cbo	OMe	O*t* Bu	Taschner *et al.* (*2283*)
Cbo	OMe	OCH$_2$CN	Losse *et al.* (*1456*)
Cbo	OMe	NHNH$_2$	Klieger and Gibian (*1259*)
Cbo	OMe	OPhNO$_2$	Goodman *et al.* (*834*); Losse *et al.* (*1456*)
Cbo	OMe	OPh	Klieger and Gibian (*1260*)
Cbo	OMe	SEt	Klieger and Gibian (*1259*)
Cbo	OMe	SPh	Losse *et al.* (*1456*)
Cbo	OPhNO$_2$	OBzl	Losse *et al.* (*1456*)
Cbo	OPhNO$_2$	OPhNO$_2$	Jones *et al.* (*1154*)
Cbo	SPh	OBzl	Weygand and Hunger (*2488*); Losse *et al.* (*1456*)
ClCbo	OCH$_2$CN	OMe	Kisfaludy (*1246*)
Form	OBzl	NH$_2$	Borek and Waelsch (*306*)
Phth (DL)	OBzl	NH$_2$	King *et al.* (*1240*)
Tos	NHNH$_2$	NH$_2$	Swan (*2246*)
Tos	NHNH$_2$	O*t* Bu	Taschner *et al.* (*2283*)
Tos	OMe	NH$_2$	Rudinger (*1846*); Swan (*2246*)
TFA	Cl	OEt	Weygand and Geiger (*2481*)
Trit	OBzl	OBzl	Amiard *et al.* (*38*); Coutsogeorgopoulos and Zervas (*531*)
Trit	OMe	NH$_2$	Amiard and Heymès (*36*)
Trit	OMe	OBzl	Amiard *et al.* (*38*)

TABLE VII (Continued)

(f) *N*-Protected glutamine α-derivatives

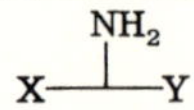

Amino substituent (X)	γ-Carboxyl substituent (Z)	α-Carboxyl substituent (Y)	References
BOC	NH_2	$NHNH_2$	Schröder (*1968*)
BOC	NH_2	NHNH-Cbo	Hofmann *et al.* (*1020b*)
BOC	NH_2	OMe	Schröder (*1968*)
Cbo	NH_2	OBzl	Bergmann *et al.* (*205*); Weygand and Hunger (*2488*); Sondheimer and Semeraro (*2163*)
Cbo	NH_2	OCH_2CN	Schneider (*1950*); Liefländer (*1425*)
Cbo	NH_2	$NHNH_2$	Sondheimer and Holley (*2161*); Liefländer (*1424*)
Cbo	NH_2	OMe	Liefländer (*1424*); Taschner and Wasielewski (*2279*)
Cbo	NH-Xanthyl	OMe	Akabori *et al.* (*27*)
Cbo	NH_2	$OPhNO_2$	Jones *et al.* (*1154*); Bodanszky and du Vigneaud (*273*)
Cbo	NH_2	SEt	Klieger and Gibian (*1259*)
Form	NH_2	OBzl	Sondheimer and Semeraro (*2163*)
MeOCbo	NH_2	$OPhNO_2$	Schröder and Klieger (*1975*)
Tos	NH_2	OEt	Zaoral and Rudinger (*2654*)
Tos	NH_2	OMe	Taschner and Wasielewski (*2279*); Zaoral and Rudinger (*2654*)
Tos (D)	NH_2	OMe	Zaoral and Rudinger (*2654*)
Trit	NH_2	OBzl	Coutsogeorgopoulos and Zervas (*531*)

(g) Glutamic acid α-derivatives

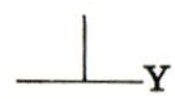

α-Carboxyl substituent (Y)	Salt	References
NH_2	–	Kraml and Bouthillier (*1301*); Bergmann and Zervas (*193*); Ressler (*1805*); Abderhalden and Nienburg (*6*); Swan and du Vigneaud (*2247*); Chambers and Carpenter (*478*); Amiard and Heymès (*36*)
NH_2	HBr	Klieger and Gibian (*1259*)
NH_2	HCl	Chambers and Carpenter (*478*)
OBzl	–	Coutsogeorgopoulos and Zervas (*531*); Sachs and Brand (*1889*)
OBzl (D)	–	Sachs and Brand (*1889*)
OEt	–	Le Quesne and Young (*1369*)
$NHNH_2$	–	Le Quesne and Young (*1370*)
OMe	–	Klieger and Gibian (*1260*)
OMe (D)	–	Bruckner *et al.* (*423*)
$OBzlNO_2$	HBr	Klieger and Gibian (*1260*); Goodman and Stueben (*837*)
SPh	HBr	Sachs and Waelsch (*1893*)

TABLE VII (Continued)

(h) Glutamic acid γ-derivatives

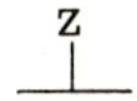

γ-Carboxyl substituent (Z)	Salt	References
NH_2	–	Coutsogeorgopoulos and Zervas (*531*); King and Kidd (*1241*); Katsoyannis and du Vigneaud (*1184*); Rudinger *et al.* (*1859*); Fruton (*770*); Bergmann *et al.* (*205*)
NH_2 (D)	–	Rudinger and Czurbová (*1852*)
OBzl	–	Heyns and Legler (*996*); Bruckner *et al.* (*428*); Clayton *et al.* (*498*); Hanby *et al.* (*931*); Guttmann and Boissonnas (*894*); Blout and Karlson (*259*)
OBzl (D)	–	Bruckner *et al.* (*418*)
O*t*Bu	–	Schröder and Klieger (*1975*)
OEt	–	Prawda (*1769*); Hegedüs (*972*); Hanby *et al.* (*931*); Miller and Waelsch (*1560*); Bergmann and Zervas (*195*)
OEt	HCl	Coleman (*510*); Rowlands and Young (*1843*); Green and Stahmann (*854*); Hegedüs (*972*); Bergmann and Zervas (*195*)
OMe	–	Biernat *et al.* (*241*); Hanby *et al.* (*931*); Goodman *et al.* (*834*)
OMe	HCl	Boissonnas *et al.* (*296*); Coleman (*510*); Kovács *et al.* (*1293*)
OMe (D)	–	Goodman *et al.* (*834*)
OMe (D)	HCl	Bruckner and Kajtár (*417*)

(i) Glutamic acid α,γ-bis derivatives

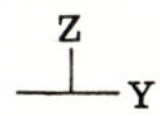

γ-Carboxyl substituent (Z)	α-Carboxyl substituent (Y)	Salt	References
NH_2	O*t*Bu	H_3PO_3	Anderson and Callahan (*48*)
NH_2	OMe	HCl	Sondheimer and Holley (*2161*)
NH_2	$OBzlNO_2$	HBr	Sondheimer and Semeraro (*2163*)
OBzl	OBzl	$PhSO_3H$	Shields *et al.* (*2103*)
OBzl	OBzl	HCl	Sachs and Brand (*1889*); Fodor *et al.* (*734*); Shields *et al.* (*2103*); Bruckner *et al.* (*431*)
OBzl (D)	OBzl	HCl	Sachs and Brand (*1889*); Bruckner *et al.* (*431*)
OBzl	OBzl	Tos-OH	Taschner and Wasielewski (*2278*); Izumiya and Makisumi (*1116*); Zervas *et al.* (*2670*)
OBzl	O*t*Bu	HCl	Roeske (*1862*)
OBzl	$OPhNO_2$	HBr	Goodman and Stueben (*837*)
O*t*Bu	OBzl	HCl	Schröder and Klieger (*1975*)
O*t*Bu	O*t*Bu	–	Anderson and Callahan (*48*)
O*t*Bu	O*t*Bu	HCl	Taschner *et al.* (*2265*)
O*t*Bu	OMe	HCl	Klieger and Gibian (*1260*)
OEt	NH_2	HCl	Angier *et al.* (*57*)
OEt	OEt	HCl	Goodman *et al.* (*834*); Schneider (*1950*)
OMe	O*t*Bu	HCl	Taschner *et al.* (*2265*)
OMe	OPh	HCl	Klieger and Gibian (*1260*)

[a] DCHA, dicyclohexylamine.

TABLE VIII

Protected Aspartic Acid

(a) *N*-Protected aspartic acid

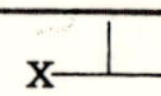

N-Substituent (X)	References
Cbo	Swallow *et al.* (*2244*); Lutz *et al.* (*1474*); Bergmann and Zervas (*193*); Karlson *et al.* (*1174*)
Cbo (D)	Stoffel (*2217*); Karlson *et al.* (*1174*)
ClCbo	Kisfaludy and Dualszky (*1247*)
MeOCbo	Weygand and Hunger (*2487*)
NO_2Cbo (DL)	Gish and Carpenter (*804*)
Phth	King and Kidd (*1242*); Balenović *et al.* (*101*)
Tos	McChesney and Swann (*1478*); Ressler (*1805*)

(b) *N*-Protected aspartic acid α-derivatives

N-Substituent (X)	α-Carboxyl substituent (Y)	References
Cbo	NH_2	Bergmann and Zervas (*193*); Lutz *et al.* (*1474*); Le Quesne and Young (*1371*)
Cbo	OBzl	Marks and Neuberger (*1505*); Bryant *et al.* (*436*); Bergmann *et al.* (*205*)
Cbo (D)	OBzl	Stoffel (*2217*)
Cbo	OEt	Le Quesne and Young (*1371*)
Cbo	OMe	Wünsch and Zwick (*2598a*)
Cbo	OBzlNO_2	Schröder and Klieger (*1974*)
Phth	NH_2	Tannenbaum (*2262*)
Tos	NH_2	Ressler (*1805*)
TFA	NH_2	Weygand *et al.* (*2490*)
TFA	OEt	Weygand *et al.* (*2490*)
TFA	OMe	Liwschitz *et al.* (*1435*)

(c) *N*-Protected aspartic acid β-derivatives

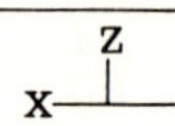

N-Substituent (X)	β-Carboxyl substituent (Z)	Salt	References
BOC	O*t*Bu	–	Schröder and Klieger (*1974*)
BOC	O*t*Bu	DCHA	Wünsch and Zwick (*2598a*)
Cbo	OBzl	–	Mazur (*1515*); Skeggs *et al.* (*2134*); Bryant *et al.* (*436*); Berger and Katchalski (*178*); Swallow *et al.* (*2244*); Izumiya *et al.* (*1124*); Frankel and Berger (*759*)
Cbo	O*t*Bu	–	Schwyzer and Dietrich (*2007*); Wünsch and Zwick (*2598*)
Cbo	OMe	–	Schwarz *et al.* (*1993*)
MeOCbo	O*t*Bu	–	Schröder and Klieger (*1974*)
MZ	OMe	–	Schwyzer *et al.* (*2037*)
Phth	O*t*Bu	–	Schröder and Klieger (*1974*); Wünsch and Zwick (*2598a*)
PZ	OMe	–	Surbeck-Wegmann (*2236*)
THF	O*t*Bu	DCHA	Wünsch and Zwick (*2598a*)

TABLE VIII (Continued)

(d) *N*-Protected asparagine

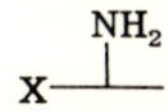

N-Substituent (X)	β-Carboxyl substituent	References
BOC	NH_2	Schröder and Klieger (*1974*)
Cbo (D)	NH_2	Berlingozzi *et al.* (*209*); Stoffel (*2217*)
Cbo	NH_2	Boissonnas *et al.* (*296*); Bergmann and Zervas (*193*); Berger and Katchalski (*178*); Zaoral and Rudinger (*2654*)
ClCbo	NH_2	Kisfaludy and Dualszky (*1247*)
Form	NH_2	Sondheimer and Semeraro (*2163*)
MeOCbo	NH_2	Schröder and Klieger (*1974*)
PZ	NH_2	CIBA Patent (*2700*)
POC	NH_2	McKay and Albertson (*1487*)
Phth	NH_2	Nefkens *et al.* (*1600*); King and Kidd (*1242*)
Tos	NH_2	Zaoral and Rudinger (*2654*); Ressler (*1805*)
TFA	NH_2	Weygand and Adermann (*2478*); Schallenberg and Calvin (*1923*)
Trit	NH_2	Stelakatos *et al.* (*2203*); Amiard and Heymès (*36*); Zervas and Theodoropoulos (*2668*)

(e) *N*-Protected aspartic acid α, β-derivatives

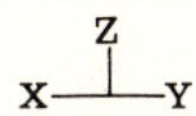

N-Substituent (X)	β-Carboxyl substituent (Z)	α-Carboxyl substituent (Y)	References
Cbo	OBzl	NH_2	Frankel and Berger (*759*); Bernhard *et al.* (*210*)
Cbo	OBzl	OBzl	Frankel and Berger (*759*); Stoffel (*2217*); Berger and Katchalski (*178*); Mazur (*1515*)
Cbo (D)	OBzl	OBzl	Karlson *et al.* (*1174*)
Cbo	OBzl	$OPhNO_2$	Guttmann (*890*)
Cbo	OBzl	$SPhNO_2$	Bryant *et al.* (*436*)
Cbo	OBzl	CH_2NPhth	Nefkens (*1598*)
Cbo	O*t*Bu	OBzl	Schwyzer and Dietrich (*2007*)
Cbo	O*t*Bu	OEt	Wünsch and Zwick (*2598*)
Cbo	O*t*Bu	$OPhNO_2$	Wünsch and Zwick (*2598a*); Bernardi *et al.* (*209b*)
Cbo	O*t*Bu	ONSu	Wünsch and Zwick (*2598a*); Hofmann *et al.* (*1020c*)
Cbo	Cl	OBzl	Bergmann *et al.* (*205*)
Cbo	OMe	NH_2	Bergmann and Zervas (*193*)
Cbo	OMe	$NHNH_2$	Schröder and Klieger (*1974*)
MeOCbo	O*t*Bu	$OPhNO_2$	Schröder and Klieger (*1974*)
Phth	O*t*Bu	$OPhNO_2$	Schröder and Klieger (*1974*)
Phth	OEt	OEt	King and Kidd (*1242*); Balenoviċ *et al.* (*101*)
TFA	Cl	OEt	Weygand *et al.* (*2490*)
Trit	OBzl	OBzl	Velluz *et al.* (*2369*)

TABLE VIII (Continued)

(f) *N*-Protected asparagine α-derivatives

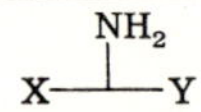

N-Substituent (X)	β-Carboxyl substituent	α-Carboxyl substituent (Y)	References
BOC	NH_2	$OPhNO_2$	Schröder and Klieger (*1974*)
Cbo	NH_2	OBzl	Bergmann *et al.* (*205*)
Cbo (D)	NH_2	OBzl	Stoffel (*2217*)
Cbo	NH_2	OCH_2CN	Schwyzer *et al.* (*2692*); Sondheimer and Semeraro (*2163*)
Cbo	NH_2	$NHNH_2$	Schüssler (*1987*)
Cbo	NH_2	OMe	Schüssler (*1987*); Zaoral and Rudinger (*2654*); Taschner and Wasielewski (*2277, 2279*); Bergmann and Zervas (*193*)
Cbo	NH_2	$OBzlNO_2$	Sondheimer and Semeraro (*2163*); Katsoyannis and Suzuki (*1192*)
Cbo	NH_2	$OPhNO_2$	Bodanszky and du Vigneaud (*273*)
Form	NH_2	OBzl	Sondheimer and Semeraro (*2163*)
MeOCbo	NH_2	$OPhNO_2$	Schröder and Klieger (*1974*)
Tos	NH_2	OMe	Taschner and Wasielewski (*2277, 2279*); Zaoral and Rudinger (*2654*)
TFA	NH_2	OEt	Weygand *et al.* (*2490*)
Trit	NH_2	OTrit	Amiard and Heymès (*36*)

(g) Aspartic acid α-derivatives

α-Carboxyl substituent (Y)	Salt	References
NH_2	–	Tannenbaum (*2262*); Ressler (*1805*); Bergmann and Zervas (*193*); Weygand *et al.* (*2490*)
OBzl	–	Bryant *et al.* (*436*)
OBzl	HCl	Amiard and Heymès (*36*)
OEt	–	Le Quesne and Young (*1371*)

(h) Aspartic acid β-esters

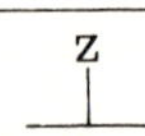

β-Carboxyl substituent (Z)	Salt	References
OBzl	–	Izumiya *et al.* (*1124*); Benoiton (*150*); Van de Linde *et al.* (*2350a*)
OBzl	HBr	Ben-Ishai and Berger (*149*)
O*t*Bu	–	Schwyzer and Dietrich (*2007*); Wünsch and Zwick (*2598*)

TABLE VIII (Continued)

(h) Aspartic acid β-esters (Continued)

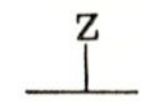

β-Carboxyl substituent (Z)	Salt	References
OEt	–	Biernat *et al.* (*241*)
OMe	–	Biernat *et al.* (*241*)
OMe	HCl	Coleman (*510*); Schwarz *et al.* (*1993*)

(i) Aspartic acid α,β-bis derivatives

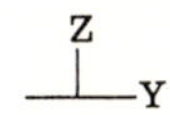

β-Carboxyl substituent (Z)	α-Carboxyl substituent (Y)	Salt	References
NH_2	NH_2	HCl	Smith and Spackman (*2152*)
NH_2	OBzl	–	Chillemi *et al.* (*490*)
NH_2 (D)	OBzl	HBr	Stoffel (*2217*)
NH_2	$OBzlNO_2$	HBr	Sondheimer and Semeraro (*2163*)
OBzl	OBzl	HBr	Ben-Ishai and Berger (*149*); Karlson *et al.* (*1174*)
OBzl (D)	OBzl	HBr	Amiard and Heymès (*36*); Karlson *et al.* (*1174*)
OBzl	OBzl	HCl	Velluz *et al.* (*2717, 2369*); Miller *et al.* (*1558*)
OBzl	OBzl	Tos-OH	Velluz *et al.* (*2717, 2369*); Zervas *et al.* (*2670*); Izumiya and Makisumi (*1116*); Mazur (*1515*)
OBzl (D)	OBzl	Tos-OH	Gibian and Schröder (*803*)
OBzl	O*t*Bu	HCl	Taschner *et al.* (*2265*); Roeske (*1862*)
OBzl	CH_2NPhth	–	Nefkens (*1598*)
O*t*Bu	OBzl	HCl	Schröder and Klieger (*1974*)
O*t*Bu	O*t*Bu	HCl	Taschner *et al.* (*2265*)
O*t*Bu	OEt	HCl	Wünsch and Zwick (*2598a*)
O*t*Bu	OMe	HCl	Wünsch and Zwick (*2598a*); Bernardi *et al.* (*209c*)
OEt	OEt	HCl	Schneider (*1950*); Balenovič *et al.* (*101*); Neumann and Smith (*1606*)
OMe	OBzl	HCl	Velluz *et al.* (*2369*)
OMe	OMe	HCl	Grassmann and Wünsch (*849*)

TABLE IX

Substituted Serine

(a) *N*-Substituted serine

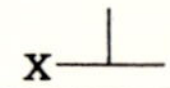

N-Substituent (X)	References
BOC (DL)	Anderson and McGregor (*49*)
Cbo	Adkins *et al.* (*22*); Guttmann and Boissonnas (*894*); Fruton (*769*); Baer and Maurukas (*85*); Moore *et al.* (*1569*)
Cbo (D)	Adkins *et al.* (*22*); Baer and Maurukas (*85*)
Cbo (DL)	Bergmann and Zervas (*193*); Baer and Maurukas (*85*)
MeOCbo	Weygand and Hunger (*2487*)
MZ	Schwyzer *et al.* (*2037*)
NO_2Cbo (DL)	Gish and Carpenter (*804*)
PZ	Schwyzer *et al.* (*2037*)
POC (DL)	McKay and Albertson (*1487*)
Phth (DL)	Nefkens *et al.* (*1600*)
Tos	Hofmann *et al.* (*1021*)
Tos (D)	Haneke (*932*)
TFA	Weygand and Rinno (*2497*)
Trit	Sheehan *et al.* (*2067*)
Trit (DL)	Amiard *et al.* (*37*)

(b) *O*-Substituted serine

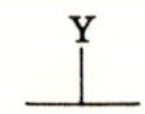

O-Substituent (Y)	References
Ac	Sheehan *et al.* (*2064*); Arakawa *et al.* (*64*)
Ac (DL)	Narita (*1596*); Sakami and Toennies (*1901*)
Ac·HCl (DL)	Fasman and Blout (*709*)
Bz	Baganz and Dransch (*91*)
Bz (DL)	Elliott (*658*); Bergmann and Miekeley (*189*)
Bzl	De Haas and Van Deenen (*579*)
Bzl (DL)	Grassmann *et al.* (*851*)
*t*Bu	Callahan *et al.* (*457*); Schröder (*1962*); Wünsch and Jentsch (*2597c*)
Cbo (DL)	Frankel and Halmann (*762*)

(c) *O*-Substituted serine esters

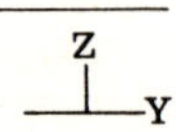

O-Substituent (Z)	Carboxyl substituent (Y)	Salt	References
Bzl (DL)	OEt	HCl	Grassmann *et al.* (*851*); Okawa (*1655*)
Bzl	OMe	HCl	Inouye and Otsuka (*1090*); Okawa (*1655*)
*t*Bu	O*t*Bu	–	Schröder (*1962*)
*t*Bu	O*t*Bu	HCl	Beyermann and Bontekoe (*228, 229*); Schröder (*1962*)
*t*Bu	OMe	–	Schröder (*1962*)
*t*Bu	OMe	HCl	Schröder (*1962*)

TABLE IX (Continued)

(d) *N, O*-Substituted serine

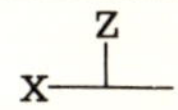

N- Substituent (X)	*O*- Substituent (Z)	References
Cbo (DL)	Ac	Frankel and Halmann (*762*)
Cbo	Bzl	Okawa (*1654*)
Cbo (DL)	Bzl	Grassmann *et al.* (*851*); Okawa (*1653*)
Cbo	O*t*Bu	Schröder (*1962*); Wünsch and Jentsch (*2597c*)
Cbo (DL)	THP	Iselin and Schwyzer (*1101*); Jones *et al.* (*1154*)
Form	Bzl	Inouye and Otsuka (*1090*)
Phth	Ac	Sheehan *et al.* (*2064*)
Phth	Bzl	De Haas and Van Deenen (*579*)
TFA	Bzl	Weygand and Rinno (*2497*)
TFA	TFA	Weygand and Rinno (*2497*)

TABLE X

Substituted Tyrosine

(a) *N*-Substituted tyrosine

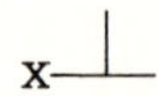

N-Substituent (X)	References
BOC	Anderson and McGregor (*49*)
Cbo	Bergmann and Zervas (*193*); Iselin and Schwyzer (*1102*); Determann *et al.* (*593*)
Form	Izumiya and Nagamatsu (*1118*); Waley and Watson (*2420*)
PZ	CIBA Patent (*2700*)
TFA	Weygand and Geiger (*2480*)
Tos	Berse *et al.* (*214*)

(b) *N*, *O*-Substituted tyrosine

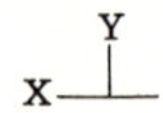

N-Substituent (X)	*O*-Substituent (Y)	References
BOC	Bzl	Schwyzer *et al.* (*2036*)
Cbo	Ac	Bergmann *et al.* (*206*); Blau and Waley (*257*)
Cbo	Bzl	Wünsch *et al.* (*2596*)
Cbo	*t* Bu	Wünsch and Jentsch (*2597c*)
Cbo	Cbo	Pannemann *et al.* (*1690*); Katchalski and Sela (*1176*)
Cbo	Tetrahydro-pyranyl	Iselin and Schwyzer (*1101*)
Cbo	Tos	Katsoyannis *et al.* (*1190*)
Form	Ac	Waley and Watson (*2420*)
Form	Me	Izumiya and Nagamatsu (*1118*)
NO_2Cbo	NO_2Cbo	Gish and Carpenter (*804*)
Tos	Tos	McChesney and Swann (*1478*)

(c) *O*-Substituted tyrosine esters

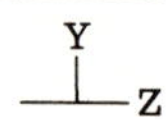

O-Substituent (Y)	Carboxyl substituent (Z)	Salt	References
Ac	OMe	HCl	Barkdoll and Ross (*108*)
Ac	$OBzlNO_2$	HCl	Schwyzer and Sieber (*2032*)
Bzl	OBzl	HCl	Schnabel (*1940*)
Bz	OEt	HCl	Harington and Pitt-Rivers (*940*)
Bzl	OMe	HCl	Wünsch *et al.* (*2596*)
t Bu	O*t* Bu	HCl	Beyermann and Bontekoe (*228*)
Cbo	Phthalimido-methyl ester	Tos-OH	Nefkens (*1598*)
Me	OEt	HCl	Jošt and Rudinger (*1159*)
Trit	OBzl	HCl	Stelakatos *et al.* (*2203*)

(d) *O*-Substituted tyrosine

O-Substituent (Y)	Salt	References
Ac	–	Schlögl *et al.* (*1933*); Kovács and Kótai (*1297*)
Ac	HCl	Barkdoll and Ross (*108*)

TABLE X (Continued)

(d) *O*-Substituted tyrosine (Continued)

O-Substituent (Y)	Salt	References
Bzl	–	Wünsch *et al.* (*2596*); Noguchi *et al.* (*1628*); Chillemi *et al.* (*490*)
t Bu	–	Wünsch and Jentsch (*2597c*)
Cbo	–	Overell and Petrow (*1672*)
Me	–	Izumiya and Nagamatsu (*1118*); Jošt and Rudinger (*1159*); Schlögl *et al.* (*1933*)
Tos	–	Katsoyannis *et al.* (*1190*); Jackson (*1128*)

(e) Carboxyl-substituted tyrosine

Carboxyl substituent (Z)	Salt	References
OBzl (D)	HCl	Erlanger *et al.* (*685*)
OBzl	–	Amiard *et al.* (*38*); Chillemi *et al.* (*490*)
OBzl	$PhSO_3H$	Gibian and Schröder (*803*)
OBzl	HCl	Erlanger and Hall (*690*)
OBzl	Tos-OH	Izumiya and Makisumi (*1117*); Zervas *et al.* (*2670*)
O *t* Bu	–	Anderson and Callahan (*48*); Roeske (*1862*)
O *t* Bu	H_3PO_3	Anderson and Callahan (*48*)
NHNH-BOC	–	Guttmann (*890*)
OMe	–	Almond and Niemann (*32*); Schwarz and Bumpus (*1992*)
OMe	HCl	Almond and Niemann (*32*); Boissonnas *et al.* (*296*); Schwarz *et al.* (*1993*)
$OPhNO_2$	HBr	Iselin and Schwyzer (*1102*)
OEt	HCl	Berse *et al.* (*214*)

TABLE XI

Substituted Cysteine

(a) *N*-Substituted cysteine

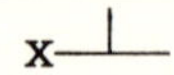

N-Substituent (X)	References
Cbo	Foye and Verderame (*753*)
Form	Foye and Verderame (*753*)
Tos	Dadić *et al.* (*558*)

(b) *N*-Protected cysteine esters

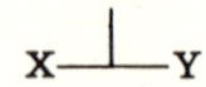

N-Substituent (X)	Carboxyl substituent (Y)	References
Cbo	OMe	Wieland *et al.* (*2531*)
Phth	OEt	Fleš *et al.* (*728*)
Tos	OBzl	Poduška and Gross (*1744*)

(c) *S*-Substituted cysteine

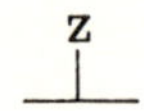

S-Substituent (Z)	References
Bzl	Wood and du Vigneaud (*2579*); Harington and Mead (*938*); Goldschmidt and Jutz (*815*); Hegedüs (*972*); Large *et al.* (*1329*)
Bzl-S-CH_2	Katsoyannis (*1183*); Pimlott and Young (*1725*)
Cbo	Berger *et al.* (*180*); Berse *et al.* (*212*)
MeOBzl	Akabori *et al.* (*27a*)
Tetrahydro-pyranyl	Holland and Cohen (*1047*)
Trit	Amiard *et al.* (*39*)

(d) *N*, *S*-Substituted cysteine

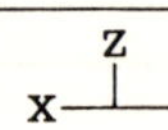

N-Substituent (X)	*S*-Substituent (Z)	References
BOC	Bzl	Anderson and McGregor (*49*)
Cbo	Bzl	Winterstein *et al.* (*2570*); Consden and Gordon (*515*); Hooper *et al.* (*1059*); Harington and Mead (*938*); Goldschmidt and Jutz (*815*)
Cbo	Bzl-S-CH_2	Pimlott and Young (*1725*)
Cbo	Cbo	Berger *et al.* (*180*)
Cbo	MeOBzl	Akabori *et al.* (*27a*)
Cbo	NO_2Bzl	Katsoyannis (*1183*)
Cbo	Tetrahydro-pyranyl	Holland and Cohen (*1047*)
ClCbo	Bzl	Kisfaludy and Dualsky (*1247*)
Form	Bzl	MacLaren *et al.* (*1491*)
Form	Bzl-S-CH_2	Katsoyannis (*1183*)
Form	Cbo	Katsoyannis (*1183*)
NO_2Cbo	Bzl	King *et al.* (*1238*); Gish and du Vigneaud (*806*); du Vigneaud *et al.* (*634*)
Phth	Bzl	Balenovič and Gašpert (*100*); King *et al.* (*1238*)
Tos	Bzl	Jošt and Rudinger (*1159*); Honzl and Rudinger (*1057*); MacLaren *et al.* (*1491*); du Vigneaud *et al.* (*626*); Poduška and Gross (*1744*)

TABLE XI (Continued)

(d) *N*, *S*-Substituted cysteine (Continued)

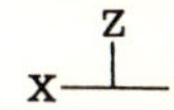

N-Substituent (X)	*S*-Substituent (Z)	References
TFA	Bzl	Weygand and Geiger (*2480*)
Trit	Bzl	Chillemi *et al.* (*490*)
Trit	Trit (diethylammonium salt)	Amiard *et al.* (*39*)

(e) *N*, *S*-Substituted cysteine derivatives

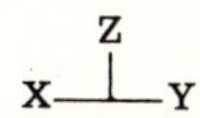

N-Substituent (X)	*S*-Substituent (Z)	Carboxyl substituent (Y)	References
Cbo	Bzl	NH_2	Stapleton and Swan (*2192*); Hegedüs (*972*)
Cbo	Bzl	Bzl	Ben-Ishai and Berger (*149*); Taschner and Wasielewski (*2278*)
Cbo	Bzl	OCH_2CN	Iselin *et al.* (*1099*); Lautsch *et al.* (*1341*)
Cbo	Bzl	3, 5-Dimethylpyrazole	Ried and Franz (*1815*)
Cbo	Bzl	OEt	Harington and Pitt-Rivers (*940*)
Cbo	Bzl	$NHNH_2$	Izumiya and Greenstein (*1113*); Hegedüs (*972*); Harington and Pitt-Rivers (*940*)
Cbo	Bzl	OMe	Hegedüs (*972*)
Cbo	Bzl	$OPhNO_2$	Bodanszky (*265*); Goodman and Stueben (*837*); Stapleton and Swan (*2192*); Wieland and Heinke (*2533*)
Cbo	$BzlNO_2$	$OPhNO_2$	Katsoyannis and Suzuki (*1192*)
ClCbo	Bzl	OCH_2CN	Kisfaludy (*1246*)
Phth	Bzl	OEt	Fleš *et al.* (*728*)
Tos	Bzl	OEt	Berse *et al.* (*214*)
Tos	Bzl	$NHNH_2$	MacLaren *et al.* (*1491*)
Tos	Bzl	OMe	MacLaren *et al.* (*1491*)
Trit	Trit	OMe	Zervas and Theodoropoulos (*2668*)

(f) *S*-Substituted cysteine esters

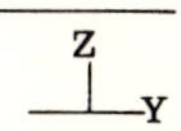

Carboxyl substituent (Y)	*S*-Substituent (Z)	Salt	References
NH_2	Bzl	–	Ressler (*1802*)
OBzl	Bzl	$PhSO_3H$	Lutz *et al.* (*1474*)
OBzl	Bzl	HBr	Ben-Ishai and Berger (*149*)
OBzl	Bzl	HCl	Hooper *et al.* (*1059*); Leach and Lindley (*1354*); Katsoyannis *et al.* (*1191*); MacLaren *et al.* (*1491*)

TABLE XI (Continued)

(f) *S*-Substituted cysteine esters (Continued)

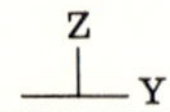

Carboxyl substituent (Y)	*S*-Substituent (Z)	Salt	References
OBzl	Bzl	Tos-OH	Hooper *et al.* (*1059*); MacLaren *et al.* (*1491*); Zervas *et al.* (*2670*)
OBzl	Cbo	–	Vollmann (*2396*)
O*t*Bu	Bzl	–	Taschner *et al.* (*2265*)
OEt	Bzl	–	Izumiya and Greenstein (*1113*); Losse and Moschall (*1458*); Harington and Pitt-Rivers (*940*)
OEt	$BzlNO_2$	–	Berse *et al.* (*212*)
OMe	Bzl	HCl	Schneider (*1950*); Boissonnas *et al.* (*296*); MacLaren *et al.* (*1491*)
OMe	Bzl-S-CH_2	–	Pimlott and Young (*1725*)
OMe	Cbo	–	Vollmann (*2396*)
OMe	Tetrahydro-pyranyl	–	Holland and Cohen (*1047*)

(g) Cysteine esters

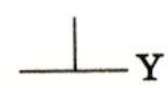

Carboxyl substituent (Y)	Salt	References
OBzl	HCl	Erlanger and Hall (*690*); Harington and Mead (*938*)
OMe	HCl	Zervas and Theodoropoulos (*2668*)
$OPhNO_2$	HBr	Goodman and Stueben (*837*)

V. THE SYNTHESIS OF CYCLIC PEPTIDES

The cyclic homomeric peptides may be divided into homodetic and heterodetic cyclic peptides. Cyclopeptides of the first group contain peptide bonds exclusively; in the second group ring closure was brought about via one or several other functional groups.

A. HOMODETIC CYCLIC PEPTIDES

The synthesis of a homodetic cyclic peptide requires the formation of a peptide bond between an amino and a carboxyl group of the same molecule. The activation required for this reaction is in principle not different from that required for the synthesis of a linear peptide. It is essential to work at the right dilution to suppress polycondensation as much as possible in favor of the formation of the intramolecular peptide bond. Boissonnas and Schumann (*302*) have tried to calculate the required dilution (Table XII, page 277). For practical purposes a ten- to a hundredfold dilution is applied. These calculations do not take into consideration molecular forces and mobility. Yet the yields of cyclic peptides are far below 50% in most cases. The cyclic dipeptides (diketopiperazines) are an exception; these compounds are formed in almost quantitative yield at much higher concentrations. The reason for this is the formation of the energetically favored six-membered ring. Schramm and Thumm (*1958*) conducted investigations on the ring formation at extremely high concentrations.

The necessity for working at high dilutions represents an inconvenience with regard to the amount of material that can be handled in the very large volumes of solvent. Therefore the concentration is frequently kept constant while the solution of the reactive compound is added dropwise at a slow rate to the reaction mixture. The dilutions reported in the literature vary within a wide range. The following values may serve as a point of reference: immediate addition of the entire amount of reactive compound, 2–4 mmoles/liter; 10 or more mmoles/

liter, depending on the drop rate if the method of slow dropwise addition is applied.

A linear cyclizable peptide may be synthesized either by leaving the amino group unprotected and activating the carboxyl group, or by removing the amino-protecting group while the carboxyl group is activated. In either case these steps are taken after proper dilution. The second approach is applied more frequently. The majority of synthetic cyclic peptides have so far been prepared via activated esters, i.e. the amino-protecting group was removed from an *N*-protected activated peptide ester. The amino group enters into salt formation as it is released, thus acquiring immediate new protection. The combination of the cyanomethyl ester with the trityl group (*1341, 2018, 2030, 2031, 2517*), or of the *p*-nitrophenyl ester with the trityl (*2026, 2028, 2029*), *tert*-butyloxycarbonyl (*377, 2033*), or preferably carbobenzoxy group (*1341, 1871, 2010, 2031, 2106*), has been used most frequently. Finally, the amino group is liberated by the addition of a base, usually pyridine (*377, 1341, 2010, 2018, 2031, 2106, 2517, 2533*) or magnesium carbonate in suspension (*1214, 1215*). The azide method has also been applied to cyclizations. The azide of a carbobenzoxy peptide was used as starting material. After dilution, the amino-protecting group was removed by catalytic hydrogenation (*2568*). Since a hydrazide group reacts more rapidly with nitrous acid than an amino group, it is possible to activate the hydrazide of a free peptide. The amino group is protected by salt formation under the required acidic reaction conditions. Liberation is brought about by the addition of an aqueous sodium bicarbonate solution (*2003, 2072, 2615, 2623, 2624, 2634*). The conversion of a free peptide to the acid chloride hydrochloride also provides a derivative suitable for cyclization. In this case the amino group was liberated by the addition of triethylamine to a solution of the peptide in dimethylformamide (*1870*). This method proved to be very useful for the synthesis of cyclic peptolides. The cyclization of peptides via the mixed anhydride method produced less satisfactory results (*302*) due to difficulties caused by the poor solubility of the precursors of the cyclic products. By contrast, it was possible to cyclize peptolides and *O*-peptides via the mixed anhydride method, since they are readily soluble in organic solvents due to the presence of ester bonds. The formation of the mixed anhydride is faster than the reaction of ethyl chlorocarbonate with the amino group. Free peptides can be cyclized with carbodiimides in aqueous methanol or ethanol (*1574, 2545*) or in dimethylformamide (*1284a, 1285*), or with water-soluble carbodiimides in a purely aqeuous solution (*2071*). Other reagents useful for cyclization are ethoxyacetylene (*1574, 1575, 2656a*) and tetraethyl pyrophosphite (*1872, 1872a*).

Several of the systematic investigations cited were not conducted on peptides of α-amino acids. Zahn *et al.* (*2615, 2622, 2623, 2624, 2634*) and Rothe *et al.* (*1866–1868, 1870–1872*) synthesized predominantly high-membered cyclic peptides of ε-aminocaproic acid (cyclodipeptide through cyclooctapeptide). Rothe (*1867*) synthesized the cyclic dipeptides of ω-aminovaleric acid and ω-aminoenanthic acid as well as the cyclic di-, tri-, and tetrapeptide of β-alanine. Zahn *et al.* (*2615, 2622, 2624*) have compared and discussed the infrared spectra and Kratky X-ray pictures of the homologous cyclic oligoamides of ε-aminocaproic acid.

To isolate a cyclic peptide from its reaction mixture, the aqueous or aqueous alcoholic mixture is passed over an acidic and basic ion-exchange resin to adsorb the polar by-products. Countercurrent distribution (*1215*) and chromatography on cellulose columns (*1285*) were also applied to the purification of cyclic peptides. The identification of the final product must be supplemented by a determination of the molecular weight. Methods applied to the molecular weight determination are isothermic distillation in trifluoroacetic acid (*1285, 2018, 2533*) or formic acid (*2545*) and cryoscopic methods. For the latter approach the use of dimethyl sulfoxide (*2031, 2035*) and hexahydro-*p*-aminobenzoic acid (*1865, 1868*) has been suggested in addition to the commonly employed phenol and camphor.

An interesting observation made during the cyclization of peptides is the so-called "doubling reaction." This effect had been noted during the synthesis of the first cyclic peptide (*105, 2065*). Schwyzer *et al.* (*2003, 2004a, 2010, 2018, 2020, 2031, 2035*) studied the phenomenon in detail. If a linear cyclizable peptide with an odd number of amino acid residues, e.g. a tripeptide, is subjected to cyclization, the corresponding cyclic tripeptide is not obtained but rather, through duplication, a cyclic hexapeptide (cf. Table XIII, page 278). Therefore the structures of all compounds described in the past as cyclic tripeptides are subject to doubt [Boissonnas and Schumann (*302*), Winitz and Fruton (*2568*), Brockmann *et al.* (*400;* cf. *397*), Smith (*2155*)], the more so since the synthesis of several of these compounds was repeated by the method of activated esters and the resulting products have been identified clearly as cyclic hexapeptides (*1872a, 2004a, 2010, 2031*).

The possible occurrence of the doubling reaction during the cyclization of peptides necessitates discussion of the basic question of possible ring sizes. The smallest cyclic peptide, a cyclodipeptide (diketopiperazine), represents a six-membered ring. Due to free rotation about the bond direction, rings of larger size normally show no strain. Theoretically, therefore, all cyclic forms of peptides higher than dipeptides should be possible. The doubling reaction demonstrates that this is not the case. Thus it appears that the phenomenon is caused by the configuration of the amide bond. Due to the presence of the mesomeric forms [1], the amide bond is planar (*104, 936*) and present in the energetically favored *trans* configuration:

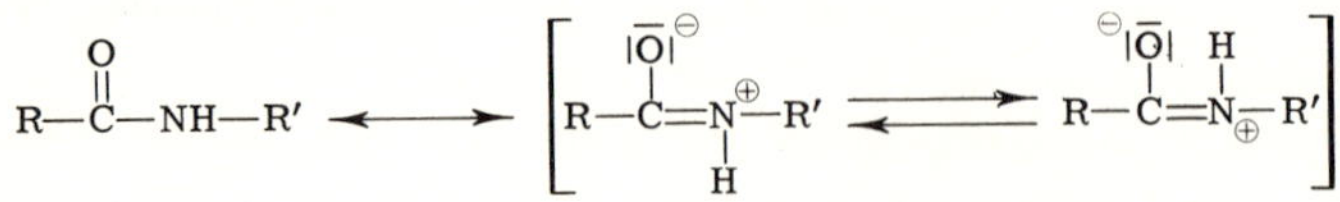

[1]

It can be demonstrated with the help of models that, while this *trans* configuration is preserved, the formation of cyclic peptides is possible only from tetrapeptides (*2018*). Schwyzer *et al.* (*2018*) succeeded indeed in synthesizing a well-defined cyclotetraglycine. Kenner *et al.* (*1214*) were unsuccessful in their attempts to synthesize cyclotetrapeptides from H-Gly-L-Leu-Gly-L-Leu-OH and H-L-Leu-Gly-L-Leu-Gly-SPhNO$_2$, respectively. Schwyzer and T. K. Aung (*2003*) compared the infrared spectra of diketopiperazines and cyclohexapeptides and found the presence of the *trans* configuration likely. In the case of diketopiperazine formation, the role of the energetically less favored *cis* configuration is suppressed by the strong tendency to form a six-membered ring. As is realized, these considerations do not sufficiently explain the occurrence of the doubling reaction, because it is observed not only with tripeptides but also with pentapeptides (*2030, 2033*), although not in the case of tetra- or hexapeptides.

According to Schwyzer *et al.* (*2004a, 2035*), three factors are responsible for the doubling reaction that occurs during cyclizations. These authors discuss steric factors, among them the activation energy, that is required to form the *cis*-peptide bond. Association phenomena due to one or several favorable steric arrangements of two peptide molecules enhance the chances of the occurrence of the doubling reactions (*1703*). In the case of the synthesis of gramicidin S from two pentapeptide units, a transition state, aided by the presence of L-proline and D-phenylalanine, sets the stage for the conformation present in gramicidin S, thus facilitating the doubling reaction. In this connection it is worthy of mention that all attempts to prepare an all-L-gramicidin S were unsuccessful (*1997*). Supposedly the average distance between the *N*-terminal amino and the C-terminal carboxyl group is also a contributing factor as was demonstrated, for instance, by the incorporation of D-amino acids. Similar observations were made during the synthesis of the LL- and LD-forms of *cyclo*-(Gly-Leu-Gly-Leu-Gly) and *cyclo*-(Gly-Leu-Leu-Gly-Gly) (*936, 1214, 1215*). The yields of the LD-compounds were considerably higher than those of the LL-compounds. Schwyzer and T. K. Aung (*2003*) arrived at the same conclusions, furnishing proof by comparison with sterically homogeneous compounds that in the course of the cyclization of H-Gly-DL-Phe-Gly-OR and H-Gly-DL-Phe-Gly-Gly-DL-Phe-Gly-OR, respectively, the LD-compound was formed preferentially.

According to our knowledge today, normal tripeptides as a rule dimerize due to their configuration and association. Pentapeptides, due to the possible association, may show the doubling reaction. However, since configurational influences of adverse effects are no longer present, monomeric cyclization is entirely possible. Further investigations are required to demonstrate the general validity of this principle. The observations with cyclizations of heptapeptides will be of great interest.

B. HETERODETIC CYCLIC PEPTIDES

1. Cyclic Disulfides

Most important biochemically among cyclic disulfides are the compounds with a disulfide bridge formed between two mercaptoamino acids of the same peptide chain (oxytocin, vasopressin). In the field of cyclic bisdisulfides—two disulfide bridges connecting different peptide chains (insulin)—efforts aimed at the total synthesis have been successful. The principles of disulfide formation will be discussed with regard to preparative methods and the aspects of dilution, ring size, and possible structures.

The protecting groups of an amino-, mercapto-, and carboxyl-protected linear peptide that is to be oxidized to a cyclic disulfide are removed subsequently (*1329*) or simultaneously (*971, 1139, 1140, 1329, 1340*), provided a suitable combination has been chosen initially. The solution of the free peptide will be diluted, either immediately or after purification via a mercaptide. The oxidation is carried out preferentially by passing air or oxygen through the solution. Potassium ferricyanide (*1060, 1061, 1138*) or hydrogen peroxide (*856*) is also employed for this purpose. The pH of the solution is important for the type of oxidation products formed (*857, 1669*); it also influences the yield (*2405*). The linear bismercaptans for the synthesis of the peptide hormones oxytocin and vasopressin are generally oxidized at a pH value of 6.8. The dilutions employed by various authors differ widely.

During the oxidation cyclic dimers of parallel [3a] or antiparallel [3b] structure and cyclic or linear higher polymers are formed in addition to the desired cyclic monomer [2]. The various products were separated by multiplicative distribution (*1300, 1329*) or by preparative paper chromatography (*1300*). The formation of higher polymers is minimal in dilute solution. The formation of the monomer [2] and the dimers [3a] and [3b] is dependent not only upon the concentration (*2405*) but also upon the distance between the two mercaptoamino acids

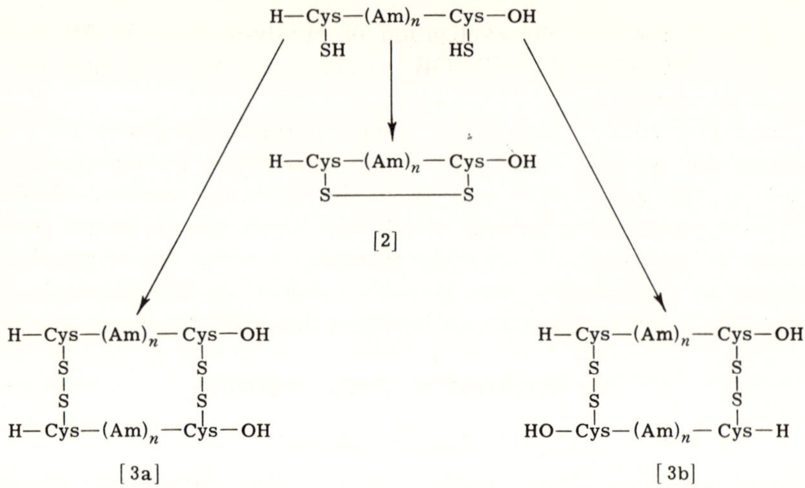

in the peptide chains (*971*). Table XIV (page 283) shows the formation of the various products in relation to the distance between the cysteine residues in H-Cys-(Gly)$_n$-Cys-OH. These data demonstrate a close analogy to the doubling reaction occurring during the synthesis of homodetic cyclic peptides. Shorter peptide chains tend to this doubling effect in either case. The dimer is formed exclusively if the ring-forming monomer contains about 10 atoms (homodetic tripeptide with nine members in the ring; heterodetic with $n = 1$, 11 members). The dimerization leads to the formation of the very stable rings of 18 (homodetic) and 22 (heterodetic) ring members, respectively. In the case of somewhat longer chains, both possibilities are open (homodetic: pentapeptide of 15 ring members; heterodetic with $n = 2$ or 3, 14 and 17 ring members, respectively). In both cases the optimum of monomer formation is reached with the hexapeptide (homodetic: 18 ring members; heterodetic: 20 ring members). The peptide hormones oxytocin and vasopressin are representatives of such cyclic disulfides with 20 ring members. Most likely the formation of cyclic bisdisulfides, as in the case of the doubling reaction of homodetic cyclic peptides, is caused by the preferred association of two molecules in pleated sheets of antiparallel orientation (*1703*). The formation of cyclic bisdisulfides from peptide chains of parallel orientation [IIIa] has not been observed (*971;* cf. *2405*). The optical rotations of the cyclic disulfides in Table XIV reveal a striking decrease of the $[\alpha]_D$ value in going from the hexapeptide ($n = 4$) to the heptapeptide ($n = 5$) (*1139, 1140*). Conformational changes have been implied to explain this observation. Investigations on the relationship of optical rotation and conformation in relation to the chain length of polypeptides were also carried out with glutamic acid peptides (cf. page 185).

The presence of the monomeric or dimeric form can be decided by a molecular weight determination [cryoscopic, isopiestic, and 2,4-dinitrophenyl (*1329*) methods]. Frequently the values obtained are too low; this is due to the presence of salts (*1300*). Heaton *et al.* (*971*) and Izumiya and Greenstein (*1113*) have devised suitable methods to distinguish in the case of dimerization between the parallel or antiparallel structure.

All basic questions were elucidated by the investigation of peptides containing glycine as the only amino acid in addition to cysteine. Lautsch *et al.*, however, synthesized a series of disulfides containing lysine, histidine, or tyrosine (*1339, 1341, 1345;* cf. Table XV, page 284). These disulfides consist of 26- and 20-membered rings, lysine, histidine, and tyrosine, respectively, being placed in 1,4-position. The same authors have also described a cyclic disulfide with a side chain, namely, S,S′-dehydro-H-L-Cys-L-Tyr-Gly-L-Cys-L-Tyr-Gly-OH (*1341*). In this case both tyrosine residues are also in 1,4-position.

2. Peptide Lactones (Cyclic Esters)

As in the case of cyclic disulfides, the synthesis of peptide lactones may lead to the formation of the cyclic monomeric [4] or the dimeric form [5]. The dimeric form can exist only in the antiparallel structure. Both forms are known as natural products [monomer: etamycin (22 ring

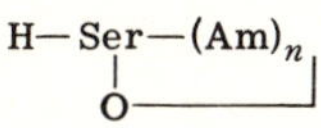

[4]

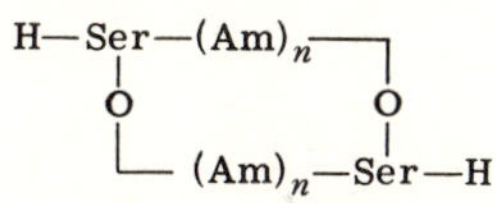

[5]

members); staphylomycin factor S (19 members); ostreogrycin (19 members); esperin (11 members); dimer: echinomycin (26 members)]. There is unfortunately a lack of systematic investigations dealing with such questions as ring size, the doubling reaction, etc., for simple cases.

The only peptide lactones that have been synthesized are related to actinomycin (*371, 381, 382a,b,c*).

TABLE XII

Dilutions Calculated for Open Peptides of Various Ring Size

Peptide	Dilution (mmoles/liter)	Peptide	Dilution (mmoles/liter)
Dipeptide	1000	Pentapeptide	68
Tripeptide	300	Hexapeptide	40
Tetrapeptide	130	Decapeptide	9

TABLE XIII

Cyclic Peptides[a]

Cyclic peptides[b]	Starting material and method of protecting group removal[c]	Cyclization conditions	Yield (%)	References
Cyclic tri- and tetrapeptides				
Gly-DL-Pro-Gly[d]	H-tripeptide-OEt HCl; Chloroform/NH_3	Ethanol, triethylamine (4 months at room temp.)	3	*(2155)*
Gly-Gly-Gly-Gly	Trit-tetrapeptide-OCH_2CN; HCl/acetonitrile	Pyridine (95°C)	12	*(2018)*
	Cbo-tetrapeptide-SCH_2COOH; HBr/glacial acetic acid	Pyridine (95°C)	13	*(2018)*
Cyclic pentapeptides				
Gly-L-Leu-Gly-L-Leu-Gly	Cbo-pentapeptide-$SPhNO_2$; HBr/glacial acetic acid	Dissolved in H_2O, diluted with $MgCO_3$ suspension in H_2O	41	*(2014, 2015)*
Gly-L-Leu-Gly-D-Leu-Gly	Cbo-pentapeptide-$SPhNO_2$; HBr/glacial acetic acid	Dissolved in H_2O, diluted with $MgCO_3$ suspension in H_2O	55	*(2014)*
Gly-L-Leu-L-Leu-Gly-Gly	Cbo-pentapeptide-$SPhNO_2$; HBr/glacial acetic acid	Dissolved in H_2O, diluted with $MgCO_3$ suspension in H_2O	12	*(936)*
	H-pentapeptide-OH	Dicyclohexylcarbodiimide, H_2O/methanol (4 days, 20°C)	13	*(936)*
Gly-L-Leu-D-Leu-Gly-Gly	Cbo-pentapeptide-$SPhNO_2$; HBr/glacial acetic acid	Dissolved in H_2O, diluted with $MgCO_3$ suspension in H_2O	39	*(936)*
	H-pentapeptide-OH	Dicyclohexylcarbodiimide, H_2O/methanol (4 days, 20°C)	39	*(936)*
Gly-Sar-L-Lys(Cbo)-D-Val-L-Pro	BOC-pentapeptide-$OPhNO_2$; HCl/methylene chloride	Dimethylformamide/pyridine (60°C)	42	*(377)*
Cyclic hexapeptides				
Gly-Gly-Gly-Gly-Gly-Gly	Cbo-tripeptide-SCH_2COOH; HBr/glacial acetic acid (DR)	Pyridine (60°C)	69	*(2018)*
	Trit-tripeptide-OCH_2CN; HCl/acetonitrile (DR)	Pyridine (95°C)	33	*(2018)*
	H-tripeptide-$NHNH_2$ (DR)	HCl/$NaNO_2$; $H_2O/NaHCO_3$ (originally described as cyclic tripeptide; later identified as cyclic hexapeptide)	42	*(2065, 2072)*

	H-tripeptide-Cl HCl	(DR)	Dimethylformamide/ triethylamine (0°C)	11	(*1870*)
	H-tripeptide-OH		Tetraethyl pyrophosphite, diethyl phosphite (140°C)	40	(*1872*)
DL-Ala-Gly-Gly-DL-Ala-Gly-Gly	H-tripeptide-OH	(DR)	Tetraethyl pyrophosphite (140°C)	40	(*1872*)
(a) Gly-DL-Val-Gly-Gly-DL-Val-Gly	Cbo-hexapeptide-SPh HBr/glacial acetic acid		Pyridine (80°C)	18	(*2533*)
	H-hexapeptide-OH		Dicyclohexylcarbodiimide, H_2O/methanol	45	(*2545*)
(b) Val-Gly-Gly-Val-Gly-Gly	H-tripeptide-Cl HCl (L or DL?)	(DR)	Dimethylformamide/ triethylamine (0°C)	6	(*1870*)
	H-tripeptide-OH	(DR)	Tetraethyl pyrophosphite, diethyl phosphite (140°C)	47	(*1872*)
Gly-L-Leu-Gly-Gly-L-Leu-Gly	Cbo-hexapeptide-$SPhNO_2$ HBr/glacial acetic acid		Dissolved in H_2O, diluted with $MgCo_3$ suspension in H_2O	20	(*1214*)
	Cbo-tripeptide-$SPhNO_2$ HBr/glacial acetic acid	(DR)	Dissolved in H_2O, diluted with $MgCO_3$ suspension in H_2O	11	(*1214*)
	Cbo-hexapeptide-$OPhNO_2$ HBr/glacial acetic acid		Pyridine (80°C)	68	(*2010*)
	Cbo-tripeptide-$OPhNO_2$ HBr/glacial acetic acid	(DR)	Pyridine (80°C)	20	(*2010*)
	H-hexapeptide-OH (L or DL?)		Ethoxyacetylene in methanol (40°-45°C)	11.2	(*1574*)
	H-hexapeptide-OH (L or DL?)		Dicyclohexylcarbodiimide, ethanol/H_2O 4:1	47	(*1574*)
(a) Gly-L-Phe-Gly-Gly-L-Phe-Gly	Cbo-tripeptide-$OPhNO_2$ HBr/glacial acetic acid	(DR)	Dimethylformamide and pyridine (60°C)	30	(*1346, 2106*)
	Cbo-hexapeptide-$OPhNO_2$ HBr/glacial acetic acid		Dimethylformamide and pyridine (60°C)	–	(*1346, 2106*)
	H-hexapeptide-OH (L or DL?)		Ethoxyacetylene in aq. methanol	48	(*1575*)
(b) L-Phe-Gly-Gly-L-Phe-Gly-Gly	PZ-hexapeptide-NHNH·BOC cat. H_2 and HCl/methanol		HCl/$NaNO_2$; H_2O/$NaHCO_3$ (48 hr at 0°C)	13	(*2003*)
(a) Gly-D-Phe-Gly-Gly-D-Phe-Gly	Cbo-tripeptide-$OPhNO_2$ HBr/glacial acetic acid	(DR)	Dimethylformamide and pyridine (60°C)	26	(*1346, 2106*)
(b) D-Phe-Gly-Gly-D-Phe-Gly-Gly	PZ-hexapeptide-NHNH·BOC cat. H_2 and HCl/methanol		HCl/$NaNO_2$; H_2O/$NaHCO_3$ (48 hr at 0°C)	17	(*2003*)
(a) Gly-L-Phe-Gly-Gly-D-Phe-Gly	Cbo-hexapeptide-$OPhNO_2$ HBr/glacial acetic acid		Dimethylformamide and pyridine (60°C)	–	(*1346, 2106*)
(b) L-Phe-Gly-Gly-D-Phe-Gly-Gly	PZ-hexapeptide-NHNH·BOC cat. H_2 and HCl/methanol		HCl/$NaNO_2$; H_2O/$NaHCO_3$ (48 hr at 0°C)	23	(*2003*)

TABLE XIII (Continued)

Cyclic peptides [b]	Starting material and method of protecting group removal [c]	Cyclization conditions	Yield (%)	References
Cyclic hexapeptides (Continued)				
Gly-Gly-L-Phe-Gly-L-Phe-Gly	Cbo-hexapeptide-$OPhNO_2$ HBr/glacial acetic acid	$HCl/NaNO_2$; $H_2O/NaHCO_3$ (48 hr at 0°C)	–	*(1346, 2106)*
(a) Gly-DL-Phe-Gly-Gly-DL-Phe-Gly	Cbo-tripeptide-O-$C_6H_5SO_2CH_3$ Hydrogenation, methanol, Pd-charcoal (DR)	Pyridine (80°C)	45	*(2031)*
	Cbo-tripeptide-$OPhNO_2$ (DR)	Pyridine (80°C)	38	*(2031)*
	Trit-tripeptide-OCH_2CN (DR)	Pyridine (80°C)	20	*(2031)*
(b) Gly-Gly-DL-Phe-Gly-Gly-DL-Phe	Cbo-tripeptide-O-$C_6H_5SO_2CH_3$ (DR) Hydrogenation, methanol, Pd-charcoal	Pyridine (80°C)	25	*(2031)*
	Cbo-hexapeptide-O-$C_6H_5SO_2CH_3$ Hydrogenation, methanol, Pd-charcoal	Pyridine (80°C)	13	*(2031)*
(a) Gly-Phe-Gly-Gly-Phe-Gly	H-hexapeptide-OH (L or DL?)	Dicyclohexylcarbodiimide, aq. methanol (3 days at −3°C; 1 week at 20°C)	~40	*(2656a)*
(b) Phe-Gly-Gly-Phe-Gly-Gly	H-tripeptide-OH (L or DL?)	Ethoxyacetylene, aq. methanol (1 month at room temperature)	~50	*(2656a)*
(a) Gly-L-Pro-Gly-Gly-L-Pro-Gly	Cbo-tripeptide-$OPhNO_2$	Pyridine (75°C)	49	*(2004a)*
	HBr/glacial acetic acid	Dimethylformamide (100°C)	88	*(1872a)*
	H-tripeptide-OH	Phosphite method/ diethyl phosphite (100°C)	55	*(1872a)*
	PZ-hexapeptide-O-$C_6H_5SO_2CH_3$ cat. H_2 in methanol	Dimethylformamide, acetic acid pyridine (90°C)	48	*(2004a)*
	Cbo-tripeptide-O-$C_6H_5SO_2CH_3$ cat. H_2 in methanol	Cyclization during hydrogenation	50	*(2004a)*
	Cbo-tripeptide-O-$C_6H_5SO_2CH_3$ cat. H_2 in methanol/HCl	Dimethylformamide, acetic acid pyridine (90°C)	–	*(2004a)*
(b) Gly-Gly-L-Pro-Gly-Gly-L-Pro	H-tripeptide-$OPh(NO_2)_2$	Dimethylformamide (50°C)	73	*(1872a)*
	H-tripeptide-OH	Phosphite method/ diethyl phosphite (100°C)	48	*(1872a)*

OBenz OBenz Gly-L-Glu-Gly-Gly-L-Glu-Gly	Cbo-tripeptide-OPhNO$_2$ HBr/glacial acetic acid	(DR)	H$_2$O/NaHCO$_3$ and extraction with ethyl acetate; free peptide by HBr/glacial acetic acid (45 min, 55°C)	–	(*1346*)
OMe OMe Gly-L-Glu-Gly-Gly-L-Glu-Gly	Cbo-tripeptide-OPhNO$_2$ HBr/glacial acetic acid	(DR)	H$_2$O/NaHCO$_3$ and extraction with ethyl acetate	26	(*1346*)
Tos Tos Gly-Lys-Gly-Gly-Lys-Gly	H-hexapeptide-OH (L or DL?)		Ethoxyacetylene, aq. methanol	30	(*1575*)
	H-tripeptide-OH (L or DL?)	(DR)	Ethoxyacetylene, aq. methanol	10	(*1575*)
(a) Gly-L-Tyr-Gly-Gly-L-Tyr-Gly	Trit-tripeptide-OCH$_2$CN HCl/acetonitrile	(DR)	Pyridine (90°C)	21	(*1341*)
(b) Gly-Gly-L-Tyr-Gly-Gly-L-Tyr	Trit-tripeptide		Pyridine (90°C)	20	(*1341*)
Benz L-Tyr-Gly-Gly-Gly-L-His-Gly			Ethyl-γ-dimethylaminopropyl-carbodiimide; dimethyl-formamide; free peptide by Na in liquid NH$_3$	58	(*1285*)
Benz Benz Gly-Gly-L-Tyr-Gly-Gly-L-His	Trit-hexapeptide-OCH$_2$CN Trifluoroacetic acid/ glacial acetic acid		Pyridine (85°C); free peptide by Na in liquid NH$_3$	57	(*2517*)
Bzl Gly-L-Tyr-Gly-Gly-L-His-Gly	H-hexapeptide-OH		*N*-Ethyl-*N'*-(3-dimethylamino-propyl)-carbodiimide; dimethyl-formamide (7 days at room temperature); free peptide by hydrogenation	–	(*1284a*)

TABLE XIII (Continued)

Cyclic peptides[b]	Starting material and method of protecting group removal[c]	Cyclization conditions	Yield (%)	References
Cyclic hexapeptides (Continued)				
Gly-D-Tyr-Gly-Gly-L-His(Bzl)-Gly	H-hexapeptide-OH (DL-L-peptide)	*N*-Ethyl-*N'*-(3-dimethylaminopropyl)-carbodiimide; dimethylformamide (7 days at room temperature); free peptide by hydrogenation; separation of D-L-cyclic peptide by chromatography	–	*(1284a)*
Gly-L-His-L-Ser-Gly-L-His-L-Ser	H-tripeptide-OH (DR)	1-Cyclohexyl-3-(dimethylaminopropyl)-carbodiimide; H_2O	35	*(2071)*
Val-Lys(Tos)-Leu-Val-Lys(Tos)-Leu	H-hexapeptide-OH (L or DL?)	Ethoxyacetylene, aq. methanol	20	*(1575)*

[a] For cyclic natural products, see Volume II, Chapter III.
[b] Order of amino acids corresponds to linear sequence used for cyclization. (a) and (b): Same cyclic peptides from different linear peptide sequences.
[c] DR = doubling reaction.
[d] Recently identified as cyclic hexapeptide (cf. *2004a*).

TABLE XIV

Formation of Products in Relation to Distance between Cysteine Residues

n	Monomeric compound ring size	Cyclized products, yield (%)[a]				$[\alpha]_D$	Ref.
		[2]	[3a]	[3b]	Higher polymers		
0	8	35	-	20	+	- 29° (*N* HCl)	(*2405*)
1	11	-	-	80	+		(*971*)
2	14	15	-	20	25		(*971*)
3	17	40	-	55	-		(*971*)
4	20	90	-	-	-	- 25° (H_2O)	(*971*)
5	23	57	Dimer and polymer are formed in low yield			- 4.5° (H_2O)	(*1140*)
6	26	47				- 4.3° (H_2O)	(*1140*)
7	29	Unknown	Unknown	Unknown	Unknown	- 1.4° (H_2O)	(*1341*)

[a] Key to symbols: -, no indications for formation;
+, formation proved, but not isolated.

TABLE XV

Cyclic Disulfides

Cyclic disulfide	Method for synthesis of cyclizable compound[a]	Cyclization conditions		Result	Ref.
		pH	Dilution (mmoles/liter)		
(1) H-Cys-Cys-OH S—S	A	6.5	–	Main product monomeric, additional by-products	(*1113*)
	A	8.5	–	Main product dimeric, additional by-products	(*1113*)
	A	(Various pH values)	(Various concentrations)	Formation of main and by-products depends upon pH value and concentration	(*2405*)
	B	8.5	75, 20, 2	Mixture from which 20% of a dimeric compound was isolated	(*971*)
(2) H-Cys-Gly-Cys-OH S———S	C	8.5	4	Mixture of different compounds	(*971*)
	C	8.5	18	Dimer only isolated	(*971*)
(3) H-Cys-Gly-Gly-Cys-OH S————S	C	8.5	13	Mixture from which 15% of a monomeric and 20% of a dimeric compound were isolated	(*971*)
(4) H-Cys-Tyr-Gly-Cys-Tyr-Gly-OH S————S	D	6.2	0.1	40% (no structural data)	(*1341*)
(5) H-Cys-Gly-Gly-Gly-Cys-OH S—————S	B	8.5	15	40% monomeric and 55% dimeric compounds	(*971*)

(6) H-Cys-Gly-Gly-Gly-Gly-Cys-OH S——S	B	8.5	14	90% (even at higher concentrations only the monomeric compound was obtained)	(*971*, *1329*)
	B (Isolation of the free peptide as Hg salt)	5.6	1	No data	(*1336*, *1340*)
(7) H-Cys-His-Gly-Gly-His-Cys-OH S——S	C (Isolation of the free peptide as Ag salt)	–	–	No data	(*1341*)
(8) H-Cys-Tyr-Gly-Gly-Tyr-Cys-OH S——S	C (Isolation of the free peptide as Ag salt)	6.2	0.5	42%	(*886*, *1339*, *1346*)
(9) H-Cys-Gly-Gly-Gly-Gly-Gly-Cys-OH S——S	C	8.6	12, 10	60%	(*1140*)
	C (Isolation of the free peptide as Hg salt)	–	0.5	No data	(*1347*)
(10) H-Cys-Gly-Gly-Gly-Gly-Gly-Gly-Cys-OH S——S	C	8.5	–	50%	(*1140*)
	C	–	0.5	No data	(*1347*)
(11) H-Cys-Gly-Gly-L-Lys-Gly-Gly-L-Lys-Cys-OH S——S	E	6.5	0.3	No data	(*1341*, *1983*)
(12) H-Cys-Gly-Gly-Gly-Gly-Gly-Gly-Gly-Cys-OH S——S	C (Isolation of the free peptide as Hg salt)	6.5	0.1	94%	(*1341*)
(13) H-Cys-Ala-Ala-Ala-Ala-Cys-OH S——S	C (Isolation of the free peptide as Hg salt)	6.5	0.5	37%	(*106*)

TABLE XV (Continued)

Cyclic disulfide	Method for synthesis of cyclizable compound[a]	Cyclization conditions: pH	Cyclization conditions: Dilution (mmoles liter)	Result	Ref.
(14) H-Cys-Ala-Ala-Ala-Ala-Ala-Cys-OH (S—S bridge between the two Cys)	C (Isolation of the free peptide as Hg salt)	6.5	0.2	17%	*(106)*

[a] Key to methods for syntheses:

Method	Protecting groups: NH_2	Protecting groups: COOH	Protecting groups: SH	Removal of the blocking groups
A	Cbo–	–OEt	Bzl	Hydrolysis of –OEt, removal of Cbo– with HBr in glacial acetic acid, purification by conversion in the free peptide and the hydrochloride, removal of S–Bzl with Na in liquid NH_3, isolation as Hg mercaptide
B	Cbo–	–OBzl	Bzl	Removal of Cbo– and –OBzl with HBr in glacial acetic acid, cleavage of S–Bzl with Na in liquid NH_3
C	Cbo–	–OBzl	Bzl	Simultaneous cleavage of all three protecting groups with Na in liquid NH_3
D	Cbo–	–OEt	Bzl	Hydrolysis of OEt, cleavage of Cbo– and S–Bzl with Na in liquid NH_3, isolation of the free peptide as Ag mercaptide
E	Cbo–,N^{ϵ}: HOC	–OMe	Bzl	Cleavage of Cbo, HOC, and OMe with HBr in glacial acetic acid (1 hour at 60°C), S–Bzl cleavage in Na in liquid NH_3

VI. DEPSIPEPTIDES

Shemyakin *et al.* (*2082*) have suggested the term *depsipeptide* for heterodetic peptides possessing one or several ester bonds as hetero link. The rather diversified group of depsipeptides contains as homomeric peptides the *O*-peptides of hydroxyamino acids. This type of *O*-peptide may be linear or, as in the case of peptide lactones, cyclic. The heteromeric peptides in the class of depsipeptides contain hydroxy acids as hetero component. Following a proposal by Russel and Brown (*1884*), these peptides are called peptolides. The hydroxy acids in peptolides—these peptides may also occur in linear or cyclic form—are arranged in alternating or nonalternating sequence.

The fact that the introduction of the term depsipeptide disrupts the classification of homomeric and heteromeric peptides resulted from preparative problems similar to those encountered with the synthesis of *O*-peptides and peptolides.

A. O-PEPTIDES

The synthesis of *O*-peptides was most actively pursued by various Russian schools. The synthetic compounds contained to a large extent racemic amino acids and as amino-protecting groups the acetyl or benzoyl residue, which cannot be removed. The compounds were used mainly in studies on the stability of the ester bond and its reactions, and as substrates for enzymatic reactions.

The anhydride method was employed to form the ester bond (ethyl chlorocarbonate/triethylamine in methylene chloride or chloroform at −10°C) (*319, 327, 328*). The results, however, were not always satisfactory. Impure products were frequently obtained; dehydration of the hydroxy acid to the acrylic acid derivative has been considered a possible cause (*319*). Better results were obtained when the acid chloride method was employed (*77, 322*). Azlactones, which do not react with hydroxy acids, were isolated as by-products (*322*). *N,N′*-Dicyclohexylcarbodiimide in pyridine or acetone (24 hours, 20°C) is also applicable (*2057*).

Several *O*-aminoacyl serine amides and threonine amides were synthesized via the phosgene and benzenesulfonyl chloride methods (*510, 561*) to be used as model compounds in studies concerned with the aminoacyl insertion. So far, the longest *O*-peptides—to be applied to the total synthesis of actinomycins—were prepared via the mixed anhydride with isovaleric acid (*371*) and via the carbonyldiimidazole method (*381*).

The ester bond in *O*-peptides differs from the alkyl ester bond of amino acids by a greater reactivity. The *O*-peptide ester bond can be hydrolyzed selectively under mild conditions in the presence of a methyl or ethyl ester. The hydrolysis of *O*-peptidylserine is faster than that of *O*-aminoacyl serine. The bond is susceptible to cleavage with trypsin or pancreatin (*78, 315, 320*). Vigorous reaction conditions are required for the ammonolysis and aminolysis (*313, 320*). The aminolysis of *O*-aminoacyl serine, but not of *O*-peptidylserine, is catalyzed by chymotrypsin (*320, 321*). In a phosphate buffer of pH 7.9 an *N*-protected *O*-aminoacyl residue of an *N*-protected hydroxyamino acid will be transferred to an added amino acid or peptide ester [1]; the *N*-terminal amino group acts as acceptor if it is not protected [2]:

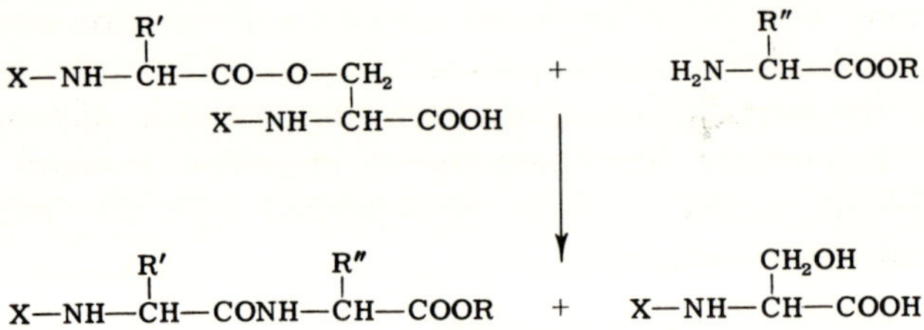

[1]

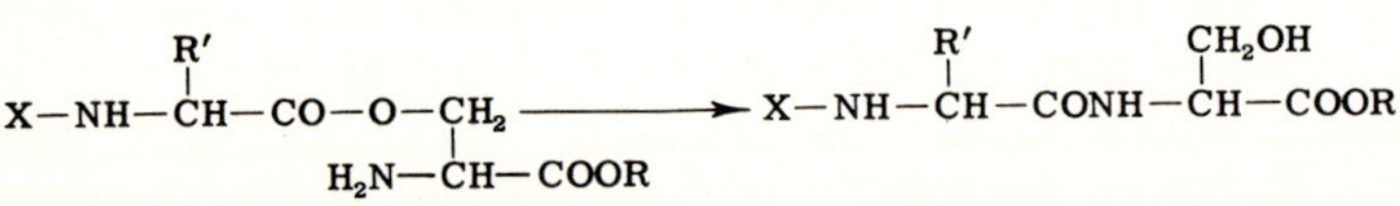

[2]

The enzymatically catalyzed transpeptidation is stereospecific (*318, 326*). An *O*-DL-aminoacyl serine will yield an *N*-L-aminoacyl serine. *O*-Aminoacyl glycolic acid shows the same behavior. *N*-Protected amino acid ethyl ester, however, do not react under the same conditions (*314, 317, 324*).

B. *S*-PEPTIDES

The mercapto group of cysteine can be acylated. If the acyl component is an amino acid, an *S*-peptide will be obtained [Shchukina *et al.* (*2056*) refer to this class of compounds as thiodepsipeptides].

The first compounds of this kind were prepared by heating a mixture of the amino acid chloride hydrochloride and cysteamine hydrochloride to 70°–110°C (*2527*), or by the reaction of the hydrochloride of a phenyl thiolester of an amino acid with cysteamine hydrochloride in methanol, aqueous tetrahydrofuran, or water only (*2528*). Glutathione was acylated successfully by this method (*1893, 2528*). Foye and Verderame (*753*) have applied the acid chloride method and Shchukina *et al.* (*2056*) the carbodiimide method in pyridine. In both cases the reactions were carried out at room temperature. Since *N*-protected derivatives are applied to these syntheses, it is possible to cleave selectively one protecting group prior to additional coupling reactions, provided the right choice of protecting groups has been made initially. Thus, the carbobenzoxy group can be removed selectively from S-(Cbo-Phe)-N-Form-Cys-OH by treatment with hydrogen bromide in glacial acetic acid (*753*).

Due to the presence of the thiolester bond in *S*-peptides the carboxyl group in *S*-peptides is activated. Wieland *et al.* (*2528*) have described the rearrangement of *S*-glycylcysteamine [3] to *N*-glycylcysteamine [4].

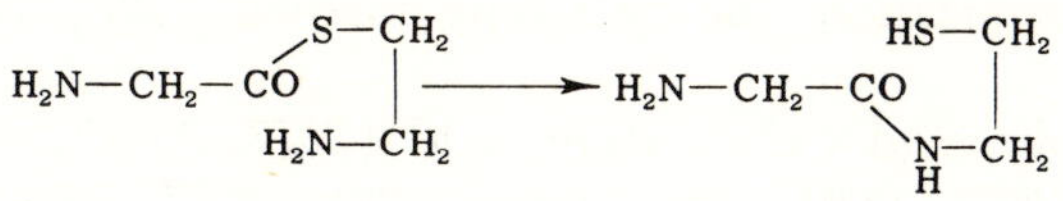

[3] [4]

This S → N shift occurs also if the amino group of cysteamine is acylated (*2528*). One of the acyl groups of the bisacyl imide formed is liberated and acylates the free amino group [5]:

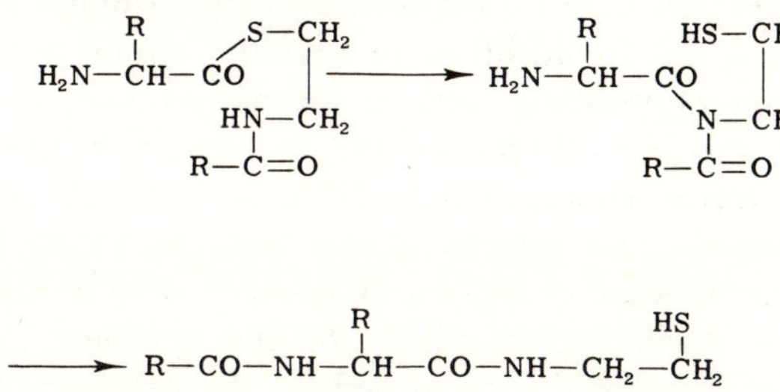

[5]

If the two acyl groups are equivalent, both are eliminated and a mixture of two dipeptidylcysteamines is formed from the original *S*-aminoacyl-*N*-aminoacyl cysteamine (*2540*).

C. PEPTOLIDES

Shemyakin *et al.* (*2077, 2091*), Schwyzer and Carrion (*2004*), Gibian and Lübke (*800, 801, 802*), Schröder and Lübke (*1977, 1979*), Quitt *et al.* (*1779b*), Losse and Bachmann (*1452a*), and Kupryszewski (*1310, 1313*) have investigated the synthesis of linear and cyclic peptolides.

There are two approaches to the synthesis of peptolides. Compounds of the aminoacyl hydroxy acid type or hydroxyacyl amino acids may be employed. The first approach is of much greater importance since compounds of this type behave like dipeptides when applied to the synthesis of higher peptolides.

1. Synthesis of the Aminoacyl Hydroxy Acid Type

Many methods for the formation of the ester bond between an amino acid and a hydroxy acid (peptolide bond) have been investigated. The benzenesulfonyl chloride, in particular as modified by Pless (*1738*), proved to be very useful (*1732, 1979, 2077, 2091*). The acid chloride method (*329, 2091*) or phosgene (*357*) and thionyl chloride (*1466*) are used less frequently for the anhydride formation. Mixed anhydrides with alkyl carbonate react only in poor yield (*319, 320, 327*), since the alcohol released during the reaction will also be esterified (*357*). Carbonyldiimidazole was used successfully (*1469, 1979*); *N,N'*-dicyclohexylcarbodiimide (*2077*), however, did not give satisfactory results. Due to the danger of racemization, the azlactone method (*327*) is not generally applicable.

Free glycolic acid (*327*), lactic acid (*2077, 2091*), or α-hydroxyisovaleric acid (*2077, 2091*), the benzyl ester (*2077, 2091*) and the *tert*-butyl ester (*1732, 2086, 2087*) of α-hydroxyisovaleric acid and *N'*-protected hydrazides of α-hydroxyisovaleric acid (*1469, 1979*) were used as hydroxy acid derivatives. The *tert*-butyl ester of α-hydroxyisovaleric acid can be obtained only via the *O*-carbobenzoxy (*2087*) or *O*-acetyl derivative (*1732*) by esterification with isobutylene and subsequent catalytic hydrogenation or hydrolysis. In addition to glycine, amino acids present in naturally occurring peptolides, such as valine, leucine, *N*-methylvaline, *N*-methylleucine, and *N*-methylisoleucine, were used in the form of their carbobenzoxy-, *p*-nitrocarbobenzoxy-, phthalyl-, or *tert*-butyloxycarbonyl-protected derivatives. The nitroso group was shown to be especially useful as amino-protecting group for *N*-methylamino acids (*1779b*).

Another route leading particularly to the synthesis of derivatives of aminoacyl glycolic acid is represented by the reaction of *N*-protected amino acids with derivatives of diazoacetic acid [6] (*802, 2004*):

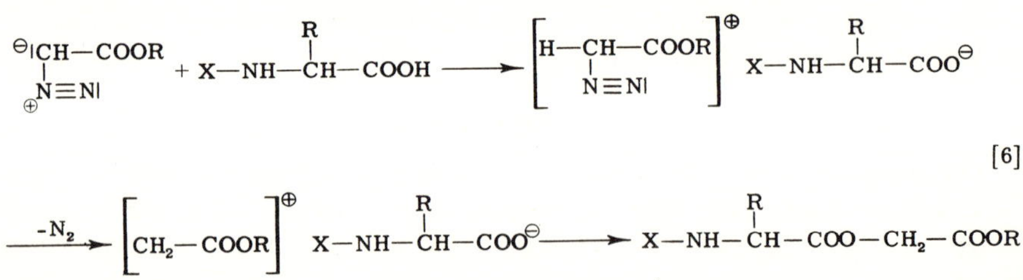

[6]

Compounds of the aminoacyl hydroxyacyl amino acid type are obtained via the Passerini reaction (cf. *2345*) by allowing *N*-protected amino acids to react with carbonyl compounds and α-isocyanocarboxylic acid esters [7] (*2345*):

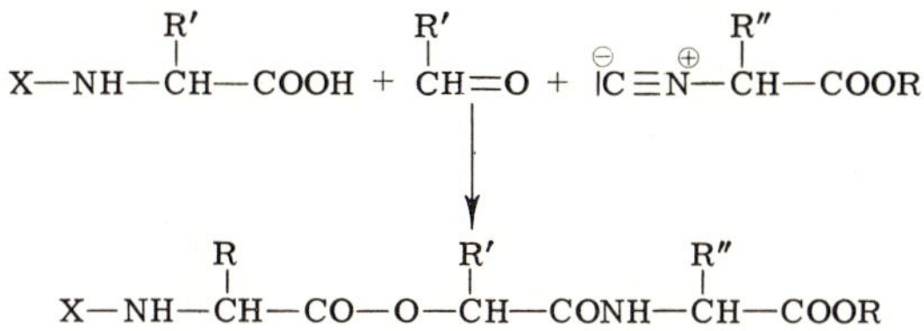

[7]

In the course of this reaction the center of asymmetry of the hydroxy acid is formed and only racemic mixtures can be expected. Peptolides with glycolic acid and those containing lactic acid were obtained by the reaction of *N*-protected amino acids with *tert*-butyl esters of α-bromocarboxylic acids (*1310, 1313*).

2. Synthesis of the Hydroxyacyl Amino Acid Type

Derivatives of the hydroxyacyl amino acid type were synthesized almost exclusively with the unprotected hydroxyl group. The azide (*1469, 1979*), carbodiimide (*1469*), and cyanomethyl ester methods (*1468, 1977*)—although only poorly activating—were used for the activation of the hydroxy acid. These syntheses were carried out with α-hydroxyisovaleric acid, which was coupled to such amino acids as glycine, valine, and leucine or to peptides composed of these amino acids. All amino components were applied as methyl, benzyl, or *tert*-butyl ester. *O*-Carbobenzoxy-α-hydroxyisovaleric acid was coupled via the acid chloride method to *N*-methylamino acids (*2086, 2087*). Hydroxyacyl amino acids are also accessible via halogen exchange (*519*) or by sustitution of the amino group by the hydroxyl group via the diazo compound (*554, 557, 2538*). These reactions are endangered, however, by the possible occurrence of Walden inversion and racemization, respectively.

3. Synthesis of Higher Peptolides

Almost all syntheses of higher peptolides have made use of aminoacyl hydroxy acids as starting material. The reaction scheme presents the various possible routes. Each of these compounds may be used again as starting material for further syntheses. Steps [8] and [9] of the synthesis require a certain combination of protecting groups to make possible the selective cleavage of the amino- or carboxyl-protecting groups without affecting the ester bond between the amino and the hydroxy acid.

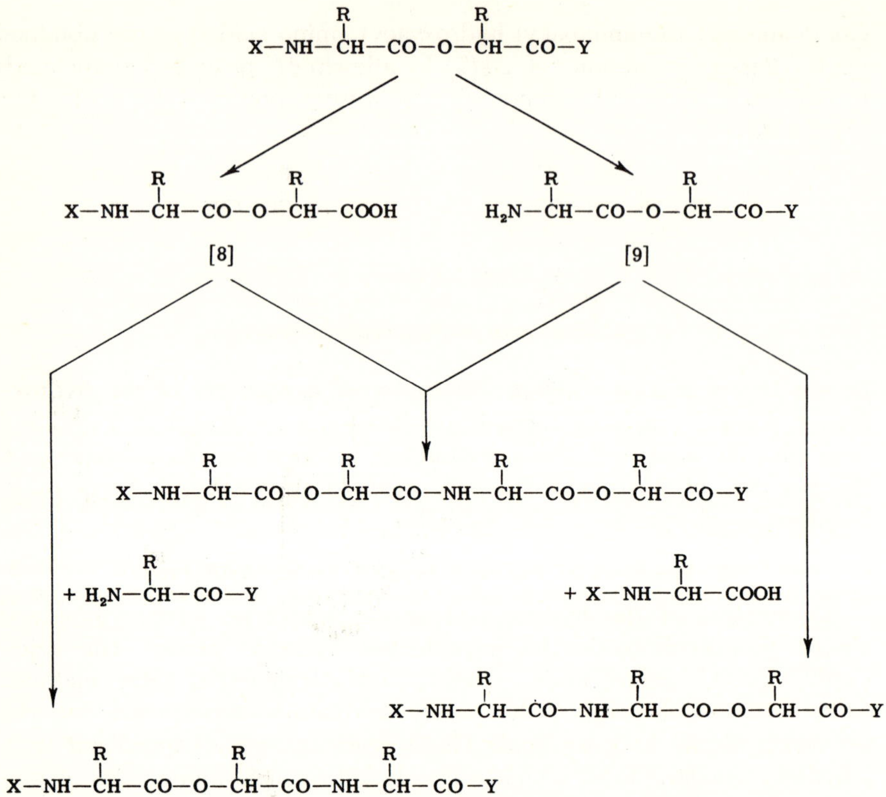

For the synthesis of *N*-protected aminoacyl hydroxy acids, the benzyl or *p*-nitrobenzyl ester (-Y) that can be cleaved by hydrogenolysis was used in combination with amino-protecting groups (X-) that cannot be removed by hydrogenolysis, such as the *tert*-butyloxycarbonyl (*802*), phthalyl (*802, 2091*), formyl (*802*), and nitroso (*1779b*) groups, or the *tert*-butyl ester, which being subject to acid hydrolysis was combined with the carbobenzoxy (*1310, 1313*), *p*-nitrocarbobenzoxy (*1732, 2086, 2087*), or phthalyl group (*1310, 1313*). This class of compounds can also be prepared directly by esterification of *N*-protected amino acids with free hydroxy acids (*2091*).

The synthesis of aminoacyl hydroxy acid esters [9] made use of amino-protecting groups (X-) that can be cleaved by acid hydrolysis, such as the *tert*-butyloxycarbonyl (*802*), formyl (*802*), carbobenzoxy (*802, 2091*), trityl (*802*), and nitroso (*1779b*) groups, or the phthalyl group (*2091*) that can be removed by hydrazinolysis. In all cases the carboxyl group was protected by the benzyl or *p*-nitrobenzyl ester. Colored protecting groups, such as the *p*-(*p*-methoxyphenylazo)-carbobenzoxy and *p*-phenylazocarbobenzoxy groups, were applied together with the *p*-nitrobenzyl ester (*2004*) which is stable to hydrogen bromide, or the *p*-nitrocarbobenzoxy group—to be removed by catalytic hydrogenation—was combined with the *tert*-butyl ester (*1732, 2086, 2087*).

The following coupling methods were applied: mixed anhydrides (*802, 1313, 1979, 2091*), activated esters (*802, 2077*), the acid chloride (*1732, 1779b, 2091*) and carbodiimide (*802*) methods as well as the azide method with *N'*-protected hydrazides (*1469, 1979*).

4. Synthesis of Cyclic Peptolides

Linear peptolides have so far, been cyclized exclusively by peptide bond formation. The acid chloride procedure (*2091*) is presently the method of choice. All peptolide-like natural products have thus far been synthesized via this method (*1732, 1779a, 2078a, 2083, 2084, 2086, 2087, 2090*). The azide method (*1980*) and the formation of mixed anhydrides with monoesters of carbonic acid (*2086*) are also applicable. The doubling reaction was also encountered in the case of peptolides. In addition to the dimerization of tripeptolides, the doubling reaction was observed for tetrapeptides (*1670a,b, 2004a*). [Cf. (*1452b,c, 1978, 2078*) for reviews on the chemistry of peptolides.]

VII. PEPTOIDS

A. PHOSPHOPEPTIDES

Various proteins, such as the casein of milk and vitellin, vitellenin, and phosphovitin of eggs, contain considerable amounts of chemically bound phosphorus. None of these phosphoproteins has been isolated in a pure state. Perlmann (*1711*) has reviewed the earlier literature on phosphoproteins. The tryptic or peptic digestion of α- or β-casein produces the so-called phosphopeptones, which contain 19–22 amino acid residues and 3–5 moles of phosphoric acid (*873, 1691, 1692, 2220*). The phosphoric acid is covalently bound via the hydroxyl groups of serine and threonine.

Phosphatidylserine, -threonine, and -hydroxyproline (serine cephalin, etc.) form a separate class of *O*-phosphorylhydroxyamino acids. Compounds of this type have been isolated, among other sources, from beef brain or muscles of tuna fish. In all probability the phosphatidylhydroxyamino acids are chemically bound to peptide chains, thus being part of true lipopeptides.

In addition to the O—P bond, which is present in *O*-phosphorylserine and -threonine and in the phosphatidylhydroxyamino acids, the presence of the N—P bond in amino acids was proven with the isolation of phosphorylarginine from muscles of invertebrates.

1. O-Phosphorylamino Acids and Peptides

The hydroxyl group of hydroxyamino acids can be phosphorylated by treatment with a mixture of phosphoric acid and phosphorus pentoxide. This method, however, gives only poor yields (*1379, 1740*). Better results were obtained upon reacting carbobenzoxytyrosine ethyl ester with phosphorus oxychloride in pyridine. Decarbobenzoxylation with phosphonium iodide in glacial acetic acid and alkaline hydrolysis led to the formation of *O*-phosphoryl-L-tyrosine in good yield. Di- and tripeptides of *O*-phosphoryl-L-tyrosine and glycine were synthesized via the same method (*1763*). In general, however, the reaction is carried out with aryl- or alkylphosphoryl chloride. In cases where diphenylphosphoryl chloride was applied (*1792, 1793*), the carbobenzoxy-*O*-(diphenylphosphoryl)-hydroxyamino acid ester [1] formed can be decarbobenzoxylated by treatment with hydrogen bromide in glacial acetic acid [2]. Alkaline hydrolysis or hydrazinolysis may lead to β-elimination and formation of diphenylphosphoric acid [3] and carbobenzoxyaminoacrylic acid [4]. The tendency to β-elimination is less pronounced in the case of compounds not substituted at the amino group.

The application of benzyl esters of carbobenzoxylated hydroxyamino acids offers certain advantages since, after the reaction with diphenylphosphoryl chloride, all protecting groups can be removed by catalytic hydrogenation in the presence of platinum oxide in glacial acetic acid [5] (*743, 1793*). Treatment with palladium black in ethanol leads only to the cleavage of the carbobenzoxy group [6]. The compound was also subjected successfully to selective hydrogenolysis (cf. *2305*), resulting in the formation of the *O*-monophenyl derivative [7]. Di-*p*-nitrobenzylphosphoryl chloride and tetra-*p*-nitrobenzyl pyrophosphate, which was prepared from di-*p*-nitrobenzylphosphoric acid by activation with *N,N′*-dicyclohexylcarbodiimide, were also applied to phosphorylation (*739, 2305, 2306a*). The di-*p*-nitrobenzylphosphoryl derivatives are distinguished by easier crystallization and hydrogenolytic cleavage. Treatment with sodium iodide in boiling acetone makes possible the selective cleavage of one *p*-nitrobenzyl group (*2304*). *O*-Pyrophosphorylserine peptides are also known (*78a*).

Fölsch (*738, 740, 741, 742*) synthesized di- and tripeptides of *O*-phospho-L- and -DL-serine containing glycine, L-leucine, L-aspartic acid, and L-glutamic acid for the purpose of studying in detail the chemical and biological properties of *O*-phosphopeptides, their behavior towards phosphatases and proteolytic enzymes, and their metal binding properties. Diphenylphosphoryl chloride was preferred for the phosphorylation. Dibenzylphosphoryl chloride has also found application.

Theodoropoulos *et al.* (*2296, 2297, 2306*) demonstrated in the case of α-carbethoxy-L-Lys-*O*-phosphoryl-L-Ser-Gly-OH that the lysylserine bond is not cleaved by trypsin, which is in contrast to the corresponding *O*-acetyl and *O*-diisopropyl compounds (cf. *745, 746*). Peptides containing *O*-phosphoserine are dephosphorylated by alkaline phosphatase from intestines (*1532, 2296*). *O*-Monophenylphosphoserine- and *O*-diphenylphosphoserine-containing peptides are dephosphorylated only very slowly or not at all (*1532*). Fölsch *et al.* (*744*) investigated the acid hydrolysis of *O*-phosphoryl-DL-serylglycine and glycyl-*O*-phosphoryl-DL-serine as well as sequence inversions in dilute hydrochloric acid. Jones and Lipkin (*1152*) synthesized a serine- and ethanolamine-con-

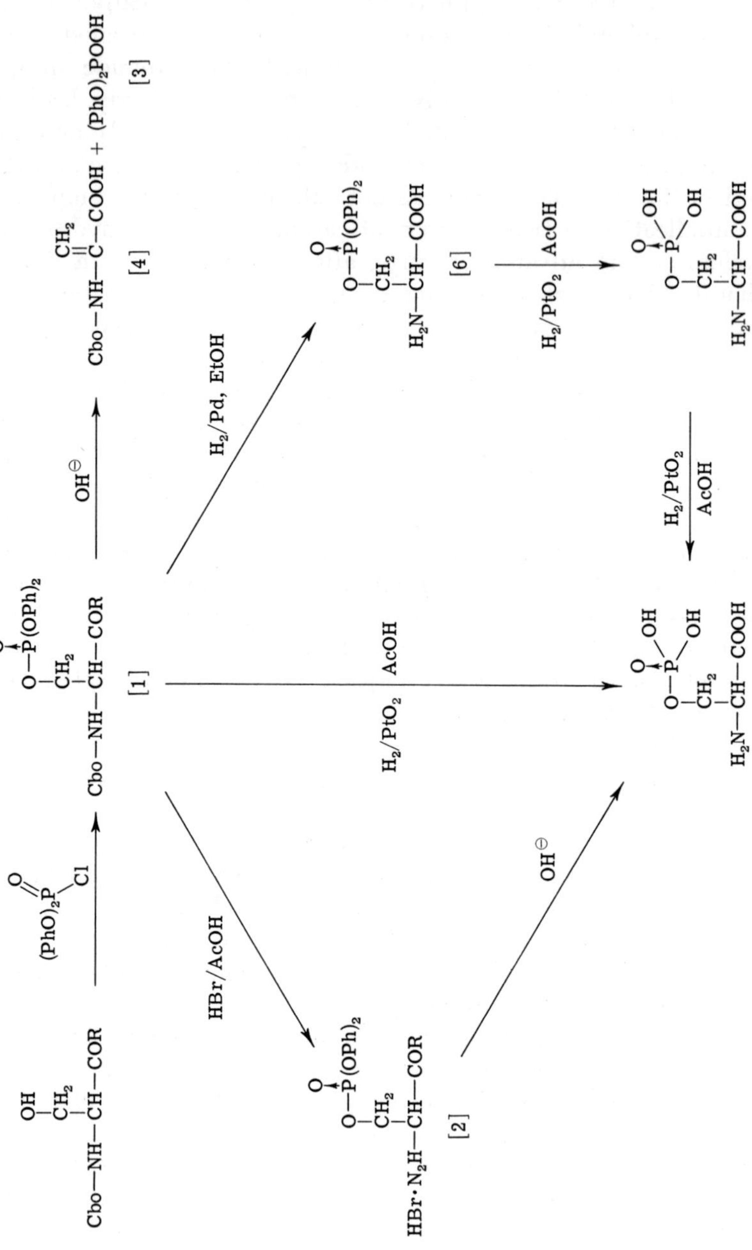
(PhO)$_2$P(O)Cl
OH$^{\ominus}$
H$_2$/Pd, EtOH
HBr/AcOH
H$_2$/PtO$_2$ AcOH
H$_2$/PtO$_2$ AcOH
H$_2$/PtO$_2$ AcOH
OH$^{\ominus}$
[1]
[2]
[3]
[4]
[5]
[6]
[7]

taining diester of phosphoric acid, namely, (2-aminoethyl)-(L-2-amino-2′-carboxyethyl)-phosphoric acid, which had been isolated originally from muscles of the sea turtle. Rosenberg *et al.* (*1841*) isolated L-threonine ethanolamine phosphate from trout (*Salmo irideus*) and lungfish (*Neoceratodus*).

2. O-Phosphatidyl Amino Acids and Peptides

Folch (*735, 736*) assigned the tentative structure of *O*-(oleylstearoylglycerylphosphoryl)-L-serine to a phosphatidylserine isolated from beef brain. The position of the phosphoric acid and fatty acid residues as well as the configuration of the phosphatidyl moiety remained uncertain. The fact that the established structures of lecithins, cephalins, and plasmalogens are represented by L-α-isomers prompted Baer and Maurukas (*85*) to synthesize *O*-(L-α-distearoylphosphatidyl)-L-serine for comparison with the naturally occurring compound. α,β-Distearoyl-

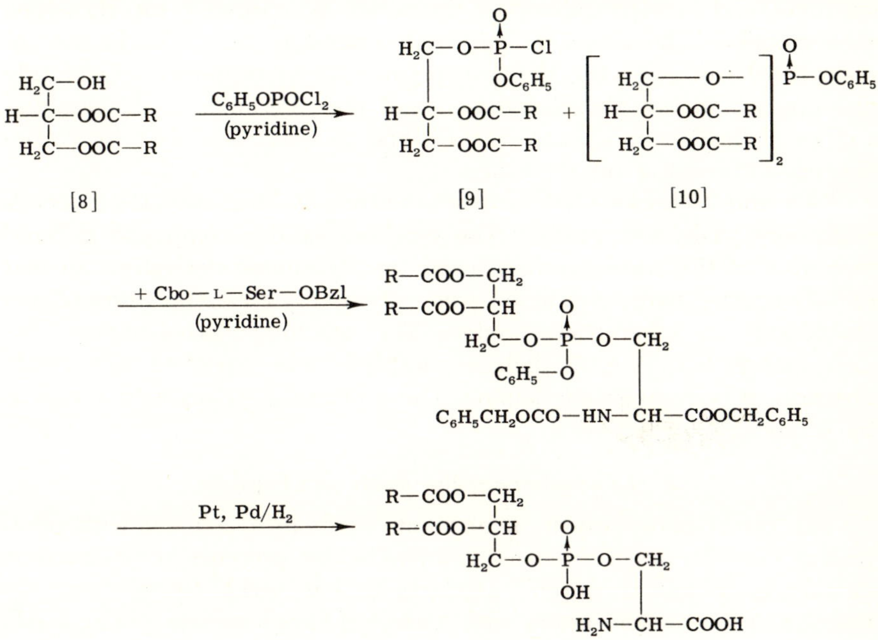

[11]

glycerine [8] was allowed to react with phenyl dichlorophosphate in anhydrous chloroform in the presence of pyridine to form a mixture of distearoyl-L-α-glycerylphenyl chlorophosphate [9] and tetrastearoyl-bis-(L-α-glyceryl)-phenyl phosphate [10]. Without further purification, the compound was esterified with carbobenzoxy-L-serine benzyl ester. L-α-

Distearoylphosphatidyl-L-serine [11] was isolated after removal of the protecting group by catalytic hydrogenation. The compound obtained from the natural product by reduction shows the same optical activity as the synthetic material. Therefore oleylstearoylglycerylphosphoryl-L-serine isolated from beef brain must be of the α-structure and possess the L-configuration. The relative position of the fatty acid residues remains open. Following the same synthetic route, Baer *et al.* (*87*) were able to synthesize *O*-(distearoyl-L-α-glycerylphosphoryl)-L-serylglycylglycine. Upon treatment with diazomethane the phosphatidyl peptide was cleaved to form L-α-distearoylglycerylphosphoric acid dimethyl ester. D-Acetonylglycerol was used as starting material for the preparation of L-α-glycerylphosphoryl-L-serine via an analogous procedure (*83*).

With the isolation of *O*-(palmitoyloleoylglycerylphosphoryl)-L-threonine from tuna fish, a representative of the phosphatidylthreonines has become known (*1086a*). Baer and Eckstein (*84*) synthesized *O*-(distearoyl-L-α-glycerylphosphoryl)-L-threonine as a model for this type of compound. The synthetic approach resembles very closely that of phosphatidylserine, with the exception that phosphorus oxychloride was applied to the phosphorylation and the intermediary distearoyl-L-α-glyceryl dichlorophosphate was allowed to react directly with carbobenzoxy-L-threonine benzyl ester.

Baer and Zschocke (*89*) have synthesized *O*-(distearoyl-L-α-glycerylphosphoryl)-L-hydroxyproline. The synthesis of this compound differed from that of the corresponding serine and threonine derivatives in that carbobenzoxy-L-hydroxyproline was phosphorylated by phosphorus oxychloride in ether/triethylamine. The resulting carbobenzoxy-L-hydroxyproline benzyl ester dichlorophosphate was esterified with D-α,β-distearin in benzene/triethylamine. The protecting groups were removed by catalytic hydrogenation.

3. *N*-Phosphorylamino Acids and Peptides

The direct phosphorylation of amino acids by treatment with phosphorus oxychloride in aqueous solution in the presence of magnesium dioxide gives rise to impure products in low yield. By contrast, the reaction of amino acid esters with dialkyl chlorophosphate (*1430, 2407*) or diphenyl chlorophosphate (*2040*) poses no problems and leads to the formation of well-defined compounds, some of them being cystalline. Hydrogenolysis of *N*-diethylphosphoryl- or diisopropylphosphorylamino acid benzyl esters yielded the corresponding *N*-dialkylphosphorylamino acids (*1430*). Hydrogenolysis of the *N*-diphenylphosphorylamino acid ester produced *N*-phosphorylamino acid esters. *N*-Diphenylphosphorylamino acid esters cannot be hydrolyzed by treatment with alkali (*2040*). The acylation of amino acid methyl (ethyl) esters with dibenzyl chlorophosphate in the presence of triethylamine and the subsequent catalytic hydrogenation proceed more smoothly (*1417*). Attempts to hydrolyze dibenzylphosphorylamino acid esters and thus another approach to the preparation of *N*-phosphorylamino acids also did not meet with success. Li and Eakin (*1418*), however, were able to obtain compounds of this type by employing amino acid benzyl esters.

Zervas and Katsoyannis (*2661*) were able to demonstrate in several cases that *N*-diphenylphosphorylamino acid esters can be hydrolyzed by treatment with barium hydroxide for 4 hours at 25°C. One phenyl group is removed simultaneously. The cleavage of the second phenyl group can be brought about by exposure of the compound to strong alkali at 100°C. These conditions do not affect the P—N bond. Di-*p*-nitrobenzylphosphoryl- and diiodobenzylphosphorylamino acid esters or peptide esters are more suitable ompounds for the synthesis of *N*-phosphoamino acids or peptides. After brief alkaline hydrolysis, the *p*-nitrobenzyl, or *p*-iodobenzyl residue can be cleaved by catalytic hydrogenation in alcoholic alkali.

N-Phosphoamino acids and peptides are stable under alkaline conditions; they are less stable at neutral pH and labile in acids. These compounds will be hydrolyzed quantitatively within a few minutes at pH 4 and room temperature. Treatment with 2 *N* hydrochloric acid at higher temperatures, in the case of *N*-diisopropylphosphoryl-DL-serine and -threonine, caused cleavage of the isopropyl groups and N → O migration of the phosphoryl residues (*1726*).

B. GLYCOPEPTIDES

The carbohydrate content of glycoproteins amounts to a few per cent. In contrast to other types of compounds, the carbohydrate component is attached to the protein by way of a homopolar bond [cf. (*1982*) on nomenclature and classification]. Glycopeptides were isolated after partial chemical or enzymatic hydrolysis of, for example, γ-globulins or ovalbumins from various species (*1166, 1167, 1306, 1363, 1629, 1630, 1631, 1635*).

The type of bond linking the carbohydrate of glycoproteins to the peptide chain has not yet been firmly established. Among others, ester bonds between the hydroxyl groups of sugars and the free carboxyl groups of amino acids, as well as peptide bonds between amino groups of hexosamines or glycosylamines, respectively, and carboxyl groups of amino acids, have been considered. *O*-Glycosidic bonds with the hydroxyl group of serine or threonine and *N*-glycosidic bonds with the amino groups of amino acids and the resulting Amadori compounds have also been discussed. Another type of compound can be derived from carbohydrates bearing carboxyl groups, i.e. gluconyl- and glucuronyl amino acids (cf. *1426, 1982*). Model compounds were synthesized from, preferably, β-D-glucose or D-glucosamine and an amino acid or a di- or tripeptide in order to answer these questions.

1. N-Aminoacyl Glucosamine or N-Peptidylglucosamine

1,3,4,6-Tetra-*O*-acetyl-D-glucosamine is preferred for the synthesis of *N*-aminoacyl D-glucosamines [12]. Carbobenzoxyamino acids are em-

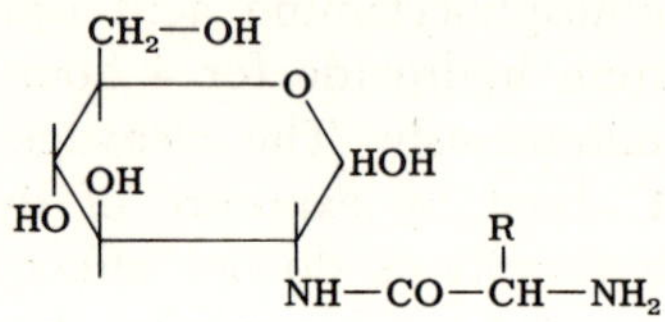

[12]

ployed and the coupling reaction is carried out via the carbodiimide (*1153, 1154, 1426, 1427, 1505, 1553*), acid chloride (*194, 614*), anhydride (*1199, 1426*), and *p*-nitrophenyl ester (*1154*) methods. The azide method, due to a rearrangement to the isocyanate, leads to the formation of urea derivatives (*1749*). Opening of carbobenzoxyglutamic acid anhydride results in the formation of the glutamyl α-derivative (*614, 1553*). The free γ-carboxyl group was coupled successfully to dibenzylglutamate (*1553*). The reaction of carbobenzoxyglutamic acid or -L-aspartic acid—activated via the *N,N'*-dicyclohexylcarbodiimide method—with 2 equivalents of tetraacetyl-D-glucosamine led to the formation of carbobenzoxy-L-glutamyl-α,γ- or carbobenzoxy-L-aspartyl-α,β-bis(tetraacetyl-D-glucosamine) (*1553*).

The acetyl groups of the sugar moiety can be removed by alkaline hydrolysis together with the amino acid ω-esters (*1426, 1427, 1428, 1429, 1553*). To prevent transpeptidations (cf. *1426*) in the cases of glutamyl-α-or-γ- and aspartyl-α-or-β-tetra-*O*-acetyl-D-glucosamine, the hydrolysis is carried out subsequent to the catalytic hydrogenolysis of the carbobenzoxy group by treating the compound with 2.5 *N* sulfuric acid for 45–60 hours at room temperature (*1427, 1429, 1505*). These conditions eliminate dehydrations that are possible during the removal of the *O*-acetyl groups (*1505*). Cleavage of the sensitive glutamyl-γ-bond occurred to only a minor extent. Acid hydrolysis of the acetyl group by treatment with hydrochloric acid/methanol is accompanied by etherification of the glucosidic hydroxyl group and formation of *N*-aminoacyl *O*-methyl-D-glucosaminide (*1199*). Protection of the hydroxyl function by the tetrahydropyranyl group proved to be advantageous in the case of the synthesis of seryl and threonyl compounds (*1153, 1154*). The compounds can be deacylated selectively by treatment with sodium or barium methylate (*1154, 1199*). The carbobenzoxy group is removed by catalytic hydrogenation. If the formyl group is applied to the protection of the amino group, the blocking group can be cleaved oxidatively by treatment with hydrogen peroxide in tetrahydrofuran while the sugar residues are not attacked (*1553*). Aminoacyl D-glucosamine and aminoacyl tetra-*O*-acetyl-D-glucosamine are isolated as hydrochlorides (*1153, 1154, 1426*) or oxalates (*1267*). These compounds are stable only in the form of their salts. Carbobenzoxyamino acids with unprotected carbohydrate components were coupled via the anhydride (*1199, 1428*) and carbodiimide (*1153, 1266, 1266a, 1267, 1727*) methods.

Aminoacyl derivatives (*222, 223*) are also obtained from α-azide carboxylic acid chlorides and tetraacetyl-D-glucosamine and subsequent catalytic hydrogenation.

2. *N*-Aminoacyl Glucopyranosylamine

N-Aminoacyl β-D-glucopyranosylamines [13] have been synthesized

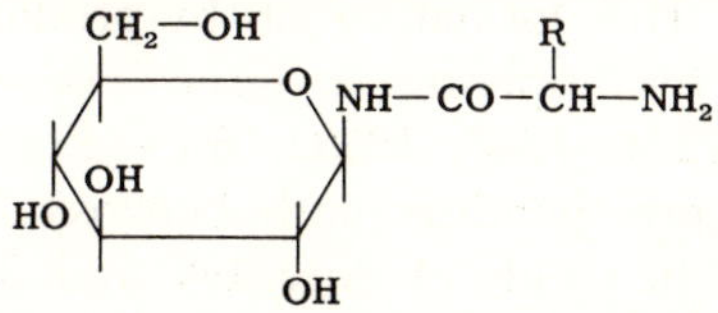

[13]

from carbobenzoxy amino acids and 4,6-*O*-benzylideneglucopyranosylamine via the anhydride method (*531*) or via the carbodiimide method (*1505*), using *2,3,4,6*-tetra-*O*-acetylglucopyranosylamine as starting material. To remove the protecting groups, the benzylidene-protected derivatives were hydrogenated catalytically which simultaneously removed the carbobenzoxy group. The acetylated compounds required alkaline hydrolysis conditions and subsequent hydrogenolysis. *N*-Aminoacyl glucopyranosylamines are cleaved by acid. Upon treatment with alkali, ammonia is formed in addition to the amino acid (*1505*).

3. 6-*O*-Aminoacyl-D-Glucose or 6-*O*-Peptidyl-D-Glucose

6-*O*-Aminoacyl D-glucose [14] can be prepared from carbobenzoxyamino acids and 1,2,3,5-di-*O*-benzylidene-protected sugars via the

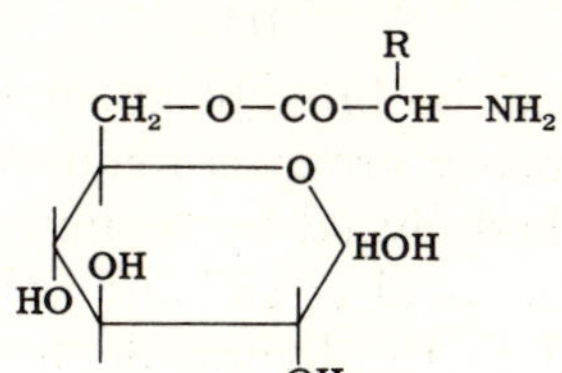

[14]

diethyl chlorophosphite (*1659*) or carbodiimide method (*2241, 2242, 2346*). Catalytic hydrogenation leads to the formation of the free compound. Treatment with aqueous acetic acid at 100°C produces 6-*O*-carbobenzoxyaminoacyl D-glucose. *N*-Protected amino acid amides (*2241, 2242*) are obtained by ammonolysis of these compounds. 6-*O*-Aminoacyl glucose or 6-*O*-peptidylglucose is decomposed at room temperature (*1659, 2241*). In aqueous solution these compounds are stable only at pH values of 2 to 3. The stability is strongly influenced by the type of amino acid or peptide (*586*).

Upon application of the carbodiimide method, unprotected sugars are acylated preferentially at the primary hydroxyl group in position 6. In addition the 3-*O*-acyl and, presumably, 3,6- or 2,6-*O,O'*-bisacyl derivatives are also formed. 1,2,5,6-Di-*O*-isopropylideneglucose was used as starting material for the synthesis of a model compound, namely, 3-*O*-carbobenzoxyglycyl-D-glucose, via the carbodiimide method followed by acid hydrolysis. The formation of by-products can be prevented to a great extent by applying a large excess of the sugar component (*587, 588, 1264, 1265, 1266*). 6-*O*-(Formylalanyl) methyl-α-D-glucoside and the corresponding carbobenzoxyalanyl compound, however, were obtained in yields of only 14% and 30%, respectively, when formylalanine or carbobenzoxyalanine was allowed to react with the methylglucoside (*1264, 1265*). 6-*O*-Peptidyl derivatives can be synthesized after removal of the carbobenzoxy group by an additional reaction with a carbobenzoxyamino acid via the carbodiimide method (*588*).

4. 1-O-Hydroxyamino Acid Glycosides

Acetylated 1-halogen-1-deoxy sugar derivatives are allowed to react with *N*-protected amino acid esters to form the *O*-glycosidic bond with the hydroxyl groups of serine or threonine [15] (*1154, 1544*).

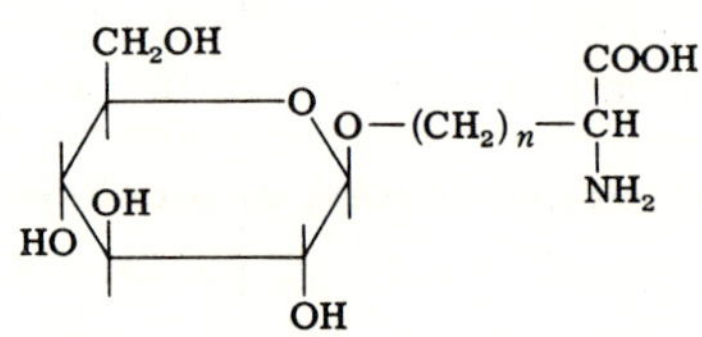

[15]

Selective deacetylation can be brought about by treatment with barium methylate. The carbobenzoxy group is removed by catalytic hydrogenation and the esters are hydrolyzed by treatment with alkali (*1154*). On the other hand, the amino acid ester bond of the fully protected derivative may be converted to the hydrazide to carry out coupling reactions with proteins via the azide method (*1544*).

5. Formation of Carbohydrate Amino Acids by Amadori or Ketosyl Rearrangement

Glucose reacts with amino acids to form an *N*-glucoside [16] that undergoes Amadori rearrangement, resulting in the formation of 1-*N*-amino acid 1-deoxyfructose ("fructose amino acid") [17]. Analogously,

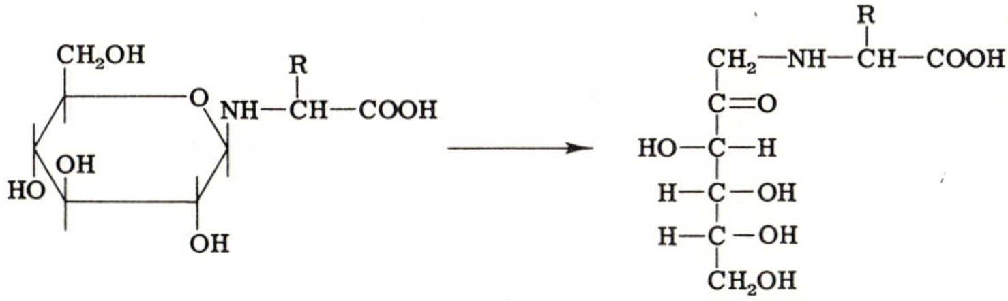

[16] [17]

2-*N*-amino acid 2-deoxyglucose ("glucose amino acid") [18] is formed from fructose by way of ketosylamine rearrangement:

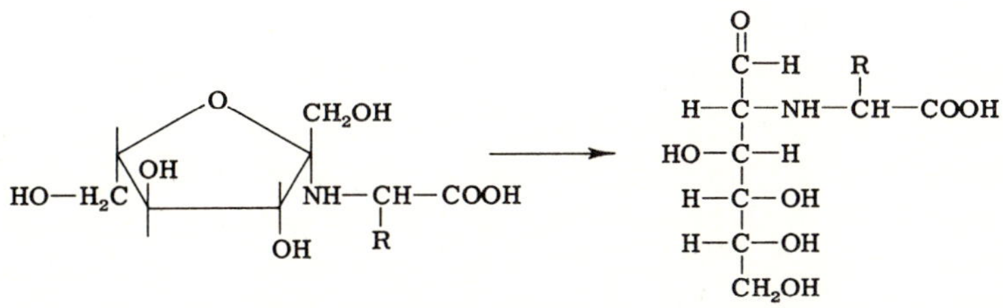

[18]

ω-Amino acids react more readily than α-amino acids (*994*). Therefore the ε-amino group of lysine will react sooner than the α-amino group. The guanido group of arginine does not react (*998*), nor does the indole ring of tryptophan or the imidazole ring of histidine (*998a*).

The steric influence of di- and tripeptides is noticeable. Peptides with N-terminal leucine react more slowly than those with N-terminal glycine (*999*). The purest products are obtained with 4,6-benzylidene-protected sugars (*1545*). Glucuronic acids are also capable of entering into this reaction (*1000, 1001*). "Sugar amino acids" are cleaved only partially by dilute organic acids, since these acids are catalysts for the formation of these compounds. The occurrence of cleavage or rearrangement depends upon the pH value (*994*). The glucoside or ketosylamine formed as intermediate normally rearranges spontaneously, since the carboxyl group of the amino acid provides a sufficient number of protons for the catalysis (*1000*). The application of the sodium salts of amino acids permits the isolation of the glucoside (*1257*).

6. Carbohydrate Guanyl Peptides and Proteins

1-Cyanoamido compounds of monosaccharides [19] or sugar isothiourea-*S*-alkyl ethers [20] react with free amino groups of amino acids or proteins to form carbohydrate guanylamino acids or proteins [21] (*1542, 1543a, 1546, 1549, 1551*). These investigations—in contrast to the

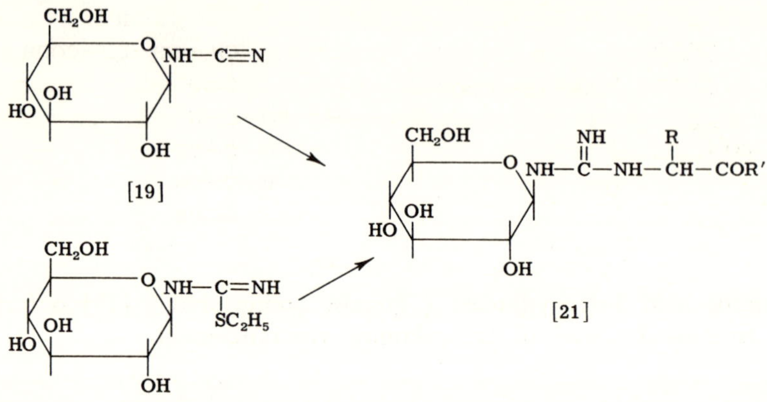

[19] [20] [21]

carbohydrate amino acid compounds discussed earlier—were extended to proteins and synthetic polyamino acids (*1546, 1550, 1551, 1554*), where about 50% of the ε-amino groups of the lysine residues in proteins were found to react (*1542, 1550*). Proteins without free primary amino groups do not react (*1550*), although histidine itself forms an N^{α},N^{im}-bis derivative (*1543a*). The guanyl derivatives do not give a well-defined Sakaguchi reaction (*1550*).

7. Gluconyl and Glucuronyl Peptides

Acetylated aldonic acid chlorides react with amino acid esters in the presence of triethylamine or with free amino acids in aqueous alkaline solution to form pentaacetyl-D-gluconyl-, pentaacetyl-D-galactonyl-, tetraacetyl-D-arabonyl-, and triacetyl-D-threonylamino acid esters or amino acids (*613*). *N,N'*-Dicyclohexylcarbodiimide can also be used as coup-

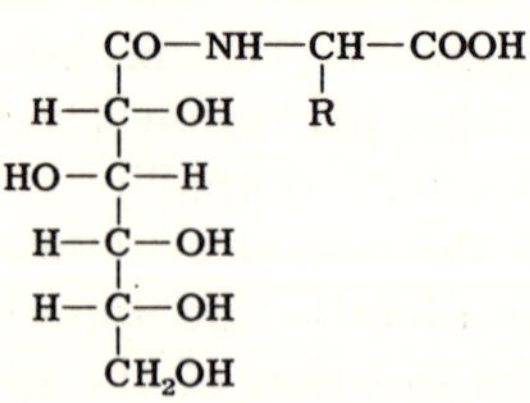

[22]

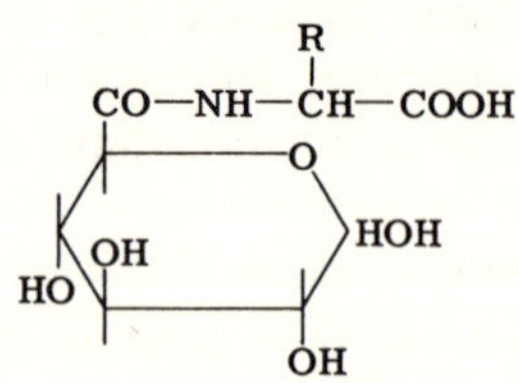

[23]

ling reagent. Protection of the hydroxyl groups of the carbohydrate is not absolutely required (*1085*). D-Gluconylamino acid esters or amino acids are also accessible by aminolysis of gluconic acid lactone (*1952*). The pentaacetylgluconylamino acid esters or corresponding peptide esters can be deacetylated selectively by treatment with barium or sodium methylate. Alkaline hydrolysis leads to the formation of free gluconylamino acids [22].

An interesting polysaccharide protein compound [23] was synthesized by Micheel and Alfes (*1543*). D-Glucose reacts with D-glucuronic acid to form poly-D-glucosido-D-glucuronide. The compound is present predominantly as the lactone. The lactone was opened to form the hydrazide by treatment with hydrazine. Application of the azide method made possible the coupling to bovine serum albumin.

C. NUCLEOPEPTIDES

Following a proposal by von Euler and Hasselquist (*2398*), compounds derived from amino acids or peptides and nucleotides or nucleosides are called nucleoamino acids or nucleopeptides. The isolation of different peptidyl nucleotides from animal tissue, yeasts, and bacteria (e.g. *477, 570, 571, 1499, 2562*) stimulated the synthesis of model substances with different bonds between amino acids and nucleotides or nucleosides.

1. O-Aminoacyl Nucleotides and Nucleosides

Aminoacyl ribonucleic acids are of great importance in the biosynthesis of proteins. The compounds contain an ester bond between the amino acid and hydroxyl group of ribose (cf. *758*). It has been established that the 3′-hydroxyl group participates in this bond (*2163a*). *O*-Aminoacyl nucleotides [24] can be synthesized from amino acid phenyl thiolesters and, for example, adenylic acid (adenosyl-5′-phosphate) (*2536*, *2544*, *2612*). Removal of the phosphoric acid with phosphatase produces aminoacyl adenosine [25], which is identical with the cleavage product obtained from aminoacyl ribonucleic acid upon treatment with ribonuclease (*2612*).

The great reactivity of the amino acid—ribose ester bond was originally believed to be the result of a phosphoric acid—amino acid anhydride (*2544*). Proof for the presence of an ester bond was furnished by oxidation with iodic acid (*758, 2536*). Zachau and Karau (*2613*) synthesized amino acid ester model compounds in order to study the reactivity of the amino acid—ribose ester bond. Carbobenzoxyamino acids were coupled via the *N,N′*-dicyclohexylcarbodiimide method to *cis*- and *trans*-3,4-dihydroxytetrahydrofuran [26].

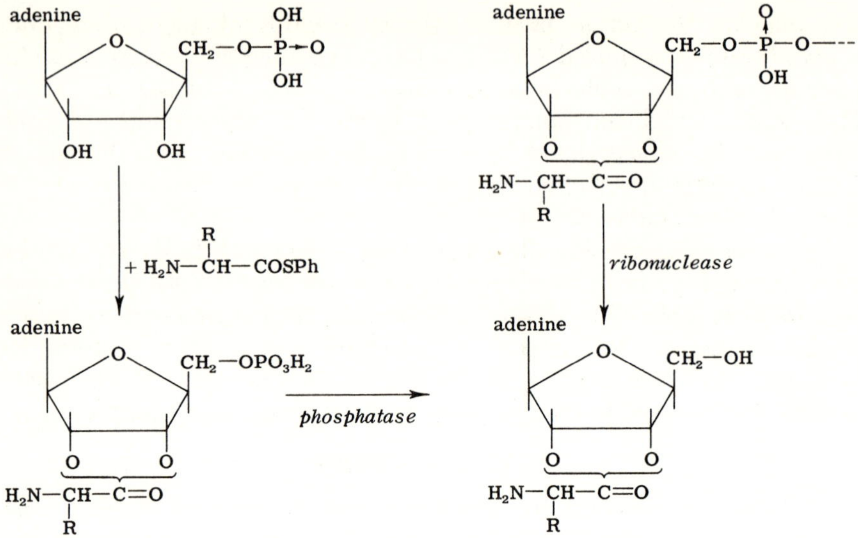

[24] [25]

Shabarova *et al.* (*2048*) synthesized 2′(or 3′),5′-*O*-bis-(carbobenzoxy-aminoacyl)-adenosine from adenosine and an excess of carbobenzoxy-

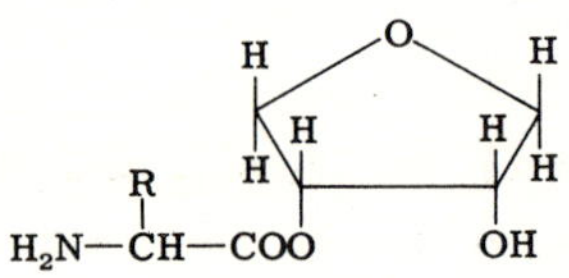

[26]

amino acids by the carbodiimide method. 5′-*O*-Carbobenzoxyaminoacyl adenosine was obtained when 2′,3′-isopropylidene-protected adenosine was used as starting material. In the presence of chymotrypsin (pH 8; 20°C) the *N*-protected amino acid residues are transferred to amino acid esters.

2. *P*-Amino Acid Nucleotides

An amino acid may be bound to the phosphoric acid residue of the nucleotide in the form of an anhydride [27] or as an amide [28a] [28b]. A compound of the anhydride type was obtained by Shabarova *et al.* (*2046*) from adenosyl-5′-phosphoric acid monoester monochloride and the silver salt of an amino acid. The hydroxyl groups of the ribose residue were protected by acetyl groups. Harris and MacWilliam (*942*) allowed unprotected adenosine-5′-phosphate or uridine-5′-phosphate to react via the carbodiimide method with *N*-protected peptides. This reaction can also be carried out with peptides with unprotected amino groups in a mix-

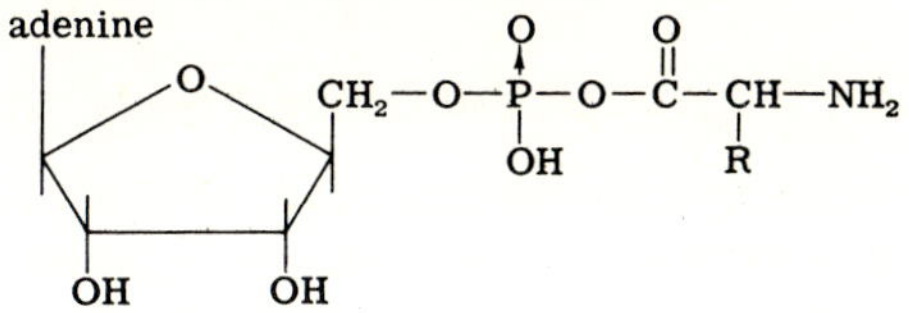

[27]

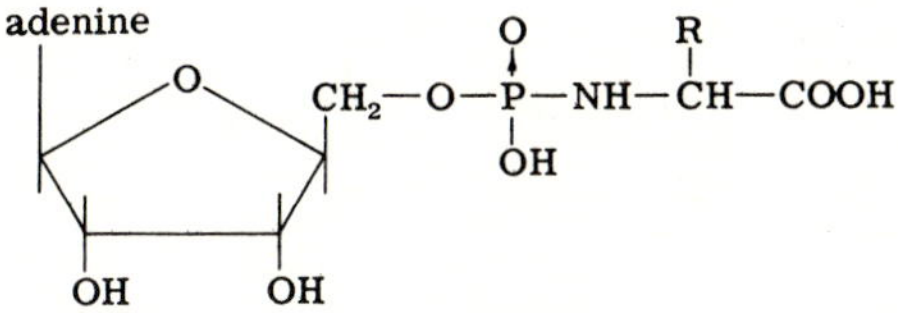

[28a]

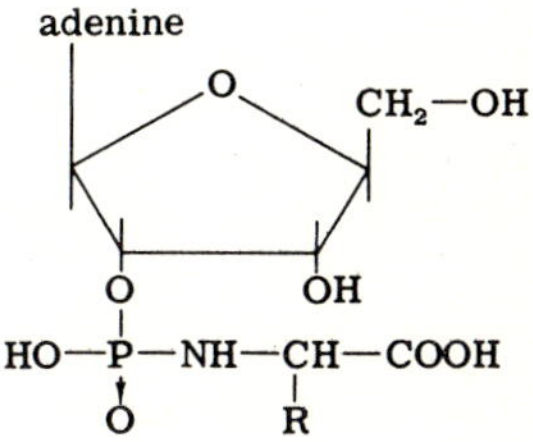

[28b]

ture of pyridine, 1 *N* hydrochloric acid, and water. Wieland and Jaenicke (*2536*) were able to couple 2-azidoisocaproic acid to adenylic acid also via the carbodiimide method. The mixed anhydride of adenylic acid and leucine was obtained by catalytic hydrogenation. Under physiological conditions—water, pH 7.5—the anhydride reacts with amino acids to form peptides (*2046, 2536*).

Compounds of the amide type [28a] [28b] were prepared from acetyl-protected adenosylphosphoric acid monoester monochloride and 3′,5′-diacetyladenosyl-2-phosphoric acid monoester monochloride, respectively (*2046, 2112*).

3. *N*-Aminoacyl Nucleosides

The acid chloride (*2397, 2398*) and carbodiimide (*2047*) methods were applied to the aminoacylation of purine and pyrimidine bases of nucleosides. The aminoacyl group is present also in these compounds in an activated form and can be transferred to other amino groups [29] in the presence of chymotrypsin (pH 8; 20°C) (*2048*):

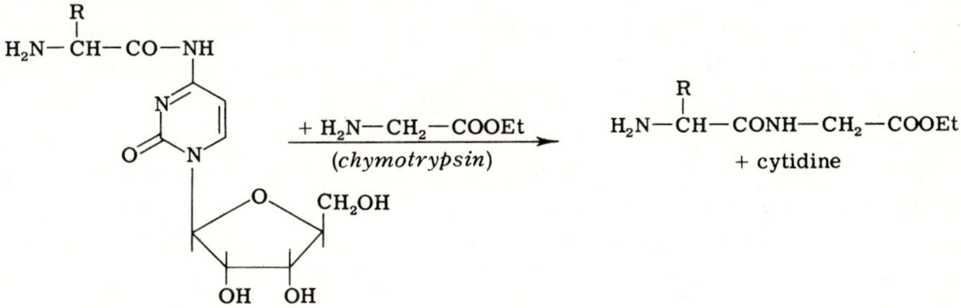

[29]

D. CHROMOPEPTIDES

Analogously to the chromoproteins, chromopeptides are compounds in which a chromophore is attached to a peptide chain by way of a covalent bond. The actinomycins, for example, are naturally occurring chromopeptides. Brockmann *et al.* (see Volume II, Chapter II,C,1,a,δ) synthesized a series of simple chromopeptides (actinocinyl amino acids and peptides) [30] in the course of the structural elucidation and total

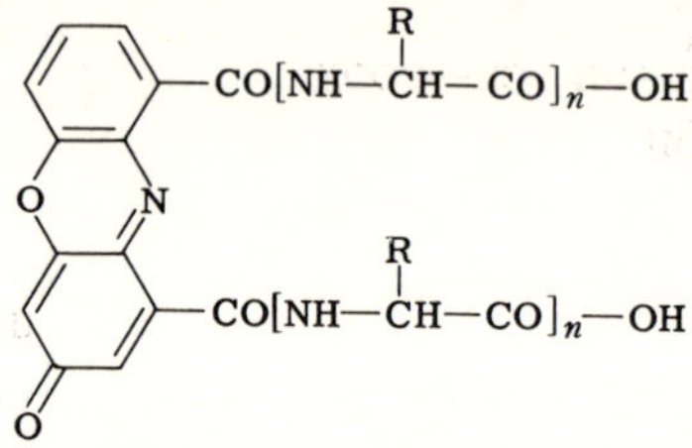

[30]

synthesis of actinomycins. The carbodiimide method was preferentially applied to these syntheses. In the case of the cytoporphyrin peptides, isolated by Tuppy *et al.* (*1689, 2327, 2331, 2332, 2336*) in the course of the structural elucidation of cytochrome c of different animal species (cf. *34, 620, 621, 1501–1504, 1514, 2333*), the peptide chain is connected with the chromophore via the mercapto groups of two cysteine residues. Lautsch *et al.* (*1332, 1336, 1338, 1341, 1343, 1344*), intrigued by the great importance of cytochrome c in biological processes, synthesized model substances of enzymes from porphyrins and amino acids, peptides, or polyamino acids.

Mesoporphyrin IX bisamino acid esters [31] ($n = 1$, OR = OEt) are accessible via the acid chloride, azide, or isocyanate method (*1337*). Pheophorbide a monoamino acid ester (*1337*) was also prepared via the isocyanate method. Protohemin IX bisamino acid esters were synthesized via the *N,N'*-dicyclohexylcarbodiimide method. Alkaline hydrolysis led to the formation of the free acids. The reaction with tri- or hexapeptide benzyl esters ($n = 3$ or 6, OR = OBzl) proceeded most smoothly via the phosphazo method. Hydrolysis of the benzyl ester and removal of the *im*-benzyl residue in the case of histidine-containing peptides can be brought about by catalytic hydrogenation or by treatment with sodium in liquid ammonia (*1341, 1344, 2561*). Compounds of the mesohemin bispeptide type containing histidine in the peptide chain were investigated for their ability to form complexes with the imidazole ring of histidine and the iron atom of hemin (*1344, 2561;* cf. *1460* on the activa-

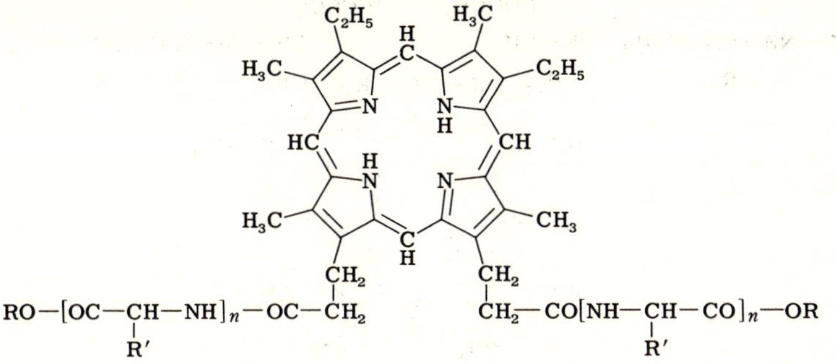

[31]

tion of hemin by histidine). *N*-Carboxylic acid anhydrides of phenylalanine, glutamine esters, or mixtures of the two react with mesoporphyrin IX bishydrazide or bisazide as initiator to form mesoporphyrin polyamino acids with a hydrazido or ureido linkage [32] [33] (*1333, 1334, 1342, 1343*).

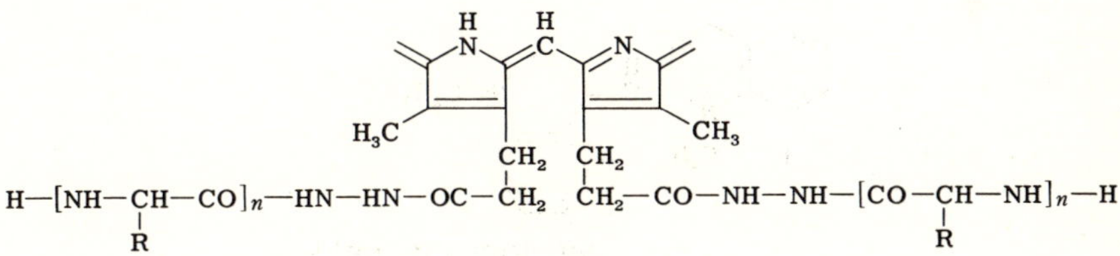

[32]

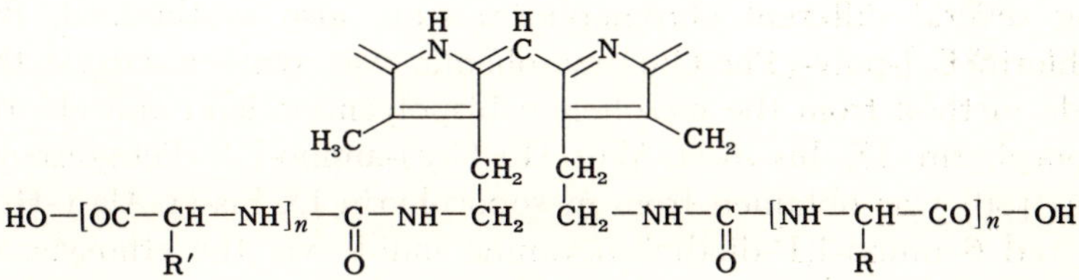

[33]

Another type of porphyrinamino acid or -polyamino acid can be obtained from cytoporphyrin tetramethyl ester (2,4-α′-dicysteinylmesoporphyrin tetramethyl ester) and *N*-protected amino acids via the anhydride method (*811*) or with amino acid *N*-carboxylic anhydrides [34] (*811, 1334, 1338*). The mesoporphyrin or cytoporphyrin derivatives were converted to hemins and investigated for their properties as oxidase models (*410, 411, 1331, 1332, 1334, 1336, 1338, 1344*).

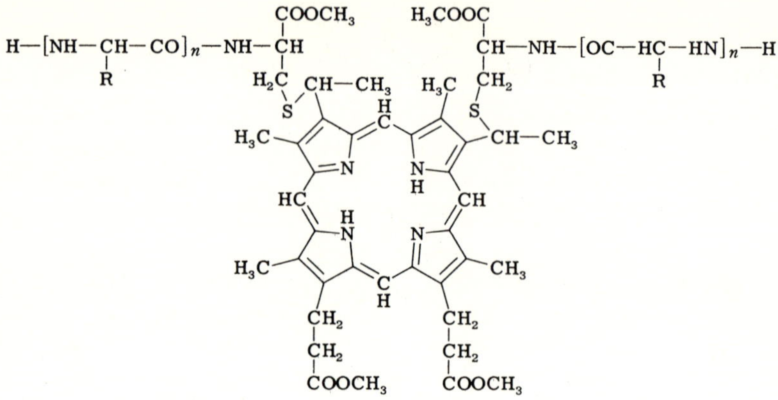

[34]

In continuation of these studies, other prosthetic groups from the chlorophyll and cyanine series were coupled to peptide chains, e.g. isocyanine-poly-DL-alanine ethyl iodide [35] (*1335*). Compounds con-

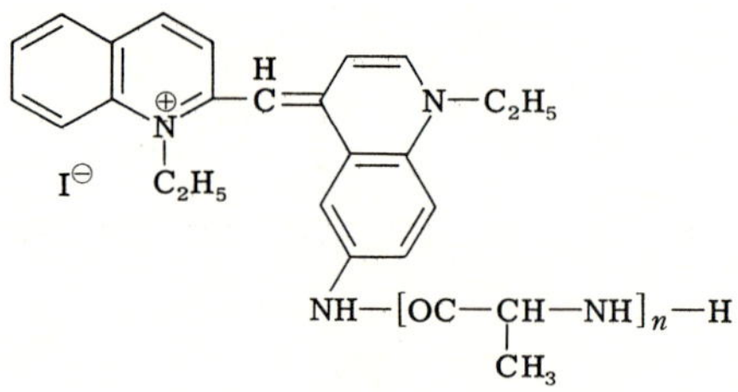

[35]

taining several different chromophores were also synthesized. Bis-[*N*-(*6*)-chlorin E_4]-poly-(Phe-Glu)-mesohemin was synthesized via the anhydride method from the mesohemin bispolyamino acid and chlorin E_4; mesoporphyrin IX bis-[6-(L-Ala-L-His-Gly)-amino-1,1′-diethylisocyanine iodide] [36] was obtained from mesoporphyrin IX bis-(L-Ala-L-His-Gly-OH) and 6-amino-1,1′-diethylisocyanine iodide via the ethoxyacetylene method (*1341*).

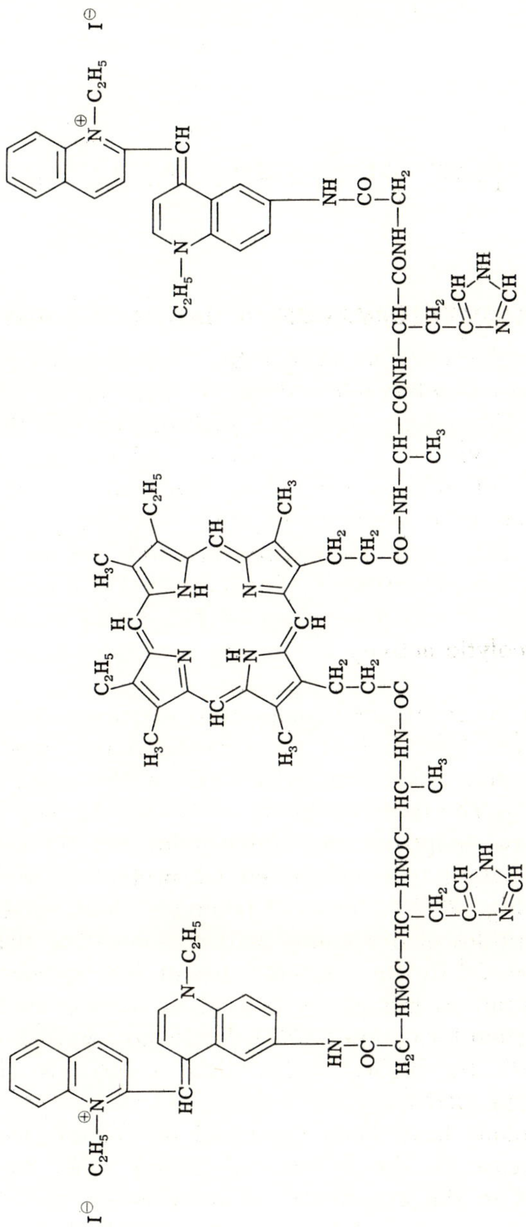

[36]

VIII. THE PLASTEIN REACTION

A. PLASTEIN FORMATION OF NATURAL PEPTIDES

Certain proteolytic enzymes initiate the formation of high molecular, protein-like substances when allowed to act upon concentrated solutions of enzymatically cleaved proteins. The investigations of this enzymatic peptide synthesis (plastein reaction) were until recently, restricted to peptide mixtures obtained by proteolytic digestion of proteins. Investigations carried out with synthetic peptides have been published only during the last few years. Chymotrypsin and pepsin are plastein-active enzymes. The pH optimum for the plastein formation is pH 7 (*2284*) in the case of chymotrypsin and pH 4 (*2378*) in the case of pepsin. The proteolytic activity of the enzymes is minimal at these pH values.

Differences in the amino acid composition of plastein from that of the applied hydrolyzate have led to the conclusion that not all peptides present in the reaction mixture participate in the plastein formation (*1068, 2285, 2378*). The terminal amino acids seem to play an important role (*2380*). Shorter peptides and amino acids are not plastein-active. The plastein formation is accompanied by proteolysis if the peptides applied are too long (*2378*). The most favorable chain length appears to be offered by peptides of 5–8 amino acid residues. The plastein formation is also influenced by the concentration of the hydrolyzate (*2378*). High salt concentrations inhibit the proteolytic activity of the enzymes, thus favoring plastein formation (*2286*). Yields are approximately 15–20% (*2285, 2378, 2380*); the highest yields, close to 50%, are obtained with hydrolyzed plastein (*2378*).

Different opinions have been expressed on the mechanism of this reaction. A decrease in the *N*-terminal amino acids was sometimes observed and led to the assumption of dehydration (*2378, 2380*). On the other hand, the number of *N*-terminal amino acids was reported to remain constant (*451, 1068*); this would require the assumption of a transpeptidation.

The molecular weights of plasteins obtained with chymotrypsin vary from 250,000 to 500,000. Some of these molecular weights are higher than those of the native proteins that were originally hydrolyzed (*2285*). The pepsin-catalyzed plastein formation produces considerably lower molecular weights; reported values range from 500 to several thousand (*1068*).

The investigations described were carried out with rather complex mixtures of plastein-active hydrolyzates. Virtanen (*2378*) reported for the first time plastein formation from a fraction of a plastein hydrolyzate isolated by electrophoresis; the yield of plastein formed was 44%. Wieland *et al.* (*2530*) and Determann *et al.* (*591, 592*) isolated homogeneous plastein-active peptides from Witte peptone, thus providing a basis for exact investigations of the plastein formation. Furthermore, sequence determinations furnished valuable information for the synthesis of suitable plastein-active peptides. Enzymatic degradation carried out with amino peptidase and carboxypeptidase as well as partial enzymatic hydrolysis carried out with chymotrypsin revealed the following structures:

H-Met-Thr-Leu-Leu-(Leu,Thr,Val)-Thr-Ser-Met-Leu-OH
H-Tyr-Ileu-(Glu,Gly,Phe)-(Glu,Ileu)-Phe-OH
H-Tyr-(Asp,Val)-Glu-Glu-Ileu-OH
H-Ileu-Asp-Val-Glu-Glu-Ileu-OH

B. SYNTHESIS OF PLASTEIN-ACTIVE PEPTIDES

Determann and Wieland (*590*) took the results of the sequence analysis of peptides isolated from Witte peptone as a guideline in their efforts to synthesize plastein-active peptides. To synthesize H-Tyr-Ileu-Gly-Glu-Phe-OH (Fig. 1), H-Ileu-Gly-OMe (A 2–3) was coupled with dicarbobenzoxytyrosine via the anhydride method as well as with carbobenzoxytyrosine hydrazide via the azide method. The tripeptide was hydrolyzed by treatment with acetone/1 *N* sodium hydroxide. Under these conditions, the *O*-carbobenzoxy group is removed from the dicarbobenzoxy compound (C 1–3). H-Glu-Phe-OH(C 4–5) was obtained from Cbo-Glu(OBzl)-Phe-OBzl by catalytic hydrogenation and from Tos-Glu-Phe-OH (prepared from tosylpyroglutamic acid chloride and phenylalanine followed by cleavage of the pyrrolidone ring upon exposure to base) (B 4–5) by treatment with sodium in liquid ammonia. The pentapeptide (D 1–5) was prepared via the anhydride method (ethyl chlorocarbonate/triethylamine in tetrahydrofuran) in a salt coupling reaction (0.5 *N* sodium hydroxide). Without purification the peptide was decarbobenzoxylated by treatment for 10 minutes with 5% hydrogen bromide solution in glacial acetic acid. The end product was purified by continuous paper electrophoresis (E 1–5).

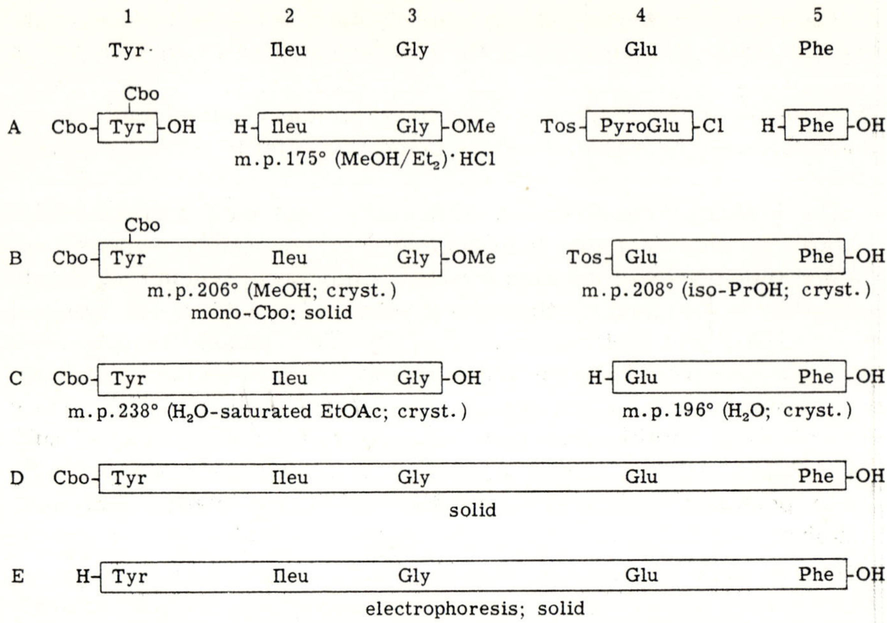

FIG. 1. Synthesis of H-Tyr-Ileu-Gly-Glu-Phe-OH [Determann and Wieland (*590*)].

Determann *et al.* (*593*) synthesized 9 analogues of plastein-active peptides:*

H-L-Tyr-L-Leu-Gly-L-Glu-L-Phe-OH [1]

H-L-Tyr-L-Leu-Gly-D-Glu-L-Phe-OH [2]

H-L-Tyr-L-Leu-Gly-L-Val-L-Phe-OH [3]

H-L-Tyr-L-Leu-Gly-L-Lys-L-Phe-OH [4]

H-L-Tyr-L-Ileu-Gly-L-Lys-L-Phe-OH [5]

H-L-Tyr-L-Lys-Gly-L-Glu-L-Phe-OH [6]

H-L-Tyr-L-Lys-Gly-L-Lys-L-Phe-OH [7]

H-L-Tyr-L-Leu-Gly-L-Glu-L-Tyr-OH [8]

H-L-Tyr-L-Leu-Gly-L-Glu(NH_2)-L-Tyr-OH [9]

In almost all peptides isoleucine was exchanged for leucine, in two cases [6] [7] lysine was the substituting amino acid. The aromatic amino acids at both terminals were not exchanged. In only two cases [8] and

* Substituting amino acid residues as underlined.

[9] was C-terminal phenylalanine replaced by tyrosine. Other exchanges involved glutamic acid, which was replaced by the D-form [2], the γ-amide [9], a neutral amino acid, namely, valine [3], and a basic amino acid, namely, lysine [4], [5], [7]. Basically the same approach was applied throughout to the synthesis of these peptides.

The plastein reaction was carried out in the following manner: the solution of 100 mg peptide in 0.1–0.2 ml water was adjusted to pH 4 and incubated at 37°C for 15–24 hours with less than 1 mg pepsin. The enzyme was destroyed by heating to 80°C and the plastein formed was isolated by centrifugation. In the case of pepsin plastein precipitates as it is formed. In the case of chymotrypsin (pH 8.6) only viscous solutions are obtained. In the latter case protein-like precipitations are caused by the addition of trifluoroacetic acid. Plasteins formed by the action of pepsin display only a very low molecular weight. The average degree of polymerization amounts to about 2.5 as calculated from optical end-group determinations of the dinitrophenyl derivatives. These observations were explained by the poor solubility of plasteins.

All synthetic peptides were plastein-active. From this the authors concluded that at least the endo-amino acids are not of great significance. It is also not important whether the peptide is neutral, basic, or acidic. It is remarkable that basic plasteins are also insoluble in acids. All plasteins form readily soluble sodium salts. Determann *et al.* (*589b*) have also studied the influence of chain length and N- and C-terminal amino acids on the condensation reaction. The following peptides were synthesized:

H-Ala-Leu-Gly-Glu-Phe-OH [10]

H-Tyr-Leu-Gly-Glu-Ala-OH [11]

H-Ala-Leu-Gly-Glu-Ala-OH [12]

H-Tyr-Leu-Ala-Gly-Glu-Phe-OH [13]

H-Tyr-Leu-Glu-Phe-OH [14]

H-Tyr-Glu-Phe-OH [15]

With regard to terminal amino acids the aromatic residue at the carboxyl end is essential, since [10] is active while [11] and [12] are inactive. The minimal chain length required is represented by the tetrapeptide [14]. Elongation of the chain is possible [13]. The tripeptide [15] is not plastein-active.

After the successful synthesis of plastein-active peptides, more reliable statements on the mechanism of the plastein reaction became possible (*590, 593*). Thus, at least the plastein reaction catalyzed by pepsin resembles a dehydration, and this seems to apply with a high degree of certainty to the reaction catalyzed by chymotrypsin. The isolated plasteins contain the same amino acids in the same ratios as the starting peptides. If mixtures of pure peptides are employed, the same ratios as those of the original mixtures are found to be present in the recovered plasteins. Except for a small amount of unreacted peptide, no by-products were found in the mother liquor after the isolation of the plastein.

IX. SOLID PHASE PEPTIDE SYNTHESIS

Recently, an entirely new technique for the synthesis of polypeptides was developed by Merrifield (*1535a*). *N*-Protected amino acids are condensed with an amino acid or peptide bound via the carboxyl group to a *solid resin particle*. Reagents and by-products are removed solely by filtration. Thus purification operations and crystallizations subsequent to each coupling step and after removal of the amino-protecting groups are eliminated. This is a simple and time-saving technique. The rather broad applicability of the technique was demonstrated by the synthesis of bradykinin and methionyllysylbradykinin.

The resin employed is a copolystyrene-2% divinylbenzene chloromethylated (22% of the aromatic rings) by treatment with chloromethyl methyl ether. The chloromethylated polymer (*1535b, 1535c*) or a nitrochloromethylated polymer (1.03 nitro groups per aromatic ring introduced by nitration) is esterified by treatment with a carbobenzoxy- or still better a *tert*-butyloxycarbonylamino acid (boiling in ethanol in the presence of triethylamine). The amount of *N*-protected amino acid bound to the resin is determined by amino acid analysis subsequent to acid hydrolysis. The carbobenzoxy group is then removed by treatment with hydrogen bromide in glacial acetic acid and the *tert*-butyloxycarbonyl group by exposure to 1 *N* HCl in acetic acid (30 minutes at room temperature). The resin itself or the amino acid—resin bond is not affected under the conditions of protecting group removal. In the following *peptide-forming* step approximately 3.5-fold excess of for example *tert*-butyloxycarbonylamino acid in dimethylformamide is allowed to react in the presence of *N,N'*-dicyclohexylcarbodiimide with the resin-bound amino acid. After a reaction time of 2 hours at room temperature, excess reagents are removed by washing with suitable solvents. After renewed removal of the protecting group from the *tert*-BOC-dipeptide resin the resulting product may be applied again to the next peptide-forming step. Since a relatively large excess of *N*-protected amino acid is applied to the peptide-forming step it is not essential any longer to acetylate (acetic anhydride/triethylamine, 2 hours) unreacted amino groups of the amino acid resin. According to investigations conducted by Bodanszky and Sheehan (*280a*), the *p*-nitrophenyl esters of *tert*-butyloxycarbonylamino acids are very well suited to reactions with amino acid and peptide resins. In the case of the *p*-nitrophenyl ester equivalent amounts of the reactants are applied. Reactions are completed within a few hours at room temperature.

The peptide is liberated from the resin either by treatment for 1 hour at 25°C with 2 *N* aqueous sodium hydroxide in ethanol (*1535a*) or by bubbling hydrogen bromide for 90 minutes through a suspension of the peptide resin in trifluoroacetic acid (*1535c*). It is also possible to cleave under amide formation the amino acid—(or peptide)—resin bond by ammonolysis (*280a*).

Employing *tert*-butyloxycarbonylamino acids, Merrifield (*1535b, 1535c*) succeeded in the course of 8 days in synthesizing via the solid-phase method biologically fully active bradykinin in an over-all yield of 32%. In the case of methionyllysylbradykinin the undecapeptide resin contained residues of nitroarginine which—following the *liberation step*—were hydrogenated over Pd-black despite the presence of methionine. Based upon the *tert*-butyloxycarbonyl nonapeptide resin the yield—inclusive of column chromatography—of methionyllysylbradykinin was 65% (*1535d*).

X. PROBLEMS OF RACEMIZATION

The degree of racemization of a chemical reaction and the optical purity of the compound obtained are not necessarily related. In spite of a considerable degree of racemization, fractionation in the course of isolation, purification, and crystallization may produce an optically pure product. In order to make a statement on the steric course of a reaction, different methods of measuring the extent of racemization must be applied than those used to check the optical purity of the isolated compound.

A. METHODS FOR THE DETERMINATION OF RACEMIZATION

1. Physical Methods

a. The Optical Rotation

The steric homogeneity of a compound is frequently tested by synthesis via two different routes and comparison of the specific rotation (cf. *1099, 1185*). Since the error of measurement is hardly better than 1°–2°, the accuracy of the method is not very great. This applies in particular to diastereoisomers since the difference in optical rotation is much smaller than in the case of optical antipodes. These differences are strongly influenced by the solvent used for the measurement of the rotation. A striking example is the fully protected ACTH sequence, Cbo-Phe-Arg(NO_2)-Try-Gly-OMe. The optically pure compound has a specific rotation of —18.5° in methanol and also in dimethylformamide. An impure product containing approximately 10% of the diastereoisomeric L-D-L-form also shows a rotation of —18.6° in dimethylformamide. Only in methanol, where the rotation is —10.9°, will the presence of the diastereoisomers be noticed (*2019*).

The tendency to racemization as exhibited by coupling methods under various conditions was determined by measuring the specific rotation together with the melting point and mixed melting point as well as the infrared spectrum. The use of Ac-Leu-Gly-OEt (L-form: $[\alpha]_D = -56.0°$, $c = 1.2$ in ethanol, melting point 100°–101°C; DL-form: melting point 120°–120.5°C) permits only an approximate classification into: less than 30% racemization, 30–70%, and more than 70% (Young test) (*2141, 2608*), since it is very difficult to crystallize partially racemized products. Bz-Leu-Gly-OEt (L-form: $[\alpha]_D = -34.0°$, $c = 2.4$ in ethanol, melting point 156.6°–157.5°C; DL-form: melting point 146°C) is a more suitable peptide, since normally also the crude products are obtained in chemically pure form and in crystalline state (*2566, 2566a*). The specific rotation is used as additional confirmation of the results obtained by test methods, such as countercurrent distribution or fractional crystallization, that depend upon separation of the reaction products and isolation of the L- and DL-forms (*47, 704*).

b. Fractional Crystallization

Very rarely is it possible to separate the pure form from the racemate by fractional crystallization. Cbo-Gly-L-Phe-Gly-OEt proved to be a suitable test peptide for systematic investigations (Anderson test, *47, 53*). Fractional crystallization from a 2% ethanolic solution causes the precipitation of the racemate. Crude products can be fractionated directly without preceding purification. Therefore this method was frequently applied to the determination of the steric course of peptide coupling reactions (*45, 235, 1690, 1700, 2068, 2363, 2581*). Cbo-Gly-Leu-Gly-OMe is another test peptide that is used in a similar way (*2363*).

c. Chromatographic Separation

Attempts to separate diastereoisomers by paper chromatography have so far been successful in only a few cases (*1009, 1821, 2264a, 2266a, 2275*). Gas chromatography, however, is capable of separating diastereoisomeric dipeptides very well in the form of their trifluoroacetyl methyl esters. The high sensitivity of this method makes possible its application to crude products. Thus, gas chromatography has become the method of choice for quantitative determination of the degree of racemization of a peptide-coupling reaction (*2477, 2495*).

d. Fractionation by Countercurrent Distribution

Resolution of diastereoisomers by countercurrent distribution was applied to Cbo-Gly-Ala-Phe-Gly-OH in order to study systematically the degree of racemization (*497, 704*). Diastereoisomeric oxytocin containing D-leucine instead of L-leucine in position 8 was separated in this manner from oxytocin (*1947*).

2. Biological Methods

Proof of the steric homogeneity of a physiologically active polypeptide by biological comparison with the naturally occurring product must be considered doubtful, since it was shown for a diastereoisomeric partial sequence of α-MSH (*1415*) that differences in activity do not necessarily occur.

The presence of certain amino acids is essential for several bacterial strains. A method for the determination of the optical purity was developed (*619, 2049*) on the basis of the possibly quantitative growth of Lactobacillus strains (*2052*).

3. Enzymatic Methods

Almost all enzymatic methods are applicable to the determination of the steric homogeneity only of isolated compounds, since one must be sure that no contaminations are present that would inhibit the action of the enzyme.

a. Amino Acid Oxidase

After total chemical hydrolysis of a peptide, the free amino acids can be degraded with L-amino acid oxidase; the remaining D-amino acids are determined by chromatography (*1821*). The degradation may also be carried out with D-amino acid oxidase by measuring the volume of oxygen evolved. One can account for the possible racemization during total hydrolysis (*1603, 2253*) by subjecting a test mixture of amino acids to the conditions of hydrolysis (*1821*).

b. Hydrolysis with Stereospecific Enzymes

Among the many proteolytic enzymes, leucine aminopeptidase, carboxypeptidase A and B, trypsin, and chymotrypsin play an important role in the determination of steric homogeneity. Leucine aminopeptidase is capable of stepwise elimination of N-terminal amino acids. The presence of a free carboxyl group is not essential. Carboxypeptidase A eliminates amino acids—except the basic ones—at the carboxyl terminal. Basic C-terminal amino acids are cleaved by carboxypeptidase B (*331, 737, 2147*); the terminal amino group may be substituted. These degradations present the most important method for the determination of optical purity of free peptides and synthetic intermediates (peptide esters with free amino groups or *N*-protected peptide acids). The color yield produced by the amino acids on chromatograms is the limiting factor for the sensitivity of this method; it is not always sufficiently high. In the case of leucine aminopeptidase and carboxypeptidase A, the rate of hydrolysis is influenced by the side chains of the amino acids forming the peptide bond to be cleaved. The terminal amino acid exerts the stronger influence. Amino acids with long or aromatic side chains are readily cleaved. Glycine is removed very slowly and the aminoacyl-proline bond will not be cleaved at all (cf. *2678*). If a given sequence contains a substituted ω-function, the cleavage will be slower but not prevented entirely.

Trypsin and chymotrypsin are enzymes of a higher degree of substrate specificity that are used for the determination of steric homogeneity. Trypsin cleaves exclusively at the carboxyl group of arginine and lysine. Cleavage does not occur or is very slow if these amino acids are *N*-terminal or next to this position. ω-Substituted amino acids and the Lys-Pro and Arg-Pro bonds are not cleaved. Nonspecific hydrolysis occurs very rarely (cf. *2678*). Chymotrypsin cleaves mainly at the carboxyl group of aromatic amino acids but also, nonspecifically, peptide bonds of other amino acids, e.g. frequently leucine. The influence of the molecule in its entirety on the specificity is more pronounced than in the case of trypsin (cf. *2678*).

The importance of these two enzymes rests predominantly upon their application to the structural elucidation of proteins. Trypsin and chymotrypsin are applicable to tests of steric homogeneity when, in the course of the synthesis, the choice of the "dissection point" may have given rise to racemization of a basic or aromatic amino acid (*209b, 1821, 1904*), or in cases where, prior to the application of another method, it appears to be desirable to work with smaller fragments (*1821*).

The sequence in question and the specificity of the enzyme determine the application of the particular enzymatic method. Frequently the final establishment of optical purity may require the application of a combination of different methods. An example of this type was furnished by Riniker and Schwyzer (*1821*) in the case of studies on synthetic Val^5-angiotensin II-Asp^1-β-amide (Fig. 2):

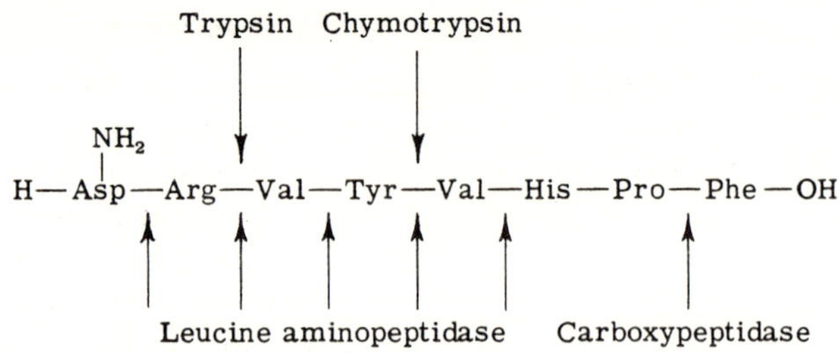

FIG. 2. Enzymatic degradation of angiotensin II-Asp^1-β-amide for detection of racemization [Riniker and Schwyzer (*1821*)].

Trypsin hydrolyzes quantitatively the Arg^{2}-Val^{3} bond, chymotrypsin the Tyr^{4}-Val^{5} bond. Carboxypeptidase removes only the C-terminal phenylalanine, leucine aminopeptidase (prolidase-free) the first five amino acids. The L-configuration of proline appeared to be confirmed by the choice of the approach to synthesis as well as by the ready cleavage of the Pro-Phe bond. The configuration of the histidine residue in position 6 was established after chymotryptic digestion, separation by countercurrent distribution, and isolation of the C-terminal tetrapeptide H-Val-His-Pro-Phe-OH and its acid hydrolysis. Due to steric hindrance caused by the presence of a valine residue (cf. *2253*), the total hydrolysis required conditions (64 hours, 105°C; or 24 hours, 115°C) that resulted in partial racemization. The degree of this racemization was determined by comparison with a test mixture of equimolar amounts of the amino acids present in the peptide. This mixture was subjected to the conditions of hydrolysis, and histidine was found to be racemized to the extent of 8–10%. Considering this amount of racemization, the degradation with L-amino acid oxidase established the L-configuration of histidine, thus proving that the synthetic peptide was present in the all-L-form.

The steric homogeneity of the isolated end product by no means excludes racemization, since diastereoisomers that have been formed may eventually have been removed in the course of purification. Indeed, it was possible to isolate a hexapeptide, namely, Cbo-Val-Tyr-Val-His-Pro-Phe-OMe, that was contaminated by 30% of the D-Tyr diastereoisomer. This diastereoisomer was isolated after removal of the carbobenzoxy group and hydrolysis of the methyl ester. The free peptide was incubated with chymotrypsin, which digested only the all-L-form, and subsequently subjected to countercurrent distribution. In addition to the stability to chymotrypsin digestion, the D-configuration of the tyrosine residue was confirmed by degradation with L-amino acid oxidase after total hydrolysis.

B. THEORY OF RACEMIZATION

Coupling reactions and operations carried out under alkaline conditions present the greatest danger of racemization during the synthesis of a peptide. Racemization under alkaline conditions is believed to be the result of withdrawal of a proton from the asymmetric C atom [1] (*1603*):

$$\text{R—NH—}\underset{\text{H}}{\overset{\text{R}'}{\text{C}}}\text{—CO—R}'' \rightleftharpoons \text{R—NH—}\overset{\text{R}'}{\underset{\ominus}{\text{C}}}\text{—CO—R}'' + \text{H}^{\oplus} \qquad [1]$$

The stability of the C—H bond is strongly influenced by the substituent R. Acyl residues enhance the chances of cleavage via possible tautomeric forms [2] (*1603*):

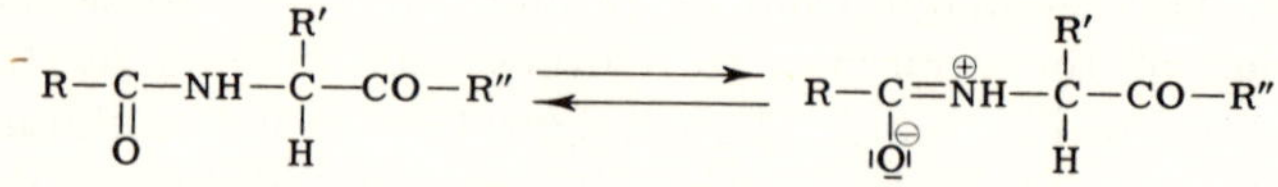

[2]

Mesomeric forms within the substituent R, such as those of the urethan-type protecting groups [3], tend to stabilize the hydrogen atom (*1570*):

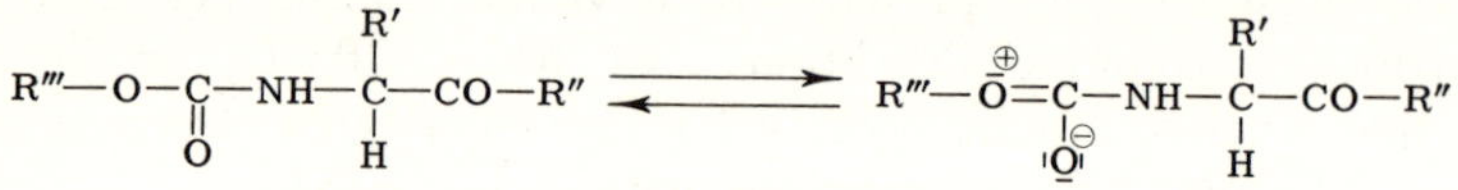

[3]

This theory is supported by the behavior of acylamino and carbobenzoxyamino acids in alkaline solution. Intermediary azlactone formation is the most likely course that racemization takes when occurring during a condensation reaction. From dipeptides on, azlactone formation is always possible [4]; *N*-protected amino acids, however, require protection in the form of an acyl group [5]. Symmetrical or mixed anhydrides

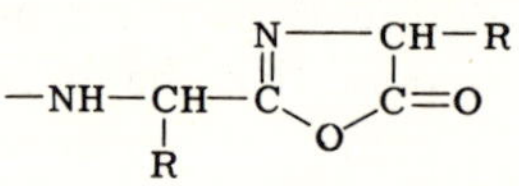

[4]

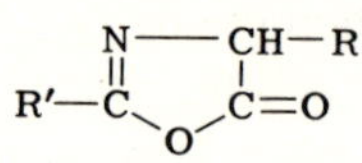

[5]

of acylamino acids form azlactones very readily. Racemization must therefore be expected from all methods proceeding via anhydride intermediates. According to Siemion *et al.* (*2118*), azlactone formation may occur in the case of the alkylcarbonic acid anhydride, phosphorus oxychloride, carbodiimide, ethoxyacetylene, and tosylchloride methods, but not with the imidazolide and *p*-nitrophenyl ester methods.

Additional functional groups in amino acids may participate in reactions that result in racemization. Besides having an influence on the electron distribution and thus affecting the stability of the C^{α}—H bond, the β-hydroxy or β-mercapto group is capable of erasing the center of asymmetry by way of β-elimination (*1491, 2005*) [6]. Therefore racemization is frequently observed in the case of peptides containing

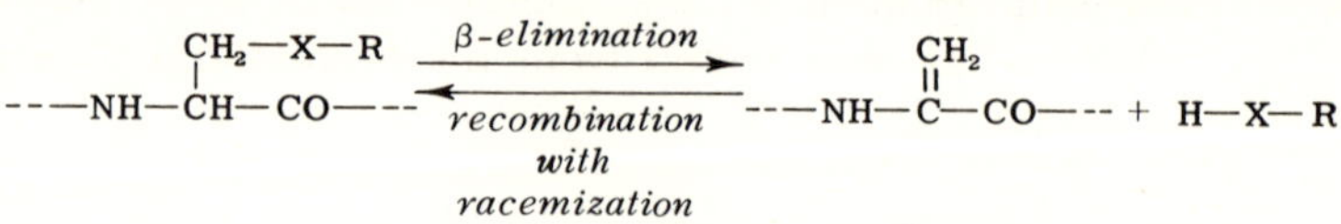

X = S or O
R = Bzl or H

[6]

serine or S-benzylcysteine upon alkaline hydrolysis or hydrazinolysis, and thus finds an explanation (*44, 1491, 1946;* cf. *268, 2497*). The great tendency to racemization of nitroarginine (*1023, 2019*) and tyrosine (*1079, 1992*) is probably due to the inductive effect of the ω-groups. This theory is supported by the fact that racemization occurs in cases where azlactone formation is unlikely or impossible, e.g. carbobenzoxy-amino acids and phthalylamino acids, respectively (*1421a, 1421b, 1421c*). Systematic investigations, carried out in connection with the synthesis of angiotensin on Cbo-Arg(NO_2)-Val-Tyr-Ileu-OMe, have revealed that the stepwise addition of amino acids is the safest approach to the synthesis of a sterically pure product of a sequence so greatly endangered by racemization (*1992*). On the other hand, an optically pure product of the ACTH sequence, Phe-Arg-Tyr-Gly, was synthesized under a set of special conditions coupling an *N*-protected peptide acid containing C-terminal arginine (*1023, 1170, 2019*).

In the following section the tendency to racemization of the most important coupling methods is reviewed for different reaction conditions [cf. Smart *et al.* (*2141*), Williams and Young (*2566*), and Weygand *et al.* (*2495*)]. Systematic investigations have been considered almost exclusively, since they permit a strict comparison of the coupling methods carried out under analogous reaction conditions. Racemization is influenced very strongly by the reaction conditions employed. Polar solvents, high temperatures, and the presence of salts favor racemization (*704, 2363, 2495, 2566*). Racemization has so far been observed to occur almost exclusively at the C-terminal amino acid of the carboxyl component (*1099*); the nature of this amino acid therefore, plays an important role also. The application of carboxyl-terminal glycine and proline is advantageous. An unfavorable influence is exerted by sterically hindered amino acids.

C. TENDENCY TO RACEMIZATION OF THE MOST IMPORTANT COUPLING METHODS

Azide method: Very little or no racemization, almost completely independent of reaction conditions. The most sensitive technique, i.e. gas chromatography, does not detect racemization (*2495, 2566*).

p-Nitrophenyl ester method: No racemization under optimum conditions (application of free esters in nonpolar solvents). Racemization occurs to the extent of 20% or more under less favorable conditions (presence of ester hydrochlorides, tertiary bases, polar solvents) (*2566, 2566a*). *N*-Protected amino acid *p*-nitrophenyl esters are readily racemized under alkaline conditions (*839*). Furthermore, *N*-protected peptide *p*-nitrophenyl esters cannot be synthesized directly but require detoured synthetic approaches (cf. page 101).

N,N'-Dicyclohexylcarbodiimide method: Almost free of racemization under optimum conditions (temperatures of —5°C, free esters, nonpolar solvents) (*47, 2068, 2073, 2495*). Solvents exert considerable influence, dimethylformamide being very unfavorable. Racemization exceeding the 50% level was observed at room temperature (*47, 2495, 2566*).

Alkylcarbonic acid anhydride method: Racemization was observed even under optimum reaction conditions (pure esters, nonpolar solvents) (*2052, 2566*). The extent of racemization is strongly influenced by solvents, not so much by the presence of salts (*2366, 2495*).

Tetraethyl pyrophosphite method: Minor but noticeable tendency to racemization where the method is carried out as "standard" or "amide" procedure (*45, 2566*). Racemization of up to 50% upon application of ester hydrochlorides or tertiary bases and when the method is used in the form of its "anhydride" version (*45*).

Phosphazo method: In general, racemization to a great extent (*852, 2141, 2566*). Operations carried out at room temperature produce no better results (*2141*).

Carbonyldiimidazole method: Practically no (*1700*) racemization or to a minor extent (*2495*) at low temperatures as long as free esters are applied. Strong tendency to racemization at room temperature and with ester hydrochlorides or tertiary bases (*2495*).

Phenylisoxazolium method: Widely differing reports in the literature (*2495, 2581*). Definitely influenced by the solvent, acetonitrile being more disadvantageous than nitromethane (*2566, 2566a*).

Ethoxyacetylene method: Minor tendency to racemization with free esters. Application of ester hydrochlorides without the addition of a tertiary base causes racemization of up to 50% (*1690*).

Cyanomethyl ester method: No racemization at temperatures that are not too high (ethyl acetate, below 70°C) (*1099, 2566*). At higher temperatures racemization of up to 20% (*2495*).

Carbonylditriazole method: Very little tendency to racemization. No racemization with carbonylditriazole under conditions where carbonyldiimidazole causes racemization to the extent of 17% (*235*).

This compilation confirms that, with the exception of the azide method, the application of all other coupling methods may cause racemization. It is also relatively safe to apply the *p*-nitrophenyl ester and—under certain conditions—the carbodiimide or imidazolide method. The application of all other methods should be restricted to the condensation of *N*-protected amino acids where, in general, racemization is not observed.

BIBLIOGRAPHY

1. Abadi, D. M., *Dissertation Abstr.* **19,** 3107 (1959).
2. Abderhalden, E., *Z. Physiol. Chem.* **52,** 207 (1907).
3. Abderhalden, E., and Bahn, A., *Z. Physiol. Chem.* **219,** 72 (1933).
4. Abderhalden, E., and Bahn, A., *Z. Physiol. Chem.* **234,** 181 (1935).
5. Abderhalden, E., and Hanson, H., *Fermentforschung* **15,** 382 (1938).
6. Abderhalden, E., and Nienburg, H., *Z. Physiol. Chem.* **219,** 155 (1933).
7. Abderhalden, E., and Nienburg, H., *Fermentforschung* **13,** 573 (1933).
8. Abderhalden, E., and Suzuki, S., *Z. Physiol. Chem.* **176–177,** 101 (1928).
9. Abraham, E. P., "Biochemistry of Some Peptide and Steroid Antibiotics," Wiley, New York, 1957.
10. Abraham, E. P., and Newton, G. G. F., *Biochem. J.* **62,** 658 (1956).
11. Acher, R., *Ann. Rev. Biochem.* **29,** 547 (1960).
12. Acher, R., and Chauvet, J., *Biochim. Biophys. Acta* **12,** 487 (1953).
13. Acher, R., Chauvet, J., Chauvet, M. T., and Crepy, D., *Biochim. Biophys. Acta* **51,** 419 (1961).
14. Acher, R., Chauvet, J., Chauvet, M. T., and Crepy, D., *Biochim. Biophys. Acta* **58,** 624 (1962).
14a. Acher, R., Chauvet, J., Chauvet, M. T., and Crepy, D., *Biochim. Biophys. Acta* **90,** 613 (1964).
15. Acher, R., Chauvet, J., and Lenci, M.-T., *Bull. Soc. Chim. Biol.* **40,** 2005 (1958).
16. Acher, R., Chauvet, J., and Lenci, M.-T., *Compt. Rend.* **248,** 1435 (1959).
17. Acher, R., Chauvet, J., and Lenci, M.-T., *Biochim. Biophys. Acta* **31,** 545 (1959).
18. Acher, R., Chauvet, J., and Lenci, M.-T., *Biochim. Biophys. Acta* **38,** 344 (1960).
19. Acher, R., Chauvet, J., Lenci, M.-T., Morel, F., and Maetz, J., *Biochim. Biophys. Acta* **42,** 379 (1960).
20. Adams, E., Davis, N. C., and Smith, E. L., *J. Biol. Chem.* **199,** 845 (1952).
21. Adams, E., Davis, N. C., and Smith, E. L., *J. Biol. Chem.* **208,** 573 (1954).
22. Adkins, H., Roos, R. M., Schroeder, D. C., Mahoney, C. L., and Gilbert, W. W., *J. Am. Chem. Soc.* **76,** 147 (1954).
23. Ahlfeld, S., Kuk-Meiri, S., and Lichtenstein, N., *Bull. Res. Council. Israel Sect. A* **84,** 115 (1959).
24. Ainsworth, G. C., Brown, A. M., and Brownlee, G., *Nature* **160,** 263 (1947).
25. Akabori, S., Ohno, K., and Narita, K., *Bull. Chem. Soc. Japan* **25,** 214 (1952).

26. Akabori, S., Okawa, K., and Sakiyama, F., *Nature* **181,** 772 (1958).
27. Akabori, S., Sakakibara, S., and Shimonishi, Y., *Bull. Chem. Soc. Japan* **34,** 739 (1961).
27a. Akabori, S., Sakakibara, S., Shimonishi, Y., and Nobuhara, Y., *Bull. Chem. Soc. Japan* **37,** 433 (1964).
28. Akabori, S., Sakakibara, S., and Shina, S., *Bull. Chem. Soc. Japan* **31,** 784 (1958).
29. Albertson, N. F., *Org. Reactions* **12,** 157 (1962).
30. Albertson, N. F., and McKay, F. C., *J. Am. Chem. Soc.* **75,** 5325 (1953).
31. Almond, H. R., Jr., Manning, D. T., and Niemann, C., *Biochemistry* **1,** 243 (1962).
32. Almond, H. R., Jr., and Niemann, C., *Biochemistry* **1,** 12 (1962).
33. Alvarez, H., Cibils, L. A., and Gonzáles-Panizza, V. H., *in* "Oxytocin" (R. Caldeyro-Barcia and H. Heller, eds.), p. 203. Macmillan (Pergamon), New York, 1961.
34. Ambler, R. P., *Biochem. J.* **82,** 30P (1962).
35. Amiard, G., and Goffinet, B., *Bull. Soc. Chim. France* p. 1133 (1957).
36. Amiard, G., and Heymès, R., *Bull. Soc. Chim. France* p. 1373 (1957).
37. Amiard, G., Heymès, R., and Velluz, L., *Bull. Soc. Chim. France* p. 191 (1955).
38. Amiard, G., Heymès, R., and Velluz, L., *Bull. Soc. Chim. France* p. 97 (1956).
39. Amiard, G., Heymès, R., and Velluz, L., *Bull. Soc. Chim. France* p. 698 (1956).
39a. Anand, N., and Koshla, M. C., *J. Sci. Ind. Res.* (*India*) **23,** 8 (1964).
40. Anastasi, A., and Erspamer, V., *Arch. Biochem. Biophys.* **101,** 56 (1963).
41. Anatol, J., and Torelli, V., *Bull. Soc. Chim. France* p. 1446 (1954).
42. Anderson, G. W., *J. Am. Chem. Soc.* **75,** 6081 (1953).
43. Anderson, G. W., *Chimia* (*Aarau*) **14,** 371 (1960).
44. Anderson, G. W., *Ann. N. Y. Acad. Sci.* **88,** 676 (1960).
45. Anderson, G. W., Blodinger, J., and Welcher, A. D., *J. Am. Chem. Soc.* **74,** 5309 (1952).
46. Anderson, G. W., Blodinger, J., Young, R. W., and Welcher, A. D., *J. Am. Chem. Soc.* **74,** 5304 (1952).
47. Anderson, G. W., and Callahan, F. M., *J. Am. Chem. Soc.* **80,** 2902 (1958).
48. Anderson, G. W., and Callahan, F. M., *J. Am. Chem. Soc.* **82,** 3359 (1960).
49. Anderson, G. W., and McGregor, A. C., *J. Am. Chem. Soc.* **79,** 6180 (1957).
50. Anderson, G. W., McGregor, A. C., and Young, R. W., *J. Org. Chem.* **23,** 1236 (1958).
51. Anderson, G. W., and Paul, R., *J. Am. Chem. Soc.* **80,** 4423 (1958).
52. Anderson, G. W., Welcher, A. D., and Young, R. W., *J. Am. Chem. Soc.* **73,** 501 (1951).
53. Anderson, G. W., and Young, R. W., *J. Am. Chem. Soc.* **74,** 5307 (1952).
53a. Anderson, G. W., Zimmerman, J. E., and Callahan, F. M., *J. Am. Chem. Soc.* **85,** 3039 (1963).
53b. Anderson, G. W., Zimmerman, J. E., and Callahan, F. M., *J. Am. Chem. Soc.* **86,** 1839 (1964).
53c. Anderson, J. C., Barton, M. A., Hardy, P. M., Kenner, G. W., MacLeod, J. K., Preston, J., Shepard, R. C., and Morley, J. S., Proc. 7th European Symp. on Peptides, Budapest, 1964, *Acta Chim. Acad. Sci. Hung.* (in press).

53d. Anderson, J. C., Barton, M. A., Gregory, R. A., Hardy, P. M., Kenner, G. W., MacLeod, J. K., Preston, J., and Sheppard, R. C., *Nature* **204,** 933 (1964).
54. Andrade, S. O., Diniz, C. R. and Rocha e Silva, M., *Arch. Intern. Pharmacodyn.* **95,** 100 (1953).
55. Andrade, S. O., and Rocha e Silva, M., *Biochem. J.* **64,** 701 (1956).
56. Andreeva, N. S., Iverenova, W. I., Kosarenko, T. D., Poroshin, K. T., Shibnev, V. A., and Shutskever, N. E., *Izv. Akad. Nauk SSSR* p. 376 (1958).
57. Angier, R. B., Waller, C. W., Hutchings, B. L., Boothe, J. H., Mowat, J. H., Semb, J., and SubbaRow, Y., *J. Am. Chem. Soc.* **72,** 74 (1950).
58. Antonio, A., and Rocha e Silva, M., *Circulation Res.* **11,** 910 (1962).
58a. Antonov, V. K., Agadzhanyan, Ts.E., Telesnina, T. R., and Shemyakin, M. M., *Tetrahedron Letters* p. 727 (1964).
59. Antonov, V. K., Ravdel, G. A., and Shemyakin, M. M., *Chimia (Aarau)* **14,** 374 (1960).
59a. Antonov, V. K., Shkrob, A. M., Shchelokov, V. I., and Shemyakin, M. M., *Tetrahedron Letters* p. 1353 (1963).
59b. Antonov, V. K., Shkrob, A. M., and Shemyakin, M. M., *Tetrahedron Letters* p. 439 (1963).
59c. Antonov, V. K., Shkrob, A. M., and Shemyakin, M. M., *in* "Peptide Symposium," Oxford, 1962 (G. T. Young, ed.), p. 221. Macmillan (Pergamon), New York, 1963.
60. Aonuma, S., Hama, T., Honda, H., and Sumita, S., *Yakugaku Kenkyu* **33,** 156 (1960).
61. Applewhite, T. H., and Niemann, C., *J. Org. Chem.* **23,** 143 (1958).
62. Arai, M., Karasawa, K., Nakamura, S., Yonehara, H., and Umezawa, H., *J. Antibiot. (Tokyo) Ser. A* **11,** 14 (1958).
63. Arakawa, K., and Bumpus, F. M., *J. Am. Chem. Soc.* **83,** 728 (1961).
64. Arakawa, K., Smeby, R. R., and Bumpus, F. M., *J. Am. Chem. Soc.* **84,** 1424 (1962).
65. Arens, J. F., *Rec. Trav. Chim.* **74,** 769 (1955).
66. Arens, J. F., *Chimia (Aarau)* **14,** 366 (1960).
67. Arman, C. G., van *in* "Polypeptides which Stimulate Plain Muscle" (J. R. Gaddum, ed.), p. 103. Livingstone, Edinburgh and London, 1955.
68. Armstrong, D., Jepson, J. B., Keele, C. A., and Stewart, J. W., *J. Physiol. (London)* **135,** 350 (1957).
69. Armstrong, D., Keele, C. A., Jepson, J. B., and Stewart, J. W., *Nature* **174,** 791 (1954).
70. Arnold, Z., *Collection Czech. Chem. Commun.* **24,** 4048 (1959).
71. Arnold, Z., and Šorm, F., *Collection Czech. Chem. Commun.* **23,** 452 (1958).
72. Arnstein, H. R. V., *Biochem. J.* **68,** 333 (1958).
73. Arnstein, H. R. V., and Morris, D., *Biochem. J.* **76,** 318 (1960).
74. Arnstein, H. R. V., and Morris, D., *Biochem. J.* **76,** 357 (1960).
75. Arnstein, H. R. V., Morris, D., and Toms, E. J., *Biochim. Biophys. Acta* **35,** 561 (1959).
75a. Arquilla, E. R., *Ciba Found. Colloq. Endocrinol.* **14,** 146 (1962).
75b. Arold, H., and Dornberger, J., *J. Prakt. Chem.* [4] **21,** 53 (1963).
75c. Arold, H., and Hartmann, M., *J. Prakt. Chem.* [4] **21,** 59 (1963).
75d. Arold, H., Proc. 7th European Peptidsymposium, Budapest, 1964, *Acta Chim. Acad. Sci. Hung.* (in press).
76. Asami, R., Shimo, K., and Sato, K., *Nippon Kagaku Zasshi* **81,** 462 (1960).

77. Avaeva, S. M., and Botvinik, M. M., *Zh. Obshch. Khim.* **26,** 2329 (1956).
78. Avaeva, S. M., and Botvinik, M. M., *Zh. Obshch. Khim.* **26,** 3066 (1956).
78a. Avaeva, S., Fölsch, G., Strid, L., and Mellander, O., *Acta Chem. Scand.* **17,** 2718 (1963).
79. Axelrod, A. E., Trakatellis, A. C., and Hofmann, K., *Nature* **197,** 146 (1963).
80. Bachelard, H. S., and Trikojus, V. M., *J. Chem. Soc.* p. 4541 (1958).
81. Back, N., Guth, P. S., and Munson, A. E., *Ann. N. Y. Acad. Sci.* **104,** 53 (1963).
82. Baddiley, J., and Mathias, A. P., *J. Chem. Soc.* p. 2803 (1954).
83. Baer, E., Buchnea, D., and Stancer, H. C., *J. Am. Chem. Soc.* **81,** 2166 (1959).
84. Baer, E., and Eckstein, F., *J. Biol. Chem.* **237,** 1449 (1962).
85. Baer, E., and Maurukas, J., *J. Biol. Chem.* **212,** 25 (1955).
86. Baer, E., Maurukas, J., and Clarke, D. D., *Can. J. Chem.* **34,** 1182 (1956).
87. Baer, E., Maurukas, J., and Clarke, D. D., *J. Biol. Chem.* **228,** 181 (1957).
88. Baer, E., and Stedman, R. J., *Can. J. Biochem. Physiol.* **37,** 583 (1959).
89. Baer, E., and Zschocke, A., *J. Biol. Chem.* **236,** 1273 (1961).
90. Baernstein, H. D., *J. Biol. Chem.* **106,** 451 (1934).
91. Baganz, H., and Dransch, G., *Chem. Ber.* **94,** 2597 (1961).
92. Baganz, H., and Peissker, H., *Chem. Ber.* **90,** 2944 (1957).
93. Baglioni, C., and Ingram, V. M., *Biochim. Biophys. Acta* **48,** 253 (1961).
94. Bailey, J. L., *Nature* **164,** 889 (1949).
95. Bailey, J. L., *J. Chem. Soc.* p. 3461 (1950).
96. Bailey, J. L., *Biochem. J.* **60,** 173 (1955).
97. Bailin, G., and Lukton, A., *J. Org. Chem.* **27,** 684 (1962).
98. Bajusz, S., Lénárd, K., Kisfaludy, L., Medzihradszky, K., and Bruckner, V., *Acta Chim. Acad. Sci. Hung.* **30,** 239 (1962).
99. Balenović, K., *J. Org. Chem.* **18,** 297 (1953).
100. Balenović, K., and Gašpert, B., *Chem. Ind.* (*London*) p. 115 (1957).
101. Balenović, K., Gašpert, B., and Štimac, N., *Croat. Chem. Acta* **29,** 93 (1957).
102. Ball, S., Boothroyd, B., Lees, K. A., Raper, A. H., and Smith, E. L., *Biochem. J.* **68,** 24P (1958).
103. Ballard, D. G. H., Bamford, C. H., and Weymouth, F. J., *Nature* **174,** 173 (1954).
104. Bamford, C. H., Elliott, A., and Hanby, W. E., "Synthetic Polypeptides," Academic Press, New York, 1956.
105. Bamford, C. H., and Weymouth, F. J., *J. Am. Chem. Soc.* **77,** 6368 (1955).
106. Bandel, W., Dissertation, Freie Universität Berlin, 1960.
106a. Banerjee, A. B., and Bose, S. K., *Nature* **200,** 471 (1963).
107. Barber, H. J., and Wragg, W. R., *Nature* **158,** 514 (1946).
108. Barkdoll, A. E., and Ross, W. F., *J. Am. Chem. Soc.* **66,** 951 (1944).
109. Barnafi, L., Rosas, R., de la Lastra, M., and Croxatto, H., *Am. J. Physiol.* **198,** 255 (1960).
110. Barnafi, L., Rosas, R., Pereda, T., and Croxatto, H., *Am. J. Physiol.* **202,** 593 (1962).
111. Barras, B. C., and Elmore, D. T., *J. Chem. Soc.* p. 3134 (1957).
112. Barras, B. C., and Elmore, D. T., *J. Chem. Soc.* p. 4830 (1957).
113. Barrett, R. J., Friesen, H., and Astwood, E. B., *J. Biol. Chem.* **237,** 432 (1962).
113a. Barth, A., and Losse, G., *Z. Naturforsch.* **19b,** 264 (1964).

113b. Barthe, P., Desaulles, P. A., Schär, B., and Staehelin, M., *Nature* **202,** 908 (1964).
114. Bartlett, M. F., Jöhl, A., Roeske, R., Stedman, R. J., Stewart, F. H. C., Ward, D. N., and du Vigneaud, V., *J. Am. Chem. Soc.* **78,** 2905 (1956).
115. Bartz, Q. R., Standiford, J., Mold, J. D., Johannessen, D. W., Ryder, A., Maretzki, A., and Haskell, T. H., *Antibiot. Ann.* p. 777 (1954–1955).
116. Battersby, A. R., and Craig, L. C., *J. Am. Chem. Soc.* **73,** 1887 (1951).
117. Battersby, A. R., and Craig, L. C., *J. Am. Chem. Soc.* **74,** 4019 (1952).
118. Battersby, A. R., and Craig, L. C., *J. Am. Chem. Soc.* **74,** 4023 (1952).
119. Battersby, A. R., and Reynolds, J. J., *J. Chem. Soc.* p. 524 (1961).
120. Battersby, A. R., and Robinson, J. C., *J. Chem. Soc.* p. 259 (1955).
121. Battersby, A. R., and Robinson, J. C., *J. Chem. Soc.* p. 2076 (1956).
122. Baudet, P., and Borecka, I., *Ann. Chim. (Rome)* **53,** 53 (1963).
123. Baudet, P., and Cherbuliez, E., *Helv. Chim. Acta* **43,** 904 (1960).
124. Baudet, P., Rössler, R., and Cherbuliez, E., *Helv. Chim. Acta* **43,** 1795 (1960).
125. Bayer, J., Dualszky, S., and Kisfaludy, L., *J. Chromatog.* **6,** 155 (1961).
126. Beecham, A. F., *J. Am. Chem. Soc.* **76,** 4613 (1954).
127. Beecham, A. F., *J. Am. Chem. Soc.* **76,** 4615 (1954).
128. Beecham, A. F., *Chem. Ind. (London)* p. 1120 (1955).
129. Beecham, A. F., *J. Am. Chem. Soc.* **79,** 3257 (1957).
130. Beecham, A. F., *J. Am. Chem. Soc.* **79,** 3262 (1957).
131. Beecham, A. F., *Australian J. Chem.* **16,** 160 (1963).
131a. Beecham, A. F., *Australian J. Chem.* **16,** 889 (1963).
132. Beglinger, U., Dissertation, University of Basel, 1959.
133. Behrens, O. K., and Bergmann, M., *J. Biol. Chem.* **129,** 587 (1939).
134. Behrens, O. K., and Bromer, W. W., *Vitamins Hormones* **16,** 263 (1958).
135. Behrens, O. K., and du Vigneaud, V., *J. Biol. Chem.* **120,** 517 (1937).
136. Bell, P. H., *J. Am. Chem. Soc.* **76,** 5565 (1954).
137. Bell, P. H., Bone, J. F., English, J. P., Fellows, C. E., Howard, K. S., Rogers, M. M., Shepherd, R. G., and Winterbottom, R., *Ann. N. Y. Acad. Sci.* **51,** 897 (1949).
138. Bell, P. H., Howard, K. S., Shepherd, R. G., Finn, B. M., and Meisenhelder, J. H., *J. Am. Chem. Soc.* **78,** 5059 (1956).
139. Beller, F. K., Krumholz, K. H., and Zeininger, K., *Acta Endocrinol.* **29,** 1 (1958).
140. Benedict, R. G., and Langlykke, A. F., *J. Bacteriol.* **54,** 24 (1947).
141. Benesch, R., and Benesch, R. E., *J. Am. Chem. Soc.* **78,** 1597 (1956).
142. Benesch, R., and Benesch, R. E., *Proc. Natl. Acad. Sci. U. S.* **44,** 848 (1958).
143. Benesch, R. E., and Benesch, R., *Biochim. Biophys. Acta* **23,** 658 (1957).
144. Benesch, R. E., and Benesch, R., *J. Am. Chem. Soc.* **80,** 1666 (1958).
145. Benfey, B. J., and Purvis, J. L., *J. Am. Chem. Soc.* **77,** 5167 (1955).
146. Benfey, B. G., and Purvis, J. L., *Biochem. J.* **62,** 588 (1956).
147. Ben-Ishai, D., *J. Org. Chem.* **19,** 62 (1954).
148. Ben-Ishai, D., *J. Am. Chem. Soc.* **79,** 5736 (1957).
149. Ben-Ishai, D., and Berger, A., *J. Biol. Chem.* **17,** 1564 (1952).
150. Benoiton, L., *Can. J. Chem.* **40,** 570 (1962).
151. Benoiton, L., Birnbaum, S. M., Winitz, M., and Greenstein, J. P., *Arch. Biochem. Biophys.* **81,** 434 (1959).
151a. Benoiton, L., Hanson, R. W., and Rydon, H. N., *J. Chem. Soc.* p. 824 (1964).

152. Benoiton, L., and Rydon, H. N., *J. Chem. Soc.* p. 3328 (1960).

153. Bentler, M., and Netter, H., *Z. Physiol. Chem.* **295**, 362 (1953).

154. Beraldo, W. T., *Am. J. Physiol.* **171**, 371 (1952).

155. Beraldo, W. T., *in* "Polypeptides which Stimulate Plain Muscle" (J. R. Gaddum, ed.), p. 58. Livingstone, Edinburgh and London, 1955.

156. Beránková, Z., Rychlík, I., Jošt, K., Rudinger, J., and Šorm, F., *Collection Czech. Chem. Commun.* **26**, 2673 (1961).

157. Beránková, Z., Rychlík, I., and Šorm, F., *Experientia* **15**, 298 (1959).

158. Beránková, Z., Rychlík, I., and Šorm, F., *Collection Czech. Chem. Commun.* **25**, 2575 (1960).

159. Beránková, Z., Rychlík, I., and Šorm, F., *Collection Czech. Chem. Commun.* **26**, 1708 (1961).

160. Berde, B., "Recent Progress in Oxytocin Research," Thomas, Springfield, Illinois, 1959.

160a. Berde, B., "Pharmacologie des hormones neurohypophysaires et de leurs analogues synthetiques," 105 pp. Masson, Paris, 1963.

160b. Berde, B., Boissonnas, R. A., Huguenin, R. L., and Stürmer, E., *Experientia* **20**, 42 (1964).

161. Berde, B., and Cerletti, A., *Gynaecologia* **144**, 275 (1957).

162. Berde, B., and Cerletti, A., *Acta Endocrinol.* **27**, 314 (1958).

163. Berde, B., and Cerletti, A., *Acta Endocrinol.* **34**, 543 (1960).

164. Berde, B., and Cerletti, A., *Helv. Physiol. Pharmacol. Acta* **19**, 135 (1961).

165. Berde, B., and Cerletti, A., *Helv. Physiol. Pharmacol. Acta* **19**, C7 (1961).

166. Berde, B., Cerletti, A., and Konzett, H., *in* "Oxytocin" (R. Caldeyro-Barcia and H. Heller, eds.), p. 247. Macmillan (Pergamon), New York, 1961.

167. Berde, B., Doepfner, W., and Konzett, H., *Brit. J. Pharmacol.* **12**, 209 (1957).

168. Berde, B., Huguenin, R., and Stürmer, E., *Experientia* **18**, 444 (1962).

169. Berde, B., and Konzett, H., *Med. Exptl.* **2**, 317 (1960).

170. Berde, B., and Saameli, K., *Acta Endocrinol.* **32**, 391 (1959).

171. Berde, B., and Saameli, K., *Nature* **191**, 83 (1961).

172. Berde, B., Weidmann, H., and Cerletti, A., *Helv. Physiol. Pharmacol. Acta* **19**, 285 (1961).

173. Berenbom, M., and White, J., *J. Am. Chem. Soc.* **71**, 2246 (1949).

174. Berg, C. P., Rose, W. C., and Marvel, C. S., *J. Biol. Chem.* **85**, 207 (1929–1930).

174a. Bergamaschi, M., and Glässer, A. H., *Circulation Res.* **8**, 329 (1963).

175. Bergel, F., and Stock, J. A., *Proc. Chem. Soc.* p. 60 (1957).

176. Bergel, F., and Wade, R., *J. Chem. Soc.* p. 941 (1959).

177. Bergen, S. S., Jr., and van Itallie, T. B., *Metab. Clin. Exptl.* **9**, 132 (1960).

178. Berger, A., and Katchalski, E., *J. Am. Chem. Soc.* **73**, 4084 (1951).

179. Berger, A., Kurtz, J., and Katchalski, E., *J. Am. Chem. Soc.* **76**, 5552 (1954).

180. Berger, A., Noguchi, J., and Katchalski E., *J. Am. Chem. Soc.* **78**, 4483 (1956).

181. Bergmann, H., *Wien. Med. Wochschr.* **110**, 148 (1960).

182. Bergmann, M., Brand, E., and Weinmann, F., *Z. Physiol. Chem.* **130/131**, 1 (1923).

183. Bergmann, M., and Fruton, J. S., *J. Biol. Chem.* **117**, 189 (1937).

184. Bergmann, M., and Fruton, J. S., *J. Biol. Chem.* **118**, 405 (1937).

185. Bergmann, M., Fruton, J. S., and Pollok, H., *J. Biol. Chem.* **127**, 643 (1939).

186. Bergmann, M., and Köster, H., *Z. Physiol. Chem.* **159**, 179 (1926).

187. Bergmann, M., and Köster, H., *Z. Physiol. Chem.* **167,** 91 (1927).
188. Bergmann, M., and Michaelis, G., *Ber. Deut. Chem. Ges.* **63,** 987 (1930).
189. Bergmann, M., and Miekeley, A., *Z. Physiol. Chem.* **140/141,** 128 (1924).
190. Bergmann, M., Stern, F., and Witte, C., *Ann. Chem. Liebigs* **449,** 277 (1926).
191. Bergmann, M., and Zervas, L., *Z. Physiol. Chem.* **175,** 154 (1928).
192. Bergmann, M., and Zervas, L., *Biochem. Z.* **203,** 280 (1928).
193. Bergmann, M., and Zervas, L., *Ber. Deut. Chem. Ges.* **65,** 1192 (1932).
194. Bergmann, M., and Zervas, L., *Ber. Deut. Chem. Ges.* **65,** 1201 (1932).
195. Bergmann, M., and Zervas, L., *Z. Physiol. Chem.* **221,** 51 (1933).
196. Bergmann, M., and Zervas, L., *Z. Physiol. Chem.* **224,** 11 (1934).
197. Bergmann, M., Zervas, L., and du Vigneaud, V., *Ber. Deut. Chem. Ges.* **62,** 1905 (1929).
198. Bergmann, M., Zervas, L., and Fruton, J. S., *J. Biol. Chem.* **111,** 225 (1935).
199. Bergmann, M., Zervas, L., and Fruton, J. S., *J. Biol. Chem.* **115,** 593 (1936).
200. Bergmann, M., Zervas, L., Fruton, J. S., Schneider, F., and Schleich, H., *J. Biol. Chem.* **109,** 325 (1935).
201. Bergmann, M., Zervas, L., and Greenstein, J. P., *Ber. Deut. Chem. Ges.* **65,** 1692 (1932).
202. Bergmann, M., Zervas, L., and Rinke, H., *Z. Physiol. Chem.* **224,** 40 (1934).
203. Bergmann, M., Zervas, L., Rinke, H., and Schleich, H., *Z. Physiol. Chem.* **224,** 26 (1934).
204. Bergmann, M., Zervas, L., and Ross, W. F., *J. Biol. Chem.* **111,** 245 (1935).
205. Bergmann, M., Zervas, L., and Salzmann, L., *Ber. Deut. Chem. Ges.* **66,** 1288 (1933).
206. Bergmann, M., Zervas, L., Salzmann, L., and Schleich, H., *Z. Physiol. Chem.* **224,** 17 (1934).
207. Bergmann, M., Zervas, L., and Schleich, H., *Z. Physiol. Chem.* **224,** 45 (1934).
208. Bergmann, M., Zervas, L., Schleich, H., and Leinert, F., *Z. Physiol. Chem.* **212–213,** 72 (1932).
209. Berlingozzi, S., Adembri, G., and Ghelardoni, M., *Gazz. Chim. Ital.* **88,** 9 (1958).
209a. Bernardi, L., Bosisio, G., Chillemi, F., de Caro, G., de Castiglione, R., Erspamer, V., Glaesser, A., and Goffredo, O., *Experientia* **20,** 306 (1964).
209b. Bernardi, L., Bosisio, G., Goffredo, O., and de Castiglione, R., *Experientia* **20,** 490 (1964).
209c. Bernardi, L., Bosisio, G., de Castiglione, R., Goffredo, O., and Chillemi, F., *Gazz. Chim. Ital.* **94,** 853 (1964).
210. Bernhard, S. A., Berger, A., Carter, J. H., Katchalski, E., Sela, M., and Shalitin, Y., *J. Am. Chem. Soc.* **84,** 2421 (1962).
211. Berridge, N. J., Newton, G. G. F., and Abraham, E. P., *Biochem. J.* **52,** 529 (1952).
212. Berse, C., Boucher, R., and Piché, L., *J. Org. Chem.* **22,** 805 (1957).
213. Berse, C., Boucher, R., and Piché, L., *Can. J. Chem.* **37,** 1733 (1959).
214. Berse, C., Massiah, T., and Piché, L., *J. Org. Chem.* **26,** 4514 (1961).
214a. Berse, C., Massiah, T., and Piché, L., *Can. J. Chem.* **41,** 2767 (1963).
215. Berse, C., and Piché, L., *J. Org. Chem.* **21,** 808 (1956).
216. Berse, C., Piché, L., Lachance, L., and Laflamme, G., *J. Org. Chem.* **27,** 3489 (1962).
217. Berse, C., Piché, L., and Uchiyama, A., *Can. J. Chem.* **38,** 1946 (1960).

218. Berson, S. A., and Yalow, R. S., *Science* **139,** 844 (1963).
219. Berson, S. A., Yalow, R. S., Baumann, A., Rothschild, M. A., and Newerly, K., *J. Clin. Invest.* **35,** 170 (1956).
220. Berson, S. A., Yalow, R. S., and Volk, B. W., *J. Lab. Clin. Med.* **49,** 331 (1957).
220a. Berthet, J., *in* "Comparative Endocrinology" (U. S. von Euler, and H. Heller, eds.), Vol. 1, p. 410. Academic press, New York, 1963.
221. Berthet, J., Sutherland, E. W., and Rall, T. W., *J. Biol. Chem.* **229,** 351 (1957).
222. Bertho, A., Schmidt, I., and Strecker, E., *Ann. Chem. Liebigs* **651,** 185 (1962).
223. Bertho, A., and Strecker, E., *Ann. Chem. Liebigs* **607,** 194 (1957).
224. Beyerman, H. C., *Rec. Trav. Chim.* **30,** 556 (1961).
225. Beyerman, H. C., and Bontekoe, J. S., *Rec. Trav. Chim.* **79,** 1044 (1960).
226. Beyerman, H. C., and Bontekoe, J. S., *Rec. Trav. Chim.* **79,** 1050 (1960).
227. Beyerman, H. C., and Bontekoe, J. S., *Rec. Trav. Chim.* **79,** 1165 (1960).
228. Beyerman, H. C., and Bontekoe, J. S., *Proc. Chem. Soc.* p. 249 (1961).
229. Beyerman, H. C., and Bontekoe, J. S., *Rec. Trav. Chim.* **81,** 691 (1962).
230. Beyerman, H. C., and Bontekoe, J. S., *Rec. Trav. Chim.* **81,** 699 (1962).
231. Beyerman, H. C., Bontekoe, J. S., and Koch, A. C., *Rec. Trav. Chim.* **78,** 935 (1959).
232. Beyerman, H. C., Bontekoe, J. S., and Koch, A. C., *Rec. Trav. Chim.* **79,** 105 (1960).
233. Beyerman, H. C., Bontekoe, J. S., and Koch, A. C., *Rec. Trav. Chim.* **79,** 1034 (1960).
234. Beyerman, H. C., Bontekoe, J. S., and Koch, A. C., *Rec. Trav. Chim.* **79,** 1039 (1960).
235. Beyerman, H. C., and Massen van der Brink, W., *Rec. Trav. Chim.* **80,** 1372 (1961).
235a. Beyerman, H. C., and Massen van der Brink, W., *Proc. Chem. Soc. (London)* p. 266 (1963).
236. Bezas, B., and Zervas, L., *J. Am. Chem. Soc.* **83,** 719 (1961).
237. Bhoola, K. D., Calle, J. D., and Schachter, M., *J. Physiol. (London)* **151,** 35P (1960).
238. Bhoola, K. D., Calle, J. D., and Schachter, M., *J. Physiol. (London)* **159,** 167 (1961).
239. Bhoola, K. D., Collier, H. O. J., Schachter, M., and Shorley, P. G., *Brit. J. Pharmacol.* **19,** 190 (1962).
240. Bianchi, A., and de Schaepdryver, A. F., *Arch. Intern. Pharmacodyn.* **124,** 21 (1960).
241. Biernat, J. F., Rzeszotarska, B., and Taschner, E., *Ann. Chem. Liebigs* **646,** 125 (1961).
242. Billmann, J. H., and Harting, W. F., *J. Am. Chem. Soc.* **70,** 1473 (1948).
243. Birkhofer, L., Bierwirth, E., and Ritter, A., *Chem. Ber.* **94,** 821 (1961).
244. Birkhofer, L., and Hartwig, I., *Chem. Ber.* **89,** 1608 (1956).
245. Birkhofer, L., and Kachel, H., *Naturwissenschaften* **41,** 576 (1954).
246. Birkhofer, L., Konkol, W., and Ritter, A., *Angew. Chem.* **71,** 701 (1959).
247. Birkhofer, L., Konkol, W., and Ritter, A., *Chem. Ber.* **94,** 1263 (1961).
248. Birkhofer, L., and Modic, R., *Ann. Chem. Liebigs* **628,** 162 (1960).
249. Birkhofer, L., and Ritter, A., *Ann. Chem. Liebigs* **612,** 22 (1958).

250. Birkhofer, L., and Ritter, A., *Chem. Ber.* **93**, 424 (1960).
251. Birkhofer, L., Ritter, A., and Neuhausen, P., *Ann. Chem. Liebigs* **659**, 190 (1962).
252. Birnbaum, S. M., and Greenstein, J. P., *Arch. Biochem. Biophys.* **39**, 108 (1952).
253. Biserte, G., and Dautrevaux, M., *Compt. Rend.* **242**, 180 (1956).
254. Biserte, G., and Dautrevaux, M., *Bull. Soc. Chim. Biol.* **39**, 795 (1957).
254a. Bishop, E., and Russel, D. W., *Biochem. J.* **92**, 19p (1964).
255. Bisset, G. W., and Lewis, G. P., *Brit. J. Pharmacol.* **19**, 168 (1962).
256. Blair, A. M. J. N., *Brit. J. Pharmacol.* **18**, 255 (1962).
257. Blau, K., and Waley, S. G., *Biochem. J.* **57**, 538 (1954).
258. Bliss, E. A., Chandler, C. A., and Schoenbach, E. B., *Ann. N. Y. Acad. Sci.* **51**, 944 (1949).
259. Blout, E. R., and Karlson, R. H., *J. Am. Chem. Soc.* **78**, 941 (1956).
259a. Bock, K. D., *Ciba Found. Symp. Basel* **10**, 224 (1962).
260. Bock, K. D., Dengler, H., Krecke, H.-J., and Reichel, G., *Klin. Wochschr.* **36**, 808 (1958).
261. Bock, K. D., and Gross, F., *Circulation Res.* **19**, 1044 (1961).
262. Bock, K. D., and Krecke, H.-J., *Klin. Wochschr.* **36**, 69 (1958).
263. Bock, K. D., Krecke, H.-J., and Kuhn, H. M., *Klin. Wochschr.* **36**, 254 (1958).
264. Bodanszky, M., *Nature* **175**, 685 (1955).
265. Bodanszky, M., *Acta Chim. Acad. Sci. Hung.* **10**, 335 (1956).
266. Bodanszky, M., *Chem. Ind. (London)* p. 524 (1957).
267. Bodanszky, M., *Ann. N. Y. Acad. Sci.* **88**, 655 (1960).
268. Bodanszky, M., and Birkhimer, C. A., *Chimia (Aarau)* **14**, 368 (1960).
269. Bodanszky, M., and Birkhimer, C. A., *Chem. Ind. (London)* p. 1620 (1962).
270. Bodanszky, M., and Birkhimer, C. A., *J. Am. Chem. Soc.* **84**, 4943 (1962).
271. Bodanszky, M., and du Vigneaud, V., *J. Am. Chem. Soc.* **81**, 1258 (1959).
272. Bodanszky, M., and du Vigneaud, V., *J. Am. Chem. Soc.* **81**, 2504 (1959).
273. Bodanszky, M., and du Vigneaud, V., *J. Am. Chem. Soc.* **81**, 5688 (1959).
274. Bodanszky, M., and du Vigneaud, V., *J. Am. Chem. Soc.* **81**, 6072 (1959).
275. Bodanszky, M., and du Vigneaud, V., *Nature* **183**, 1324 (1959).
276. Bodanszky, M., and du Vigneaud, V., *Nature* **184**, 981 (1959).
277. Bodanszky, M., Meienhofer, J., and du Vigneaud, V., *J. Am. Chem. Soc.* **82**, 3195 (1960).
277a. Bodanszky, M., and Ondetti, M. A., *Antimicrobial Agents Chemotherapy* p. 360 (1963).
277b. Bodanszky, M., Ondetti, M. A., Birkhimer, C. A., and Thomas, P. L., *J. Am. Chem. Soc.* **86**, 4452 (1964).
278. Bodanszky, M., Ondetti, M. A., Rubin, B., Piala, J. J., Fried, J., Sheehan, J. T., and Birkhimer, C. A., *Nature* **194**, 485 (1962).
279. Bodanszky, M., Ondetti, M. A., Sheehan, J. T., and Lande, S., *Ann. N. Y. Acad. Sci.* **104**, 24 (1963).
280. Bodanszky, M., and Sheehan, J. T., *Chem. Ind. (London)* p. 1268 (1960).
280a. Bodanszky, M., and Sheehan, J. T., *Chem. Ind. (London)* p. 1423 (1964).
280b. Bodanszky, M., and Sheehan, J. T., *Antimicrobial Agents Chemotherapy* p. 38 (1963).
281. Bodanszky, M., Sheehan, J. T., Ondetti, M. A., and Lande, S., *J. Am. Chem. Soc.* **85**, 991 (1963).

282. Bodanszky, M., Szelke, M., Tömörkény, E., and Weisz, E., *Chem. Ind. (London)* p. 1517 (1955).
283. Bodanszky, M., Szelke, M., Tömörkény, E., and Weisz, E., *Acta Chim. Acad. Sci. Hung.* **11**, 179 (1957).
284. Boissonnas, R. A., *Helv. Chim. Acta* **34**, 874 (1951).
285. Boissonnas, R. A., *in* "Polypeptides which Affect Smooth Muscles and Blood Vessels" (M. Schachter, ed.), p. 7. Macmillan (Pergamon), New York.
286. Boissonnas, R. A., *Paper on the 8th Symp. Endocrinology, München, 1961* p. 195. Springer, Berlin, 1962.
286a. Boissonnas, R. A., *Advan. Org. Chem.* **3**, 159 (1963).
287. Boissonnas, R. A., Franz, J., and Stürmer, E., *Ann. N. Y. Acad. Sci.* **104**, 376 (1963).
288. Boissonnas, R. A., and Guttmann, S., *Helv. Chim. Acta* **43**, 190 (1960).
289. Boissonnas, R. A., *Biochem. Pharmacol.* **10**, 35 (1962).
290. Boissonnas, R. A., Guttmann, S., Berde, B., and Konzett, H., *Experientia* **17**, 377 (1961).
291. Boissonnas, R. A., Guttmann, S., Huguenin, R. L., Jaquenoud, P.-A., and Sandrin, E., *Helv. Chim. Acta* **41**, 1867 (1958).
291a. Boissonnas, R. A., Guttmann, S., Huguenin, R. L., Jaquenoud, P.-A., and Sandrin, E., *Helv. Chim. Acta* **46**, 2347 (1963).
292. Boissonnas, R. A., Guttmann, S., and Jaquenoud, P.-A., *Helv. Chim. Acta* **43**, 1349 (1960).
293. Boissonnas, R. A., Guttmann, S., and Jaquenoud, P.-A., *Helv. Chim. Acta* **43**, 1481 (1960).
294. Boissonnas, R. A., Guttmann, S., Jaquenoud, P.-A., Konzett, H., and Stürmer, E., *Experientia* **16**, 326 (1960).
295. Boissonnas, R. A., Guttmann, S., Jaquenoud, P.-A., Sandrin, E., and Waller, J.-P., *Helv. Chim. Acta* **44**, 123 (1961).
296. Boissonnas, R. A., Guttmann, S., Jaquenoud, P.-A., and Waller, J.-P., *Helv. Chim. Acta* **38**, 1491 (1955).
297. Boissonnas, R. A., Guttmann, S., Jaquenoud, P.-A., and Waller, J.-P., *Helv. Chim. Acta* **39**, 1421 (1956).
298. Boissonnas, R. A., Guttmann, S., Jaquenoud, P.-A., and Waller, J.-P., *Nature* **178**, 260 (1956).
299. Boissonnas, R. A., Guttmann, S., Waller, J.-P., and Jaquenoud, P.-A., *Experientia* **12**, 446 (1956).
300. Boissonnas, R. A., and Huguenin, R. L., *Helv. Chim. Acta* **43**, 182 (1960).
301. Boissonnas, R. A., and Preitner, G., *Helv. Chim. Acta* **36**, 875 (1953).
301a. Boissonnas, R. A., and Sandrin, E., Proc. 7th European Symp. on Peptides, Budapest, *Acta Chim. Acad. Sci. Hung.* (in press).
302. Boissonnas, R. A., and Schumann, I., *Helv. Chim. Acta* **35**, 2229 (1952).
303. Boothe, J. H., Mowat, J. H., Hutchings, B. L., Angier, R. B., Waller, C. W., Stokstad, E. L. R., Semb, J., Gazzola, A. L., and SubbaRow, Y., *J. Am. Chem. Soc.* **70**, 1099 (1948).
304. Boothe, J. H., Semb, J., Waller, C. W., Angier, R. B., Mowat, J. H., Hutchings, B. L., Stokstad, E. L. R., and SubbaRow, Y., *J. Am. Chem. Soc.* **71**, 2304 (1949).
305. Bonta, I. L., Wijmenga, H. G., and Hohensee, F., *Acta Physiol. Pharmacol. Neerl.* **10**, 114 (1961).
306. Borek, B. A., and Waelsch, H., *J. Biol. Chem.* **205**, 459 (1953).

307. Bösch, K., and Käser, O., *Schweiz. Med. Wochschr.* **86**, 229 (1963).
308. Bose, A. K., Greer, F., and Price, C. C., *J. Org. Chem.* **23**, 1335 (1958).
309. Böshagen, H., and Ullrich, J., *Chem. Ber.* **92**, 1478 (1959).
310. Bosshard, H. H., Mory, R., Schmid, M., and Zollinger, H., *Helv. Chim. Acta* **42**, 1653 (1953).
311. Bosshard, H. H., and Zollinger, H., *Helv. Chim. Acta* **42**, 1659 (1959).
312. Bossi, R., Hütter, R., Keller-Schierlein, W., Neip, L., and Zähner, H., *Helv. Chim. Acta* **41**, 1645 (1958).
313. Botvinik, M. M., *Collection Czech. Chem. Commun.* **24**, 54 (1959).
314. Botvinik, M. M., Andreeva, A. P., and Koksharova, L. M., *Collection Czech. Chem. Commun.* **27**, 2244 (1962).
315. Botvinik, M. M., and Avaeva, S. M., *Dokl. Akad. Nauk SSSR* **84**, 951 (1952).
316. Botvinik, M. M., and Avaeva, S. M., *Zh. Obshch. Khim.* **26**, 2325 (1956).
317. Botvinik, M. M., and Avaeva, S. M., *Dokl. Akad. Nauk SSSR* **112**, 1053 (1957).
318. Botvinik, M. M., and Avaeva, S. M., *Zh. Obshch. Khim.* **28**, 1534 (1958).
319. Botvinik, M. M., Avaeva, S. M., Koksharova, L. M., and Oladkina, V. A., *Zh. Obshch. Khim.* **30**, 3877 (1960).
320. Botvinik, M. M., Avaeva, S. M., Koksharova, L. M., and Oladkina, V. A., *Zh. Obshch. Khim.* **30**, 3883 (1960).
321. Botvinik, M. M., Avaeva, S. M., Konovalova, M. I., and Ostoslavskaya, V. I., *Zh. Obshch. Khim.* **27**, 1910 (1957).
322. Botvinik, M. M., Avaeva, S. M., and Mistryukov, E. A., *Zh. Obshch. Khim.* **23**, 971 (1953).
323. Botvinik, M. M., Avaeva, S. M., and Mistryukov, E. A., *Zh. Obshch. Khim.* **23**, 1716 (1953).
324. Botvinik, M. M., Avaeva, S. M., and Mistryukov, E. A., *Zh. Obshch. Khim.* **24**, 2084 (1954).
325. Botvinik, M. M., Avaeva, S. M., and Noskova, N. B., *Zh. Obshch. Khim.* **26**, 2325 (1956).
326. Botvinik, M. M., and Ostoslavskaya, V. I., *Izv. Akad. Nauk SSSR, Otd. Khim. Nauk* **123**, 285 (1958).
327. Botvinik, M. M., Ostoslavskaya, V. I., and Svanov, L. L., *Zh. Obshch. Khim.* **31**, 42 (1961).
328. Botvinik, M. M., Rodvjasniji, V. P., and Koksharova, L. M., *Zh. Obshch. Khim.* **32**, 1619 (1962).
329. Botvinik, M. M., and Severin, I. S., *Zh. Obshch. Khim.* **20**, 1062 (1950).
330. Bovarnick, M., *J. Biol. Chem.* **148**, 151 (1943).
331. Boyer, P. D., Lardy, H., and Myrbäck, K. (eds.), "The Enzymes," 2nd ed., Vol. 4. Academic Press, New York, 1960.
332. Brand, E., and Erlanger, B. F., *J. Am. Chem. Soc.* **72**, 3314 (1950).
333. Brand, E., Erlanger, B. F., Polatnick, J., Sachs, H., and Kirschbaum, D., *J. Am. Chem. Soc.* **73**, 4027 (1951).
334. Brand, E., Erlanger, B. F., and Sachs, H., *J. Am. Chem. Soc.* **74**, 1849 (1952).
335. Brand, E., Erlanger, B. F., and Sachs, H., *J. Am. Chem. Soc.* **74**, 1851 (1952).
336. Brand, E., Erlanger, B. F., Sachs, H., and Polatnick, J., *J. Am. Chem. Soc.* **73**, 3510 (1951).
337. Brandenburg, D., Dissertation, Rheinisch-Westfälische Technische Hochschule Aachen, 1961.
338. Bräuninger, H., and Mengering, S., *Pharm. Zentralhalle* **99**, 284 (1960).

339. Braun-Menéndez, E., *Pharmacol. Rev.* **8,** 25 (1956).
340. Braun-Menéndez, E., Fasciolo, J. C., Leloir, L. F., and Muñoz, J. M., *Rev. Soc. Arg. Biol.* **15,** 420 (1939).
341. Braun-Menéndez, E., Fasciolo, J. C., Leloir, L. F., and Muñoz, J. M., *J. Physiol.* (*London*) **98,** 283 (1940).
342. Braun-Menéndez, E., and Page, I. H., *Science* **127,** 242 (1958).
343. Brenner, M., *J. Cellular Comp. Physiol.* **54,** 221 (1959).
344. Brenner, M., and Burckhardt, C. H., *Helv. Chim. Acta* **34,** 1070 (1951).
344a. Brenner, M., and Curtius, H. C., *Helv. Chim. Acta* **46,** 2126 (1963).
345. Brenner, M., and Hartmann, A., *Collection Czech. Chem. Commun.* **24,** 120 (1959).
346. Brenner, M., and Hofer, W., *Helv. Chim. Acta* **44,** 1794 (1961).
347. Brenner, M., and Hofer, W., *Helv. Chim. Acta* **44,** 1798 (1961).
348. Brenner, M., and Huber, W., *Helv. Chim. Acta* **36,** 1109 (1953).
349. Brenner, M., Müller, H. R., and Pfister, R. W., *Helv. Chim. Acta* **33,** 568 (1950).
350. Brenner, M., and Pfister, R. W., *Helv. Chim. Acta* **34,** 2085 (1951).
351. Brenner, M., and Photaki, I., *Helv. Chim. Acta* **39,** 1525 (1956).
352. Brenner, M., and Rickenbacher, H. R., *Helv. Chim. Acta* **43,** 2152 (1960).
353. Brenner, M., Sailer, E., and Rüfenacht, K., *Helv. Chim. Acta* **34,** 2096 (1951).
354. Brenner, M., and Wehrmüller, J., *Helv. Chim. Acta* **40,** 2374 (1957).
355. Brenner, M., and Zimmermann, J. P., *Helv. Chim. Acta* **40,** 1933 (1957).
356. Brenner, M., and Zimmermann, J. P., *Helv. Chim. Acta* **41,** 467 (1958).
357. Brenner, M., Zimmermann, J. P., Quitt, P., Schneider, W., and Hartmann, A., *Helv. Chim. Acta* **40,** 604 (1957).
358. Brenner, M., Zimmermann, J. P., Wehrmüller, J., Quitt, P., Hartmann, A., Schneider, W., and Beglinger, U., *Helv. Chim. Acta* **40,** 1497 (1957).
358a. Brenner, M., Zimmermann, J. P., Wehrmüller, J., Quitt, P., and Photaki, I., *Experientia* **11,** 297 (1955).
359. Brewster, J. H., and Ciotti, C. J., Jr., *J. Am. Chem. Soc.* **77,** 6214 (1955).
360. Bricas, E., and Nicot-Gutton, C., *Bull. Soc. Chim. France* p. 466 (1960).
361. Bricas, E., Nicot, C., and Lederer, E., *Bull. Soc. Chim. Biol.* **44,** 1115 (1962).
362. Bricas, E., Nicot, C., and Van Heijencort, J., *Collection Czech. Chem. Commun.* **27,** 2240 (1962).
363. Bricas, E., Nicot, C., and Van Heijencort, J., *Compt. Rend.* **254,** 1685 (1962).
363a. Broghammer, H., *Klin. Wochschr.* **41,** 1097 (1963).
364. Brockmann, H., *Angew. Chem.* **66,** 1 (1954).
365. Brockmann, H., *Angew. Chem.* **72,** 939 (1960).
366. Brockmann, H., *Fortschr. Chem. Org. Naturstoffe* **13,** 1 (1960).
366a. Brockmann, H., *Naturwissenschaften* **50,** 689 (1963).
366b. Brockmann, H., *Ann. N. Y. Acad. Sci.* **89,** 323 (1960).
367. Brockmann, H., Bohnsack, G., and Süling, G. H., *Angew. Chem.* **68,** 66 (1956).
368. Brockmann, H., and Boldt, P., *Naturwissenschaften* **46,** 262 (1959).
369. Brockmann, H., and Boldt, P., *Naturwissenschaften* **50,** 19 (1963).
370. Brockmann, H., Boldt, P., and Petras, H.-S., *Naturwissenschaften* **47,** 62 (1960).
371. Brockmann, H., and Bujard, H., *Naturwissenschaften* **49,** 515 (1962).
372. Brockmann, H., and Franck, B., *Naturwissenschaften* **41,** 451 (1954).
373. Brockmann, H., and Franck, B., *Chem. Ber.* **87,** 1767 (1954).

374. Brockmann, H., and Franck, B., *Angew. Chem.* **68,** 68 (1956).
375. Brockmann, H., and Franck, B., *Angew. Chem.* **68,** 70 (1956).
376. Brockmann, H., Geeren, H., *Ann. Chem. Liebigs* **603,** 216 (1957).
377. Brockmann, H., and Graef, V., *Naturwissenschaften* **49,** 540 (1962).
378. Brockmann, H., and Gröne, H., *Chem. Ber.* **87,** 1036 (1954).
379. Brockmann, H., Gröne, H., and Pampus, G., *Chem. Ber.* **91,** 1916 (1958).
380. Brockmann, H., Grubhofer, N., Kass, W., and Kalbe, H., *Chem. Ber.* **84,** 260 (1951).
381. Brockmann, H., and Lackner, H., *Naturwissenschaften* **47,** 230 (1960).
382. Brockmann, H., and Lackner, H., *Naturwissenschaften* **48,** 555 (1961).
382a. Brockmann, H., and Lackner, H., *Naturwissenschaften* **51,** 384 (1964).
382b. Brockmann, H., and Lackner, H., *Naturwissenschaften* **51,** 407 (1964).
382c. Brockmann, H., and Lackner, H., *Naturwissenschaften* **51,** 435 (1964).
383. Brockmann, H., and Manegold, J. H., *Naturwissenschaften* **45,** 310 (1958).
384. Brockmann, H., and Manegold, J. H., *Chem. Ber.* **93,** 2971 (1960).
385. Brockmann, H., and Manegold, J. H., *Chem. Ber.* **95,** 1081 (1962).
385a. Brockmann, H., and Manegold, J. H., *Naturwissenschaften* **51,** 383 (1964).
386. Brockmann, H., and Musso, H., *Chem. Ber.* **89,** 241 (1956).
387. Brockmann, H., and Muxfeldt, H., *Angew. Chem.* **68,** 67 (1956).
388. Brockmann, H., and Muxfeldt, H., *Angew. Chem.* **68,** 69 (1956).
389. Brockmann, H., and Muxfeldt, H., *Chem. Ber.* **89,** 1379 (1956).
390. Brockmann, H., and Muxfeldt, H., *Chem. Ber.* **89,** 1397 (1956).
391. Brockmann, H., and Muxfeldt, H., *Chem. Ber.* **91,** 1242 (1958).
392. Brockmann, H., Pampus, G., and Manegold, J. H., *Chem. Ber.* **92,** 1294 (1959).
393. Brockmann, H., Pampus, G., and Mecke, R., *Chem. Ber.* **92,** 3082 (1959).
394. Brockmann, H., and Petras, H.-S., *Naturwissenschaften* **46,** 400 (1959).
395. Brockmann, H., and Petras, H.-S., *Naturwissenschaften* **48,** 218 (1961).
396. Brockmann, H., and Schmidt-Kastner, G., *Chem. Ber.* **88,** 57 (1955).
397. Brockmann, H., and Springorum, M., *Naturwissenschaften* **49,** 514 (1962).
398. Brockmann, H., and Sunderkötter, W., *Naturwissenschaften* **47,** 229 (1960).
399. Brockmann, H., Sunderkötter, W., Ohly, K. W., and Boldt, P., *Naturwissenschaften* **47,** 230 (1960).
400. Brockmann, H., Tummes, H., and von Metzsch, F.-A., *Naturwissenschaften* **41,** 37 (1954).
401. Brockmann, H., and Vohwinkel, K., *Angew. Chem.* **67,** 618 (1955).
402. Brockmann, H., and Vohwinkel, K., *Chem. Ber.* **89,** 1373 (1956).
403. Brocks, F. P., and Pickford, M., *J. Physiol. (London)* **142,** 468 (1958).
404. Broghammer, H., and Wernitsch, W., *Z. Kreislaufforsch.* **51,** 798 (1962).
405. Bromer, W. W., Sinn, L. G., and Behrens, O. K., *J. Am. Chem. Soc.* **79,** 2798 (1957).
406. Bromer, W. W., Sinn, L. G., and Behrens, O. K., *J. Am. Chem. Soc.* **79,** 2807 (1957).
407. Bromer, W. W., Sinn, L. G., Staub, A., and Behrens, O. K., *J. Am. Chem. Soc.* **78,** 3858 (1956).
408. Bromer, W. W., Staub, A., Diller, E. R., Bird, H. L., Sinn, L. G., and Behrens, O. K., *J. Am. Chem. Soc.* **79,** 2794 (1957)
409. Bromer, W. W., Staub, A., Sinn, L. G., and Behrens, O. K., *J. Am. Chem. Soc.* **79,** 2801 (1954).
410. Broser, W., and Lautsch, W., *Naturwissenschaften* **42,** 513 (1955).

411. Broser, W., and Lautsch, W., *Z. Naturforsch.* **11b**, 453 (1956).
412. Brown, G. B., and du Vigneaud, V., *J. Biol. Chem.* **140**, 767 (1941).
413. Brown, H., Sanger, F., and Kitai, R., *Biochem. J.* **60**, 556 (1955).
413a. Brown, J. J., and Peart, W. S., *Clin. Sci.* **22**, 1 (1962).
414. Brown, R., Brennan, J., and Kelley, C., *Antibiot. Chemotherapy* **12**, 482 (1962).
415. Brown, S. S., and Wade, R., *J. Chem. Soc.* p. 3280 (1962).
416. Brownlee, G., and Jones, T. S. G., *Biochem. J.* **43**, xxv (1948).
417. Bruckner, V., and Kajtár, M., *Acta Chim. Acad. Sci. Hung.* **21**, 417 (1959).
418. Bruckner, V., Kajtár, M., Kovács, J., Nagy, H., and Wein, J., *Tetrahedron* **2**, 211 (1958).
419. Bruckner, V., Kótai, A., and Kovács, K., *Acta Chim. Acad. Sci. Hung.* **21**, 427 (1959).
420. Bruckner, V., and Kovács, J., *Acta Chim. Acad. Sci. Hung.* **12**, 363 (1957).
421. Bruckner, V., Kovács, K., and Kótai, A., *Acta Chim. Acad. Sci. Hung.* **5**, 267 (1955).
422. Bruckner, V., Kovács, J., and Kovács, K., *Naturwissenschaften* **39**, 380 (1952).
423. Bruckner, V., Kovács, J., and Kovács, K., *Acta Chim. Acad. Sci. Hung.* **31**, 361 (1953).
424. Bruckner, V., Kovács, J., and Kovács, K., *Naturwissenschaften* **40**, 243 (1953).
425. Bruckner, V., Kovács, K., Kovács, J., and Kótai, A., *Experientia* **10**, 166 (1954).
426. Bruckner, V., Kovács, J., and Medzihradszky, K., *Naturwissenschaften* **42**, 96 (1955).
427. Bruckner, V., Kovács, J., Nagy, H., and Kajtár, M., *Naturwissenschaften* **41**, 528 (1954).
428. Bruckner, V., Kovács, J., Nagy, H., and Kajtár, M., *Acta Chim. Acad. Sci. Hung.* **6**, 219 (1955).
429. Bruckner, V., Medzihradszky, K., Kandel, I., and Kandel, M., *Acta Chim. Acad. Sci. Hung.* **21**, 105 (1959).
430. Bruckner, V., and Szekerke, M., *Acta Chim. Acad. Sci. Hung.* **34**, 93 (1962).
431. Bruckner, V., Szekerke, M., and Kovács, J., *Z. Physiol. Chem.* **309**, 25 (1957).
432. Brunner, H., *Arch. Exptl. Pathol. Pharmakol.* **245**, 283 (1963).
433. Brunner, H., Kuschinsky, G., Münchow, O., and Peters, G., *Klin. Wochschr.* **34**, 451 (1956).
434. Brunner, H., and Regoli, D., *Experientia* **18**, 504 (1962).
435. Brunn-Leube, I. von, and Schramm, G., *Chem. Ber.* **89**, 2045 (1956).
436. Bryant, P. M., Moore, R. H., Pimlott, P. J., and Young, G. T., *J. Chem. Soc.* p. 3868 (1959).
437. Bullock, E., and Johnson, A. W., *J. Chem. Soc.* p. 3280 (1957).
438. Bumpus, F. M., Green, A. A., and Page, I. H., *J. Biol. Chem.* **210**, 287 (1954).
439. Bumpus, F. M., Khairallah, P. A., Arakawa, K., Page, I. H., and Smeby, R. R., *Biochim. Biophys. Acta* **46**, 38 (1961).
440. Bumpus, F. M., and Page, I. H., *Science* **119**, 849 (1954).
441. Bumpus, F. M., Schwarz, H., and Page, I. H., *Circulation Res.* **4**, 488 (1956).
442. Bumpus, F. M., Schwarz, H., and Page, I. H., *Science* **125**, 886 (1957).
443. Bumpus, F. M., Schwarz, H., and Page, I. H., *Circulation* **17**, 664 (1958).
444. Bumpus, F. M., and Smeby, R. R., *Circulation* **25**, 183 (1962).

445. Bumpus, F. M., Smeby, R. R., and Page, I. H., *Circulation Res.* **9,** 762 (1961).

446. Burckhardt, D., *Helv. Physiol. Pharmacol. Acta* **20,** 135 (1962).

447. Burgers, A. C. J., *Acta Endocrinol. Suppl.* **51,** 329 (1960).

448. Burgers, A. C. J., *Endocrinology* **68,** 698 (1961).

449. Burgers, A. C. J., *Ann. N. Y. Acad. Sci.* **100,** 669 (1963).

450. Burkhardt, H.-J., and Mitchel, H. K., *Arch. Biochem. Biophys.* **94,** 32 (1961).

451. Butler, J. A. V., Dodds, E. C., Phillips, D. M. P., and Stephen, J. L. M., *Biochem. J.* **42,** 122 (1948).

452. Butz, L. W., and du Vigneaud, V., *J. Biol. Chem.* **99,** 135 (1932–1933).

453. Buzas, A., Egnell, C., and Fréon, P., *Compt. Rend.* **255,** 945 (1962).

454. Cahill, W. M., and Burton, I. F., *J. Biol. Chem.* **132,** 161 (1940).

455. Caldeyro-Barcia, R., *Acta Endocrinol.* Suppl. **47–50,** 41 (1959–1960).

456. Caldeyro-Barcia, R., and Sereno, J. A., *in* "Oxytocin" (R. Caldeyro-Barcia and H. Heller, eds.), p. 177. Macmillan (Pergamon), New York, 1961.

457. Callahan, F. M., Anderson, G. W., Paul, R., and Zimmermann, J. E., *J. Am. Chem. Soc.* **85,** 201 (1963).

457a. Camerino, B., DeCaro, G., Boissonnas, R. A., Sandrin, E., and Stürmer, E., *Experientia* **19,** 339 (1963).

458. Capri, A., and Corrado, A. P., *Experientia* **17,** 326 (1961).

459. Carayon-Gentil, A., and Van-Thoai, M. N., *Compt. Rend.* **239,** 1051 (1954).

460. Carlini, E. A., and Huggins, C. G., *Biochem. Pharmacol.* **11,** 171 (1962).

461. Carlini, E. A., Picarelli, Z. P., and Prado, J. L., *Bull. Soc. Chim. Biol.* **40,** 1825 (1958).

461a. Carnegie, P. R., *Biochem. J.* **89,** 459 (1963).

461b. Carnegie, P. R., *Biochem. J.* **89,** 471 (1963).

462. Carpenter, F. H., and Baum, W. E., *J. Biol. Chem.* **237,** 409 (1962).

463. Carpenter, F. H., and Gish, D. T., *J. Am. Chem. Soc.* **74,** 3818 (1952).

464. Carpino, L. A., *J. Am. Chem. Soc.* **79,** 98 (1957).

465. Carpino, L. A., *J. Am. Chem. Soc.* **79,** 4427 (1957).

465a. Carpino, L. A., Giza, C. A., and Carpino, B. A., *J. Am. Chem. Soc.* **81,** 955 (1959).

466. Casentini, S., de Poli, A., Hukovic, S., and Martini, L., *Endocrinology* **64,** 483 (1959).

467. Cash, W. D., *J. Org. Chem.* **26,** 2136 (1961).

468. Cash, W. D., *J. Org. Chem.* **27,** 3329 (1962).

469. Cash, W. D., McCulloch Mahaffey, L., Buck, A. S., Nettleton, D. E., Jr., Romas, C., and du Vigneaud, V., *J. Med. Pharm. Chem.* **5,** 413 (1962).

470. Cash, W. D., and Smith, B. L., *J. Biol. Chem.* **278,** 994 (1963).

470a. de Castiglione, R., Chillemi, F., Bernardi, L., and Goffredo, O., *Gazz. Chim. Ital.* **94,** 875 (1964).

471. Catch, J. R., Jones, T. S. G., and Wilkinson, S., *Biochem. J.* **43,** xvii–xviii (1948).

472. Catch, J. R., Jones, T. S. G., and Wilkinson, S., *Ann. N. Y. Acad. Sci.* **51,** 917 (1949).

473. Cecil, R., and McPhee, J. R., *Advan. Protein Chem.* **14,** 255 (1959).

474. Čehović, G., *Compt. Rend.* **254,** 1872 (1962).

475. Celmer, W. D., and Sobin, R. A., *Antibiot. Ann.* p. 437 (1954–1955).

476. Cerletti, A., Stürmer, E., and Konzett, H., *Deut. Med. Wochschr.* **86,** 678 (1961).

476a. Cerletti, A., Weber, H., and Weidmann, H., *Helv. Physiol. Acta* **21,** 394 (1963).
477. Černá, J., Grünberger, D., and Šorm F., *Collection Czech. Chem. Commun.* **26,** 1212 (1961).
478. Chambers, R. W., and Carpenter, F. H., *J. Am. Chem. Soc.* **77,** 1522 (1955).
479. Chambers, R. W., and Carpenter, F. H., *J. Am. Chem. Soc.* **77,** 1527 (1955).
480. Chan, W. Y., and du Vigneaud, V., *Endocrinology* **71,** 977 (1962).
481. Chan, W. Y., and Sawyer, W. H., *Am. J. Physiol.* **201,** 799 (1961).
482. Channing, D. M., Turner, P. B., and Young, G. T., *Nature* **167,** 487 (1951).
483. Chantrenne, H., *Biochim. Biophys. Acta* **2,** 286 (1948).
484. Charney, J., Fisher, W. P., Curran, C., Machlowitz, R. A., and Tyteli, A. A., *Antibiot. Chemotherapy* **3,** 1283 (1953).
485. Chauvet, J., Lenci, M.-T., and Acher, R., *Biochim. Biophys. Acta* **38,** 266 (1960).
486. Chauvet, J., Lenci, M.-T., and Acher, R., *Biochim. Biophys. Acta* **38,** 571 (1960).
487. Chen, C. C., Huang, W. T., and Chang, K. P., *Acta Biochim. Biophys. Sinica* **2,** 152 (1962).
487a. Chen, C.-C., Huang, W.-T., and Niu, C.-I., *Sci. Sinica* (*Peking*) **13,** 1235 (1964).
488. Chibnall, A. C., and Spahr, P. F., *Biochem. J.* **68,** 135 (1958).
489. Chibnall, A. C., and Westall, R. G., *Biochem. J.* **26,** 122 (1932).
489a. Chillemi, F., Bernardi, L., and Bosisio, G., *Gazz. Chim. Ital.* **94,** 891 (1964).
489b. Chillemi, F., and Goffredo, O., *Gazz. Chim. Ital.* **94,** 866 (1964).
490. Chillemi, D., Scarso, L., and Scoffone, E., *Gazz. Chim. Ital.* **87,** 1356 (1957).
490a. Choppin, A. R., and Rogers, J. W., *J. Am. Chem. Soc.* **70,** 2967 (1948).
490b. Chillemi, F., *Gazz. Chim. Ital.* **93,** 1079 (1963).
491. Christensen, H. N., and Riggs, T. R., *J. Biol. Chem.* **220,** 265 (1956).
492. Čihař, M., Beránková, Z., Rychlík, I., and Šorm, F., *Collection Czech. Chem. Commun.* **26,** 2632 (1961).
493. Cipera, J. D., *J. Org. Chem.* **26,** 206 (1961).
494. Cipera, J. D., and Nicholls, R. V. V., *Chem. Ind.* (*London*) p. 16 (1955).
495. Clark, L. C., Jr., Winkler, C., Collan, F., and Fox, R. P., *J. Biol. Chem.* **206,** 717 (1954).
496. Clark-Lewis, J. W., and Fruton, J. S., *J. Biol. Chem.* **207,** 477 (1954).
497. Clayton, D. W., Farrington, J. A., Kenner, G. W., and Turner, J. M., *J. Chem. Soc.* p. 1398 (1957).
498. Clayton, D. W., Kenner, G. W., and Sheppard, R. C., *J. Chem. Soc.* p. 371 (1956).
499. Cleugh, J., and Gaddum, J. H., *Experientia* **19,** 72 (1963).
500. Cliffe, E. E., and Waley, S. G., *Biochem. J.* **89,** 649 (1958).
501. Cliffe, E. E., and Waley, S. G., *Biochem. J.* **72,** 12P (1959).
502. Cliffe, E. E., and Waley, S. G., *Biochem. J.* **73,** 25P (1959).
503. Cliffe, E. E., and Waley, S. G., *Biochem. J.* **79,** 118 (1961).
504. Cliffe, E. E., and Waley, S. G., *Biochem. J.* **79,** 669 (1961).
505. Clubb, M. E., Scopes, P. M., and Young, G. T., *Chimia* (*Aarau*) **14,** 373 (1960).
506. Cohen, L. A., and Witkop, B., *Angew. Chem.* **73,** 253 (1961).
507. Colasito, D. J., Koffler, H., Reitz, H. C., and Tetrault, T. H., *Bacteriol. Proc.* (*Soc. Am. Bacteriologists*) p. 721 (1956).

508. Colasito, D. J., Koffler, H., Tetrault, P. A., and Reitz, H. C., *Can. J. Microbiol.* **1,** 685 (1955).

509. Cole, R. D., Li, C. H., Harris, J. I., and Pon, N. G., *J. Biol. Chem.* **219,** 903 (1956).

510. Coleman, D., *J. Chem. Soc.* p. 2294 (1951).

511. Collier, H. O. J., *Actualités Pharmacol.* **14,** 53 (1961).

511a. Collier, H. O. J., *Ann. N. Y. Acad. Sci.* **104,** 290 (1963).

512. Collier, H. O. J., Holgate, J. A., Schachter, M., and Shorley, P. G., *J. Physiol. (London)* **149,** 54P (1959).

513. Collier, H. O. J., Holgate, J. A., Schachter, M., and Shorley, P. G., *Brit. J. Pharmacol.* **15,** 290 (1960).

514. Collier, H. O. J., and Shorley, P. G., *Brit. J. Pharmacol.* **15,** 601 (1960).

515. Consden, R., and Gordon, A. H., *Biochem. J.* **46,** 8 (1950).

516. Consden, R., Gordon, A. H., and Martin, A. J. P., *Biochem. J.* **40,** 580 (1946).

517. Consden, R., Gordon, A. H., Martin, A. J. P., and Synge, L. M., *Biochem. J.* **40,** xliii (1946).

518. Consden, R., Gordon, A. H., Martin, A. J. P., and Synge, L. M., *Biochem. J.* **41,** 596 (1947).

518a. Constantine, J. W., *Experientia* **20,** 381 (1964).

519. Cook, A. H., Cox, S. F., and Farmer, T. H., *J. Chem. Soc.* p. 1022 (1949).

520. Cook, A. H., and Elvidge, J. A., *J. Chem. Soc.* p. 2362 (1949).

521. Cook, A. H., Heilbron, I., and Levy, A. L., *J. Chem. Soc.* p. 201 (1948).

522. Cook, A. H., and Levy, A. L., *J. Chem. Soc.* p. 637 (1950).

523. Cook, A. H., and Levy, A. L., *J. Chem. Soc.* p. 642 (1950).

524. Cook, A. H., and Levy, A. L., *J. Chem. Soc.* p. 646 (1950).

525. Cook, A. H., and Levy, A. L., *J. Chem. Soc.* p. 651 (1950).

526. Coon, J. A., *Arch. Intern. Pharmacodyn.* **62,** 79 (1939).

527. Corbaz, R., Ettlinger, L., Gäumann, E., Keller-Schierlein, W., Kradolfer, F., Neipp, L., Prelog, V., Reusser, P., and Zähner, H., *Helv. Chim. Acta* **40,** 199 (1957).

528. Corbett, C. E., Mauri, A. C., Saraiva, P. A. P., Adachi, T., Vital Brazil, O., and Pereira, S. A., *Chemotherapia* **3,** 497 (1961).

529. Cosmatos, A., Photaki, I., and Zervas, L., *Chem. Ber.* **94,** 2644 (1961).

530. Courrier, R., and Čehović, G., *Compt. Rend.* **251,** 832 (1960).

531. Coutsogeorgopoulos, C., and Zervas, L., *J. Am. Chem. Soc.* **83,** 1885 (1961).

532. Cox, R. W., Henley, E. D., Narahara, H. T., Van Arsdel, P. P., and Williams, R. H., *Endocrinology* **60,** 277 (1957).

533. Craig, L. C., *Proc. 3rd Intern. Congr. Biochem., Brussels, 1955* p. 416 (1956). Academic Press, New York.

533a. Craig, L. C., Harfenist, E. J., and Paladini, A. C., *Biochemistry* **3,** 764 (1964).

534. Craig, L. C., Hausmann, W., and Weisiger, J. R., *J. Biol. Chem.* **200,** 765 (1953).

535. Craig, L. C., and Konigsberg, W. M., *J. Org. Chem.* **22,** 1345 (1957).

536. Craig, L. C., Konigsberg, W., and Hill, R. J., *Ciba Found. Symp. Amino Acids Peptides Antimetab. Activity, 1958* p. 226. (1959).

537. Craig, L. C., Weisiger, J. R., Hausmann, W., and Harfenist, E. J., *J. Biol. Chem.* **199,** 259 (1952).

538. Cramer, F. D., and Gärtner, K. G., *Chem. Ind. (London)* p. 560 (1958).

539. Cramer, F., and Gärtner, K.-G., *Chem. Ber.* **91,** 1562 (1958).

540. Cramer, F., and Schaller, H., *Chem. Ber.* **94,** 1634 (1961).

540a. Crawhall, J. C., and Elliott, D. F., *Biochem. J.* **61,** 264 (1955).

541. Crofts, P. C., Markes, J. H. H., and Rydon, H. N., *J. Chem. Soc.* p. 4250 (1958).

542. Crofts, P. C., Markes, J. H. H., and Rydon, H. N., *J. Chem. Soc.* p. 3610 (1959).

543. Cronyn, M. W., and Jiu, J., *J. Am. Chem. Soc.* **74,** 4726 (1952).

544. Cross, B. A., *in* "Oxytocin" (R. Caldeyro-Barcia, and H. Heller, eds.), p. 24. Macmillan (Pergamon), New York, 1961.

545. Croxatto, H., *in* "Polypeptides which Stimulate Plain Muscle" (J. R. Gaddum, ed.), p. 92. Livingston, Edinburgh and London, 1955.

546. Croxatto, H., *in* "The Neurohypophysis" (H. Heller, ed.), p. 51. Butterworth, London and Washington, D. C., 1957.

547. Croxatto, H., and Barnafi, L., *Recent Progr. Hormone Res.* **16,** 263 (1960).

548. Croxatto, H., Pereda, T., Belmar, J., and Labarca, E., *Ann. N. Y. Acad. Sci.* **104,** 146 (1963).

549. Croxatto, H., Pereda, T., and Mellada, R., *Nature* **184,** 1496 (1959).

550. Csapo, A., *in* "Oxytocin" (R. Caldeyro-Barcia, and H. Heller, eds.), p. 100. Macmillan (Pergamon), New York, 1961.

551. Curragh, E. F., and Elmore, D. T., *J. Chem. Soc.* p. 2948 (1962).

552. Curtius, T., *J. Prakt. Chem.* **70,** 57 (1904).

553. Curtius, T., and Curtius, H., *J. Prakt. Chem.* **70,** 158 (1904).

554. Curtius, T., and Darapsky, A., *Ber. Deut. Chem. Ges.* **39,** 1373 (1906).

555. Curtius, T., and Lambotte, E., *J. Prakt. Chem.* **70,** 109 (1904).

556. Curtius, T., and Levy, L., *J. Prakt. Chem.* **70,** 89 (1904).

557. Curtius, T., and Thompson, J., *Ber. Deut. Chem. Ges.* **39,** 1379 (1906).

558. Dadić, M., Fleš, D., and Markovac-Prpić, A., *Croat. Chem. Acta* **33,** 73 (1961).

559. Dahlberg, B. C. G., *Acta Endocrinol.* Suppl. 51, 331 (1960).

560. Dahlberg, B. C. G., *Acta Endocrinol.* **38** Suppl. 60 (1961).

561. Dahn, H., Manassé, R., Rosenthaler, J., and Brenner, M., *Helv. Chim. Acta* **42,** 2249 (1959).

562. Dalgliesh, C. E., Johnson, A. W., Todd, A. R., and Vining, L. C., *J. Chem. Soc.* p. 2946 (1950).

563. Dane, E., Drees, F., Konrad, P., and Dockner, T., *Angew. Chem.* **74,** 873 (1962).

564. Danowski, T. S., Hofmann, K., Yajima, H., and Moses, C., *Proc. Soc. Exptl. Biol. Med.* **108,** 559 (1961).

565. Danowski, T. S., Hofmann, K., Yajima, H., and Moses, C., *Metab., Clin. Exptl.* **10,** 835 (1961).

566. Dautrevaux, M., and Biserte, G., *Compt. Rend.* **243,** 923 (1956).

567. Dautrevaux, M., and Biserte, G., *Compt. Rend. Soc. Biol.* **153,** 1346 (1959).

568. Dautrevaux, M., and Biserte, G., *Bull. Soc. Chim. Biol.* **43,** 495 (1961).

569. Davis, A. C., and Levy, A. L., *J. Chem. Soc.* p. 2419 (1951).

570. Davies, J. W., and Harris, G., *Proc. Roy. Soc.* **B151,** 537 (1960).

571. Davies, J. W., and Harris, G., *Biochim. Biophys. Acta* **45,** 28 (1960).

572. Davis, N. C., *J. Biol. Chem.* **223,** 935 (1956).

573. Davis, N. C., and Smith, E. L., *J. Biol. Chem.* **200,** 373 (1953).

574. Dababov, V. G., Kozarenko, T. D., and Shibnev, V. A., *Proc 5th Intern. Congr. Biochem. Moscow, 1961* Abstr., Sect. 1.17. Macmillan (Pergamon), New York, 1961.

575. Debabov, V. G., and Shibnev, V. A., *Izv. Akad. Nauk. SSSR Otd. Khim. Nauk* p. 35 (1962).

575a. Debabov, V. G., and Shibnev, V. A., *Izv. Akad. Nauk SSSR Otd. Khim. Nauk* p. 870 (1963).

575b. De Caro, G., *Arch. Intern. Pharmacodyn.* **146**, 27 (1963).

576. Dedman, M. L., Farmer, T. H., and Morris, C. J. O. R., *Biochem. J.* **59**, xii (1955).

577. Dedman, M. L., Farmer, T. H., and Morris, C. J. O. R., *Biochem. J.* **66**, 166 (1957).

578. Dedman, M. L., Farmer, T. H., and Morris, C. J. O. R., *Biochem. J.* **78**, 348 (1961).

579. De Haas, G. H., and Van Deenen, L. L. M., *Rec. Trav. Chim.* **81**, 215 (1962).

580. Dekanski, J., *Brit. J. Pharmacol.* **7**, 567 (1952).

581. Dekker, C. A., and Fruton, J. S., *J. Biol. Chem.* **173**, 471 (1948).

582. Dekker, C. A., Stone, D., and Fruton, J. S., *J. Biol. Chem.* **181**, 719 (1949).

583. Dekker, C. A., Taylor, S. P., Jr., and Fruton, J. S., *J. Biol. Chem.* **180**, 155 (1949).

584. Dengler, H., *Z. Ges. Exptl. Med.* **127**, 153 (1956).

585. Deodhar, S. D., *J. Exptl. Med.* **111**, 419 (1960).

586. Derevitskaya, V. A., Likhosherstov, L. M., and Kochetkov, N. K., *in* "Peptide Symposium," Oxford, 1962, (G. T. Young, ed.), p. 231. Macmillan (Pergamon), New York, 1963.

587. Derevitskaya, V. A., Likhosherstov, L. M., Kara-Murza, S. G., and Kochetkov, N. K., *Zh. Obshch. Khim.* **32**, 2134 (1962).

588. Derevitskaya, V. A., Likhosherstov, L. M., and Kochetkov, N. K., *Izv. Akad. Nauk SSSR Otd. Khim. Nauk* p. 1795 (1962).

589. Desnuelle, P., and Casal, A., *Biochim. Biophys. Acta* **2**, 64 (1948).

589a. DeTar, D. F., Honsberg, W., Honsberg, U., Wieland, A., Gouge, M., Bach, H., Tahara, A., Brinigar, W. S., and Rogers, F. F., Jr., *J. Am. Chem. Soc.* **85**, 2873 (1963).

589b. Determann, H., Bomhard, K., Köhler, R., and Wieland, T., *Helv. Chim. Acta* **46**, 2498 (1963).

589c. Determann, H., Torff, H.-J., and Zipp, O., *Ann. Chem. Liebigs* **670**, 141 (1963).

590. Determann, H., and Wieland, T., *Makromol. Chem.* **44–46**, 312 (1961).

590a. Determann, H., and Wieland, T., *Ann. Chem. Liebigs* **670**, 136 (1963).

591. Determann, H., Wieland, T., and Zipp, O., *Chimia (Aarau)* **14**, 374 (1960).

592. Determann, H., and Zipp, O., *Ann. Chem. Liebigs* **649**, 203 (1961).

593. Determann, H., Zipp, O., and Wieland, T., *Ann. Chem. Liebigs* **651**, 172 (1962).

594. Deutsch, H. F., and Diniz, C. R., *J. Biol. Chem.* **216**, 17 (1955).

595. DeWald, H. A., Behn, D. C., and Moore, A. M., *J. Am. Chem. Soc.* **81**, 4364 (1959).

596. DeWald, H. A., Behn, D. C., Morgan, M. S., Renfrew, A. G., and Moore, A. M., *J. Am. Chem. Soc.* **81**, 4367 (1959).

596a. DeWald, H. A., Craft, M. K., and Nicolaides, E. D., *J. Med. Chem.* **6**, 741 (1963).

597. DeWald, H. A., and Moore, A. M., *J. Am. Chem. Soc.* **80**, 3941 (1958).

597a. DeWald, H. A., and Nicolaides, E. D., *J. Med. Chem.* **7**, 50 (1964).

598. de Wied, D., Siderius, P., and Mirsky, I. A., *Arch. Intern. Pharmacodyn.* **133,** 50 (1961).
599. Dicker, S. E., *in* "Polypeptides which Affect Smooth Muscles and Blood Vessels" (M. Schachter, ed.), p. 79. Macmillan (Pergamon), New York, 1960.
600. Dicker, S. E., and Greenbaum, A. L., *J. Physiol.* (*London*) **141,** 107 (1958).
601. Dickison, H. L., Cull, K. M., and Tisch, D. E., *Antibiot. Ann.* p. 733 (1954–1955).
602. Diehl, J. F., Dissertation, Ruprecht-Karl Universität Heidelberg, 1957.
603. Diniz, C. R., and Carvalho, I. F., *Ann. N. Y. Acad. Sci.* **104,** 77 (1963).
604. Dixon, G. H., and Wardlaw, A. C., *Nature* **188,** 721 (1960).
605. Dixon, H. B. F., *Biochem. J.* **62,** 25P (1956).
606. Dixon, H. B. F., *Biochem. J.* **83,** 91 (1962).
607. Dixon, H. B. F., and Stack-Dunne, M. P., *Biochem. J.* **61,** 483 (1955).
608. Dixon, H. B. F., and Weitkamp, L. R., *Biochem. J.* **84,** 462 (1962).
609. Dixon, J. S., and Li, C. H., *J. Am. Chem. Soc.* **82,** 4568 (1960).
610. Dixon, J. S., and Li, C. H., *Gen. Comp. Endocrinol.* **1,** 161 (1961).
611. Dixon, J. S., Lo, T.-B., and Li, C. H., *Arch. Biochem. Biophys.* **92,** 296 (1961).
612. Doepfner, W., Stürmer, E., and Berde, B., *Endocrinology* **72,** 897 (1963).
613. Doherty, D. G., *J. Biol. Chem.* **201,** 857 (1953).
614. Doherty, D. G., Popenoe, E. A., and Link, K. P., *J. Am. Chem. Soc.* **75,** 3466 (1953).
615. Doty, P., and Lundberg, R. D., *J. Am. Chem. Soc.* **78,** 4810 (1956).
616. Dowling, J. H., Koffler, H., Reitz, H. C., Peterson, D., and Tetrault, P. A., *Science* **116,** 147 (1952).
616a. Drabarek, S., and du Vigneaud, V., *in* "Proceedings of the 6th European Symp. on Peptides, Athens, 1963" Macmillan (Pergamon), New York, (in press).
617. Drechsel, E., *J. Prakt. Chem.* **27,** 418 (1883).
618. Du, Y.-C., Zhang, Y.-S., Lu, Z.-X., and Tsou, C.-L., *Sci. Sinica* (*Peking*) **10,** 84 (1961).
619. Dunn, M. S., Camien, M. N., Rockland, L. B., Shankman, S., and Goldberg, S. C., *J. Biol. Chem.* **155,** 591 (1945).
620. Dus, K., Bartsch, R. G., and Kamen, M. D., *J. Biol. Chem.* **236,** PC47 (1961).
621. Dus, K., Bartsch, R. G., and Kamen, M. D., *J. Biol. Chem.* **237,** 3083 (1962).
622. du Vigneaud, V., *Harvey Lectures Ser.* **50,** 1 (1954).
623. du Vigneaud, V., *Chem. Soc.* (*London*) *Spec. Publ.* **2,** 49 (1955).
624. du Vigneaud, V., *Experientia* **11,** Suppl. 2, 9 (1955).
625. du Vigneaud, V., Audrieth, L. F., and Loring, H. S., *J. Am. Chem. Soc.* **52,** 4500 (1930).
626. du Vigneaud, V., Bartlett, M. F., and Jöhl, A., *J. Am. Chem. Soc.* **79,** 5572 (1957).
627. du Vigneaud, V., and Behrens, O. K., *J. Biol. Chem.* **117,** 27 (1937).
628. du Vigneaud, V., and Brown, G. B., *J. Biol. Chem.* **138,** 151 (1941).
629. du Vigneaud, V., Brown, G. B., and Bonsnes, R. W., *J. Biol. Chem.* **141,** 707 (1941).
630. du Vigneaud, V., Denning, G. S., Jr., Drabarek, S., and Chan, W. Y., *J. Biol. Chem.* **238,** PC1560 (1963).
630a. du Vigneaud, V., Denning, G. S., Jr., Drabarek, S., and Chan, W. Y., *J. Biol. Chem.* **239,** 472 (1964).

631. du Vigneaud, V., Dorfmann, R., and Loring, H. S., *J. Biol. Chem.* **98**, 577 (1932).
632. du Vigneaud, V., Dyer, H. M., and Harmon, J., *J. Biol. Chem.* **101**, 719 (1933).
633. du Vigneaud, V., Fitt, P. S., Bodanszky, M., and O'Connell, M., *Proc. Soc. Exptl. Biol. Med.* **104**, 653 (1960).
634. du Vigneaud, V., Gish, D. T., and Katsoyannis, P. G., *J. Am. Chem. Soc.* **76**, 4751 (1954).
635. du Vigneaud, V., Gish, D. T., Katsoyannis, P. G., and Hess, G. P., *J. Am. Chem. Soc.* **80**, 3355 (1958).
636. du Vigneaud, V., and Hunt, M., *J. Biol. Chem.* **125**, 269 (1938).
637. du Vigneaud, V., Lawler, H. C., and Popenoe, E. A., *J. Am. Chem. Soc.* **75**, 4880 (1953).
638. du Vigneaud, V., Loring, H. S., and Miller, G. L., *J. Biol. Chem.* **118**, 391 (1937).
638a. du Vigneaud, V., and Meyer, C. E., *J. Biol. Chem.* **98**, 295 (1932).
639. du Vigneaud, V., and Meyer, C. E., *J. Biol. Chem.* **99**, 143 (1932–1933).
639a. du Vigneaud, V., and Miller, G. L., *J. Biol. Chem.* **116**, 469 (1936).
640. du Vigneaud, V., and Patterson, W. I., *J. Biol. Chem.* **109**, 97 (1935).
641. du Vigneaud, V., Patterson, W. I., and Hunt, M., *J. Biol. Chem.* **126**, 217 (1938).
642. du Vigneaud, V., Ressler, C., Swan, J. M., Roberts, C. W., and Katsoyannis, P. G., *J. Am. Chem. Soc.* **76**, 3115 (1954).
643. du Vigneaud, V., Ressler, C., Swan, J. M., Roberts, C. W., Katsoyannis, P. G., and Gordon, S., *J. Am. Chem. Soc.* **75**, 4879 (1953).
644. du Vigneaud, V., Ressler, C., and Trippett, S., *J. Biol. Chem.* **205**, 949 (1953).
645. du Vigneaud, V., Schneider, C. H., Stouffer, J. E., Murti, V. V. S., Aroskar, J. P., and Winestock, G., *J. Am. Chem. Soc.* **84**, 409 (1962).
646. du Vigneaud, V., Winestock, G., Murti, V. V. S., Hope, D. B., and Kimbrough, R. D., Jr., *J. Biol. Chem.* **235**, PC64 (1960).
647. Dvonch, W., Shotwell, O. L., Benedict, R. G., Pridhan, T. G., and Lindenfelser, L. A., *Antibiot. Chemotherapy* **4**, 1135 (1954).
648. Dyer, H. M., *J. Biol. Chem.* **124**, 519 (1938).
649. Eastwood, F. W., Hughes, G. K., and Ritchie, E., *Australian J. Chem.* **8**, 552 (1955).
650. Eastwood, F. W., Snell, B. K., and Todd, A., *J. Chem. Soc.* p. 2286 (1960).
651. Ebata, M., *J. Biochem.* (*Tokyo*) **49**, 110 (1961).
652. Edman, P., *Acta Chem. Scand.* **4**, 277 (1950).
653. Edman, P., *Acta Chem. Scand.* **4**, 283 (1950).
654. Edman, P., *Acta Chem. Scand.* **10**, 761 (1956).
655. Ehrensvärd, G. C. H., *Nature* **159**, 500 (1947).
656. Ehringer, H., Herzog, P., and Konzett, H., *Helv. Physiol. Pharmacol. Acta* **19**, C 66 (1961).
657. Ehrlich, J., Coffey, G. L., Fisher, M. W., Galbraith, M. M., Knudsen, M. P., Sarber, R. W., Schlingman, A. S., Smith, R. M., and Weston, J. K., *Antibiot. Ann.* p. 790 (1954–1955).
658. Elliott, D. F., *J. Chem. Soc.* p. 589 (1949).
659. Elliott, D. F., *J. Chem. Soc.* p. 62 (1950).
660. Elliott, D. F., *Biochem. J.* **50**, 542 (1952).
661. Elliott, D. F., *Biochem. Pharmacol.* **10**, 25 (1962).

662. Elliott, D. F., *Ann. N. Y. Acad. Sci.* **104,** 35 (1963).

663. Elliott, D. F., *in* "Polypeptides which Affect Smooth Muscles and Blood Vessels" (M. Schachter, ed.), p. 99. Macmillan (Pergamon), New York, 1960.

663a. Elliott, D. F., *Record Chem. Progr.* (*Kresge-Hooker Sci. Lib.*) **22,** 157 (1961).

664. Elliott, D. F., Horton, E. W., and Lewis, G. P., *J. Physiol.* (*London*) **150,** 6P (1960).

665. Elliott, D. F., Horton, E. W., and Lewis, G. P., *J. Physiol.* (*London*) **153,** 473 (1960).

666. Elliott, D. F., Horton, E. W., and Lewis, G. P., *Biochem. J.* **78,** 60 (1961).

666a. Elliott, D. F., and Lewis, G. P., *Biochem. J.* (in press).

667. Elliott, D. F., Lewis, G. P., and Horton, E. W., *Biochem. Biophys. Res. Commun.* **3,** 87 (1960).

668. Elliott, D. F., Lewis, G. P., and Horton, E. W., *Biochem. J.* **74,** 15P (1960).

669. Elliott, D. F., Lewis, G. P., and Horton, E. W., *Biochem. J.* **76,** 16P (1960).

670. Elliott, D. F., Lewis, G. P., and Smyth, D. G., *Biochem. J.* **87,** 21P (1963).

671. Elliott, D. F., and Morris, D., *Chimia* (*Aarau*) **14,** 373 (1960).

672. Elliott, D. F., and Peart, W. S., *Nature* **177,** 527 (1956).

673. Elliott, D. F., and Peart, W. S., *Biochem. J.* **65,** 246 (1957).

674. Elliott, D. F., and Russell, D. W., *Biochem. J.* **66,** 49P (1957).

675. Elmore, D. T., *J. Chem. Soc.* p. 3152 (1959).

676. Elmore, D. T., and Smyth, J., *Proc. Chem. Soc.* p. 18 (1963).

677. Emery, A. R., and Gold, V., *J. Chem. Soc.* p. 1443 (1950).

678. Emery, A. R., and Gold, V., *J. Chem. Soc.* p. 1447 (1950).

679. Emery, A. R., and Gold, V., *J. Chem. Soc.* p. 1455 (1950).

680. Engel, F. L., *Vitamins Hormones* **19,** 189 (1961).

681. Erdös, E. G., Renfrew, A. G., Sloane, E. M., and Wholer, J. R., *Ann. N. Y. Acad. Sci.* **104,** 222 (1963).

682. Erdös, E. G., and Sloane, E. M., *Biochem. Pharmacol.* **11,** 585 (1962).

682a. Erdös, E. G., Wohler, J. R., and Levine, M. I., *J. Pharmacol. Exptl. Therap.* **142,** 327 (1963).

682b. Erdös, E. G., and Wohler, J. R., *Biochem. Pharmacol.* **12,** 1193 (1963).

683. Erlanger, B. F., and Brand, E., *J. Am. Chem. Soc.* **73,** 3508 (1951).

684. Erlanger, B. F., and Brand, E., *J. Am. Chem. Soc.* **73,** 4025 (1951).

685. Erlanger, B. F., Curran, W. V., and Kokowsky, N., *J. Am. Chem. Soc.* **80,** 1128 (1958).

686. Erlanger, B. F., Curran, W. V., and Kokowsky, N., *J. Am. Chem. Soc.* **81,** 3051 (1959).

687. Erlanger, B. F., Curran, W. V., and Kokowsky, N., *J. Am. Chem. Soc.* **81,** 3055 (1959).

688. Erlanger, B. F., and Goode, L., *Nature* **174,** 840 (1954).

689. Erlanger, B. F., and Goode, L., *Science* **131,** 669 (1960).

690. Erlanger, B. F., and Hall, R. M., *J. Am. Chem. Soc.* **76,** 5781 (1954).

691. Erlanger, B. F., and Kokowsky, N., *J. Org. Chem.* **26,** 2534 (1961).

692. Erlanger, B. F., Sachs, H., and Brand, E., *J. Am. Chem. Soc.* **76,** 1806 (1954).

693. Erspamer, V., *Arzneimittel-Forsch.* **2,** 253 (1952).

694. Erspamer, V., *Ann. Rev. Pharmacol.* **1,** 175 (1961).

695. Erspamer, V., and Anastasi, A., *Experientia* **18,** 59 (1962).

695a. Erspamer, V., Anastasi, A., Bertaccini, G., and Cei, J. M., *Experientia* **20,** 489 (1964).

696. Erspamer, V., Bertaccini, G., and Cei, J. M., *Experientia* **18**, 562 (1962).
697. Erspamer, V., Bertaccini, G., and Cei, J. M., *Experientia* **18**, 563 (1962).
698. Erspamer, V., and Erspamer, G. F., *Brit. J. Pharmacol.* **19**, 337 (1962).
698a. Erspamer, V., and Glaesser, A., *Brit. J. Pharmacol.* **20**, 516 (1963).
699. Evans, R. L., and Irreverre, F., *J. Org. Chem.* **24**, 863 (1959).
700. Fahnenstich, R., Dissertation, Rheinisch-Westfälische Technische Hochschule Aachen, 1961.
701. Falkenburg, H. R., Dissertation, Rheinisch-Westfälische Technische Hochschule Aachen, 1959.
702. Farber, S., *Ciba Found. Symp. Amino Acids Peptides Antimetab. Activity, 1958* p. 138 (1959).
703. Farmer, T. H., *Biochem. J.* **73**, 321 (1959).
704. Farrington, J. A., Hextall, P. J., Kenner, G. W., and Turner, J. M., *J. Chem. Soc.* p. 1407 (1957).
705. Farrington, J. A., Kenner, G. W., and Turner, J. M., *Chem. Ind. (London)* p. 601 (1955).
705a. Fasciolo, J. C., *Acta Physiol. Latinoam.* **12**, 360 (1962).
706. Fasciolo, J. C., Halvorsen, K. A., Binia, A., and Itoiz, J. E., *Acta Physiol. Latinoam.* **10**, 23 (1960).
707. Fasciolo, J. C., Halvorsen, K., Zangheri, E., and Fernández, F. O., *Compt. Rend. Soc. Biol.* **153**, 500 (1959).
707a. Fasciolo, J. C., Muñoz, A., and Sosa, H., *Brit. J. Pharmacol.* **21**, 250 (1963).
708. Fasman, G. D., *Science* **131**, 420 (1960).
709. Fasman, G. D., and Blout, E. R., *J. Am. Chem. Soc.* **82**, 2262 (1960).
710. Faust, G., and Lange, H., *J. Prakt. Chem.* [4] **11**, 153 (1960).
711. Fedoseeva, N. V., Silaev, A. B., and Andreeva, L. I., *Zh. Obshch. Khim.* **33**, 1019 (1963).
712. Ferreira, S. H., and Rocha e Silva, M., *Biochem. Pharmacol.* **11**, 1123 (1962).
713. Ferren, R. A., Miller, J. G., and Day, A. R., *J. Am. Chem. Soc.* **79**, 70 (1957).
713a. Finn, F. M., and Hofmann, K., *J. Am. Chem. Soc.* **87**, 645 (1965).
714. Fischer, E., *Ber. Deut. Chem. Ges.* **34**, 433 (1901).
715. Fischer, E., *Ber. Deut. Chem. Ges.* **39**, 2893 (1906).
716. Fischer, E., *Ber. Deut. Chem. Ges.* **40**, 1754 (1907).
717. Fischer, E., *Ber. Deut. Chem. Ges.* **40**, 3704 (1907).
718. Fischer, E., *Ber. Deut. Chem. Ges.* **48**, 93 (1915).
719. Fischer, E., and Cone, L. H., *Ann. Chem. Liebigs* **363**, 107 (1908).
720. Fischer, E., and Lipschitz, W., *Ber. Deut. Chem. Ges.* **48**, 360 (1915).
721. Fischer, E., and Raske, K., *Ber. Deut. Chem. Ges.* **41**, 893 (1908).
722. Fischer, E., and Suzuki, U., *Ber. Deut. Chem. Ges.* **38**, 4173 (1905).
723. Fischer, E., and Warburg, O., *Ber. Deut. Chem. Ges.* **38**, 3997 (1905).
724. Fischer, R. F., and Whetstone, R. R., *J. Am. Chem. Soc.* **76**, 5076 (1954).
725. Fischer, R. F., and Whetstone, R. R., *J. Am. Chem. Soc.* **77**, 750 (1955).
726. Fisher, R. B., and Zachariah, P., *Biochem. J.* **76**, 155 (1960).
727. Fleischhauer, G., *Deut. Med. Wochschr.* **85**, 1717 (1960).
728. Fleš, D., Markovac-Prpić, A., and Tomašić, V., *J. Am. Chem. Soc.* **80**, 4654 (1958).
729. Fleš, D., Markovac-Prpić, A., Tomašić, V., and Milohnoja, M., *Croat. Chem. Acta* **30**, 167 (1958).

730. Fling, M., Minard, F. N., and Fox, S. W., *J. Am. Chem. Soc.* **69,** 2466 (1947).
731. Florsheim, H. A., Makineni, S., and Shankman, S., *Arch. Biochem. Biophys.* **97,** 243 (1962).
732. Flückiger, E., *Acta Endocrinol.* Suppl. 51, 333 (1960).
733. Flückiger, E., *Arch. Exptl. Pathol. Pharmakol.* **245,** 168 (1963).
734. Fodor, P. J., Miller, A., Neidle, A., and Waelsch, H., *J. Biol. Chem.* **203,** 991 (1953).
735. Folch, J., *J. Biol. Chem.* **174,** 439 (1948).
736. Folch, J., *J. Biol. Chem.* **177,** 497 (1949).
737. Folk, J. E., and Gladner, J. A., *J. Biol. Chem.* **231,** 379 (1958).
738. Fölsch, G., *Acta Chem. Scand.* **9,** 1039 (1955).
739. Fölsch, G., *Acta Chem. Scand.* **10,** 686 (1956).
740. Fölsch, G., *Acta Chem. Scand.* **12,** 561 (1958).
741. Fölsch, G., *Acta Chem. Scand.* **13,** 1407 (1959).
742. Fölsch, G., *Acta Chem. Scand.* **13,** 1422 (1959).
743. Fölsch, G., and Mellander, O., *Acta Chem. Scand.* **11,** 1232 (1957).
744. Fölsch, G., Mellander, O., and Strid, L., *Acta Chem. Scand.* **14,** 625 (1960).
745. Fölsch, G., and Österberg, R., *J. Biol. Chem.* **234,** 2298 (1959).
746. Fölsch, G., and Österberg, R., *Acta Chem. Scand.* **15,** 1963 (1961).
747. Fölsch, G., Ryhage, R., and Stenhagen, E., *Arkiv Kemi* **20,** 55 (1963).
748. Fölsch, G., and Serck-Hanssen, K., *Acta Chem. Scand.* **13,** 1243 (1959).
749. Fones, W. S., *J. Org. Chem.* **17,** 1661 (1952).
750. Fones, W. S., and Lee, M., *J. Biol. Chem.* **210,** 227 (1954).
751. Fong, C. T. O., Silver, L., Christman, D. R., and Schwartz, I. L., *Proc. Natl. Acad. Sci. U. S.* **46,** 1273 (1960).
751a. Fong, C. T. O., Silver, L., and Louie, D. D., *Biochem. Biophys. Res. Commun.* **14,** 302 (1964).
752. Forsham, P. H., DiRaimondo, V. C., Biglieri, E. G., and Li, C. H., *Metab. Clin. Exptl.* **10,** 335 (1961).
752a. Fox, S. W., Hayakawa, T., and Harada, K., *Bull. Chem. Soc. Japan* **36,** 1050 (1963).
753. Foye, W. O., and Verderame, M., *J. Am. Pharm. Assoc. Sci. Ed.* **46,** 273 (1957).
754. Fraenkel-Conrat, H., *Ann. Rev. Biochem.* **25,** 291 (1956).
755. Fraenkel-Conrat, H., and Harris, J. I., *J. Am. Chem. Soc.* **76,** 6058 (1954).
756. Fraenkel-Conrat, H., Harris, J. I., and Levy, A. L., *Methods Biochem. Anal.* **2,** 359 (1955).
757. Fraenkel-Conrat, J., and Fraenkel-Conrat, H., *Biochim. Biophys. Acta* **5,** 89 (1950).
758. Frank, W., and Zachau, H. G., *Z. Physiol. Chem.* **331,** 258 (1963).
759. Frankel, M., and Berger, A., *J. Org. Chem.* **16,** 1513 (1951).
760. Frankel, M., and Gertner, D., *J. Chem. Soc.* p. 459 (1961).
761. Frankel, M., and Gertner, D., *J. Chem. Soc.* p. 463 (1961).
762. Frankel, M., and Halmann, M., *J. Chem. Soc.* p. 2735 (1952).
763. Frankel, M., Liwschitz, Y., and Amiel, Y., *J. Am. Chem. Soc.* **75,** 330 (1953).
763a. Frankel, M., and Moses, P., *Tetrahedron* **9,** 289 (1960).
764. Franz, J., Boissonnas, R. A., and Stürmer, E., *Helv. Chim. Acta* **44,** 881 (1961).
765. Freudenberg, K., and Wegmann, T., *Z. Physiol. Chem.* **233,** 159 (1935).

766. Frey, E. K., Kraut, H., and Werle, E., "Kallikrein (Padutin)." Enke, Stuttgart 1950.
767. Frey, H., Meyer, H. J., and Nievergelt, J., *Schweiz. Med. Wochschr.* **92,** 785 (1962).
768. Frey, J., Kerp, L., and Reichardt, W., *Klin. Wochschr.* **37,** 325 (1959).
769. Fruton, J. S., *J. Biol. Chem.* **146,** 463 (1942).
770. Fruton, J. S., *J. Biol. Chem.* **165,** 333 (1946).
771. Fruton, J. S., *J. Am. Chem. Soc.* **70,** 1280 (1948).
772. Fruton, J. S., *Advan. Protein Chem.* **5,** 1 (1949).
773. Fruton, J. S., and Bergmann, M., *J. Biol. Chem.* **127,** 627 (1939).
774. Fruton, J. S., and Bergmann, M., *J. Biol. Chem.* **145,** 253 (1942).
775. Fruton, J. S., Hearn, W. R., Ingram, V. M., Wiggans, D. S., and Winitz, M., *J. Biol. Chem.* **204,** 891 (1953).
776. Fruton, J. S., Smith, V. A., and Driscoll, P. E., *J. Biol. Chem.* **173,** 457 (1948).
777. Fujiwara, S., Morinaga, S., and Narita, K., *Bull. Chem. Soc. Japan* **35,** 438 (1962).
778. Furuyama, T., Sakakibara, S., and Akabori, S., *Bull. Chem. Soc. Japan* **35,** 171 (1962).
799. Fusari, S. A., Haskell, T. H., Frohardt, R. P., and Bartz, Q. R., *J. Am. Chem. Soc.* **76,** 2881 (1954).
780. Gaddum, J. H., and Horton, E. W., *Brit. J. Pharmacol.* **14,** 117 (1959).
781. Gäumann, E., Roth, S., Ettlinger, L., Plattner, P. A., and Nager, U., *Experientia* **3,** 202 (1947).
782. Gause, G. F., and Brashnikova, M. G., *Lancet* II, 715 (1944).
783. Gause, G. F., Brashnikova, M. G., and Lissovskaya, N. P., *Am. Rev. Soviet Med.* **2,** 134 (1944).
784. Gavrilov, N. I., Grigor'eva, I. P., Akimova, L. N., and Erokhin, V. K., *Zh. Obshch. Khim.* **31,** 739 (1961).
785. Gawron, O., and Draus, F., *J. Org. Chem.* **23,** 1040 (1958).
786. Gawron, O., and Draus, F., *J. Org. Chem.* **24,** 1392 (1959).
786a. Gazis, E., Bezas, B., Stelekatos, G. C., and Zervas, L., *in* "Peptide Symposium" (G. T. Young, ed.), p. 17. Macmillan (Pergamon), New York, 1963.
787. Geiger, R., Sturm, K., and Siedel, W., *Chem. Ber.* **96,** 1080 (1963).
787a. Geiger, R., Sturm, K., Vogel, G., and Siedel, W., *Z. Naturforsch.* **19b,** 858 (1964).
787b. Geiger, R., Sturm, K., and Siedel, W., *Chem. Ber.* **97,** 1207 (1964).
788. Gersmeyer, E. F., and Spitzbarth, H., *Klin. Wochschr.* **39,** 1227 (1961).
789. Gersmeyer, E. F., Spitzbarth, H., and Weyland, H., *Klin. Wochschr.* **36,** 1061 (1958).
790. Geschwind, I. I., *in* "Columbia University Conference on Comparative Endocrinology," p. 421. Wiley, New York, 1958.
791. Geschwind, I. I., and Li, C. H., *Biochim. Biophys. Acta* **15,** 442 (1954).
792. Geschwind, I. I., and Li, C. H., *J. Am. Chem. Soc.* **79,** 615 (1957).
793. Geschwind, I. I., Li, C. H., and Barnafi, L., *J. Am. Chem. Soc.* **78,** 4494 (1956).
794. Geschwind, I. I., Li, C. H., and Barnafi, L., *J. Am. Chem. Soc.* **79,** 620 (1957).
795. Geschwind, I. I., Li, C. H., and Barnafi, L., *J. Am. Chem. Soc.* **79,** 1003 (1957).

796. Geschwind, I. I., Li, C. H., and Barnafi, L., *J. Am. Chem. Soc.* **79,** 6394 (1957).

797. Gibian, H., *Z. Naturforsch.* **16b,** 18 (1961).

798. Gibian, H., and Klieger, E., *Angew. Chem.* **72,** 708 (1960).

799. Gibian, H., and Klieger, E., *Ann. Chem. Liebigs* **640,** 145 (1961).

800. Gibian, H., and Lübke, K., *Chimia (Aarau)* **14,** 380 (1960).

801. Gibian, H., and Lübke, K., *Angew. Chem.* **72,** 523 (1960).

802. Gibian, H., and Lübke, K., *Ann. Chem. Liebigs* **644,** 130 (1961).

803. Gibian, H., and Schröder, E., *Ann. Chem. Liebigs* **642,** 145 (1961).

803a. Gillessen, D., Schnabel, E., and Meienhofer, J., *Ann. Chem. Liebigs* **667,** 164 (1963).

803b. Gillessen, D., Dissertation, Rheinisch-Westfälische Technische Hochschule Aachen, 1963.

804. Gish, D. T., and Carpenter, F. H., *J. Am. Chem. Soc.* **75,** 950 (1953).

805. Gish, D. T., and Carpenter, F. H., *J. Am. Chem. Soc.* **75,** 5872 (1953).

806. Gish, D. T., and du Vigneaud, V., *J. Am. Chem. Soc.* **79,** 3579 (1957).

807. Gish, D. T., Katsoyannis, P. G., Hess, G. P., and Stedman, R. J., *J. Am. Chem. Soc.* **78,** 5954 (1956).

808. Gjuriš, V., Heicke, B., Holtz, P., and Westermann, E., *Experientia* **18,** 385 (1962).

809. Gjuriš, V., and Westermann, E., *Arch. Exptl. Pathol. Pharmakol.* **245,** 285 (1963).

810. Gmelin, R., Strauss, G., and Hasenmaier, G., *Z. Physiol. Chem.* **314,** 28 (1959).

811. Gnichtel, H., and Lautsch, W., *Angew. Chem.* **73,** 29 (1961).

812. Goerdeler, J., and Holst, A., *Angew. Chem.* **71,** 775 (1959).

813. Goldblatt, H., *Physiol. Rev.* **27,** 120 (1947).

814. Goldschmidt, S., *Collection Czech. Chem. Commun.* **24,** 15 (1959).

815. Goldschmidt, S., and Jutz, C., *Chem. Ber.* **86,** 1116 (1953).

816. Goldschmidt, S., and Jutz, C., *Chem. Ber.* **89,** 518 (1956).

817. Goldschmidt, S., and Kraus, H. L., *Angew. Chem.* **67,** 471 (1955).

818. Goldschmidt, S., and Lautenschlager, H., *Ann. Chem. Liebigs* **580,** 68 (1953).

818a. Goldschmidt, S., Lautenschlager, W., Kolb, B., and Zumach, G., *Chem. Ber.* **97,** 2434 (1964).

819. Goldschmidt, S., and Obermeier, F., *Ann. Chem. Liebigs* **588,** 24 (1954).

820. Goldschmidt, S., and Rosculet, G., *Chem. Ber.* **93,** 2387 (1960).

821. Goldschmidt, S., and Rosculet, G., *Chimia (Aarau)* **14,** 367 (1960).

822. Goldschmidt, S., and Wick, M., *Z. Naturforsch.* **5b,** 170 (1950).

823. Goldschmidt, S., and Wick, M., *Ann. Chem. Liebigs* **575,** 217 (1952).

824. Golubow I., and du Vigneaud, V., *Proc. Soc. Exptl. Biol. Med.* **112,** 218 (1963).

825. Gomes, F. P., *Brit. J. Pharmacol.* **10,** 200 (1955).

826. Gomes, F. P., *Compt. Rend. Soc. Biol.* **151,** 812 (1957).

827. Gonin, J., Baudet, P., and Cherbuliez, E., *Pharm. Acta Helv.* **37,** 425 (1962).

827a. Goodman, M., and Boardman, F., *J. Am. Chem. Soc.* **85,** 2483 (1963).

827b. Goodman, M., Boardman, F., and Listowsky, I., *J. Am. Chem. Soc.* **85,** 2491 (1963).

828. Goodman, M., Deber, C. M., and Felix, A. M., *J. Am. Chem. Soc.* **84,** 3773 (1962).

829. Goodman, M., and Kenner, G. W., *Advan. Protein Chem.* **12,** 465 (1957).
830. Goodman, M., Listowsky, I., and Schmitt, E. E., *J. Am. Chem. Soc.* **84,** 1296 (1962).
831. Goodman, M., and Schmitt, E. E., *J. Am. Chem. Soc.* **81,** 5507 (1959).
832. Goodman, M., and Schmitt, E. E., *Ann. N. Y. Acad. Sci.* **88,** 669 (1960).
833. Goodman, M., Schmitt, E. E., and Yphantis, D. A., *J. Am. Chem. Soc.* **82,** 3483 (1960).
834. Goodman, M., Schmitt, E. E., and Yphantis, D. A., *J. Am. Chem. Soc.* **84,** 1283 (1962).
835. Goodman, M., Schmitt, E. E., and Yphantis, D. A., *J. Am. Chem. Soc.* **84,** 1288 (1962).
836. Goodman, M., and Stueben, K. C., *J. Org. Chem.* **24,** 112 (1959).
837. Goodman, M., and Stueben, K. C., *J. Am. Chem. Soc.* **81,** 3980 (1959).
838. Goodman, M., and Stueben, K. C., *J. Am. Chem. Soc.* **84,** 1279 (1962).
839. Goodman, M., and Stueben, K. C., *J. Org. Chem.* **27,** 3409 (1962).
840. Gordon, M., Miller, J. G., and Day, A. R., *J. Am. Chem. Soc.* **70,** 1946 (1948).
841. Gordon, M., Miller, J. G., and Day, A. R., *J. Am. Chem. Soc.* **71,** 1245 (1949).
842. Grady, J. E., Koffler, H., Parsons, J. L., Kobayashi, T., and Tetrault, P. A., *Federation Proc.* **17,** 233 (1958).
843. Grassmann, W., Dyckerhoff, H., and Eibeler, H., *Z. Physiol. Chem.* **189,** 112 (1930).
844. Grassmann, W., Hannig, K., Endres, H., and Riedel, A., *Z. Physiol. Chem.* **306,** 123 (1957).
845. Grassmann, W., Hannig, K., and Schleyer, M., *Z. Physiol. Chem.* **322,** 71 (1960).
846. Grassmann, W., Nordwig, A., and Hörmann, H., *Z. Physiol. Chem.* **323,** 48 (1961).
847. Grassmann, W., and Schneider, F., *Biochem. Z.* **273,** 452 (1934).
848. Grassmann, W., and Schulte-Uebbing, E., *Chem. Ber.* **83,** 244 (1950).
849. Grassmann, W., and Wünsch, E., *Chem. Ber.* **91,** 449 (1958).
850. Grassmann, W., and Wünsch, E., *Chem. Ber.* **91,** 462 (1958).
851. Grassmann, W., Wünsch, E., Deufel, P., and Zwick, A., *Chem. Ber.* **91,** 538 (1958).
852. Grassmann, W., Wünsch, E., and Riedel, A., *Chem. Ber.* **91,** 455 (1958).
853. Green, A. A., and Bumpus, F. M., *J. Biol. Chem.* **210,** 281 (1954).
854. Green, M., and Stahmann, M. A., *J. Biol. Chem.* **197,** 771 (1952).
855. Greenbaum, L. M., and Hosoda, T., *Biochem. Pharmacol.* **12,** 325 (1963).
856. Greenstein, J. P., *J. Biol. Chem.* **118,** 321 (1937).
857. Greenstein, J. P., *J. Biol. Chem.* **121,** 9 (1937).
858. Greenstein, J. P., *J. Biol. Chem.* **124,** 255 (1938).
859. Greenstein, J. P., *J. Biol. Chem.* **128,** 241 (1939).
860. Greenstein, J. P., and Klemperer, F. W., *J. Biol. Chem.* **128,** 245 (1939).
861. Greenstein, J. P., and Winitz, M., "Chemistry of the Amino Acids," Wiley, New York, 1960.
862. Gregory, J. D., and Craig, L. C., *J. Biol. Chem.* **172,** 839 (1948).
862a. Gregory, R. A., and Tracy, H. J., *Gut* **5,** 103 (1964).
862b. Gregory, H., Hardy, P. M., Jones, D. S., Kenner, G. W., and Sheppard, R. C., *Nature* **204,** 931 (1964).

863. Grindeland, R. E., Wherry, F. E., and Anderson, E., *Proc. Soc. Exptl. Biol. Med.* **110,** 377 (1962).
864. Grommers, E. P., and Arens, J. F., *Rec. Trav. Chim.* **78,** 558 (1959).
865. Gros, C., and Privat de Garilhe, M., *Compt. Rend.* **249,** 2234 (1959).
866. Gros, C., Privat de Garilhe, M., Costopanagiotis, A., and Schwyzer, R., *Helv. Chim. Acta* **44,** 2042 (1961).
867. Gross, F., *Klin. Wochschr.* **36,** 693 (1958).
868. Gross, F., *Arch. Exptl. Pathol. Pharmakol.* **245,** 196 (1963).
869. Gross, F., and Bock, K. D., *Circulation* **25,** 193 (1962).
870. Gross, F., Bock, K. D., and Turrian, H., *Helv. Physiol. Pharmacol. Acta* **19,** 42 (1961).
871. Gross, F., and Lichtlen, P., *Arch. Exptl. Pathol. Pharmakol.* **233,** 323 (1958).
872. Gross, F., and Turrian, H., *in* "Polypeptides which Affect Smooth Muscles and Blood Vessels" (M. Schachter, ed.), p. 137. Macmillan (Pergamon), New York, 1960.
873. Gründig, E., and Pantlitschko, N., *Monatsh. Chem.* **90,** 891 (1959).
874. Guhl, U., *Schweiz. Med. Wochschr.* **91,** 798 (1961).
875. Guillemin, R., *Endocrinology* **66,** 819 (1960).
876. Guillemin, R., *Pathol. Biol. Semaine Hop.* **9,** 2143 (1961).
877. Guillemin, R., *Symp. Biologisch Aktive Polypeptide, Vienna, 1962.* p. 11. Wien. med Akad. ärztl. Fortbildung, Vienna, 1962.
878. Guillemin, R., *Arch. Exptl. Pathol. Pharmakol.* **245,** 187 (1963).
878a. Guillemin, R., *Recent Progr. Hormone Res.* **20,** 89 (1964).
879. Guillemin, R., Clayton, G. W., Smith, J. D., and Lipscomb, H. S., *Endocrinology* **63,** 349 (1958).
880. Guillemin, R., Dear, W. E., Nichols, B., Jr., and Lipscomb, H. S., *Proc. Soc. Exptl. Biol. Med.* **101,** 107 (1959).
881. Guillemin, R., Jutisz, M., and Sakiz, E., *Compt. Rend.* **256,** 504 (1963).
882. Guillemin, R., and Krivoy, W. A., *Compt. Rend.* **250,** 1117 (1960).
883. Guillemin, R., and Schally, A. V., *Endocrinology* **65,** 555 (1959).
884. Guillemin, R., Schally, A., Andersen, R., Lipscomb, H., and Long, J., *Compt. Rend.* **250,** 4462 (1960).
885. Guillemin, R., Yamazaki, E., Jutisz, M., and Sakiz, E., *Compt. Rend.* **255,** 1018 (1962).
886. Günther, D., Dissertation, Freie Universität Berlin, 1958.
886a. Gut, V., and Rudinger, J., *Collection Czech. Chem. Commun.* **28,** 2953 (1963).
887. Guth, P. S., *Brit. J. Pharmacol.* **14,** 549 (1959).
888. Gutmann, H. R., and Chang, S. F., *J. Org. Chem.* **27,** 2248 (1962).
889. Guttmann, S., *Chimia (Aarau)* **14,** 368 (1960).
890. Guttmann, S., *Helv. Chim. Acta* **44,** 721 (1961).
891. Guttmann, S., *Helv. Chim. Acta* **45,** 2622 (1962).
892. Guttmann, S., *in* "Peptide Symposium," Oxford 1962 (G. T. Young, ed.), p. 41. Macmillan (Pergamon), New York, 1963.
893. Guttmann, S., Berde, B., and Stürmer, E., *Experientia* **18,** 445 (1962).
894. Guttmann, S., and Boissonnas, R. A., *Helv. Chim. Acta* **41,** 1852 (1958).
895. Guttmann, S., and Boissonnas, R. A., *Helv. Chim. Acta* **42,** 1257 (1959).
896. Guttmann, S., and Boissonnas, R. A., *Helv. Chim. Acta* **43,** 200 (1960).
897. Guttmann, S., and Boissonnas, R. A., *Experientia* **17,** 265 (1961).

898. Guttmann, S., and Boissonnas, R. A., *Angew. Chem.* **73,** 738 (1961).
899. Guttmann, S., and Boissonnas, R. A., *Helv. Chim. Acta* **44,** 1713 (1961).
899a. Guttmann, S., and Boissonnas, R. A., *Helv. Chim. Acta* **46,** 1626 (1963).
900. Guttmann, S., and Boissonnas, R. A., *Helv. Chim. Acta* **45,** 2517 (1962).
901. Guttmann, S., Jaquenoud, P.-A., Boissonnas, R. A., Konzett, H., and Berde, B., *Naturwissenschaften* **44,** 632 (1958).
901a. Guttmann, S., and Pless, J., *Chimia (Aarau)* **18,** 185 (1964).
901b. Guttmann, S., and Pless, J., Proc. 7th European Symp. on Peptides, Budapest, 1964, *Acta Chim. Acad. Sci. Hung.* (in press).
902. Guttmann, S., Pless, J., and Boissonnas, R. A., *Helv. Chim. Acta* **45,** 170 (1962).
902a. Guttmann, S., Pless, J., and Boissonnas, R. A., Proc. 7th European Symp. on Peptides, Budapest, 1964, *Acta Chim. Acad. Sci. Hung.* (in press).
903. Haas, H. J., *Chem. Ber.* **94,** 2442 (1961).
904. Haas, W. L., Dissertation, Rheinisch-Westfälische Technische Hochschule Aachen, 1959.
905. Habermann, E., *Arch. Exptl. Pathol. Pharmakol.* **236,** 492 (1959).
906. Habermann, E., *Naturwissenschaften* **47,** 111 (1960).
907. Habermann, E., *Arch. Exptl. Pathol. Pharmakol.* **240,** 552 (1961).
908. Habermann, E., *Arch. Exptl. Pathol. Pharmakol.* **241,** 202 (1961).
909. Habermann, E., *Arch. Exptl. Pathol. Pharmakol.* **241,** 543 (1961).
910. Habermann, E., *Z. Physiol. Chem.* **328,** 15 (1962).
911. Habermann, E., *Z. Physiol. Chem.* **328,** 24 (1962).
912. Habermann, E., *Ann. N. Y. Acad. Sci.* **104,** 130 (1963).
913. Habermann, E., *Arch. Exptl. Pathol. Pharmakol.* **245,** 230 (1963).
913a. Habermann, E., *Biochem. Z.* **337,** 440 (1963).
913b. Habermann, E., Klett, W., and Rosenbusch, G., *Z. Physiol. Chem.* **332,** 121 (1963).
914. Habermann, E., and Okon, M. E., *Z. Physiol. Chem.* **326,** 171 (1961).
915. Habermann, E., and Rosenbusch, G., *Arch. Exptl. Pathol. Pharmakol.* **243,** 357 (1962).
916. Hacker, E., Dissertation, Ludwig-Maximilian-Universität München, 1959.
917. Hackmann, Chr., *Deut. Med. Wochschr.* **80,** 812 (1955).
918. Haefely, W., and Hürlimann, A., *Experientia* **18,** 297 (1962).
919. Hajós, A., *Chem. Ber.* **94,** 2350 (1961).
920. Halvorsen, K. A., Fasciolo, J. C., Calvo, R., and Chionetti, I., *Compt. Rend. Soc. Biol.* **153,** 496 (1959).
921. Hamberg, U., *Biochim. Biophys. Acta* **34,** 135 (1959).
922. Hamberg, U., *Biochim. Biophys. Acta* **36,** 296 (1959).
923. Hamberg, U., *Suomen Kemistilehti* **34B,** 168 (1961).
924. Hamberg, U., *Biochem. Biophys. Res. Commun.* **7,** 95 (1962).
925. Hamberg, U., *Ann. Acad. Sci. Fennicae; Ser. AII* **113** (1962).
926. Hamberg, U., Bumpus, F. M., and Page, I. H., *Biochim. Biophys. Acta* **52,** 533 (1961).
927. Hamberg, U., and Deutsch, H. F., *Arch. Biochem. Biophys.* **76,** 262 (1958).
928. Hamberg, U., and Rocha e Silva, M., *Acta Physiol. Scand.* **30,** 215 (1954).
929. Hamberg, U., and Rocha e Silva, M., *Experientia* **13,** 489 (1957).
930. Hamberg, U., and Rocha e Silva, M., *Arch. Intern. Pharmacodyn.* **110,** 222 (1957).
931. Hanby, W. E., Waley, S. G., and Watson, J., *J. Chem. Soc.* p. 3239 (1950).

932. Haneke, H., Dissertation, Westfälische Wilhelms-Universität Münster, 1960.
932a. Hano, K., Koida, M., Kubo, K., and Yajima, H., *Biochim. Biophys. Acta* **90,** 204 (1964).
933. Hanson, H., and Illhardt, R., *Z. Physiol. Chem.* **298,** 210 (1954).
934. Hanson, H., and Kirschke, H., *Z. Vitamin-Hormon- Fermentforsch.* **11,** 343 (1960–61).
935. Hanson, H. T., and Smith, E. L., *J. Biol. Chem.* **175,** 833 (1948).
935a. Hanson, H. T., and Smith, E. L., *J. Biol. Chem.* **179,** 789 (1949).
935b. Hanson, R. W., and Rydon, H. N., *J. Chem. Soc.* p. 836 (1964).
936. Hardy, P. M., Kenner, G. W., and Sheppard, R. C., *Tetrahedron* **19,** 95 (1963).
937. Harington, C. R., and Mead, T. H., *Biochem. J.* **29,** 1602 (1935).
938. Harington, C. R., and Mead, T. H., *Biochem. J.* **30,** 1598 (1936).
939. Harington, C. R., and Moggridge, R. C. G., *J. Chem. Soc.* p. 706 (1940).
940. Harington, C. R., and Pitt-Rivers, R. V., *Biochem. J.* **38,** 417 (1944).
941. Harrington, W. F., and von Hippel, P. H., *Advan. Protein Chem.* **16,** 1 (1961).
942. Harris, G., and MacWilliam, I. C., *J. Chem. Soc.* p. 2053 (1961).
943. Harris, J. I., *Symp. Protein Struct., Paris, 1957* p. 333 (1958).
944. Harris, J. I., *Biochem. J.* **71,** 451 (1959).
945. Harris, J. I., *Nature* **184,** 167 (1959).
946. Harris, J. I., *Ciba Found. Colloq. Endocrinol.* **13,** 266 (1960).
947. Harris, J. I., *Brit. Med. Bull.* **16,** 189 (1960).
948. Harris, J. I., Cole, R. D., and Pon, N. G., *Biochem. J.* **62,** 154 (1956).
949. Harris, J. I., and Fruton, J. S., *J. Biol. Chem.* **191,** 143 (1951).
950. Harris, J. I., and Lerner, A. B., *Nature* **179,** 1346 (1957).
951. Harris, J. I., and Li, C. H., *J. Am. Chem. Soc.* **74,** 2945 (1952).
952. Harris, J. I., and Li, C. H., *J. Am. Chem. Soc.* **76,** 3607 (1954).
953. Harris, J. I., and Li, C. H., *J. Biol. Chem.* **213,** 499 (1955).
954. Harris, J. I., and Roos, P., *Nature* **178,** 90 (1956).
955. Harris, J. I., and Roos, P., *Biochem. J.* **71,** 434 (1959).
955a. Harris, J. I., and Roos, P., *Biochem. J.* **71,** 445 (1959).
956. Harris, J. I., Sanger, F., and Naughton, M. A., *Arch. Biochem. Biophys.* **65,** 427 (1956).
957. Harris, J. I., and Work, T. S., *Biochem. J.* **46,** 196 (1950).
958. Harris, J. I., and Work, T. S., *Biochem. J.* **46,** 582 (1950).
959. Haskell, T. H., Fusari, S. A., Frohardt, R. P., and Bartz, Q. R., *J. Am. Chem. Soc.* **74,** 599 (1952).
960. Haskell, T. H., Maretzki, A., and Bartz, Q. R., *Antibiot. Ann.* p. 784 (1954–1955).
961. Hassall, C. H., and John, D. I., *J. Chem. Soc.* p. 4112 (1960).
962. Hatanaka, S.-I., and Virtanen, A. I., *Acta Chem. Scand.* **16,** 514 (1962).
963. Hausmann, W., *J. Am. Chem. Soc.* **78,** 3663 (1956).
964. Hausmann, W., Weisiger, J. R., and Craig, L. C., *J. Am. Chem. Soc.* **77,** 721 (1955).
965. Hausmann, W., Weisiger, J. R., and Craig, L. C., *J. Am. Chem. Soc.* **77,** 723 (1955).
966. Häussler, R., *Chimia (Aarau)* **14,** 369 (1960).
966a. Havinga, E., Schattenkerk, C., Visser, G. H., and Kerling, K. E. T., *Rec. Trav. Chim.* **83,** 672 (1964).

967. Hayakawa, T., Nishi, N., Noguchi, J., Ikeda, S., Yamashita, T., and Isemura, T., *Nippon Kagaku Zasshi* **82**, 601 (1961).
968. Hays, E. E., and White, W. F., *Recent Progr. Hormone Res.* **10**, 265 (1954).
969. Hearn, W. R., and Hendry, R. A., *J. Am. Chem. Soc.* **79**, 5213 (1957).
970. Hearn, W. R., Weber, E. J., Woodward Randolph, P., and Barks, N. E., *Proc. Soc. Exptl. Biol. Med.* **107**, 515 (1961).
971. Heaton, G. S., Rydon, H. N., and Schofield, J. A., *J. Chem. Soc.* p. 3157 (1956).
972. Hegedüs, B., *Helv. Chim. Acta* **31**, 737 (1948).
973. Heidenreich, O., *Deut. Med. Wochschr.* **86**, 674 (1961).
974. Heidenreich, O., Kook, Y., Ling, V., and Menzel, H., *Arch. Exptl. Pathol. Pharmakol.* **239**, 328 (1960).
975. Heidenreich, O., Kook, Y., Ling, V., and Menzel, H., *Arch. Exptl. Pathol. Pharmakol.* **239**, 336 (1960).
976. Heilbron, I., *J. Chem. Soc.* p. 2099 (1949).
977. Heinemann, B., Gourevitch, A., Lein, J., Johnson, D. L., Kaplan, M. A., Vanas, D., and Hooper, I. R., *Antibiot. Ann.* p. 728 (1954–1955).
978. Helferich, B., and Böshagen, H., *Chem. Ber.* **92**, 2813 (1959).
979. Helferich, B., Schellenberg, P., and Ullrich, J., *Chem. Ber.* **90**, 700 (1957).
980. Heller, H., *in* "Oxytocin" (R. Caldeyro-Barcia and H. Heller, eds.), p. 3. Macmillan (Pergamon), New York, 1961.
981. Heller, H., *in* "Polypeptides which Affect Smooth Muscles and Blood Vessels" (M. Schachter, ed.), p. 59. Macmillan (Pergamon), New York, 1960.
982. Heller, H., *Arch. Exptl. Pathol. Pharmakol.* **245**, 142 (1963).
982a. Heller, H., *in* "Comparative Endocrinology" (U. S. von Euler and H. Heller, eds.), Vol. 1, p. 26. Academic Press, New York, 1963
983. Heller, H., and Pickering, B. T., *J. Physiol. (London)* **152**, 56P (1960).
984. Heller, H., and Pickering, B. T., *J. Physiol. (London)* **155**, 98 (1961).
985. Heller, H., Pickering, B. T., Maetz, J., and Morel, F., *Nature* **191**, 670 (1961).
986. Helmer, O. M., *Proc. Soc. Exptl. Biol. Med.* **74**, 642 (1950).
987. Helmer, O. M., *Federation Proc.* **15**, 273 (1956).
988. Helmer, O. M., *Am. J. Physiol.* **188**, 571 (1957).
988a. Helmer, O. M., and Sanders, R. M., *Am. J. Physiol.* **207**, 368 (1964).
989. Helmer, R., Tripod, J., and Studer, A., *Arch. Intern. Pharmacodyn.* **117**, 185 (1958).
990. Henriques, O. B., Fichman, M., and Beraldo, W., *Nature* **187**, 414 (1960).
991. Henriques, O. B., Picarelli, Z. P., and Ferraz de Oliveira, M. C., *Biochem. Pharmacol.* **11**, 707 (1962).
992. Heslinga, L., and Arens, J. F., *Rec. Trav. Chim.* **76**, 982 (1957).
992a. Hess, H.-J., Moreland, W. T., and Laubach, G. D., *J. Am. Chem. Soc.* **85**, 4040 (1963).
992b. Hess, H.-J., and Constantine, J. W., *J. Med. Chem.* **7**, 602 (1964).
993. Heyboer, N., Visser, G. H. and Kerling, K. E. T., *Rec. Trav. Chim.* **81**, 69 (1962).
994. Heyns, K., and Breuer, H., *Chem. Ber.* **91**, 2750 (1958).
995. Heyns, K., and Legler, G., *Z. Physiol. Chem.* **325**, 288 (1959).
996. Heyns, K., and Legler, G., *Z. Physiol. Chem.* **321**, 161 (1960).
997. Heyns, K., and Legler, G., *Z. Physiol. Chem.* **321**, 184 (1960).

998. Heyns, K., and Noack, H., *Chem. Ber.* **95,** 720 (1962).
998a. Heyns, K., and Noack, H., *Chem. Ber.* **97,** 415 (1964).
999. Heyns, K., and Rolle, M., *Chem. Ber.* **92,** 2439 (1959).
1000. Heyns, K., and Schulz, W., *Chem. Ber.* **93,** 128 (1950).
1001. Heyns, K., and Schulz, W., *Chem. Ber.* **95,** 709 (1962).
1002. Heyns, K., Walter, W., and Müller, F., *Angew. Chem.* **68,** 617 (1956).
1003. Hietala, P. K., *Acta Chem. Scand.* **14,** 212 (1960).
1004. Hill, R. L., and Smith, E. L., *Biochim. Biophys. Acta* **31,** 257 (1959).
1005. Hillmann, A., and Hillmann, G., *Z. Naturforsch.* **6b,** 340 (1951).
1006. Hillmann-Elies, A., Hillmann, G., and Jatzkewitz, H., *Z. Naturforsch.* **8b,** 445 (1953).
1007. Hillmann, G., *Z. Naturforsch.* **1,** 682 (1946).
1008. Hilton, S. M., and Lewis, G. P., *Brit. Med. Bull.* **13,** 189 (1957).
1009. Hinman, J. W., Caron, E. L., and Christensen, H. N., *J. Am. Chem. Soc.* **72,** 1620 (1950).
1010. Hiramatsu, S., *Osaka Shiritsu Daigaku Igaku Zasshi* **8,** 2017 (1959).
1011. Hirs, C. H. W., Stein, W. H., and Moore, S., *J. Biol. Chem.* **221,** 151 (1956).
1012. Hiskey, R. G., and Tucker, W. P., *J. Am. Chem. Soc.* **84,** 4789 (1962).
1013. Hiskey, R. G., and Tucker, W. P., *J. Am. Chem. Soc.* **84,** 4794 (1962).
1014. Hodgkin, D. C., and Oughton, B. M., *Biochem. J.* **65,** 752 (1957).
1015. Hoffmann, E., and Schiff-Shenhav, H., *J. Org. Chem.* **27,** 4686 (1962).
1016. Hofmann, K., *Brookhaven Symp. Biol.* **13,** 184 (1960).
1017. Hofmann, K., *Ann. N. Y. Acad. Sci.* **88,** 689 (1960).
1018. Hofmann, K., *Ann. Rev. Biochem.* **31,** 213 (1961).
1018a. Hofmann, K., *Pure Appl. Chem.* **6,** 245 (1963).
1019. Hofmann, K., and Bergmann, M., *J. Biol. Chem.* **130,** 81 (1939).
1020. Hofmann, K., and Bergmann, M., *J. Biol. Chem.* **134,** 225 (1940).
1020a. Hofmann, K., Finn, F., Haas, W., Smithers, M. J., Wolman, Y., and Yanaihara, N., *J. Am. Chem. Soc.* **85,** 833 (1963).
1020b. Hofmann, K., Haas, W., Smithers, M. J., Wells, R. D., Wolman, Y., Yanaihara, N., and Zanetti, G., *J. Am. Chem. Soc.* **87,** 620 (1965).
1020c. Hofmann K., Haas, W., Smithers, M. J., and Zanetti, G., *J. Am. Chem. Soc.* **87,** 631 (1965).
1021. Hofmann, K., Jöhl, A., Furlenmeier, A. E., and Kappeler, H., *J. Am. Chem. Soc.* **79,** 1636 (1957).
1022. Hofmann, K., Kappeler, H., Furlenmeier, A. E., Woolner, M. E., Schwartz, E. T.,. and Thompson, T. A., *J. Am. Chem. Soc.* **79,** 1641 (1957).
1022a. Hofmann, K., and Katsoyannis, P. G., *in* "The Proteins" (H. Neurath ed.), 2nd ed., Vol. 1, p. 53. Academic Press, New York, 1963.
1023. Hofmann, K., and Lande, S., *J. Am. Chem. Soc.* **83,** 2286 (1961).
1024. Hofmann, K., Lindenmann, A., Magee, M. Z., and Kahn, N. H., *J. Am. Chem. Soc.* **74,** 470 (1952).
1025. Hofmann, K., Liu, T.-Y., Yajima, H., Yanaihara, N., and Lande, S., *J. Am. Chem. Soc.* **83,** 2294 (1961).
1026. Hofmann, K., Liu, T.-Y., Yajima, H., Yanaihara, N., Yanaihara, C., and Humes, J. L., *J. Am. Chem. Soc.* **84,** 1054 (1962).
1027. Hofmann, K., Magee, M. Z., and Lindenmann, A., *J. Am. Chem. Soc.* **72,** 2814 (1950).
1028. Hofmann, K., Peckham, W. D., and Rheiner, A., *J. Am. Chem. Soc.* **78,** 238 (1956).

1029. Hofmann, K., Rheiner, A., and Peckham, W. D., *J. Am. Chem. Soc.* **75,** 6083 (1953).

1029a. Hofmann, K., Rosenthaler, J., Wells, R. D., and Yajima, H., *J. Am. Chem. Soc.* **86,** 4991 (1964).

1029b. Hofmann, K., Schmiechen, R., Smithers, M. J., Wells, R. D., Wolman, Y., and Zanetti, G., *J. Am. Chem. Soc.* **87,** 640 (1965).

1029c. Hofmann, K., Schmiechen, R., Wells, R. D., Wolman, Y., and Yanaihara, N., *J. Am. Chem. Soc.* **87,** 611 (1965).

1030. Hofmann, K., Stutz, E., Spühler, G., Yajima, H., and Schwartz, E. T., *J. Am. Chem. Soc.* **82,** 3727 (1960).

1031. Hofmann, K., Thompson, T. A., and Schwartz, E. T., *J. Am. Chem. Soc.* **79,** 6087 (1957).

1032. Hofmann, K., Thompson, T. A., Woolner, M. E., Spühler, G., Yajima, H., Cipera, J. D., and Schwartz, E. T., *J. Am. Chem. Soc.* **82,** 3721 (1960).

1033. Hofmann K., Thompson, T. A., Yajima, H., Schwartz, E. T., and Inouye, H., *J. Am. Chem. Soc.* **82,** 3715 (1960).

1034. Hofmann, K., Wells, R. D., Yajima, H., and Rosenthaler, J., *J. Am. Chem. Soc.* **85,** 1546 (1963).

1035. Hofmann, K., Woolner, M. E., Spühler, G., and Schwartz, E. T., *J. Am. Chem. Soc.* **80,** 1486 (1958).

1036. Hofmann, K., Woolner, M. E., Yajima, H., Spühler, G., Thompson, T. A., and Schwartz, E. T., *J. Am. Chem. Soc.* **80,** 6458 (1958).

1037. Hofmann, K., and Yajima, H., *J. Am. Chem. Soc.* **83,** 2289 (1961).

1038. Hofmann, K., and Yajima, H., *in* "Polyamino Acids Polypeptides and Proteins" (M. A. Stahmann, ed.), p. 21. Univ. of Wisconsin Press, Madison, Wisconsin, 1962.

1039. Hofmann, K., and Yajima, H., *Recent Progr. Hormone Res.* **18,** 41 (1962).

1040. Hofmann, K., Yajima, H., Liu, T.-Y., and Yanaihara, N., *J. Am. Chem. Soc.* **84,** 4475 (1962).

1041. Hofmann, K., Yajima, H., Liu, T.-Y., Yanaihara, N., Yanaihara, C., and Humes, J. L., *J. Am. Chem. Soc.* **84,** 4481 (1962).

1042. Hofmann, K., Yajima, H., and Schwartz, E. T., *Biochim. Biophys. Acta* **36,** 252 (1959).

1043. Hofmann, K., Yajima, H., and Schwartz, E. T., *J. Am. Chem. Soc.* **82,** 3732 (1960).

1044. Hofmann, K., Yajima, H., Yanaihara, N., Liu, T.-Y., and Lande, S., *J. Am. Chem. Soc.* **83,** 487 (1961).

1045. Hofmann, K., Yanaihara, N., Lande, S., and Yajima, H., *J. Am. Chem. Soc.* **84,** 4470 (1962).

1046. Holdstock, D. J., Mathias, A. P., and Schachter, M., *Brit. J. Pharmacol.* **12,** 149 (1957).

1047. Holland, G. F., and Cohen, L. A., *J. Am. Chem. Soc.* **80,** 3765 (1958).

1048. Hollenbach, C., *Zentr. Gynaekol.* **80,** 1750 (1958).

1048a. Hollenbach, C., *Zentr. Gynaekol.* **81,** 1980 (1959).

1049. Hollenberg, C. H., Raben, M. S., and Astwood, E. B., *Endocrinology* **68,** 589 (1961).

1050. Holley, R. W., and Holley, A. D., *J. Am. Chem. Soc.* **74,** 3069 (1952).

1051. Holley, R. W., and Holley, A. D., *J. Am. Chem. Soc.* **74,** 5445 (1952).

1052. Holley, R. W., and Sondheimer, E., *J. Am. Chem. Soc.* **76,** 1326 (1954).

1053. Holly, F. W., Peel, E. W., Luz, E. L., and Folkers, K., *J. Am. Chem. Soc.* **74**, 4539 (1952).
1054. Holton, P., *Brit. J. Pharmacol.* **3**, 328 (1948).
1054a. Holtz, P., Barck, S., and Heicke, B., *Arch. Intern. Pharmacodyn,* **140**, 79 (1962).
1055. Holtz, P., Heicke, B., and Barck, S., *Experientia* **18**, 184 (1962).
1056. Holtz, P., and Raudonat, H. W., *Arch. Exptl. Pathol. Pharmakol.* **229**, 113 (1956).
1057. Honzl, J., and Rudinger, J., *Collection Czech. Chem. Commun.* **20**, 1190 (1955).
1058. Honzl, J., and Rudinger, J., *Collection Czech. Chem. Commun.* **26**, 2333 (1961).
1059. Hooper, K. C., Rydon, H. N., Schofield, J. A., and Heaton, G. S., *J. Chem. Soc.* p. 3148 (1956).
1060. Hope, D. B., and du Vigneaud, V., *J. Biol. Chem.* **237**, 3146 (1962).
1061. Hope, D. B., Murti, V. V. S., and du Vigneaud, V., *J. Biol. Chem.* **237**, 1563 (1962).
1061a. Hope, D. B., Murti, V. V. S., and du Vigneaud, V., *J. Am. Chem. Soc.* **85**, 3686 (1963).
1062. Hopkins, F. G., *Biochem. J.* **15**, 286 (1921).
1063. Hopkins, F. G., *J. Biol. Chem.* **84**, 269 (1929).
1064. Hörmann, H., Grassmann, W., Wünsch, E., and Preller, H., *Chem. Ber.* **89**, 933 (1956).
1065. Horn, M. J., Jones, D. B., and Ringel, S. J., *J. Biol. Chem.* **138**, 141 (1941).
1066. Horn, M. J., Jones, D. B., and Ringel, S. J., *J. Biol. Chem.* **144**, 87 (1942).
1067. Horn, M. J., Jones, D. B., and Ringel, S. J., *J. Biol. Chem.* **144**, 93 (1942).
1068. Horowitz, J., and Haurowitz, F., *Biochim. Biophys. Acta* **33**, 231 (1959).
1069. Horton, E. W., *Brit. J. Pharmacol.* **14**, 125 (1959).
1070. Horton, E. W., *in* "Polypeptides which Affect Smooth Muscles and Blood Vessels" (M. Schachter, ed.), p. 263. Macmillan (Pergamon), New York, 1960.
1071. Hotchkiss, R. D., *Advan. Enzymol.* **4**, 153 (1944).
1072. Hotchkiss, R. D., and Dubos, R. J., *J. Biol. Chem.* **132**, 79 (1940).
1073. Hotchkiss, R. D., and Dubos, R. J., *J. Biol. Chem.* **136**, 803 (1940).
1074. Hotchkiss, R. D., and Dubos, R. J., *J. Biol. Chem.* **141**, 155 (1941).
1075. Huang, W.-T., Yang, C.-S., Wang, K.-Z., and Niu, C.-I., *Sci. Sinica (Peking)* **11**, 499 (1962).
1076. Huggins, C. G., and Walaszek, E. J., *Am. Heart. J.* **60**, 976 (1960).
1077. Huggins, C. G., and Walaszek, E. J., *J. Med. Pharm. Chem.* **5**, 183 (1962).
1078. Hughes, G. K., Neill, K. G., and Ritchie, E., *Australian J. Sci. Res.* **A5**, 401 (1952).
1078a. Huguenin, R. L., *Helv. Chim. Acta* **47**, 1934 (1964).
1079. Huguenin, R. L., and Boissonnas, R. A., *Helv. Chim. Acta* **44**, 213 (1961).
1080. Huguenin, R. L., and Boissonnas, R. A., *Helv. Chim. Acta* **45**, 1629 (1962).
1081. Huguenin, R. L., and Boissonnas, R. A., *Helv. Chim. Acta* **46**, 1669 (1963).
1082. Hunt, M. and du Vigneaud, V., *J. Biol. Chem.* **124**, 699 (1938).
1083. Hunt, M., and du Vigneaud, V., *J. Biol. Chem.* **127**, 43 (1939).
1084. Hunter, C. A., Jr., and Howard, W. F., *Am. J. Obstet. Gynecol.* **79**, 838 (1960).
1085. Iacobellis, M., Schroeder, L. J., and Smith, A. H., *Arch. Biochem. Biophys.* **84**, 134 (1959).

1086. Ichii, S.-I., and Witkop, B., *J. Am. Chem. Soc.* **85**, 1832 (1963).
1086a. Igarashi, H., Zama, K., and Katada, M., *Nature* **181**, 1282 (1958).
1087. Ikawa, M., and Snell, E. E., *Biochim. Biophys. Acta* **60**, 186 (1962).
1087a. Ikekawa, N., *J. Biochem. (Tokyo)* **54**, 279 (1963).
1087b. Imhof, P., Brunner, H., Quitt, J., Steinmann, B., and Jacono, A., *Schweiz. Med. Wochschr.* **94**, 1199 (1964).
1088. Ing, H. R., and Manske, R. H. F., *J. Chem. Soc.* p. 2348 (1926).
1089. Ingle, T. R., *J. Sci. Ind. Res. (India)* **17BC**, 492 (1958).
1090. Inouye, K., and Otsuka, H., *Bull. Chem. Soc. Japan* **34**, 1 (1961).
1091. Inouye, K., and Otsuka, H., *Bull. Chem. Soc. Japan* **34**, 4 (1961).
1092. Inouye, K., and Otsuka, H., *J. Org. Chem.* **26**, 2613 (1961).
1093. Inouye, K., and Otsuka, H., *J. Org. Chem.* **27**, 4236 (1962).
1094. Ippen, H., *Deut. Med. Wochschr.* **86**, 307 (1961).
1095. Irving, C. C., and Gutmann, H. R., *J. Org. Chem.* **24**, 1979 (1959).
1096. Iselin, B., *Arch. Biochem. Biophys.* **78**, 532 (1958).
1097. Iselin, B., *Helv. Chim. Acta* **44**, 61 (1961).
1098. Iselin, B., *Helv. Chim. Acta* **45**, 1510 (1962).
1099. Iselin, B., Feurer, M., and Schwyzer, R., *Helv. Chim. Acta* **38**, 1508 (1955).
1100. Iselin, B., Rittel, W., Sieber, P., and Schwyzer, R., *Helv. Chim. Acta* **40**, 373 (1957).
1101. Iselin, B., and Schwyzer, R., *Helv. Chim. Acta* **39**, 57 (1956).
1102. Iselin, B., and Schwyzer, R., *Helv. Chim. Acta* **43**, 1760 (1960).
1103. Iselin, B., and Schwyzer, R., *Helv. Chim. Acta* **44**, 169 (1961).
1104. Iselin, B., and Schwyzer, R., *Helv. Chim. Acta* **45**, 1499 (1962).
1105. Ishihara, Y., Saito, T., Ito, Y., and Fujino, M., *Nature* **181**, 1468 (1958).
1106. Ishizuka, Y., *Nippon Kagaku Zasshi* **79**, 1190 (1958).
1106a. Isono, K., and Curtis, R. W., *Phytochemistry* **3**, 277 (1964).
1107. Ito, T., and Ogawa, H., *Bull. Agr. Chem. Soc. Japan* **23**, 536 (1959).
1108. Itoh, T., *Bull. Chem. Soc. Japan* **36**, 25 (1963).
1109. Izumiya, N., *Nippon Kagaku Zasshi* **70**, 404 (1949).
1110. Izumiya, N., *Nippon Kagaku Zasshi* **70**, 447 (1949).
1111. Izumiya, N., *Nippon Kagaku Zasshi* **71**, 214 (1950).
1112. Izumiya, N., *Nippon Kagaku Zasshi* **71**, 500 (1950).
1113. Izumiya, N., and Greenstein, J. P., *Arch. Biochem. Biophys.* **52**, 203 (1954).
1114. Izumiya, N., and Fruton, J. S., *J. Biol. Chem.* **218**, 59 (1956).
1115. Izumiya, N., Fu, S.-C. J., Birnbaum, S. M., and Greenstein, J. P., *J. Biol. Chem.* **205**, 221 (1953).
1116. Izumiya, N., and Makisumi, S., *Nippon Kagaku Zasshi* **78**, 662 (1957).
1117. Izumiya, N., and Makisumi, S., *Nippon Kagaku Zasshi* **78**, 1768 (1957).
1118. Izumiya, N., and Nagamatsu, A., *Bull. Chem. Soc. Japan* **25**, 265 (1952).
1119. Izumiya, N., Okazaki, H., Matsumoto, I., and Takiguchi, H., *J. Biochem. (Tokyo)* **46**, 1347 (1959).
1120. Izumiya, N., and Shitô, M., *Nippon Kagaku Zasshi* **79**, 975 (1958).
1121. Izumiya, N., and Uchio, H., *J. Biochem. (Tokyo)* **46**, 235 (1959).
1122. Izumiya, N., and Uchio, H., *J. Biochem. (Tokyo)* **46**, 645 (1959).
1123. Izumiya, N., Uchio, H., and Kamata, T., *Bull. Chem. Soc. Japan* **33**, 66 (1960).
1124. Izumiya, N., Uchio, H., and Yamashita, T., *Nippon Kagaku Zasshi* **79**, 420 (1958).
1125. Izumiya, N., and Yamashita, T., *J. Biochem. (Tokyo)* **46**, 19 (1959).
1126. Izumiya, N., and Yamashita, T., *J. Biochem. (Tokyo)* **46**, 337 (1959).

1127. Izumiya, N., Yamashita, T., Uchio, H., and Kitagawa, K., *Arch. Biochem. Biophys.* **90,** 170 (1960).
1128. Jackson, E. L., *J. Am. Chem. Soc.* **74,** 837 (1952).
1129. Jaenike, J., and Waterhouse, C., *J. Clin. Endocrinol. Metab.* **21,** 231 (1961).
1130. Janicki, J., Skupin, J., and Zagalak, B., *Roczniki Chem.* **36,** 353 (1962).
1131. Jaquenoud, P.-A., *Chimia (Aarau)* **14,** 373 (1960).
1132. Jaquenoud, P.-A., and Boissonnas, R. A., *Helv. Chim. Acta* **42,** 788 (1959).
1133. Jaquenoud, P.-A., and Boissonnas, R. A., *Helv. Chim. Acta* **44,** 113 (1961).
1134. Jaquenoud, P.-A., and Boissonnas, R. A., *Helv. Chim. Acta* **45,** 1462 (1962).
1135. Jaquenoud, P.-A., and Boissonnas, R. A., *Helv. Chim. Acta* **45,** 1601 (1962).
1136. Jaques, R., and Meier, R., *Experientia* **16,** 371 (1960).
1137. Jaques, R., and Schachter, M., *Brit. J. Pharmacol.* **9,** 53 (1954).
1137a. Jard, S., and Morel, F., *Am. J. Physiol.* **204,** 222 (1963).
1138. Jarvis, D., Bodanszky, M., and du Vigneaud, V., *J. Am. Chem. Soc.* **83,** 4780 (1961).
1138a. Jarvis, D., and du Vigneaud, V., *Science* **143,** 545 (1964).
1139. Jarvis, D., and Rydon, H. N., *Chimia (Aarau)* **14,** 376 (1960).
1140. Jarvis, D., Rydon, H. N., and Schofield, J. A., *J. Chem. Soc.* p. 1752 (1961).
1140a. Jenny, M., Muller, A. F., and Mach, R. S., *Helv. Med. Acta* **30,** 476 (1963).
1140b. Jenny, M., Muller, A. F., and Mach, R. S., *Schweiz. Med. Wochschr.* **93,** 766 (1963).
1141. Jensen, K. B., *Brit. J. Pharmacol.* **13,** 271 (1958).
1142. Jensen, K. B., and Venneröd, A. M., *Acta Pharmacol. Toxicol.* **19,** 265 (1962).
1143. Jensen, K. B., and Venneröd, A. M., *Acta Pharmacol. Toxicol.* **19,** 275 (1962).
1144. Jensen, K. B., and Venneröd, A. M., *Acta Pharmacol. Toxicol.* **19,** 337 (1962).
1144a. Jeschkeit, M., Losse, G., and Knopf, D., *Pharmazie* **18,** 658 (1963).
1145. Jiang, R.-Q., Du, Y.-C., and Tsou, C.-L., *Sci. Sinica (Peking)* **12,** 452 (1963).
1146. Jöhl, A., Hartmann, A., and Rink, H., *Biochim. Biophys. Acta* **69,** 193 (1963).
1147. Jöhl, A., and Stoll, W. G., *Helv. Chim. Acta* **42,** 156 (1959).
1148. Jöhl, A., and Stoll, W. G., *Helv. Chim. Acta* **42,** 716 (1959).
1149. John, W. D., and Young, G. T., *J. Chem. Soc.* p. 2870 (1954).
1150. Johnson, B., Anker, H. S., and Meleney, F. L., *Science* **102,** 376 (1945).
1151. Jollès, P., Cros, N.-T.-L., Lederer, H., and Lederer, E., *Biochim. Biophys. Acta* **43,** 559 (1960).
1152. Jones, E. E., and Lipkin, D., *J. Am. Chem. Soc.* **78,** 2408 (1956).
1153. Jones, J. K. N., Millington, J. P., and Perry, M. B., *Can. J. Chem.* **40,** 2229 (1962).
1154. Jones, J. K. N., Perry, M. B., Shelton, B., and Walton, D. J., *Can. J. Chem.* **39,** 1005 (1961).
1155. Jones, K. M., and Wooley, D. W., *J. Bacteriol.* **83,** 797 (1962).
1156. Jones, T. S. G., *Biochem. J.* **43,** XXVI (1948).
1157. Jones, T. S. G., *Ann. N. Y. Acad. Sci.* **51,** 909 (1949).
1157a. Jorpes, J. E., and Mutt, V., *Acta Chem. Scand.* **15,** 1790 (1961).
1157b. Jorpes, J. E., Mutt, V., Magnusson, S., and Steele, B. B., *Biochem. Biophys. Res. Commun.* **9,** 275 (1962).
1157c. Josefsson, L., *Biochim. Biophys. Acta* **74,** 774 (1963).

1158. Josefsson, L., and Edman, P., *Biochim. Biophys. Acta* **25,** 614 (1957).

1158a. Jošt, K., Debabov, V. G., Nesvadba, H., and Rudinger, J., *Collection Czech. Chem. Commun.* **29,** 419 (1964).

1159. Jošt, K., and Rudinger, J., *Collection Czech. Chem. Commun.* **26,** 2345 (1961).

1160. Jošt, K., Rudinger, J., and Šorm, F., *Collection Czech. Chem. Commun.* **26,** 2496 (1961).

1160a. Jošt, K., Rudinger, J., and Šorm, F., *Collection Czech. Chem. Commun.* **28,** 1706 (1963).

1160b. Jošt, K., Rudinger, J., and Šorm, F., *Collection Czech. Chem. Commun.* **28,** 2021 (1963).

1161. Jung, H., *in* "Oxytocin" (R. Caldeyro-Barcia and H. Heller, eds.), p. 87. Macmillan (Pergamon), New York, 1961.

1162. Jutisz, M., and De La Llosa, P., *Annee Biol.* **37,** Fasc. 3–4, 81 (1961).

1163. Jutisz, M., Yamazaki, E., Bérault, A., Sakiz, E., and Guillemin, R., *Compt. Rend.* **256,** 2925 (1963).

1164. Kader, A. T., and Stirling, C. J. M., *Proc. Chem. Soc.* p. 363 (1962).

1165. Kahn, J. R., Skeggs, L. T., and Shumway, N. P., *Circulation* **2,** 363 (1950).

1166. Kamiyama, S., and Schmid, K., *Biochim. Biophys. Acta* **58,** 80 (1962).

1167. Kamiyama, S., and Schmid, K., *Biochim. Biophys. Acta* **63,** 266 (1962).

1168. Kaneko, T., and Inui, T., *Bull. Chem. Soc. Japan* **35,** 1145 (1962).

1169. Kaneko, T., Shiba, T., Watarai, S., Imai, S., Shimada, T., and Ueno, K., *Chem. Ind. (London)* p. 986 (1957).

1170. Kappeler, H., *Helv. Chim. Acta* **44,** 476 (1961).

1171. Kappeler, H., and Schwyzer, R., *Experientia* **16,** 415 (1960).

1172. Kappeler, H., and Schwyzer, R., *Helv. Chim. Acta* **43,** 1453 (1960).

1173. Kappeler, H., and Schwyzer, R., *Helv. Chim. Acta* **44,** 1136 (1961).

1173a. Karkun, J. N., and Landgrebe, F. W., *in* "Comparative Endocrinology" (U. S. von Euler and H. Heller, eds.), Vol. 1, p. 81. Academic Press, New York, 1963.

1173b. Karl, H. J., *Klin. Wochschr.* **41,** 633 (1963).

1174. Karlson, R. H., Norland, K. S., Fasman, G. D., and Blout, E. R., *J. Am. Chem. Soc.* **82,** 2268 (1960).

1175. Katchalski, E., *Advan. Protein Chem.* **6,** 123 (1951).

1176. Katchalski, E., and Sela, M., *J. Am. Chem. Soc.* **75,** 5284 (1953).

1177. Katchalski, E., and Sela, M., *Advan. Protein Chem.* **13,** 243 (1958).

1178. Katchalski, E., and Spitnik, P., *J. Am. Chem. Soc.* **73,** 2946 (1951).

1179. Katchalskyi, A., and Paecht, M., *Bull. Res. Council Israel* **2A,** 312 (1952).

1180. Katchalskyi, A., and Paecht, M., *J. Am. Chem. Soc.* **76,** 6042 (1954).

1181. Katsoyannis, P. G., *J. Am. Chem. Soc.* **79,** 109 (1957).

1182. Katsoyannis, P. G., *J. Polymer Sci.* **49,** 51 (1961).

1183. Katsoyannis, P. G., *J. Am. Chem. Soc.* **83,** 4053 (1961).

1183a. Katsoyannis, P. G., *Diabetes* **13,** 339 (1964).

1184. Katsoyannis, P. G., and du Vigneaud, V., *J. Am. Chem. Soc.* **76,** 3113 (1954).

1185. Katsoyannis, P. G., and du Vigneaud, V., *J. Am. Chem. Soc.* **78,** 4482 (1956).

1186. Katsoyannis, P. G., and du Vigneaud, V., *Arch. Biochem. Biophys.* **78,** 555 (1958).

1187. Katsoyannis, P. G., and du Vigneaud, V., *J. Biol. Chem.* **233,** 1352 (1958).

1188. Katsoyannis, P. G., and du Vigneaud, V., *Nature* **184,** 1465 (1959).

1189. Katsoyannis, P. G., Fukuda, K., and Tometsko, A., *J. Am. Chem. Soc.* **85,** 1681 (1963).

1189a. Katsoyannis, P. G., Fukuda, K., Tometsko, A., Suzuki, K., and Tilak, M., *J. Am. Chem. Soc.* **86,** 930 (1964).

1190. Katsoyannis, P. G., Gish, D. T., and du Vigneaud, V., *J. Am. Chem. Soc.* **79,** 4516 (1957).

1191. Katsoyannis, P. G., Gish, D. T., Hess, G. P., and du Vigneaud, V., *J. Am. Chem. Soc.* **80,** 2558 (1958).

1192. Katsoyannis, P. G., and Suzuki, K., *J. Am. Chem. Soc.* **83,** 4057 (1961).

1193. Katsoyannis, P. G., and Suzuki, K., *J. Am. Chem. Soc.* **84,** 1420 (1962).

1194. Katsoyannis, P. G., and Suzuki, K., *J. Am. Chem. Soc.* **85,** 1679 (1963).

1194a. Katsoyannis, P. G., and Suzuki, K., *J. Am. Chem. Soc.* **85,** 2659 (1963).

1195. Katsoyannis, P. G., Suzuki, K., and Tometsko, A., *J. Am. Chem. Soc.* **85,** 1139 (1963).

1195a. Katsoyannis, P. G., Tometsko, A., and Fukuda, K., *J. Am. Chem. Soc.* **85,** 2863 (1963).

1195b. Katsoyannis, P. G., and Tilak, M., *J. Am. Chem. Soc.* **85,** 4028 (1963).

1196. Katz, E., Waldron, C. R., and Meloni, M. L., *J. Bacteriol.* **82,** 600 (1961).

1197. Katz, Y. L., and Goldblatt, H., *J. Exptl. Med.* **78,** 67 (1943).

1198. Kauzmann, W., *in* "Sulfur in Protein" (R. Benesch *et al.*, eds.), p. 93. Academic Press, New York, 1959.

1199. Kaverzneva, D., and Konovalova, M. I., *Izv. Akad. Nauk SSSR, Otd. Khim. Nauk* p. 124 (1963).

1200. Ke, L.-T., Kung, Y.-T., Huang, W.-T., Chi, K.-Y., and Niu, C.-I., *Sci. Sinica (Peking)* **11,** 337 (1962).

1200a. Ke, L.-T., Kung, Y.-T., Wang, K.-Z., and Niu, C.-I., *Sci. Sinica (Peking)* **13,** 1435 (1964).

1201. Keele, C. A., *in* "Polypeptides which Affect Smooth Muscles and Blood Vessels" (M. Schachter, ed.), p. 253. Macmillan (Pergamon), New York, 1960.

1202. Keenan, C. W., and McDowell, W. J., *J. Am. Chem. Soc.* **75,** 6348 (1953).

1203. Keil, B., and Šorm, F., *Collection Czech. Chem. Commun.* **27,** 1310 (1962).

1204. Keil, B., Zikán, J., Rexová, L., and Šorm, F., *Collection Czech. Chem. Commun.* **27,** 1678 (1962).

1205. Keller, H., Netter, H., and Niemann, B., *Z. Physiol. Chem.* **313,** 244 (1958).

1206. Keller-Schierlein, W., Mihailović, M. L., and Prelog, V., *Helv. Chim. Acta* **42,** 305 (1959).

1207. Keller-Schierlein, W., and Prelog, V., *Helv. Chim. Acta* **40,** 205 (1957).

1207a. Kelly, R. B., *J. Org. Chem.* **28,** 453 (1963).

1208. Kendall, E. C., McKenzie, B. F., and Mason, H. L., *J. Biol. Chem.* **84,** 657 (1929).

1209. Kendall, E. C., Mason, H. L., and McKenzie, B. F., *J. Biol. Chem.* **87,** 55 (1930).

1210. Kenner, G. W., *Chem. Ind.* p. 15 (1951).

1211. Kenner, G. W., *Chem. Soc. (London) Spec. Publ.* **2,** 103 (1955).

1212. Kenner, G. W., *Angew. Chem.* **71,** 741 (1959).

1213. Kenner, G. W., and Stedman, R. J., *J. Chem. Soc.* p. 2069 (1952).

1214. Kenner, G. W., Thomson, P. J., and Turner, J. M., *J. Chem. Soc.* p. 4148 (1958).

1215. Kenner, G. W., and Turner, J. M., *Chem. Ind. (London)* p. 602 (1955).

1216. Kenny, A. J., *Brit. Med. Bull.* **16,** 202 (1960).

1217. Khairallah, P. A., Bumpus, F. M., Page, I. H., and Smeby, R. R., *Nature* **196,** 1059 (1962).

1218. Khairallah, P. A., Bumpus, F. M., Page, I. H., and Smeby, R. R., *Science* **140,** 672 (1963).

1219. Khairallah, P. A., and Page, I. H., *Am. J. Physiol.* **200,** 51 (1961).

1220. Khairallah, P. A., and Page, I. H., *Ann. N. Y. Acad. Sci.* **104,** 212 (1963).

1221. Khokhlov, A. S., Mamiofe, S. M., and Sinitsyna, Z. T., *Antibiotiki* **3,** 6 (1958).

1222. Khokhlov, A. S., Silaev, A. B., Stepanov, V. M., Yulikova, E. P., Troshko, E. V., Levin, E. D., Mamiofe, S. M., Sinitsyna, Z. T., Chi, C.-T., Solov'eva, N. K., K'inskaya, S. A., Rossovskaya, V. S., Dmitrieva, V. S., Semenov, S. M., Veǐls, R. A., Berezina, E. K., and Rubtsova, L. K., *Antibiotiki* **5,** 3 (1960).

1223. Khorana, H. G., *Chem. Rev.* **53,** 145 (1953).

1224. Khorana, H. G., *Chem. Ind.* (*London*) p. 1087 (1955).

1224a. Khosla, M. C., Garg, H. G., and Anand, N., *J. Sci. Ind. Res.* (*India*) **21B,** 318 (1962).

1224b. Khosla, M. C., and Anand, N., *Indian J. Chem.* **1,** 49 (1963).

1225. Khurgin, Y. I., and Dmitrieva, M. G., *Dokl. Akad. Nauk SSSR* **143,** 629 (1962).

1226. Khurgin, Y. I., and Dmitrieva, M. G., *Collection Czech. Chem. Commun.* **27,** 2235 (1962).

1227. Kidd, D. A., and King, F. E., *Nature* **162,** 776 (1948).

1228. Kienhuis, H., Van de Linde, A., Van der Holst, J. P. J., and Verweij, A., *Rec. Trav. Chim.* **80,** 1278 (1961).

1229. Kihara, H., Ikawa, M., and Snell, E. E., *J. Biol. Chem.* **236,** 172 (1961).

1230. Kihara, H., and Snell, E. E., *J. Biol. Chem.* **235,** 1409 (1960).

1231. Kimbrough, R. D., Jr., Cash, W. D., Branda, L. A., Chan, W. Y., and du Vigneaud, V., *J. Biol. Chem.* **238,** 1411 (1963).

1232. Kimbrough, R. D., Jr., and du Vigneaud, V., *J. Biol. Chem.* **236,** 778 (1961).

1233. King, F. E., Clark-Lewis, J. W., Kidd, D. A. A., and Smith, G. R., *J. Chem. Soc.* p. 1039 (1954).

1234. King, F. E., Clark-Lewis, J. W., and Smith, G. R., *J. Chem. Soc.* p. 1044 (*1954*).

1235. King, F. E., Clark-Lewis, J. W., and Smith, G. R., *J. Chem. Soc.* p. 1046 (1954).

1236. King, F. E., Clark-Lewis, J. W., and Swindin, W. A., *J. Chem. Soc.* p. 2259 (1959).

1237. King, F. E., Clark-Lewis, J. W., and Wade, R., *J. Chem. Soc.* p. 880 (1957).

1238. King, F. E., Clark-Lewis, J. W., and Wade, R., *J. Chem. Soc.* p. 886 (1957).

1239. King, F. E., Clark-Lewis, J. W., Wade, R., and Swindin, W. A., *J. Chem. Soc.* p. 873 (1957).

1240. King, F. E., Jackson, B. S., and Kidd, D. A. A., *J. Chem. Soc.* p. 243 (1951).

1241. King, F. E., and Kidd, D. A. A., *J. Chem. Soc.* p. 3315 (1949).

1242. King, F. E., and Kidd, D. A. A., *J. Chem. Soc.* p. 2976 (1951).

1243. King, L. C., and Suydam, F. H., *J. Am. Chem. Soc.* **74,** 5499 (1952).

1244. King, T. P., and Craig, L. C., *J. Am. Chem. Soc.* **77,** 6624 (1955).

1245. King, T. P., and Craig, L. C., *J. Am. Chem. Soc.* **77,** 6627 (1955).

1245a. Kiryushkin, A. A., Ovchinnikov, Y. A., and Shemyakin, M. M., *Tetrahedron Letters* p. 3313 (1964).

1246. Kisfaludy, L., *Acta Chim. Acad. Sci. Hung.* **24,** 309 (1960).
1247. Kisfaludy, L., and Dualszky, S., *Acta Chim. Acad. Sci. Hung.* **24,** 301 (1960).
1248. Kisfaludy, L., Dualszky, S., and Bayer, J., *Chimia (Aarau)* **14,** 368 (1960).
1249. Kisfaludy, L., Dualszky, S., Medzihradszky, K., Bajusz, S., and Bruckner, V., *Acta Chim. Acad. Sci. Hung.* **30,** 473 (1962).
1249a. Kisfaludy, L., and Löw, M., *in* "Peptide Symposium," Oxford 1962 (G. T. Young, ed.), p. 93. Macmillan, (Pergamon), New York, 1963.
1250. Kitaoka, H., Sakakibara, S., and Tani, H., *Bull. Chem. Soc. Japan* **31,** 802 (1958).
1251. Kjaer, A., *Acta Chem. Scand.* **3,** 1087 (1949).
1252. Kjaer, A., and Larsen, P. O., *Acta Chem. Scand.* **13,** 1565 (1959).
1252a. Kjaer, A., and Larsen, P. O., *Acta Chem. Scand.* **15,** 750 (1961).
1253. Klaus, D., Kaffarnik, H., and Pfeil, H., *Klin. Wochschr.* **41,** 376 (1963).
1254. Klaus, D., Kaffarnik, H., and Pfeil, H., *Klin. Wochschr.* **41,** 380 (1963).
1255. Klee, W., and Brenner, M., *Helv. Chim. Acta* **44,** 2151 (1961).
1256. Kleeman, C. R., and Cutler, R. E., *Ann. Rev. Physiol.* **25,** 385 (1963).
1257. Klemer, A., and Micheel, F., *Chem. Ber.* **89,** 1242 (1957).
1258. Klieger, E., and Gibian, H., *Ann. Chem. Liebigs* **649,** 183 (1961).
1259. Klieger, E., and Gibian, H., *Ann. Chem. Liebigs* **651,** 194 (1962).
1260. Klieger, E., and Gibian, H., *Ann. Chem. Liebigs* **655,** 195 (1962).
1261. Klieger, E., and Schröder, E., *Ann. Chem. Liebigs* **661,** 193 (1963).
1262. Klieger, E., Schröder, E., and Gibian, H., *Ann. Chem. Liebigs* **640, 157** (1961).
1263. Klippel, R., König, J., *Arzneimittel-Forsch.* **6,** 489 (1956).
1263a. Kloss, G., and Schröder, E., *Z. Physiol. Chem.* **336,** 248 (1964).
1264. Kochetkov, N. K., Derevitskaya, V. A., and Likhosherstov, L. M., *Chem. Ind. (London)* p. 1532 (1960).
1265. Kochetkov, N. K., Derevitskaya, V. A., and Likhosherstov, L. M., *Zh. Vses. Khim. Obshchestva im. D. I. Mendeleeva* **6,** 228 (1961).
1266. Kochetkov, N. K., Derevitskaya, V. A., Likhosherstov, L. M., Molodtsov, N. V., and Kara-Murza, S. G., *Tetrahedron* **18,** 273 (1962).
1266a. Kochetkov, N. K., Derevitskaya, V. A., and Molodtsov, N. V., *Chem. Ind. (London)* p. 1159 (1959).
1267. Kochetkov, N. K., Derevitskaya, V. A., and Molodtsov, N. V., *Zh. Obshch. Khim.* **32,** 2500 (1962).
1268. Koffler, H., *Science* **130,** 1419 (1959).
1269. Koffler, H., and Kobayashi, T., *Proc. 4th Intern. Congr. Biochem., Vienna 1958* Abstr. Sect. 1, p. 63 (1959). Macmillan (Pergamon), New York.
1270. Kögl, F., and Akkermann, A. M., *Rec. Trav. Chim.* **65,** 216 (1946).
1271. Kollonitsch, J., Gábor, V., and Hajós, A., *Nature* **177,** 841 (1956) .
1272. Kollonitsch, J., Gábor, V., and Hajós, A., *Chem. Ber.* **89,** 2293 (1956).
1273. Kollonitsch, J., Hajós, A., and Gábor, V., *Chem. Ber.* **89,** 2288 (1956).
1274. Kollonitsch, J., and Vita, J., *Nature* **178,** 1307 (1956).
1275. Konigsberg, W., and Craig, L. C., *J. Org. Chem.* **27,** 934 (1962).
1276. Konigsberg, W., Hill, R. J., and Craig, L. C., *J. Org. Chem.* **26,** 3867 (1961).
1276a. Kontos, H. A., Magee, J. H., Shapiro, W., and Patterson, J. L., Jr., *Circulation Res.* **14,** 351 (1964).
1277. Konzett, H., *Helv. Physiol. Pharmacol. Acta* **15,** 419 (1957).
1278. Konzett, H., *Arch. Exptl. Pathol. Pharmakol.* **245,** 153 (1963).

1279. Konzett, H., and Berde, B., *Brit. J. Pharmacol.* **14,** 133 (1959).
1280. Konzett, H., Berde, B., and Cerletti, A., *Schweiz. Med. Wochschr.* **86,** 226 (1956).
1281. Konzett, H., and Boissonnas, R. A., *Experientia* **16,** 456 (1960).
1282. Konzett, H., and Stürmer, E., *Brit. J. Pharmacol.* **15,** 544 (1960).
1283. Konzett, H., and Stürmer, E., *Nature* **188,** 998 (1960).
1284. Konzett, H., Stürmer, E., and Cerletti, A., *Clin. Terap.* **20,** 21 (1961).
1284a. Kopple, K. D., Jarabak, R. R., and Bhatia, P. L., *Biochemistry* **2,** 958 (1963).
1285. Kopple, K. D., and Nitecki, D. E., *J. Am. Chem. Soc.* **83,** 4103 (1961).
1286. Kopple, K. D., and Renick, R. J., *J. Org. Chem.* **23,** 1565 (1958).
1286a. Kornguth, M. L., Neidle, A., and Waelsch, H., *Biochemistry* **2,** 740 (1963).
1287. Korzybski, T., and Kurylowicz, W., "Antibiotika, Herkunft, Arten, Eigenschaften," Fischer, Jena, 1961.
1288. Kossel, A., and Kennaway, E. L., *Z. Physiol. Chem.* **72,** 486(1911).
1289. Kotaki, A., *J. Biochem.* (*Tokyo*) **50,** 256 (1961).
1290. Kotaki, A., *J. Biochem.* (*Tokyo*) **51,** 301 (1962).
1291. Kotaki, A., Kurioka, U., and Satake, K., *J. Biochem.* (*Tokyo*) **51,** 375 (1962).
1292. Kovács, J., and Bruckner, V., *J. Chem. Soc.* p. 4255 (1952).
1293. Kovács, J., Bruckner, V., and Kovács, K., *J. Chem. Soc.* p. 145 (1953).
1293a. Kovács, J., and Ghatak, U. R., *Chem. Ind.* (*London*) p. 913 (1963).
1293b. Kovács, J., Kovács, H. N., and Ballina, R., *J. Am. Chem. Soc.* **85,** 1839 (1963).
1294. Kovács, J., Kovács, H. N., Könyves, I., Császár, J., Vajda,, T., and Mix, H., *J. Org. Chem.* **26,** 1084 (1961).
1295. Kovács, J., Medzihradszky, K., and Bruckner, V., *Naturwissenschaften* **41,** 450 (1954).
1296. Kovács, J., Medzihradszky, K., and Bruckner, V., *Acta Chim. Acad. Sci. Hung.* **6,** 183 (1955).
1297. Kovács, K., and Kótai, A., *Acta Chim. Acad. Sci. Hung.* **5,** 313 (1955).
1298. Kovalcik, V., *Nature* **196,** 174 (1962).
1299. Koyama, Y., Kurosasa, A., Tsuchiya, A., and Takakuta, K., *J. Antibiot.* (*Japan*) **3,** 457 (1950).
1300. Kraege, H. J., Dissertation, Freie Universität Berlin, 1958.
1301. Kraml, M., and Bouthillier, L. P., *Can. J. Chem.* **33,** 1630 (1955).
1302. Krivoy, W. A., and Guillemin, R., *Endocrinology* **69,** 170 (1961).
1303. Krivoy, W. A., and Guillemin, R., *Experientia* **18,** 20 (1962).
1304. Krivoy, W. A., Lane, M., Childers, H. E., and Guillemin, R., *Experientia* **18,** 521 (1962).
1305. Kuether, C. A., and Haney, M. E., Jr., *Science* **121,** 65 (1955).
1306. Kuhn, R., and Ekong, D., *Chem. Ber.* **96,** 683 (1963).
1307. Kuhn, R., and Haas, H. J., *Angew. Chem.* **67,** 785 (1955).
1308. Kunde, J., and Zahn, H., *Ann. Chem. Liebigs* **646,** 137 (1961).
1308a. Kung, Y.-T., Ke, L.-T., and Niu, C.-I., *Sci. Sinica* (*Peking*) **12,** 1321 (1963).
1308b. Kung, Y.-T., Ke, L.-T., Niu, C.-T., and Hu, S.-C., *Sci. Sinica* (*Peking*) **13,** 1245 (1964).
1309. Kupryszewski, G., *Roczniki Chem.* **35,** 595 (1961).
1310. Kupryszewski, G., *Roczniki Chem.* **36,** 1593 (1962).
1311. Kupryszewski, G., *Roczniki Chem.* **37,** 593 (1963).
1312. Kupryszewski, G., and Formela, M., *Roczniki Chem.* **35,** 1533 (1961).
1313. Kupryszewski, G., and Formela, M., *Roczniki Chem.* **37,** 161 (1963).

1314. Kupryszewski, G., and Kaczmarek, M., *Roczniki Chem.* **35,** 931 (1961).
1315. Kupryszewski, G., and Sokolowska, T., *Acta Biochim. Polon.* **4,** 85 (1957).
1315a. Kuranova, S. I. P., and Smirnov, N. V., *Izv. Akad. Nauk SSSR, Otd. Khim. Nauk* p. 1148 (1963).
1315b. Kurahashi, K., *J. Biochem.* (*Tokyo*) **56,** 101 (1964).
1316. Kurihara, T., and Suzuki, K., *Ann. Rept. Tohoku Coll. Pharm.* **3,** 1 (1956).
1317. Kurtz, A. C., *J. Biol. Chem.* **122,** 477 (1937).
1318. Kurtz, A. N., and Niemann, C., *J. Am. Chem. Soc.* **83,** 3309 (1961).
1318a. LaBella, F. S., *Can. J. Physiol. Pharmacol.* **42,** 75 (1964).
1319. Lackner, H., Dissertation, Georg-August-Universität Göttingen, 1961.
1320. Lamfrom, H., and Nielsen, S. O., *J. Am. Chem. Soc.* **79,** 1966 (1957).
1321. Lande, S., *Dissertation Abstr.* **22,** 52 (1961).
1322. Lande, S., *J. Org. Chem.* **27,** 4558 (1962).
1323. Landgrebe, F. W., and Waring, H., *Quart. J. Exptl. Physiol.* **33,** 1 (1944).
1324. Landmann, W. A., Drake, M. P., and White, W. F., *J. Am. Chem. Soc.* **75,** 4370 (1953).
1324a. Landon, J., James, V. H. T., and Stamp, T. C. B., *Biochem. J.* **90,** 20p (1964).
1325. Landy, M., Warren, C. H., Rosenman, S. B., and Colio, L. G., *Proc. Soc. Exptl. Biol. Med.* **67,** 539 (1948).
1326. Langenbeck, W., and Kresse, P., *J. Prakt, Chem.* [4] **2,** 261 (1955).
1327. Laragh, J. H., *Circulation* **25,** 203 (1962).
1328. Laragh, J. H., *Circulation* **25,** 1015 (1962).
1329. Large, D. G., Rydon, H. N., and Schofield, J. A., *J. Chem. Soc.* p. 1749 (1961).
1330. Larsen, P. O., *Acta Chem. Scand.* **16,** 1511 (1962).
1331. Lautsch, W., and Broser, W., *Z. Naturforsch.* **13b,** 48 (1958).
1332. Lautsch, W., Broser, W., Bandel, W., Biedermann, W., Gehrmann, W., Schröder, E., Gnichtel, H., Zehmisch, I., Kurth, G., Krüger, R., and Kraege, H.-J., *Kolloid-Z.* **138,** 129 (1954).
1333. Lautsch, W., Broser, W., Biedermann, W., Döring, U., and Zoschke, H., *Kolloid-Z.* **125,** 72 (1952).
1334. Lautsch, W., Broser, W., Biedermann, W., and Gnitchtel, H., *Angew. Chem.* **66,** 123 (1954).
1335. Lautsch, W., Broser, W., and Gödicke, V., *Z. Naturforsch.* **12b,** 303 (1957).
1336. Lautsch, W., Broser, W., Höfling, E., Gnichtel, H., Schröder, E., Krüger, R., Woldt, J., Schulz, G., Wiechert, R., Bandel, W., Kurth, G., Kraege, H.-J., Gehrmann, W., Prater, K., Parsiegla, G., Pasedag, R., and Hunger, W., *Kolloid-Z.* **144,** 82 (1955).
1337. Lautsch, W., Gehrmann, W., Pasedag, R., and Prater, K., *Chem. Ber.* **90,** 470 (1957).
1338. Lautsch, W., Gnichtel, H., Gnichtel, J., and Höfling, E., *Kolloid-Z.* **141,** 132 (1955).
1339. Lautsch, W., and Günther, D., *Naturwissenschaften* **44,** 492 (1957).
1340. Lautsch, W., and Kraege, H.-J., *Chem. Ber.* **89,** 737 (1956).
1341. Lautsch, W., Pasedag, R., Sommer, I., Julius, H. J., Boedefeld, E., Boroschewski, G., Gante, J., Shingte R., Rauhut, H., Grimm, W., Rimpler, M., Schulz, G., Paetzke, I., and Heinicke, D., *Chimia* (*Aarau*) **13,** 129 (1959).
1342. Lautsch, W., and Schröder, E., *Z. Naturforsch.* **9b,** 277 (1954).
1343. Lautsch, W., and Schröder, E., *Monatsh. Chem.* **88,** 432 (1957).

1344. Lautsch, W., Schröder, E., Krüger, R., Broser, W., Becker, K., Sommer, I., Pasedag, R., Kurth, G., Müller, H., Jehring, H., Wiechert, R., Heinicke, D., Rauhut, R., Grimm, W., Wiemer, B., Zschenderlein, P., Kraege, H.-J., Bandel, W., Günther, D., Schulz, G., and Gnichtel, H., *Kolloid-Z.* **161**, 1 (1958).
1345. Lautsch, W., and Schulz, G., *Naturwissenschaften* **45**, 1 (1958).
1346. Lautsch, W., Shingte, R., Heinicke, D., Vollmann, D., Wieczorek, H., Günther, D., and Uhde, W., *Kolloid-Z.* **183**, 38 (1962).
1347. Lautsch, W., Wiechert, R., Gnichtel, H., Schuchardt, G., Kraege, H.-J., Singewald, C., Broser, W., Becker, H., Rauhut, H., and Grimm, W., *Oesterr. Chemiker-Z.* **58**, 33 (1957).
1348. Lavine, T. F., *J. Biol. Chem.* **113**, 583 (1936).
1348a. Law, H. D., *Advan. Med. Chem.* (in press).
1349. Law, H. D., and du Vigneaud, V., *J. Am. Chem. Soc.* **82**, 4579 (1960).
1350. Law, H. D., Millar, I. T., and Springall, H. D., *Proc. Chem. Soc.* p. 198 (1958).
1351. Law, H. D., Millar, I. T., and Springall, H. D., *J. Chem. Soc.* p. 279 (1961).
1352. Leach, S. J., and Lindley, H., *Trans. Faraday Soc.* **49**, 915 (1953).
1353. Leach, S. J., and Lindley, H., *Trans. Faraday Soc.* **49**, 921 (1953).
1354. Leach, S. J., and Lindley, H., *Australian J. Chem.* **7**, 173 (1954).
1354a. Lebovitz, H. E., and Engel, F. L., *Endocrinology* **73**, 573 (1963).
1355. Lechat, P., *Therapie* **15**, 412 (1960).
1356. Lecomte, J., Petit, J. M., Mélon, I., Troquet, J., and Marcelle, R., *Arch. Intern. Pharmacodyn.* **137**, 232 (1962).
1356a. Lee, T. H., and Buettner-Janusch, V., *J. Biol. Chem.* **238**, 2012 (1963).
1357. Lee, T. H., and Lerner, A. B., *J. Biol. Chem.* **221**, 943 (1956).
1358. Lee, T. H., and Lerner, A. B., *J. Am. Chem. Soc.* **81**, 6084 (1959).
1359. Lee, T. H., Lerner, A. B., and Buettner-Janusch, V., *Ciba Found. Colloq. Endocrinol.* **13**, 251 (1960).
1360. Lee, T. H., Lerner, A. B., and Buettner-Janusch, V., *J. Biol. Chem.* **236**, 1390 (1961).
1361. Lee, T. H., Lerner, A. B., and Buettner-Janusch, V., *J. Biol. Chem.* **236**, 2970 (1961).
1362. Lee, T. H., Lerner, A. B., and Buettner-Janusch, V., *Ann. N. Y. Acad. Sci.* **100**, 658 (1963).
1363. Lee, Y. C., and Montgomery, R., *Arch. Biochem. Biophys.* **97**, 9 (1962).
1364. Lembeck, F., and Zetler, G., *Intern. Rev. Neurobiol.* **4**, 159 (1962).
1365. Lentz, K. E., Skeggs, L. T., Jr., Woods, K. R., Kahn, J. R., and Shumway, N. P., *J. Exptl. Med.* **104**, 183 (1956).
1366. Léonis, J., Li, C. H., and Chung, D., *J. Am. Chem. Soc.* **81**, 419 (1959).
1367. Leplawy, M. T., Jones, D. S., Kenner, G. W., and Sheppard, R. C., *Tetrahedron* **11**, 39 (1960).
1368. Le Quesne, W. J., and Young, G. T., *Nature* **163**, 604 (1949).
1369. Le Quesne, W. J., and Young, G. T., *J. Chem. Soc.* p. 1954 (1950).
1370. Le Quesne, W. J., and Young, G. T., *J. Chem. Soc.* p. 1959 (1950).
1371. Le Quesne, W. J., and Young, G. T., *J. Chem. Soc.* p. 24 (1952).
1372. Le Quesne, W. J., and Young, G.-T., *J. Chem. Soc.* p. 594 (1952).
1373. Lerner, A. B., *Nature* **184**, 674 (1959).
1374. Lerner, A. B., and Case, J. D., *J. Invest. Dermatol.* **32**, 211 (1959).
1375. Lerner, A. B., and Lee, T. H., *J. Am. Chem. Soc.* **77**, 1066 (1955).

1376. Lerner, A. B., Lee, T. H., Wright, M. R., and McGuire, J. S., *Acta Endocrinol.* Suppl. 47–50, 73 (1959–1960).
1377. Lerner, A. B., and McGuire, J. S., *Nature* **189,** 176 (1961).
1378. Lerner, A. B., and Wright, M. R., *Methods of Biochem. Anal.* **8,** 295 (1960).
1379. Levene, P. A., and Schormüller, A., *J. Biol. Chem.* **105,** 547 (1934).
1380. Levi, I., Weed, J. W. R., LaFlamme, G., and Koller, A. E., *Can. J. Chem.* **39,** 2491 (1961).
1381. Levin, Y., Berger, A., and Katchalski, E., *Biochem. J.* **63,** 308 (1956).
1382. Levine, S., *J. Am. Chem. Soc.* **76,** 1382 (1954).
1383. Lewis, G. P., *J. Physiol.* (*London*) **140,** 285 (1958).
1384. Lewis, G. P., *J. Physiol.* (*London*) **147,** 458 (1959).
1385. Lewis, G. P., *Nature* **188,** 999 (1960).
1386. Lewis, G. P., *Physiol. Rev.* **40,** 647 (1960).
1387. Lewis, G. P., *Nature* **192,** 596 (1961).
1388. Lewis, G. P., "The Scientific Basis of Medicine," p. 242. Annual Reviews, Palo Alto, California, 1962.
1389. Lewis, G. P., *Ann. N. Y. Acad. Sci.* **104,** 236 (1963).
1390. Lewis, G. P., and Work, T., *J. Physiol.* (*London*) **135,** 7P (1957).
1391. Lewis, J. C., and Snell, N. S., *J. Am. Chem. Soc.* **73,** 4812 (1951).
1392. Li, C. H., *Advan. Protein Chem.* **11,** 101 (1956).
1393. Li, C. H., *Advan. Protein Chem.* **12,** 269 (1957).
1394. Li, C. H., *Symp. Protein Struct. Paris, 1957* p. 302 (1958).
1395. Li, C. H., *Science* **129,** 969 (1959).
1396. Li, C. H., *Lab. Invest.* **8,** 574 (1959).
1397. Li, C. H., *J. Biol. Chem.* **235,** 1383 (1960).
1398. Li, C. H., *Vitamins Hormones* **19,** 313 (1961).
1399. Li, C. H., *Surv. Biol. Progr.* **4,** 93 (1962).
1400. Li, C. H., *Recent Progr. Hormone Res.* **18,** 1 (1962).
1400a. Li, C. H., *Nature* **201,** 924 (1964).
1401. Li, C. H., and Bertsch, L., *J. Biol. Chem.* **235,** 2638 (1960).
1401a. Li, C. H., Chung, D., and Ramachandran, J., *J. Am. Chem. Soc.* **86,** 2715 (1964).
1402. Li, C. H., Chung, D., Ramachandran, J., and Gorup, B., *J. Am. Chem. Soc.* **84,** 2460 (1962).
1403. Li, C. H., Cole, D., Chung, D., and Léonis, J., *J. Biol. Chem.* **227,** 207 (1957).
1404. Li, C. H., and Dixon, J. S., *Science* **124,** 934 (1956).
1405. Li, C. H., Dixon, J. S., and Chung, D., *J. Am. Chem. Soc.* **80,** 2587 (1958).
1406. Li, C. H., Dixon, J. S., and Chung, D., *Biochim. Biophys. Acta* **46,** 324 (1961).
1407. Li, C. H., Fonss-Bech, P., Geschwind, I. I., Hayashida, T., Hungerford, G. F., Lostroh, A. J., Lyons, W. R., Moon, H. D., Reinhardt, W. O., and Sideman, M., *J. Exptl. Med.* **105,** 335 (1957).
1408. Li, C. H., Geschwind, I. I., Cole, R. D., Raacke, I. D., Harris, J. I., and Dixon, J. S., *Nature* **176,** 687 (1955).
1409. Li, C. H., Geschwind, I. I., Dixon, J. S., Levy, A. L., and Harris, J. I., *J. Biol. Chem.* **213,** 171 (1955).
1410. Li, C. H., Geschwind, I. I., Levy, A. L., Harris, J. I., Dixon, J. S., Pon, N. G., and Porath, J. O., *Nature* **173,** 251 (1954).
1411. Li, C. H., Gorup, B., Chung, D., and Ramachandran, J., *J. Org. Chem.* **28,** 178 (1963).

1412. Li, C. H., Liu, W.-K., and Dixon, J. S., *Arch. Biochem. Biophys.* Suppl. 1, 327 (1962).

1413. Li, C. H., Meienhofer, J., Schnabel, E., Chung, D., Lo, T.-B., and Ramachandran, J., *J. Am. Chem. Soc.* **82,** 5760 (1960).

1414. Li, C. H., Meienhofer, J., Schnabel, E., Chung, D., Lo, T.-B., and Ramachandran, J., *J. Am. Chem. Soc.* **83,** 4449 (1961).

1415. Li, C. H., Schnabel, E., and Chung, D., *J. Am. Chem. Soc.* **82,** 2062 (1960).

1415a. Li, C. H., Ramachandran, J., and Chung, D., *J. Am. Chem. Soc.* **85,** 1895 (1963).

1415b. Li, C. H., Ramachandran, J., and Chung, D., *J. Am. Chem. Soc.* **86,** 2711 (1964).

1415c. Li, C. H., Ramachandran, J., Chung, D., and Gorup, B., *J. Am. Chem. Soc.* **86,** 2703 (1964).

1415d. Li, C. H., Ramachandran, J., and Chung, D., *Proc. 6th Intern. Congr. Biochem., New York, 1964,* Abstract II, p. 165 (1964).

1416. Li, C. H., Schnabel, E., Chung, D., and Lo, T.-B., *Nature* **189,** 143 (1961).

1416a. Li, C. H., Loh, T.-P., Chang, M.-C., Wu, S.-L., and Hsing, C.-Y., *Acta Chim. Sinica* **29,** 133 (1963).

1417. Li, S. O., *J. Am. Chem. Soc.* **74,** 5959 (1952).

1418. Li, S. O., and Eakin, R. E., *J. Am. Chem. Soc.* **77,** 1866 (1955).

1419. Liberek, B., *Chem. Ind. (London)* p. 987 (1961).

1420. Liberek, B., *Bull. Acad. Polon. Sci., Ser. Sci. Chim.* **10,** 227 (1962).

1421. Liberek, B., *Bull. Acad. Polon. Sci., Ser. Sci. Chim.* **10,** 407 (1962).

1421a. Liberek, B., *Tetrahedron Letters* p. 925 (1963).

1421b. Liberek, B., *Tetrahedron Letters* p. 1103 (1963).

1421c. Liberek, B., and Grzonka, Z., *Tetrahedron Letters* p. 159 (1964).

1422. Liberek, B., and Nowicka, A., *Bull. Acad. Polon. Sci., Ser. Sci. Chim.* **10,** 413 (1962).

1422a. Liberek, B., Nowicka, A., and Grzonka, Z., *Tetrahedron Letters* p. 1479 (1963).

1423. Lichtlen, P., Schaub, F., and Bühlmann, A., *Schweiz. Med. Wochschr.* **92,** 639 (1962).

1424. Liefländer, M., Dissertation, Georg-August-Universität Göttingen, 1958.

1425. Liefländer, M., *Z. Physiol. Chem.* **320,** 35 (1960).

1426. Liefländer, M., *Z. Physiol. Chem.* **329,** 1 (1962).

1427. Liefländer, M., *Naturwissenschaften* **49,** 394 (1962).

1428. Liefländer, M., *Naturwissenschaften* **49,** 541 (1962).

1429. Liefländer, M., and Thomas, K., *Z. Physiol. Chem.* **331,** 154 (1963).

1430. Lies, T., Plapinger, R. E., and Wagner-Jauregg, T., *J. Am. Chem. Soc.* **75,** 5755 (1953).

1431. Light, A., and du Vigneaud, V., *Proc. Soc. Exptl. Biol. Med.* **98,** 692 (1958).

1432. Lindenmann, A., Khan, N. H., and Hofmann, K., *J. Am. Chem. Soc.* **74,** 476 (1952).

1433. Livermore, A. H., and Muecke, E. C., *Nature* **173,** 265 (1954).

1434. Liwschitz, Y., Edlitz-Pfeffermann, Y., and Lapidoth, Y., *J. Am. Chem. Soc.* **78,** 3069 (1956).

1435. Liwschitz, Y., Irsay, R. D., and Vincze, A. I., *J. Chem. Soc.* p. 1308 (1959).

1436. Liwschitz, Y., Nemes, E., and Levi, E., *J. Org. Chem.* **27,** 3555 (1962).

1437. Liwschitz, Y., Rabinsohn, Y., and Singermann, A., *J. Chem. Soc.* p. 3726 (1962).

1438. Liwschitz, Y., Vincze, A. I., and Nemes, E., *Bull. Res. Council Israel* **9A,** 49 (1960).
1439. Liwschitz, Y., and Zilkha, A., *J. Am. Chem. Soc.* **76,** 3698 (1954).
1440. Liwschitz, Y., and Zilkha, A., *J. Am. Chem. Soc.* **77,** 1265 (1955).
1441. Liwschitz, Y., and Zilkha, A., *J. Chem. Soc.* p. 4394 (1957).
1442. Liwschitz, Y., Zilkha, A., and Amiel, Y., *J. Am. Chem. Soc.* **78,** 3067 (1956).
1443. Liwschitz, Y., Zilkha, A., Borensztain, H., and Frankel, M., *J. Org. Chem.* **21,** 1531 (1956).
1444. Liwschitz, Y., Zilkha, A., and Nemes, E., *Bull. Res. Council Israel* **10A,** 107 (1961).
1445. Lo, T.-B., *J. Biochem.* (*Tokyo*) **52,** 409 (1962).
1446. Lo, T.-B., Dixon, J. S., and Li, C. H., *Biochim. Biophys. Acta* **53,** 584 (1961).
1447. Lockhart, I. M., and Abraham, E. P., *Biochem. J.* **58,** 633 (1954).
1448. Lockhart, I. M., and Abraham, E. P., *Biochem. J.* **62,** 645 (1956).
1449. Lockhart, I. M., Abraham, E. P., and Newton, G. G. F., *Biochem. J.* **61,** 534 (1955).
1450. Long, J. M., Krivoy, W. A., and Guillemin, R., *Endocrinology* **69,** 176 (1961).
1451. Loring, H. S., and du Vigneaud, V., *J. Biol. Chem.* **111,** 385 (1935).
1452. Losse, G., *Chem. Ber.* **90,** 1919 (1957).
1452a. Losse, G., and Bachmann, G., *Chem. Ber.* **97,** 2671 (1964).
1452b. Losse, G., and Bachmann, G., *Z. Chem.* **4,** 204 (1964).
1452c. Losse, G., and Bachmann, G., *Z. Chem.* **4,** 241 (1964).
1453. Losse, G., and Augustin, M., *Chem. Ber.* **91,** 2418 (1958).
1453a. Losse, G., Barth, A., and Jasche, K., *J. Prakt. Chem.* [4] **21,** 32 (1963).
1454. Losse, G., Barth, A., and Langenbeck, W., *Chem. Ber.* **95,** 918 (1962).
1455. Losse, G., and Demuth, E., *Chem. Ber.* **94,** 1762 (1961).
1455a. Losse, G., Jeschkeit, H., and Knopf, D., *Chem. Ber.* **97,** 1789 (1964).
1456. Losse, G., Jeschkeit, H., and Langenbeck, W., *Z. Chem.* **1,** 279 (1961).
1457. Losse, G., Jeschkeit, H., and Langenbeck, W., *Chem. Ber.* **96,** 204 (1963).
1458. Losse, G., and Moschall, G., *J. Prakt. Chem.* [4] **7,** 38 (1958).
1459. Losse, G., and Müller, G., *Chem. Ber.* **94,** 2768 (1961).
1460. Losse, G., and Müller, G., *Z. Physiol. Chem.* **327,** 205 (1962).
1460a. Losse, G., and Nadolski, D., *J. Prakt. Chem.* [4] **24,** 118 (1964).
1461. Losse, G., and Weddige, H., *Ann. Chem. Liebigs* **636,** 144 (1960).
1462. Losse, G., and Weddige, H., *Angew. Chem.* **72,** 323 (1960).
1463. Losse, G., and Zönnchen, W., *Angew. Chem.* **72,** 385 (1960).
1464. Losse, G., and Zönnchen, W., *Ann. Chem. Liebigs* **636,** 140 (1960).
1465. Lu, Z.-X., and Tsao, T.-C., *Sci. Sinica* (*Peking*) **12,** 83 (1963).
1466. Lübke, K., Dissertation, Freie Universität Berlin, 1961.
1466a. Lübke, K., Hempel, R., and Schröder, E., *Experientia* **21,** 84 (1965).
1467. Lübke, K., and Schröder, E., *Z. Naturforsch.* **16b,** 765 (1961).
1468. Lübke, K., and Schröder, E., *Z. Naturforsch.* **16b,** 847 (1961).
1469. Lübke, K., and Schröder, E., *Ann. Chem. Liebigs* **665,** 205 (1963).
1469a. Lübke, K., and Schröder, E., *Ann. Chem. Liebigs* **681,** 250 (1965).
1470. Lübke, K., Schröder, E., Schmiechen, R., and Gibian, H., *Ann. Chem. Liebigs* **679,** 195 (1964).
1471. Lucas, F., Shaw, J. T. B., and Smith, S. G., *Biochem. J.* **66,** 468 (1957).
1472. Lucas, F., Shaw, J. T. B., and Smith, S. G., *Advan. Protein Chem.* **13,** 107 (1958).

1473. Ludueña, F. P., *Rev. Soc. Arg. Biol.* **16,** 358 (1940).
1474. Lutz, W. B., Ressler, C., Nettleton, D. E., Jr., and du Vigneaud, V., *J. Am. Chem. Soc.* **81,** 167 (1959).
1475. McCann, S. M., *Endocrinology* **60,** 664 (1957).
1476. McCann, S. M., and Fruit, A., *Proc. Soc. Exptl. Biol. Med.* **96,** 566 (1957).
1477. McCann, S. M., and Haberland, P., *Proc. Soc. Exptl. Biol. Med.* **102,** 319 (1959).
1478. McChesney, E. W., and Swann, W. K., Jr., *J. Am. Chem. Soc.* **59,** 1116 (1937).
1479. McCubbin, J. W., and Page, I. H., *Circulation Res.* **2,** 35 (1954).
1480. McCubbin, J. W., Page, I. H., and Bumpus, F. M., *Circulation Res.* **5,** 458 (1957).
1480a. Macdonald, C. G., and Shannon, J. S., *Tetrahedron Letters* p. 3113 (1964).
1481. McDonald, R. K., and Weise, V. K., *Proc. Soc. Exptl. Biol. Med.* **52,** 481 (1956).
1482. McDonald, R. K., Weise, V. K., and Patric, R. W., *Proc. Soc. Exptl. Biol. Med.* **53,** 348 (1956).
1483. McGregor, W. H., and Carpenter, F. H., *J. Org. Chem.* **26,** 1849 (1961).
1484. McGuire, J. S., and Lerner, A. B., *Ann. N. Y. Acad. Sci.* **100,** 622 (1963).
1485. McHale, D., Mamalis, P., and Green, J., *J. Chem. Soc.* p. 2847 (1960).
1486. McKay, A. F., *Chem. Rev.* **51,** 301 (1952).
1487. McKay, F. C., and Albertson, N. F., *J. Am. Chem. Soc.* **79,** 4686 (1957).
1488. MacLaren, J. A., *Proc. Intern. Wool. Textile Res. Conf. Australia* Vol. C, p. 168 (1955). CSIRO, Victoria, Australia.
1489. MacLaren, J. A., *Australian J. Chem.* **11,** 360 (1958).
1490. MacLaren, J. A., *Experientia* **17,** 346 (1961).
1491. MacLaren, J. A., Savige, W. E., and Swan, J. M., *Australian J. Chem.* **11,** 345 (1958).
1492. McLeod, C., *J. Bacteriol.* **56,** 749 (1948).
1493. Magee, M. Z., and Hofmann, K., *J. Am. Chem. Soc.* **71,** 1515 (1949).
1494. Majumdar, S. K., *Indian J. Appl. Chem.* **22,** 228 (1959).
1495. Majumdar, S. K., and Bose, S. K., *Ann. Biochem. Exptl. Med. (Calcutta)* **15,** 127 (1955).
1496. Majumdar, S. K., and Bose, S. K., *Nature* **181,** 134 (1958).
1497. Majumdar, S. K., and Bose, S. K., *Biochem. J.* **74,** 596 (1960).
1497a. Makhlouf, G. M., McManus, J. P. A., and Card, W. I., *Lancet* **II,** 485 (1964).
1497b. Makhlouf, G. M., McManus, J. P. A., and Card, W. I., *Gut* **5,** 379 (1964).
1498. Mamiofe, S. M., and Sinitsyna, Z. T., *Antibiotiki* **6,** 551 (1961).
1499. Mandelstam, P., Loercher, L., and Strominger, J. L., *J. Biol. Chem.* **237,** 2683 (1962).
1499a. Marchiori, F., Rocchi, R., and Scoffone, E., *Gazz. Chim. Ital.* **93,** 834 (1963).
1500. Mardashev, S. R., and Yurkevich, A. M., *Zh. Obshch. Khim.* **29,** 4056 (1959).
1501. Margoliash, E., *J. Biol. Chem.* **237,** 2161 (1962).
1502. Margoliash, E., Kimmel, J. R., Hill, R. L., and Schmidt, W. R., *J. Biol. Chem.* **237,** 2148 (1962).
1503. Margoliash, E., and Smith, E. L., *J. Biol. Chem.* **237,** 2151 (1962).
1504. Margoliash, E., Smith, E. L., Kreil, G., and Tuppy, H., *Nature* **192,** 1121 (1961).
1505. Marks, G. S., and Neuberger, A., *J. Chem. Soc.* p. 4872 (1961).

1506. Marsh, R. E., Corey, R. B., and Pauling, L., *Biochim. Biophys. Acta* **16,** 1 (1955).
1507. Marshall, R., Winitz, M., Birnbaum, S. M., and Greenstein, J. P., *J. Am. Chem. Soc.* **79,** 4538 (1957).
1508. Martini, L., and Pecile, A., *J. Endocrinol.* **24,** xi (1962).
1508a. Marumo, S., and Curtis, R. W., *Phytochemistry* **1,** 245 (1962).
1509. Mäsiar, P., *Collection Czech. Chem. Commun.* **27,** *1598* (1962).
1510. Masson, G. M. C., del Greco, F., Corcoran, A. C., and Page, I. H., *Proc. Soc. Exptl. Biol. Med.* **83,** 631 (1953).
1511. Mathias, A. P., and Schachter, M., *Brit. J. Pharmacol.* **13,** 326 (1958).
1512. Matschinsky, F., Meyer, U., and Wieland, O., *Biochem. Z.* **333,** 48 (1960).
1513. Matschinsky, F., and Wieland, O., *Biochem. Z.* **333,** 33 (1960).
1514. Matsubara, H., and Smith, E. L., *J. Biol. Chem.* **237,** PC 3575 (1962).
1515. Mazur, R. H., *Can. J. Chem.* **40,** 1098 (1962).
1515a. Mazur, R. H., *J. Org. Chem.* **28,** 2498 (1963).
1516. Mechanic, G. L., and Levy, M., *J. Am. Chem. Soc.* **81,** 1889 (1959).
1517. Medzihradszky, K., *Collection Czech. Chem. Commun.* **24,** 107 (1959).
1518. Medzihradszky, K., *Chimia (Aarau)* **14,** 375 (1960).
1519. Medzihradszky, K., and Bajusz, S., *in* "Peptide Symposium," Oxford, 1962 (G. T. Young, ed.), p. 49. Macmillan (Pergamon), New York, 1963.
1520. Medzihradszky, K., and Bruckner, G., *Magy. Kem. Folyoirat* **64,** 307 (1958).
1521. Medzihradszky, K., Bruckner, V., Katjár, M., Löw, M., Bajusz, S., and Kisfaludy, L., *Acta Chim. Acad. Sci. Hung.* **30,** 105 (1962).
1522. Medzihradszky-Schweiger, H., *Acta Chim. Acad. Sci. Hung.* **34,** 213 (1962).
1522a. Meienhofer, J., *Chemiker-Zt.* **87,** 788 (1963).
1522b. Meienhofer, J., and Brinkhoff, O., *Nature* **199,** 1095 (1963).
1522c. Meienhofer, J., *Z. Naturforsch.* **19b,** 114 (1964).
1523. Meienhofer, J., and du Vigneaud, V., *J. Am. Chem. Soc.* **82,** 2279 (1960).
1524. Meienhofer, J., and du Vigneaud, V., *J. Am. Chem. Soc.* **82,** 6336 (1960).
1525. Meienhofer, J., and du Vigneaud, V., *J. Am. Chem. Soc.* **83,** 142 (1961).
1526. Meienhofer, J., and Li, C. H., *J. Am. Chem. Soc.* **84,** 2434 (1962).
1527. Meienhofer, J., and Li, C. H., *Biochim. Biophys. Acta* **54,** 351 (1961).
1527a. Meienhofer, J., Schnabel, E., Bremer, H., Brinkhoff, O., Zabel, R., Sroka, W., Klostermeyer, H., Brandenburg, D., Okuda, T., and Zahn, H., *Z. Naturforsch.* **18b,** 1120 (1963).
1527b. Meienhofer, J., and Zahn, H., *Angew. Chem.* **75,** 1126 (1963).
1528. Meier, R., Gross, F., Tripod, J., and Turrian, H., *Experientia* **13,** 361 (1957).
1529. Mejer, S., *Roczniki Chem.* **36,** 1115 (1962).
1530. Mejer, S., Kocor, M., and Taschner, E., *Roczniki Chem.* **32,** 277 (1958).
1531. Mélon, J., and Lecomte, J., *Compt. Rend. Soc. Biol.* **155,** 947 (1961).
1532. Mellander, O., Bennich, H., and Fölsch, G., *Acta Soc. Med. Upsalien.* **64,** 303 (1959).
1533. Melville, J., *Biochem. J.* **29,** 179 (1935).
1534. Méndez-Bauer, C. J., Carballo, M. A., Cabot, H. M., Negreiros De Paiva, C. E., and Gonzáles-Panizza, V. H., *in* "Oxytocin" (R. Caldeyro-Barcia and H. Heller, eds.), p. 325. Macmillan (Pergamon), New York, 1961.
1535. Merrifield, R. B., *J. Biol. Chem.* **232,** 43 (1958).
1535a. Merrifield, R. B., *J. Am. Chem. Soc.* **85,** 2149 (1964).
1535b. Merrifield, R. B., *J. Am. Chem. Soc.* **86,** 304 (1964).
1535c. Merrifield, R. B., *Biochemistry* **3,** 1385 (1964).

1535d. Merrifield, R. B., *J. Org. Chem.* **29,** 3100 (1964).
1536. Merrifield, R. B., and Woolley, D. W., *Arch. Biochem. Biophys.* **56,** 265 (1955).
1537. Merrifield, R. B., and Woolley, D. W., *J. Am. Chem. Soc.* **78,** 358 (1956).
1538. Merrifield, R. B., and Woolley, D. W., *J. Am. Chem. Soc.* **78,** 4646 (1956).
1539. Merrifield, R. B., and Woolley, D. W., *J. Am. Chem. Soc.* **80,** 6635 (1958).
1540. Mertz, D. P., *Arch. Exptl. Pathol. Pharmakol.* **239,** 410 (1960).
1541. Mertz, D. P., *Arch. Exptl. Pathol. Pharmakol.* **244,** 405 (1963).
1541a. Mertz, D. P., *Arch. Exptl. Pathol. Pharmakol.* **246,** 338 (1964).
1542. Micheel, F., *J. Polymer. Sci.* **12,** 577 (1954).
1543. Micheel, F., and Alfes, H., *Makromol. Chem.* **48,** 33 (1961).
1543a. Micheel, F., Berlenbach, W., and Weichbrodt, K., *Chem. Ber.* **85,** 189 (1952).
1544. Micheel, F., Emde, H., and Dörner, H., *Z. Physiol. Chem.* **275,** 258 (1942).
1545. Micheel, F., and Frowein, A., *Chem. Ber.* **92,** 304 (1959).
1546. Micheel, F., and Habendorff, R., *Chem. Ber.* **90,** 1590 (1957).
1547. Micheel, F., and Haneke, H., *Chem. Ber.* **92,** 309 (1959).
1548. Micheel, F., and Haneke, H., *Chem. Ber.* **95,** 1009 (1962).
1549. Micheel, F., and Heesing, A., *Ann. Chem. Liebigs* **604,** 34 (1957).
1550. Micheel, F., and Herold, B., *Z. Physiol. Chem.* **293,** 187 (1953).
1551. Micheel, F., and Hülsmann, H. L., *Chem. Ber.* **93,** 13 (1960).
1552. Micheel, F., and Meckstroth, W., *Chem. Ber.* **92,** 1675 (1959).
1553. Micheel, F., Ostmann, E.-A., and Alfes, F., *Tetrahedron* **18,** 1155 (1962).
1554. Micheel, F., and Schmidt, K., *Makromol. Chem.* **3,** 210 (1949).
1555. Micheel, F., and Thomas, S., *Chem. Ber.* **90,** 2906 (1957).
1556. Mikeš, O., Schuh, V., and Šorm, F., *Collection Czech. Chem. Commun.* **24,** 583 (1959).
1557. Mikeš, O., and Šorm, F., *Collection Czech. Chem. Commun.* **24,** 1897 (1959).
1558. Miller, A., Neidle, A., and Waelsch, H., *Arch. Biochem. Biophys.* **56,** 11 (1955).
1559. Miller, G. L., Behrens, O., and du Vigneaud, V., *J. Biol. Chem.* **140,** 411 (1941).
1560. Miller, H. K., and Waelsch, H., *Arch. Biochem. Biophys.* **35,** 176 (1952).
1561. Miller, H. K., and Waelsch, H., *J. Am. Chem. Soc.* **74,** 1092 (1952).
1562. Milne, H. B., Halver, J. E., Ho, D. S., and Mason, M. S., *J. Am. Chem. Soc.* **79,** 637 (1957).
1563. Milne, H. B., and Peng, C.-H., *J. Am. Chem. Soc.* **79,** 639 (1957).
1564. Milne, H. B., and Peng, C.-H., *J. Am. Chem. Soc.* **79,** 645 (1957).
1564a. Mitchell, J. C., *Arch. Dermatol.* **88,** 48 (1963).
1565. Miyanoki, Y., *J. Biochem. (Tokyo)* **13,** 389 (1931).
1566. Miyao, K., *Bull. Agr. Chem. Soc. Japan* **19,** 86 (1955).
1567. Miyao, K., *Bull. Agr. Chem. Soc. Japan* **24,** 23 (1960).
1568. Moloney, P. J., and Coval, M., *Biochem. J.* **59,** 179 (1955).
1568a. Montervino, C., Motolese, M., and Jacobelli, A., *Ann. Endocrinol. (Paris)* **24,** 1055 (1963).
1569. Moore, J. A., Dice, J. R., Nicolaides, E. D., Westland, R. D., and Wittle, E. L., *J. Am. Chem. Soc.* **76,** 2884 (1954).
1570. Morawiecki, A., Siemion, I. Z., and Nowak, K., *Roczniki Chem.* **36,** 983 (1962).
1571. Morel, F., and Jard, S., *Am. J. Physiol.* **204,** 227 (1963).

1572. Moriya, H., Pierce, J. V., and Webster, M. E., *Ann. N. Y. Acad. Sci.* **104,** 172 (1963).

1573. Morozova, E. A., and Zhenodarova, S. M., *Zh. Obshch. Khim.* **28,** 1661 (1958).

1574. Morozova, E. A., and Zhenodarova, S. M., *Dokl. Akad. Nauk SSSR* **125,** 93 (1959).

1575. Morozova, E. A., and Zhenodarova, S. M., *Zh. Obshch. Khim.* **31,** 45 (1961).

1576. Morris, C. J., Thompson, J. T., Asen, A., and Irreverre, F., *J. Biol. Chem.* **237,** 2180 (1962).

1577. Morris, D., *Biochem. J.* **76,** 349 (1960).

1578. Morito, T., *J. Agr. Chem. Soc. Japan* **36,** 18 (1962).

1579. Morito, T., *J. Agr. Chem. Soc. Japan* **36,** 24 (1962).

1580. Morito, T., *J. Agr. Chem. Soc. Japan* **36,** 393 (1962).

1581. Morito, T., *J. Agr. Chem. Soc. Japan* **36,** 398 (1962).

1582. Mowat, J. H., Gazzola, A. L., Hutchings, B. L., Boothe, J. H., Waller, C. W., Angier, R. B., Semb, J., and SubbaRow, Y., *J. Am. Chem. Soc.* **71,** 2308 (1949).

1583. Mowat, J. H., Hutchings, B. L., Angier, R. B., Stokstad, E. L. R., Boothe, J. H., Waller, C. W., Semb, J., and SubbaRow, Y., *J. Am. Chem. Soc.* **70,** 1096 (1948).

1584. Müller, H. A., *in* "Oxytocin" (R. Caldeyro-Barcia and H. Heller, eds.), p. 137. Macmillan (Pergamon), New York, 1961.

1585. Mulrow, P. J., Ganong, W. F., and Boryczka, A., *Proc. Soc. Exptl. Biol. Med.* **112,** 7 (1963).

1586. Munsick, R. A., Sawyer, W. H., and van Dyke, H. B., *Endocrinology* **63,** 688 (1958).

1587. Muramatsu, I., *Nippon Kagaku Zasshi* **82,** 83 (1961).

1588. Murase, Y., Okawa, K., and Akabori, S., *Bull. Chem. Soc. Japan* **33,** 123 (1960).

1589. Murray, F. J., and Tetrault, P. A., *Bacteriol. Proc.* (*Soc. Am. Bacteriologists*) **1,** 20 (1948).

1590. Murray, F. J., Tetrault, P. A., Kaufmann, O. W., Koffler, H., Peterson, D. H., and Colingsworth, D. R., *J. Bacteriol.* **57,** 305 (1949).

1591. Nagai, Y., *J. Biochem.* (*Tokyo*) **50,** 486 (1961).

1592. Nagai, Y., *J. Biochem.* (*Tokyo*) **50,** 493 (1961).

1593. Nagai, Y., and Noda, H., *Biochim. Biophys. Acta* **34,** 298 (1959).

1594. Nagai, Y., Sakakibara, S., Noda, H., and Akabori, S., *Biochim. Biophys. Acta* **37,** 567 (1960).

1594a. Nagamatsu, A., Watanabe, M., and Okuma, T., *Kyushu J. Med. Sci.* **13,** 99 (1962).

1595. Narahara, H. T., and Williams, R. H., *Endocrinology* **60,** 285 (1957).

1596. Narita, K., *J. Am. Chem. Soc.* **81,** 1751 (1959).

1597. Nefkens, G. H. L., *Nature* **185,** 309 (1960).

1598. Nefkens, G. H. L., *Nature* **193,** 974 (1962).

1598a. Nefkens, G. H. L., and Nivard, R. J. F., *Rec. Trav. Chim.* **83,** 199 (1964).

1599. Nefkens, G. H. L., and Tesser, G. I., *J. Am. Chem. Soc.* **83,** 1263 (1961).

1600. Nefkens, G. H. L., Tesser, G. I., and Nivard, R. J. F., *Rec. Trav. Chim.* **79,** 688 (1960).

1600a. Nefkens, G. H. L., Tesser, G. I., and Nivard, R. J. F., *Rec. Trav. Chim.* **82,** 941 (1963).

1601. O'Neill, J. J., Veitch, F. P., and Wagner-Jauregg, T., *J. Org. Chem.* **21,** 363 (1956).
1601a. Nesvadba, H., Honzl, J., and Rudinger, J., *Collection Czech. Chem. Commun.* **28,** 1691 (1963).
1602. Nesvadba, H., and Young, G. T., *Tetrahedron Letters* p. 361 (1963).
1603. Neuberger, A., *Advan. Protein Chem.* **4,** 297 (1948).
1604. Neuberger, A., and Sanger, F., *Biochem. J.* **37,** 515 (1943).
1605. Neumann, H., Levin, Y., Berger, A., and Katchalski, E., *Biochem. J.* **73,** 33 (1959).
1606. Neumann, R. E., and Smith, E. L., *J. Biol. Chem.* **193,** 97 (1951).
1607. Newton, B. A., *Bacteriol. Rev.* **20,** 14 (1956).
1608. Newton, G. G. F., and Abraham, E. P., *Biochem. J.* **47,** 257 (1950).
1609. Newton, G. G. F., and Abraham, E. P., *Nature* **171,** 606 (1953).
1610. Newton, G. G. F., and Abraham, E. P., *Biochem. J.* **53,** 597 (1953).
1611. Newton, G. G. F., and Abraham, E. P., *Biochem. J.* **53,** 604 (1953).
1612. Newton, G. G. F., and Abraham, E. P., *Biochem. J.* **62,** 651 (1956).
1613. Nicol, D. S. H. W., *Biochim. Biophys. Acta* **34,** 257 (1959).
1614. Nicol, D. S. H. W., *Biochem. J.* **75,** 395 (1960).
1615. Nicol, D. S. H. W., and Smith, L. F., *Nature* **187,** 483 (1960).
1615a. Nicolaides, E. D., Craft, M. K., and DeWald, H. A., *J. Med. Chem.* **6,** 524 (1963).
1616. Nicolaides, E. D., and DeWald, H. A., *J. Org. Chem.* **26,** 3872 (1961).
1616a. Nicolaides, E. D., DeWald, H. A., *J. Org. Chem.* **28,** 1926 (1963).
1617. Nicolaides, E. D., DeWald, H. A., and Craft, M. K., *Ann. N. Y. Acad. Sci.* **104,** 15 (1963).
1617a. Nicolaides, E. D., DeWald, H. A., and Craft, M. K., *J. Med. Chem.* **6,** 739 (1963).
1618. Nicolaides, E. D., DeWald, H. A., and McCarthy, D. A., *Biochem. Biophys. Res. Commun.* **6,** 210 (1961).
1619. Nicolaides, E. D., DeWald, H. A., Shorley, P. G., and Collier, H. O. I., *Nature* **187,** 773 (1960).
1620. Nicolet, B. H., *J. Biol. Chem.* **88,** 389 (1930).
1621. Nicot, C., and Bricas, E., *Compt. Rend.* **256,** 1391 (1963).
1622. Niu, C.-I., and Fraenkel-Conrat, H., *J. Am. Chem. Soc.* **77,** 5882 (1955).
1623. Niu, C.-I., Ke, L. T., Chi, K. Y., Chen, C. C., and Kung, Y. T., *Sci. Sinica (Peking)* **12,** 327 (1963).
1623a. Niu, C.-I., Kung, Y.-T., Huang, W.-T., Ke, L.-T., Chen, C.-C., Chen, Y.-C., Du, Y.-C., Jiang, R.-Q., Tsou, C.-L., Hu, S.-C., Chu, S.-Q., and Wang, K.-Z., *Sci. Sinica (Peking)* **13,** 1343 (1964).
1624. Noda, Y., *Nippon Kagaku Zasshi* **79,** 662 (1958).
1625. Noda, Y., *Nippon Kagaku Zasshi* **80,** 411 (1959).
1626. Noguchi, J., Hayakawa, T., and Hiraoka, M., *Nippon Kagaku Zasshi* **82,** 604 (1961).
1627. Noguchi, J., Hayakawa, T., Nishimura, M., Ishizaki, S., Shimotani, T., Ebata, M., Saitô, T., and Nakayama, N., *Nippon Kagaku Zasshi* **77,** 463 (1956).
1628. Noguchi, J., Saitô, T., and Hayakawa, T., *Nippon Kagaku Zasshi* **80,** 82 (1959).
1629. Nolan, C., *Dissertation Abstr.* **22,** 1388 (1961).
1630. Nolan, C., and Smith, E. L., *J. Biol. Chem.* **237,** 446 (1962).

1631. Nolan, C., and Smith, E. L., *J. Biol. Chem.* **237**, 453 (1962).
1632. Nordwig, A., Hörmann, H., Kühn, K., and Grassmann, W., *Z. Physiol. Chem.* **326**, 242 (1961).
1632a. Nowak, K., *Roczniki Chem.* **38**, 129 (1964).
1633. Nowak, K., and Siemion, I. Z., *Roczniki Chem.* **35**, 153 (1961).
1634. Nowak, K., Siemion, I. Z., and Siemieniewski, H., *Roczniki Chem.* **36**, 557 (1962).
1635. Nuenke, R. H., *Dissertation Abstr.* **22**, 54 (1961).
1636. Nyman, M. A., and Herbst, R. M., *J. Org. Chem.* **15**, 108 (1950).
1637. Oda, T., Kinoshita, M., Yamanaka, O., and Ueda, F., *J. Pharm. Soc. Japan* **74**, 1243 (1954).
1638. Oda, T., and Ueda, F., *J. Pharm. Soc. Japan* **74**, 1246 (1954).
1639. Oertel, G. W., *Angew. Chem.* **70**, 51 (1960).
1640. Ogawa, H., and Ito, T., *J. Agr. Chem. Soc. Japan* **24**, 191 (1951).
1641. Ogawa, H., and Ito, T., *J. Agr. Chem. Soc. Japan* **26**, 432 (1952).
1642. Ōhira, T., *Bull. Agr. Chem. Soc. Japan* **16**, 10 (1940).
1643. Ōhira, T., *Bull. Agr. Chem. Soc. Japan* **16**, 71 (1940).
1644. Ōhira, T., *Bull. Agr. Chem. Soc. Japan* **18**, 74 (1942).
1645. Ōhira, T., Imai, K., and Kidō, H., *Bull. Agr. Chem. Soc. Japan* **28**, 765 (1954).
1646. Ōhira, T., Imai, K., and Kidō, H., *Bull. Agr. Chem. Soc. Japan* **29**, 237 (1955).
1647. Ōhira, T., Kidō, H., and Imai, K., *Bull. Agr. Chem. Soc. Japan* **29**, 520 (1955).
1648. Ohno, K., *J. Biochem. (Tokyo)* **40**, 621 (1953).
1649. Okabe, K., *J. Antibiot. (Tokyo) Ser. A* **12**, 90 (1959).
1650. Okabe, K., Yonehara, H., and Umezawa, H., *J. Antibiot. (Tokyo) Ser. A* **12**, 192 (1959).
1651. Okabe, S., *Osaka Shiritsu Daigaku Igaku Zasshi* **9**, 5017 (1960).
1652. Okawa, K., *Nippon Kagaku Zasshi* **75**, 1199 (1954).
1653. Okawa, K., *Bull. Chem. Soc. Japan* **29**, 486 (1956).
1654. Okawa, K., *Bull. Chem. Soc. Japan* **29**, 488 (1956).
1655. Okawa, K., *Bull. Chem. Soc. Japan* **30**, 976 (1957).
1656. Okawa, K., *Bull. Chem. Soc. Japan* **31**, 88 (1958).
1656a. Okawa, K., and Hase, S., *Bull. Chem. Soc. Japan* **36**, 754 (1963).
1657. Okoshi, S., Kitano, N., Tomoda, T., Usui, M., Takashio, M., Suzuki, N., and Konoshi, T., *J. Antibiot. (Tokyo) Ser. A.* **13**, 137 (1960).
1658. Okuda, K., Lin, C.-S., and Winnick, T., *Nature* **195**, 1067 (1962).
1659. Okui, S., and Suzuki, S., *J. Pharm. Soc. Japan (Yakugaku Zasshi)* **79**, 471 (1959).
1660. Olmsted, F., and Page, I. H., *Am. J. Physiol.* **203**, 951 (1962).
1661. Olofson, R. A., Dissertation, Harvard University, Cambridge, Massachusetts, 1961.
1662. Olofson, R. A., *Dissertation Abstr.* **12**, 1413 (1961).
1663. Ondetti, M. A., *J. Med. Chem.* **6**, 10 (1963).
1664. Ondetti, M. A., and Bodanszky, M., *Chem. Ind. (London)* p. 697 (1962).
1665. Ono, T., *J. Japan. Biochem. Soc.* **27**, 278 (1955).
1666. Otani, S., Nippon Academy Conference of Antibiotic Substances, Japan, p. 3997 (1957); *Chem. Abstr.* **52**, 16704 (1958).
1667. Otano, S., and Saito, Y., *Proc. Japan Acad.* **30**, 991 (1954).

1667a. Otani, S., and Saito, Y., *J. Biochem.* (*Tokyo*) **56,** 103 (1964).
1668. Otani, T. T., and Winitz, M., *Arch. Biochem. Biophys.* **90,** 254 (1960).
1669. Otey, M. C., and Greenstein, J. P., *Arch. Biochem. Biophys.* **53,** 501 (1954).
1669a. Otsuka, H., and Inouye, K., *Bull. Chem. Soc. Japan* **37,** 289 (1964).
1670. Ovchinnikov, Y. A., Ivanov, V. T., Kiryushkin, A. A., and Chalilanina, K. C., *Izv. Akad. Nauk SSSR Otd. Khim. Nauk* p. 578 (1963).
1670a. Ovchinnikov, Y. A., Ivanov, V. T., Kiryushkin, A. A., and Shemyakin, M. M., *in* "Peptide Symposium," Oxford, 1962 (G. T. Young, ed.), p. 207. Macmillan (Pergamon), New York, 1963.
1670b. Ovchinnikov, Y. A., Ivanov, V. T., Kiryushkin, A. A., and Shemyakin, M. M., *Dokl. Akad. Nauk SSSR* **153,** 122 (1963).
1670c. Ovchinnikov, Y. A., Ivanov, V. T., Kiryushkin, A. A., and Shemyakin, M. M., *Izv. Akad. Nauk SSSR Otd. Khim.* p. 1497 (1963).
1671. Ovchinnikov, Y. A., Kiryushkin, A. A., Ivanov, V. T., and Shemyakin, M. M., *Izv. Akad. Nauk SSSR Otd. Khim. Nauk* p. 770 (1963).
1672. Overell, B. G., and Petrow, V., *J. Chem. Soc.* p. 232 (1955).
1673. Page, I. H., *in* "Polypeptides which Stimulate Plain Muscle" (J. R. Gaddum, ed.), p. 71 Livingstone, Edinburgh and London, 1955.
1674. Page, I. H., and Bumpus, F. M., *Physiol. Rev.* **41,** 331 (1961).
1675. Page, I. H., and Bumpus, F. M., *Recent Progr. Hormone Res.* **18,** 167 (1962).
1676. Page, I. H., and Bumpus, F. M., *Clin. Pharmacol. Therap.* **3,** 758 (1962).
1677. Page, I. H., and Helmer, O. M., *J. Exptl. Med.* **71,** 29 (1940).
1678. Page, I. H., and Helmer, O. M., *J. Exptl. Med.* **71,** 495 (1940).
1679. Page, I. H., McCubbin, J. W., Schwarz, H., and Bumpus, F. M., *Circulation Res.* **5,** 552 (1957).
1680. Page, I. H., and Olmsted, F., *Am. J. Physiol.* **201,** 92 (1961).
1681. Paiva, A. C. M., thesis, Universität Sao Paolo, 1954.
1682. Paiva, A. C. M., Bandiera, T., and Prado, J. L., *Science* **120,** 611 (1954).
1683. Paiva, A. C. M., and Pavia, T. B., *Biochem. Pharmacol.* **5,** 187 (1960).
1684. Paiva, A. C. M., and Paiva, T. B., *Biochim. Biophys. Acta* **48,** 412 (1961).
1685. Paiva, A. C. M., and Paiva, T. B., *Biochim. Biophys. Acta* **56,** 339 (1962).
1686. Paiva, T. B., and Paiva, A. C. M., *J. Pharmacol. Exptl. Therap.* **130,** 177 (1960).
1687. Paiva, T. B., and Paiva, A. C. M., *Brit. J. Pharmacol.* **15,** 557 (1960).
1687a. Paiva, T. B., Paiva, A. C. M., and Scheraga, H. A., *Biochemistry* **2,** 1327 (1963).
1688. Paladini, A., and Craig, L. C., *J. Am. Chem. Soc.* **76,** 688 (1954).
1688a. Paladini, A. C., Franze de Ferandez, M. T., and Delius, A. E., *Proc. 6th Intern. Congr. Biochem. New York, 1964,* p. 173.
1689. Paléus, S., and Tuppy, H., *Acta Chem. Scand.* **13,** 641 (1959).
1690. Pannemann, H. J., Marx, A. F., and Arens, J. F., *Rec. Trav. Chim.* **78,** 487 (1959).
1691. Pantlitschko, M., and Gründig, E., *Monatsh. Chem.* **89,** 274 (1958).
1692. Pantlitschko, M., and Gründig, E., *Monatsh. Chem.* **89,** 489 (1958).
1693. Park, C. R., Morgan, H. E., Henderson, M. J., Regen, D. M., Cadenas, E., and Post, R. L., *Recent Progr. Hormone Res.* **17,** 493 (1961).
1694. Patchett, A. A., and Witkop, B., *J. Am. Chem. Soc.* **79,** 185 (1957).
1695. Patchornik, A., Berger, A., and Katchalski, E., *J. Am. Chem. Soc.* **79,** 6416 (1957).

1696. Patchornik, A., Lawson, W. B., Gross, E., and Witkop, B., *J. Am. Chem. Soc.* **82**, 5923 (1960).
1696a. Patchornik, A., and Sokolovsky, M., *J. Am. Chem. Soc.* **86**, 1206 (1964).
1697. Patterson, W. I., and du Vigneaud, V., *J. Biol. Chem.* **111**, 393 (1935).
1698. Pätzold, W., Dissertation, Rheinisch-Westfälische Hochschule Aachen, 1961.
1699. Paul, R., *J. Org. Chem.* **28**, 236 (1963).
1700. Paul, R., and Anderson, G. W., *J. Am. Chem. Soc.* **82**, 4596 (1960).
1701. Paul, R., and Anderson, G. W., *J. Org. Chem.* **27**, 2094 (1962).
1702. Paul, R., Anderson, G. W., and Callahan, F. M., *J. Org. Chem.* **26**, 3347 (1961).
1703. Pauling, L., and Corey, R. B., *Fortschr. Chem. Org. Naturstoffe* **11**, 180 (1954).
1704. Peart, W. S., *Biochem. J.* **59**, 300 (1955).
1705. Peart, W. S., *Biochem. J.* **62**, 520 (1956).
1706. Peart, W. S., *Ergeb. Physiol. Biol. Chem. Exptl. Pharmakol.* **50**, 409 (1959).
1707. Peart, W. S., *in* "Essentielle Hypertonie" (K. D. Bock and P. T. Cottier, ed.), p. 127. Springer, Berlin, 1960.
1708. Peck, R. L., Wolf, D. E., and Folkers, K., *J. Am. Chem. Soc.* **74**, 1999 (1952).
1709. Pedersen, K. O., and Synge, R. L. M., *Acta Chem. Scand.* **2**, 408 (1948).
1709a. Pereira, J. R., *Arch. Intern. Pharmacodyn.* **147**, 450 (1964).
1710. Pereira, J. R., and Blustos, R. E., *Arch. Intern. Pharmacodyn.* **141**, 551 (1963).
1711. Perlmann, G. E., *Advan. Protein Chem.* **10**, 1 (1955).
1712. Pernow, B., *Acta Physiol. Scand.* **29**, Suppl. 105, 1 (1953).
1713. Pernow, B., *Ann. N. Y. Acad. Sci.* **104**, 393 (1963).
1714. Peter, H., Dissertation, Eidgenössische Technische Hochschule Zürich 1961.
1715. Peterson, D. H., and Reineke, L. M., *J. Biol. Chem.* **181**, 95 (1949).
1716. Peterson, D. H., and Reineke, L. M., *J. Clin. Invest.* **28**, 1053 (1949).
1716a. Photaki, I., *J. Am. Chem. Soc.* **85**, 1123 (1963).
1716b. Photaki, I., and du Vigneaud, V., *in* "Proceedings of the 6th European Symp. on Peptides, Athens, 1963." Macmillan (Pergamon), New York (in press).
1716c. Phillips, J. G., and Bellamy, D., *in* "Comparative Endocrinology" (U. S. von Euler and H. Heller, eds.), Vol. 1. p. 208. Academic Press, New York, 1963.
1717. Picarelli, Z. P., Henriques, O. B., and Oliveira, M. C. F., *Experientia* **18, 77** (1962).
1718. Picarelli, Z. P., Kupper, R., Prado, E. S., and Valle, J. R., *Circulation Res.* **2**, 354 (1954).
1719. Pickering, B. T., and Heller, H., *Nature* **184**, 1463 (1959).
1720. Pickering, B. T., and Li, C. H., *Biochim. Biophys. Acta* **62**, 475 (1962).
1720a. Pickering, B. T., and Li, C. H., *Biochim. Biophys. Acta* **74**, 156 (1963).
1720b. Pickering, B. T., and Li, C. H., *Arch. Biochem. Biophys.* **104**, 119 (1964).
1721. Pickford, M., *in* "Polypeptides which Affect Smooth Muscles and Blood Vessels" (M. Schachter, ed.), p. 42. Macmillan (Pergamon), New York, 1960.
1722. Pickford, M., *in* "Oxytocin" (R. Caldeyro-Barcia and H. Heller, eds.), p. 68. Macmillan (Pergamon), New York, 1961.
1723. Pierce, J. G., Gordon, S., and du Vigneaud, V., *J. Biol. Chem.* **199**, 929 (1952).

1724. Pierce, J. V., and Webster, M. E., *Biochem. Biophys. Res. Commun.* **5**, 353 (1961).
1725. Pimlott, P. J. E., and Young, G. T., *Proc. Chem. Soc.* p. 257 (1958).
1726. Plapinger, R. E., and Wagner-Jauregg, T., *J. Am. Chem. Soc.* **75**, 5757 (1953).
1727. Platt, D., and Finn, F. M., *J. Org. Chem.* **27**, 2958 (1962).
1728. Plattner, P. A., and Nager, U., *Helv. Chim. Acta* **31**, 665 (1948).
1729. Plattner, P. A., and Nager, U., *Helv. Chim. Acta* **31**, 2192 (1948).
1730. Plattner, P. A., and Nager, U., *Helv. Chim. Acta* **31**, 2203 (1948).
1731. Plattner, P. A., Nager, U., and Boller, A., *Helv. Chim. Acta* **31**, 594 (1948).
1732. Plattner, P. A., Vogler, K., Studer, R. O., Quitt, P., and Keller-Schierlein, W., *Experientia* **19**, 71 (1963).
1733. Plattner, P. A., Vogler, K., Studer, R. O., Quitt, P., and Keller-Schierlein, W., *Helv. Chim. Acta* **46**, 927 (1963).
1734. Plentl, A. A., and Page, I. H., *J. Biol. Chem.* **147**, 135 (1943).
1735. Plentl, A. A., and Page, I. H., *J. Biol. Chem.* **158**, 49 (1945).
1736. Plentl, A. A., and Page, I. H., *J. Biol. Chem.* **163**, 49 (1946).
1737. Plentl, A. A., Page, I. H., and Davis, W. W., *J. Biol. Chem.* **147**, 143 (1943).
1738. Pless, J., Dissertation, Universität Basel, 1961.
1738a. Pless, J., *in* "Peptide Symposium" (G. T. Young, ed.), p. 69. Macmillan (Pergamon), New York, 1963.
1738b. Pless, J., and Boissonnas, R. A., *Helv. Chim. Acta* **46**, 1609 (1963).
1739. Pless, J., Stürmer, E., Guttmann, S., and Boissonnas, R. A., *Helv. Chim. Acta* **45**, 394 (1962).
1740. Plimmer, R. H. A., *Biochem. J.* **35**, 461 (1941).
1741. Poddubnaya, N. A., Aleynikova, M. J., and Egorov, T. A., *Zh. Obshch. Khim.* **30**, 3591 (1960).
1742. Poddubnaya, N. A., and Maksimov, V. I., *Zh. Obshch. Khim.* **30**, 308 (1960).
1742a. Poduska, K., Proc. 7th European Symp. on Peptides, Budapest, 1964, *Acta Chim. Acad. Sci. Hung.* (in press).
1743. Poduška, K., and Gross, H., *Chimia (Aarau)* **14**, 371 (1960).
1744. Poduška, K., and Gross, H., *Chem. Ber.* **94**, 527 (1961).
1745. Poduška, K., and Rudinger, J., *Collection Czech. Chem. Commun.* **22**, 1283 (1957).
1746. Poduška, K., and Rudinger, J., *Collection Czech. Chem. Commun.* **24**, 3449 (1959).
1747. Poduška, K., Rudinger, J., and Šorm, F., *Collection Czech. Chem. Commun.* **20**, 1174 (1955).
1748. Polglase, W. J., and Smith, E. L., *J. Am. Chem. Soc.* **71**, 3081 (1949).
1749. Popenoe, E. A., Doherty, D. C., and Link, K. P., *J. Am. Chem. Soc.* **75**, 3469 (1953).
1750. Popenoe, E. A., and du Vigneaud, V., *J. Am. Chem. Soc.* **76**, 6202 (1954).
1751. Popenoe, E. A., Lawler, H. C., and du Vigneaud, V., *J. Am. Chem. Soc.* **74**, 3713 (1952).
1752. Porath, J., Roos, P., Landgrebe, F. W., and Mitchell, G. M., *Biochim. Biophys. Acta* **17**, 598 (1955).
1753. Porath, J., and Schally, A. V., *Endocrinology* **70**, 738 (1962).
1754. Poroshin, K. T., Debabov, V. G., and Maksimov, V. I., *Izv. Akad. Nauk SSSR Otd. Khim. Nauk* p. 1134 (1961).
1755. Poroshin, K. T., Debabov, V. G., Shibnev, V. A., and Kosarenko, T. D., *Zh. Obshch. Khim.* **31**, 3006 (1961).

1756. Poroshin, K. T., Kosarenko, T. D., Shibnev, V. A., and Debabov, V. G., *Izv. Akad. Nauk. SSSR Otd. Khim. Nauk* p. 1851 (1959).
1757. Poroshin, K. T., Kosarenko, T. D., Shibnev, V. A., and Debabov, V. G., *Biokhimiya* **26**, 244 (1961).
1758. Poroshin, K. T., Shibnev, V. A., and Debabov, V. G., *Proc. 5th Intern. Congr. Biochem. Moscow, 1961,* Abstr., Sect. 2, p. 125 (1961).
1759. Poroshin, K. T., Shibnev, V. A., Debabov, V. G., and Kosarenko, T. D., *Biokhimiya* **25**, 693 (1960).
1760. Poroshin, K. T., Shibnev, V. A., and Kosarenko, T. D., *Izv. Akad. Nauk SSSR Otd. Khim. Nauk* p. 736 (1959).
1761. Poroshin, K. T., Shibnev, V. A., Kosarenko, T. D., and Debabov, V. G., *Vysokomolekul. Soedin.* **3**, 122 (1961).
1762. Poseiro, J. J., and Noriega-Guerra, L., *in* "Oxytocin" (R. Caldeyro-Barcia and H. Heller, eds.), p. 158. Macmillan (Pergamon). New York, 1961.
1763. Posternak, T., and Grafl, S., *Helv. Chim. Acta* **28**, 1258 (1945).
1764. Prado, E. S., Beraldo, W. T., and Rocha e Silva, M., *Arch. Biochem. Biophys.* **27**, 410 (1950).
1765. Prado, E. S., Prado, J. L., and Brandi, C. M. W., *Arch. Intern. Pharmacodyn.* **137**, 358 (1962).
1766. Prado, J. L., Monier, R., Prado, E. S., and Fromageot, C., *Biochim. Biophys. Acta* **22**, 87 (1956).
1767. Prado, J. L., Prado, E. S., Brandi, C. M. W., and Katchburian, A. V., *Ann. N. Y. Acad. Sci.* **104**, 186 (1963).
1768. Prado, J. L., Valle, J. R., and Picarelli, Z. P., *Acta Physiol. Latinoam.* **4**, 104 (1954).
1769. Pravda, Z., *Collection Czech. Chem. Commun.* **24**, 2083 (1959).
1770. Pravda, Z., and Rudinger, J., *Collection Czech. Chem. Commun.* **20**, 1 (1955).
1771. Prelog, V., and Wieland, P., *Helv. Chim. Acta* **29**, 1128 (1946).
1772. Privat de Garilhe, M., and Gros, C., *in* "Biologisch Aktive Peptide," 27. Tagung Deut. Pharmakol. Ges., Wien, 1962, p. 7. Eigenverlag Wiener med. Akad., Wien.
1773. Privat de Garilhe, M., and Gros, C., *Arch. Exptl. Pathol. Pharmakol.* **245**, 184 (1963).
1774. Privat de Garilhe, M., Gros, C., Chauvet, J., Fromageot, C., Miolhe-Voloss, C., and Benoit, J., *Biochim. Biophys. Acta* **29**, 603 (1958).
1775. Privat de Garilhe, M., Gros, C., Lozac'h, Y., and Garnuchot, B., *Experientia* **18**, 92 (1962).
1776. Privat de Garilhe, M., Gros, C., Porath, J., and Lindner, E.-B., *Experientia* **16**, 414 (1960).
1776a. Prout, T. E., *J. Chronic Diseases* **15**, 879 (1962).
1776b. Prout, T. E., *Metab. Clin. Exptl.* **12**, 673 (1963).
1777. Prox, A., Dissertation, Technische Hochschule München, 1962.
1778. Quastel, J. H., Stewart, C. P., and Tunnicliffe, H. E., *Biochem. J.* **17**, 586 (1923).
1778a. Quitt, P., *in* "Peptide Symposium," Oxford, 1962 (G. T. Young, ed.), p. 165. Macmillan (Pergamon), New York, 1963.
1779. Quitt, P., Hellerbach, J., and Vogler, K., *Helv. Chim. Acta* **46**, 327 (1963).
1779a. Quitt, P., Studer, R. O., and Vogler, K., *Helv. Chim. Acta* **46**, 1715 (1963).
1779b. Quitt, P., Studer, R. O., and Vogler, K., *Helv. Chim. Acta* **47**, 166 (1964).

1780. Raben, M. S., Landolt, R., Smith, F. A., Hofmann, K., and Yajima, H., *Nature* **189,** 681 (1961).
1781. Rabinowitz, J. L., *Acta Chem. Scand.* **13,** 1463 (1959).
1782. Radenhausen, R., *J. Prakt. Chem.* **52,** 446 (1895).
1783. Rall, T. W., Sutherland, E. W., and Berthet, J., *J. Biol. Chem.* **224,** 463 (1957).
1784. Ram, M., Ingle, T. R., and Damodaran, M., *J. Sci. Ind. Res.* (*India*) **17C,** 21 (1958).
1785. Ramachandran, J., and Li, C. H., *J. Org. Chem.* **27,** 4006 (1962).
1786. Ramachandran, J., and Li, C. H., *J. Org. Chem.* **28,** 173 (1963).
1787. Ramachandran, L. K., and McConnell, W. B., *Can. J. Chem.* **33,** 1638 (1955).
1788. Ramachandran, L. K., and Sastry, L. V. S., *Biochemistry* **1,** 75 (1962).
1789. Ramachandran, L. K., and Witkop, B., *J. Am. Chem. Soc.* **81,** 4028 (1959).
1790. Rasmussen, H., Schwartz, I. L., Schoessler, M. A., and Hochster, G., *Proc. Natl. Acad. Sci. U. S.* **46,** 1278 (1960).
1791. Rasmussen, H., and Craig, L., *Endocrinology* **68,** 1051 (1961).
1792. Riley, G., Turnbull, J. H., and Wilson, W., *Chem. Ind.* (*London*) p. 1181 (1953).
1793. Riley, G., Turnbull, J. H., and Wilson, W., *J. Chem. Soc.* p. 1373 (1957).
1794. Raudonat, H. W., and Rocha e Silva, M., *Arch. Exptl. Pathol. Pharmakol.* **243,** 232 (1962).
1795. Reed, L. J., Kidwai, A. R., and du Vigneaud, V., *J. Biol. Chem.* **180,** 571 (1949).
1796. Rees, P. S., Tong, D. P., and Young, G. T., *J. Chem. Soc.* p. 662 (1954).
1797. Regna, P. P., *Am. J. Med.* **18,** 686 (1955).
1798. Regna, P. P., Solomons, I. A., Forscher, B. K., and Timreck, A. E., *J. Clin. Invest.* **28,** 1022 (1948).
1798a. Regoli, D., Riniker, B., and Brunner, H., *Biochem. Pharmacol.* **12,** 637 (1963).
1799. Rehbinder, D., Löffler, G., Wieland, O., and Wieland, T., *Z. Physiol. Chem.* **331,** 132 (1963).
1800. Reich, E., Goldberg, I. H., and Rabinowitz, M., *Nature* **196,** 743 (1962).
1801. Rerup, C., *Acta Endocrinol.* **25,** 17 (1957).
1802. Ressler, C., *Proc. Soc. Exptl. Biol. Med.* **92,** 725 (1956).
1803. Ressler, C., *J. Am. Chem. Soc.* **78,** 5956 (1956).
1804. Ressler, C., *Science* **128,** 1281 (1958).
1805. Ressler, C., *J. Am. Chem. Soc.* **82,** 1641 (1960).
1806. Ressler, C., and du Vigneaud, V., *J. Am. Chem. Soc.* **76,** 3107 (1954).
1807. Ressler, C., and du Vigneaud, V., *J. Am. Chem. Soc.* **79,** 4511 (1957).
1808. Ressler, C., and Rachele, J. R., *Proc. Soc. Exptl. Biol. Med.* **98,** 170 (1958).
1809. Ressler, C., and Ratzkin, H., *J. Org. Chem.* **26,** 3356 (1961).
1810. Rzeszotarska, B., Taschner, E., and Wasielewski, C., *Collection Czech. Chem. Commun.* **27,** 2238 (1962).
1811. Richards, F. M., and Vithayathil, P. J., *J. Biol. Chem.* **234,** 1459 (1959).
1812. Rieche, A., and Gross, H., *Chem. Ber.* **92,** 83 (1959).
1813. Ried, W., and Beck, B. M., *Ann. Chem. Liebigs* **646,** 96 (1961).
1814. Ried, W., and Czack, A., *Ann. Chem. Liebigs* **642,** 133 (1961).
1815. Ried, W., and Franz, G., *Ann. Chem. Liebigs* **644,** 141 (1961).
1816. Ried, W., and Marquard, K., *Ann. Chem. Liebigs* **642,** 141 (1961).
1817. Ried, W., and Schleimer, B., *Ann. Chem. Liebigs* **619,** 43 (1958).

1818. Riegel, B., and du Vigneaud, V., *J. Biol. Chem.* **112,** 149 (1935–1936).
1819. Rinderknecht, H., Thomas, D., and Aslin, S., *Helv. Chim. Acta* **41,** 1 (1958).
1820. Riniker, B., Brunner, H., and Schwyzer, R., *Angew. Chem.* **74,** 469 (1962).
1821. Riniker, B., and Schwyzer, R., *Helv. Chim. Acta* **44,** 658 (1961).
1822. Riniker, B., and Schwyzer, R., *Helv. Chim. Acta* **44,** 674 (1961).
1823. Riniker, B., and Schwyzer, R., *Helv. Chim. Acta* **44,** 677 (1961).
1824. Riniker, B., and Schwyzer, R., *Helv. Chim. Acta* **44,** 685 (1961).
1824a. Riniker, B., and Schwyzer, R., *Helv. Chim. Acta* **47,** 2357 (1964).
1825. Rittel, W., *Helv. Chim. Acta* **45,** 2465 (1962).
1826. Rittel, W., Iselin, B., Kappeler, H., Riniker, B., and Schwyzer, R., *Angew. Chem.* **69,** 179 (1957).
1827. Rittel, W., Iselin, B., Kappeler, H., Riniker, B., and Schwyzer, R., *Helv. Chim. Acta* **40,** 614 (1957).
1828. Roberts, C. W., *J. Am. Chem. Soc.* **76,** 6203 (1954).
1829. Roberts, C. W., and du Vigneaud, V., *J. Biol. Chem.* **204,** 871 (1953).
1829a. Rocchi, R., Marchiori, F., and Scoffone, E., *Gazz. Chim. Ital.* **93,** 823 (1963).
1830. Rocha e Silva, M., *in* "Polypeptides which Stimulate Plain Muscle" (J. R. Gaddum, ed.), p. 57. Livingston, Edinburgh and London, 1955.
1831. Rocha e Silva, M., *21st Intern. Congr. Physiol. Sci., Buenos Aires, 1959, Symp. Spec. Lectures* p. 50 (1959).
1832. Rocha e Silva, M., *in* "Polypeptides which Affect Smooth Muscles and Blood Vessels" (M. Schachter, ed.), p. 210. Macmillan (Pergamon), New York, 1960.
1833. Rocha e Silva, M., *Biochem. Pharmacol.* **8,** 177 (1961).
1834. Rocha e Silva, M., *Biochem. Pharmacol.* **10,** 3 (1962).
1835. Rocha e Silva, M., *Acta Physiol. Latinoam.* **12,** 66 (1962).
1836. Rocha e Silva, M., *Ann. N. Y. Acad. Sci.* **104,** 190 (1963).
1836a. Rocha e Silva, M., *in* "Comparative Endocrinology" (U. S. von Euler and H. Heller, eds.), Vol. 2, p. 64. Academic Press, New York, 1963.
1836b. Roch e Silva, M., *Actualites Pharmacol.* **15,** 163 (1963).
1837. Rocha e Silva, M., Beraldo, W. T., and Rosenfeld, G., *Am. J. Physiol.* **156,** 261 (1949).
1838. Rocha e Silva, M., and Holzhacker, E. L., *Arch. Intern. Pharmacodyn.* **122,** 168 (1959).
1838a. Rocha e Silva, M., and Leme, J. G., *Med. Exptl.* **8,** 287 (1963).
1839. Rocha e Silva, M., Jr., and Malnic, G., *Med. Exptl.* **7,** 329 (1962).
1840. Roper, J. A., and McIlwain, H., *Biochem. J.* **42,** 485 (1948).
1841. Rosenberg, H., Ennor, A. H., Hagerman, D. D., and Sugai, S., *Biochem. J.* **84,** 536 (1962).
1841a. Ross, D. L., Skinner, C. G., and Shive, W., *J. Med. Pharm. Chem.* **3,** 519 (1961).
1842. Rosenberg, L. L., Evans, E. S., and Simpson, M. E., *Endocrinology* **68,** 1 (1961).
1843. Rowlands, D. A., and Young, G. T., *J. Chem. Soc.* p. 3937 (1952).
1844. Rowlands, D. A., and Young, G. T., *Biochem. J.* **65,** 516 (1957).
1845. Rudinger, J., *Collection Czech. Chem. Commun.* **16,** 615 (1951).
1846. Rudinger, J., *Collection Czech. Chem. Commun.* **19,** 365 (1954).
1847. Rudinger, J., *Collection Czech. Chem. Commun.* **19,** 375 (1954).
1848. Rudinger, J., *Collection Czech. Chem. Commun.* **24,** 95 (1959).
1849. Rudinger, J., *Angew. Chem.* **71,** 742 (1959).

1850. Rudinger, J., *Record Chem. Progr.* (*Kresge-Hooker Sci. Lib.*) **23,** 3 (1962).
1851. Rudinger, J., *Acta Biol. Med. Ger.* Suppl. **2,** 38 (1963).
1851a. Rudinger, J., *in* "Polypeptide Symposium," Oxford, 1962 (G. T. Young, ed.), p. 133. Macmillan (Pergamon), New York, 1963.
1852. Rudinger, J., and Czurbová, H., *Collection Czech. Chem. Commun.* **19,** 386 (1954).
1852a. Rudinger, J., and Farkašová, H., *Collection Czech. Chem. Commun.* **28,** 2941 (1963).
1853. Rudinger, J., and Honzl, J., *Chimia* (*Aarau*) **14,** 367 (1960).
1854. Rudinger, J., Honzl, J., and Zaoral, M., *Collection Czech. Chem. Commun.* **21,** 202 (1956).
1855. Rudinger, J., Honzl, J., and Zaoral, M., *Collection Czech. Chem. Commun.* **21,** 770 (1956).
1855a. Rudinger, J., and Jošt, K., *Experientia* **20,** 570 (1964).
1856. Rudinger, J., and Krejčí, I., *Experientia* **18,** 585 (1962).
1857. Rudinger, J., Krupička, J., Zaoral, M., and Černík, V., *Collection Czech. Chem. Commun.* **25,** 3338 (1960).
1858. Rudinger, J., Poduška, K., and Zaoral, M., *Collection Czech. Chem. Commun.* **25,** 2022 (1960).
1859. Rudinger, J., Poduška, K., Zaoral, M., and Jošt, K., *Collection Czech. Chem. Commun.* **24,** 2013 (1959).
1860. Rudinger, J., and Pravda, Z., *Collection Czech. Chem. Commun.* **23,** 1947 (1958).
1861. Rudinger, J., and Šorm, F., *Collection Czech. Chem. Commun.* **16,** 214 (1951).
1862. Roeske, R. W., *Chem. Ind.* (*London*) p. 1121 (1959).
1863. Roeske, R., Stewart, F. H. C., Stedman, R. J., and du Vigneaud, V., *J. Am. Chem. Soc.* **78,** 5883 (1956).
1864. Ross, C. A., Ludden, C. T., and Stone, C. A., *Proc. Soc. Exptl. Biol. Med.* **105,** 558 (1960).
1865. Rothe, I., and Rothe, M., *Chem. Ber.* **88,** 284 (1955).
1866. Rothe, M., *J. Polymer Sci.* **30,** 227 (1958).
1867. Rothe, M., *Acta Chim. Acad. Sci. Hung.* **18,** 449 (1959).
1868. Rothe, M., *Collection Czech. Chem. Commun.* **24,** 148 (1959).
1869. Rothe, M., *Angew. Chem.* **74,** 725 (1962).
1870. Rothe, M., Brünig, H., and Eppert, G., *J. Prakt. Chem.* [4] **8,** 323 (1959).
1871. Rothe, M., and Kunitz, F. W., *Ann. Chem. Liebigs* **609,** 88 (1957).
1872. Rothe, M., Rothe, I., Brünig, H., and Schwenke, K.-D., *Angew. Chem.* **71,** 700 (1959).
1872a. Rothe, M., Steffen, K.-D., and Rothe, I., *Angew. Chem.* **75,** 1206 (1963).
1873. Rudman, D., *J. Lipid Res.* **4,** 119 (1963).
1874. Rudman, D., Brown, S. J., and Malkin, M. F., *Endocrinology* **72,** 527 (1963).
1875. Rudman, D., Hirsch, R. L., Kendall, F. E., Seidman, F., and Brown, S. J., *Recent Progr. Hormone Res.* **18,** 89 (1962).
1876. Ruggli, P., Ratti, R., and Henzi, E., *Helv. Chim. Acta* **7,** 332 (1929).
1877. Rühlmann, K., *J. Prakt. Chem.* [4] **9,** 86 (1959).
1878. Rühlmann, K., *J. Prakt. Chem.* [4] **9,** 315 (1959).
1879. Rühlmann, K., *Chem. Ber.* **94,** 1876 (1961).
1880. Rühlmann, K., and Kaufmann, U., *Ann. Chem. Liebigs* **656,** 22 (1962).
1881. Rühlmann, K., and Michael, G., *Z. Naturforsch.* **15b,** 811 (1960).

1882. Russell, D. W., *Biochim. Biophys. Acta* **45,** 411 (1960).
1883. Russell, D. W., *J. Chem. Soc.* p. 753 (1962).
1883a. Russell, D. W., Macdonald, C. G., and Shannon, J. S., *Tetrahedron Letters* p. 2759 (1964).
1884. Russell, D. W., and Brown, M. E., *Biochim. Biophys. Acta* **38,** 382 (1960).
1885. Rydon, H. N., *Angew. Chem.* **71,** 741 (1959).
1886. Rydon, H. N., and Smith, P. G. W., *J. Chem. Soc.* p. 3642 (1956).
1887. Ryle, A. P., and Sanger, F., *Biochem. J.* **58,** V (1954).
1888. Sachs, H., and Brand, E., *J. Am. Chem. Soc.* **75,** 4608 (1953).
1889. Sachs, H., and Brand, E., *J. Am. Chem. Soc.* **75,** 4610 (1953).
1890. Sachs, H., and Brand, E., *J. Am. Chem. Soc.* **76,** 1811 (1954).
1891. Sachs, H., and Brand, E., *J. Am. Chem. Soc.* **76,** 1815 (1954).
1892. Sachs, H., and Brand, E., *J. Am. Chem. Soc.* **76,** 3601 (1954).
1893. Sachs, H., and Waelsch, H., *J. Am. Chem. Soc.* **77,** 660 (1955).
1894. Saffran, M., *Can. J. Biochem. Physiol.* **37,** 319 (1959).
1895. Saffran, M., and Schally, A. V., *Can. J. Biochem. Physiol.* **33,** 408 (1955).
1896. Saffran, M., and Schally, A. V., *Endocrinology* **56,** 523 (1955).
1897. Saffran, M., Schally, A. V., and Benfey, B. G., *Endocrinology* **57,** 439 (1955).
1898. Saito, Y., *Osaka Shiritsu Daigaku Igaku Zasshi* **7,** 499 (1958).
1899. Saito, Y., *Osaka Shiritsu Daigaku Igaku Zasshi* **7,** 503 (1958).
1900. Sakakibara, S., and Nagai, Y., *Bull. Chem. Soc. Japan* **33,** 1537 (1960).
1901. Sakami, W., and Toennies, G., *J. Biol. Chem.* **144,** 203 (1942).
1902. Sakiyama, F., *Bull. Chem. Soc. Japan* **35,** 1943 (1962).
1903. Sakiyama, F., Okawa, K., Yamakawa, T., and Akabori, S., *Bull. Chem. Soc. Japan* **31,** 926 (1958).
1904. Sandrin, E., and Boissonnas, R. A., *Experientia* **18,** 59 (1962).
1904a. Sandrin, E., and Boissonnas, R. A., *Helv. Chim. Acta* **46,** 1637 (1963).
1904b. Sandrin, E., and Boissonnas, R. A., *Helv. Chim. Acta* **47,** 417 (1964).
1904c. Sandrin, E, and Boissonnas, R. A., *Helv. Chim. Acta* **47,** 1294 (1964).
1905. Sanger, F., *Biochem. J.* **40,** 261 (1946).
1906. Sanger, F., *Nature* **171,** 1025 (1953).
1907. Sanger, F., *Brit. Med. Bull.* **16,** 183 (1960).
1908. Sanger, F., and Smith, L. F., *Endeavour* **16,** 48 (1957).
1909. Sanger, F., and Thompson, E. O. P., *Biochem. J.* **53,** 366 (1953).
1910. Sanger, F., and Tuppy, H., *Biochem. J.* **49,** 463 (1951).
1911. Sanger, F., and Tuppy, H., *Biochem. J.* **49,** 481 (1951).
1911a. Sarin, P. S., and Fasman, G. D., *Biochim. Biophys. Acta* **82,** 175 (1964).
1911b. Sarges, R., and Witkop, B., *J. Am. Chem. Soc.* **86,** 1861 (1964).
1911c. Sarges, R., and Witkop, B., *J. Am. Chem. Soc.* **86,** 1862 (1964).
1911d. Sarges, R., and Witkop, B., *J. Am. Chem. Soc.* (in press).
1912. Savige, W. E., *Australian J. Chem.* **14,** 664 (1961).
1913. Sawyer, W. H., *Endocrinology* **63,** 694 (1958).
1914. Sawyer, W. H., *Pharmacol. Rev.* **13,** 225 (1961).
1915. Sawyer, W. H., *Recent Progr. Hormone Res.* **17,** 437 (1961).
1916. Sawyer, W. H., Chan, W. Y., and van Dyke, H. B., *Endocrinology* **71,** 536 (1962).
1917. Sawyer, W. H., Munsick, R. A., and van Dyke, H. B., *Endocrinology* **68,** 215 (1961).
1918. Sayers, G., *Acta Endocrinol.* Suppl. 50, 25 (1960).

1919. Sayers, M. A., Sayers, G., and Woodbury, L. A., *Endocrinology* **42**, 379 (1948).
1920. Schachter, M., *in* "Polypeptides which Affect Smooth Muscles and Blood Vessels" (M. Schachter, ed.), p. 232. Macmillan (Pergamon), New York, 1960.
1921. Schachter, M., *in* "Recent Advances in Pharmacology" (J. M. Robson and R. S. Stacy, eds.), p. 156. Churchill, London, 1962.
1922. Schachter, M., *Ann. N. Y. Acad. Sci.* **104**, 108 (1963).
1922a. Schachter, M., *Ann. Rev. Pharmacol.* **4**, 281 (1964).
1923. Schallenberg, E. E., and Calvin, M., *J. Am. Chem. Soc.* **77**, 277 (1955).
1924. Schally, A. V., Andersen, R. N., Lipscomb, H. S., Long, J. M., and Guillemin, R., *Nature* **188**, 1192 (1960).
1925. Schally, A. V., Andersen, R. N., Long, J. M., and Guillemin, R., *Proc. Soc. Exptl. Biol. Med.* **104**, 290 (1960).
1925a. Schally, A. V., Bowers, C. Y., Kuroshima, A., Ishida, Y., Carter, W. H., and Redding, T. W., *Am. J. Physiol.* **207**, 378 (1964).
1926. Schally, A. V., and Guillemin, R., *Proc. Soc. Exptl. Biol. Med.* **100**, 138 (1959).
1927. Schally, A. V., and Guillemin, R., *Texas Rept. Biol. Med.* **18**, 133 (1960).
1928. Schally, A. V., and Guillemin, R., *Proc. Soc. Exptl. Biol. Med.* **112**, 1014 (1963).
1928a. Schally, A. V., and Guillemin, R., *J. Biol. Chem.* **239**, 1038 (1964).
1929. Schally, A. V., Lipscomb, H. S., and Guillemin, R., *Endocrinology* **71**, 164 (1962).
1929a. Schally, A. V., Saffran, M., and Zimmermann, B., *Biochem. J.* **70**, 97 (1959).
1929b. Schattenkerk, C., Visser, G. H., Kerling, K. E. T., and Havinga, E., *Rec. Trav. Chim.* **83**, 677 (1964).
1930. Schellenberg, P., and Ullrich, J., *Chem. Ber.* **92**, 1276 (1959).
1931. Schlögl, K., *Monatsh. Chem.* **89**, 61 (1958).
1932. Schlögl, K., and Fabitschowitz, H., *Monatsh. Chem.* **84**, 937 (1953).
1933. Schlögl, K., Wessely, F., and Wawersich, E., *Monatsh. Chem.* **84**, 706 (1953).
1934. Schlögl, K., Wessely, F., and Woidich, H., *Monatsh. Chem.* **87**, 425 (1956).
1935. Schmidt, G. M. J., Hodgkin, D. C., and Oughton, B. M., *Biochem. J.* **65**, 744 (1957).
1936. Schmidt, L., Bernauer, W., and Menzel, H., *Arzneimittel-Forsch.* **12**, 1019 (1962).
1937. Schmidt-Kastner, G., *Naturwissenschaften* **43**, 131 (1956).
1938. Schnabel, E., *Ann. Chem. Liebigs* **615**, 165 (1958).
1939. Schnabel, E., *Ann. Chem. Liebigs* **615**, 173 (1958).
1940. Schnabel, E., *Ann. Chem. Liebigs* **622**, 181 (1959).
1941. Schnabel, E., *Z. Physiol. Chem.* **314**, 114 (1959).
1942. Schnabel, E., *Ann. Chem. Liebigs* **659**, 168 (1962).
1942a. Schnabel, E., *Ann. Chem. Liebigs* **667**, 179 (1963).
1942b. Schnabel, E., *in* "Peptide Symposium," Oxford, 1962 (G. T. Young, ed.), p. 77. Macmillan (Pergamon), New York, 1963.
1942c. Schnabel, E., *Ann. Chem. Liebigs* **667**, 171 (1963).
1942d. Schnabel, E., *Z. Naturforsch.* **19b**, 120 (1964).
1942e. Schnabel, E., *Ann. Chem. Liebigs* **674**, 218 (1964).
1943. Schnabel, E., and Li, C. H., *J. Am. Chem. Soc.* **82**, 457 (1960).

1944. Schnabel, E., and Li, C. H., *J. Biol. Chem.* **235,** 2010 (1960).
1944a. Schnabel, E., and Meienhofer, J., Proc. 7th European Symp. on Peptides, Budapest, 1964, *Acta Chim. Acad. Sci. Hung.* (in press).
1945. Schnabel, E., and Zahn, H., *Monatsh. Chem.* **88,** 646 (1957).
1946. Schnabel, E., and Zahn, H., *Ann. Chem. Liebigs* **614,** 141 (1958).
1947. Schneider, C. H., and du Vigneaud, V., *J. Am. Chem. Soc.* **84,** 3005 (1962).
1948. Schneider, F., *Biochem. Z.* **291,** 328 (1937).
1949. Schneider, F., *Ann. Chem. Liebigs* **529,** 1 (1937).
1950. Schneider, F., *Z. Physiol. Chem.* **320,** 82 (1960).
1951. Schneider, F., *Z. Physiol. Chem.* **321,** 38 (1960).
1951a. Schneider, F., *Z. Physiol. Chem.* **332,** 38 (1963).
1952. Schneider, F., and Geyer, H.-U., *Z. Physiol. Chem.* **330,** 189 (1963).
1953. Schoenbach, E. B., Bryer, M. S., and Long, P. H., *Ann. N. Y. Acad. Sci.* **51,** 987 (1949).
1954. Schönenberg, H., *Deut. Med. Wochschr.* **85,** 1714 (1960).
1955. Schönenberg, H., *Antibiot. Chemotherapy* **13,** 136 (1963).
1956. Schönheimer, R., *Z. Physiol. Chem.* **154,** 203 (1926).
1957. Schott, H. F., Larkin, J. B., Rockland, L. B., and Dunn, M. S., *J. Org. Chem.* **12,** 490 (1947).
1958. Schramm, G., and Thumm, G., *Z. Naturforsch.* **3b,** 218 (1948).
1959. Schramm, G., and Wissmann, H., *Chem. Ber.* **91,** 1073 (1958).
1960. Schreiber, V., Eckertová, A., Franc, Z., Kočí, J., Rybák, M., and Kmentová, V., *Experientia* **17,** 264 (1961).
1960a. Schreiber, V., Eckertová, A., Franc, Z., Rybák, M., Gregorová, I., Kmentová, V., and Jirgl, V., *Physiol. Bohemoslov.* **12,** 1 (1963).
1961. Schreiber, V., Rybák, M., Eckertová, A., Jirgl, V., Kočí, J., Franc, Z., and Kmentová, V., *Experientia* **18,** 338 (1962).
1962. Schröder, E., *Ann. Chem. Liebigs* **670,** 127 (1963).
1963. Schröder, E., *Ann. Chem. Liebigs* **673,** 186 (1964).
1964. Schröder, E., *Ann. Chem. Liebigs* **673,** 220 (1964).
1965. Schröder, E., *Experientia* **20,** 39 (1964).
1966. Schröder, E., *Ann. Chem. Liebigs* **681,** 231 (1965).
1967. Schröder, E., *Ann. Chem. Liebigs* **681,** 241 (1965).
1968. Schröder, E., *Ann. Chem. Liebigs* (in press).
1969. Schröder, E., *In* "6th Peptid-Symposium, Athen, 1963," Macmillan (Pergamon), New York (in press).
1969a. Schröder, E., *Ann. Chem. Liebigs* **679,** 207 (1964).
1969b. Schröder, E., *Ann. Chem. Liebigs* **680,** 132 (1964).
1969c. Schröder, E., *Ann. Chem. Liebigs* **680,** 142 (1964).
1969d. Schröder, E., *Experientia* (in press).
1969e. Schröder, E., unpublished work.
1970. Schröder, E., and Gibian, H., *Z. Naturforsch.* **15b,** 814 (1960).
1971. Schröder, E., and Gibian, H., *Ann. Chem. Liebigs* **649,** 168 (1961).
1972. Schröder, E., and Gibian, H., *Ann. Chem. Liebigs* **656,** 190 (1962).
1973. Schröder, E., and Gibian, H., *Ann. Chem. Liebigs* **673,** 176 (1964).
1973a. Schröder, E., and Hempel, R., *Experientia* **20,** 529 (1964).
1973b. Schröder, E., and Hempel, R., *Ann. Chem. Liebigs* (in press).
1974. Schröder, E., and Klieger, E., *Ann. Chem. Liebigs* **673,** 208 (1964).
1975. Schröder, E., and Klieger, E., *Ann. Chem. Liebigs* **673,** 196 (1964).
1976. Schröder, E., Klieger, E., and Gibian, H., *Ann. Chem. Liebigs* **646,** 101 (1961).

1977. Schröder, E., and Lübke, K., *Ann. Chem. Liebigs* **655,** 211 (1962).
1978. Schröder, E., and Lübke, K., *Experientia* **19,** 57 (1963).
1979. Schröder, E., and Lübke, K., *in* "Peptide Symposium," Oxford, 1962 (G. T. Young, ed.), p. 195. Macmillan (Pergamon), New York, 1963.
1979a. Schröder, E., and Lübke, K., *Experientia* **20,** 19 (1964).
1979b. Schröder, E., Lübke, K., and Hempel, R., *Experientia* **21,** 70 (1965).
1980. Schröder, E., Lübke, K., Klieger, E., and Schmiechen, R., (unpublished work).
1980a. Schröder, E., Petras, H.-S., and Klieger, E., *Ann. Chem. Liebigs* **679,** 221 (1964).
1981. Schröder, E., and Schmiechen, R., *Collection Czech. Chem. Commun.* **27,** 2259 (1962).
1981a. Schuler, W., Schär, B., and Desaulles, P., *Schweiz. Med. Wochschr.* **93,** 1027 (1963).
1982. Schultze, H. E., *Deut. Med. Wochenschr.* **83,** 1742 (1958).
1983. Schulz, G., Dissertation, Freie Universität Berlin, 1958.
1984. Schumann, I., and Boissonnas, R. A., *Nature* **169,** 154 (1952).
1985. Schumann, I., and Boissonnas, R. A., *Helv. Chim. Acta* **35,** 2235 (1952).
1986. Schumann, I., and Boissonnas, R. A., *Helv. Chim. Acta* **35,** 2237 (1952).
1987. Schüssler, H., Dissertation, Rheinisch-Westfälische Technische Hochschule Aachen, 1960.
1988. Schüssler, H., and Zahn, H., *Chem. Ber.* **95,** 1076 (1962).
1989. Schwartz, B. S., Warren, M. R., Barkley, F. A., and Landis, L., *Antibiot. Ann.* p. 41 (1959–1960).
1990. Schwartz, I. L., Rasmussen, H., Schoessler, M. A., Silver, L., and Fong, C. T. O., *Proc. Natl. Acad. Sci. U. S.* **46,** 1288 (1960).
1991. Schwarz, H., and Arakawa, K., *J. Am. Chem. Soc.* **81,** 5691 (1959).
1992. Schwarz, H., and Bumpus, F. M., *J. Am. Chem. Soc.* **81,** 890 (1959).
1993. Schwarz, H., Bumpus, F. M., and Page, I. H., *J. Am. Chem. Soc.* **79,** 5697 (1957).
1994. Schwarz, H., and Page, I. H., *Federation Proc.* **13,** 293 (1954).
1995. Schwyzer, R., *Helv. Chim. Acta* **36,** 414 (1953).
1996. Schwyzer, R., *Helv. Chim. Acta* **37,** 647 (1954).
1997. Schwyzer, R., *Ciba Found. Symp. Amino Acids Peptides Antimetab. Activity* p. 171 (1958).
1998. Schwyzer, R., *Chimia (Aarau)* **12,** 53 (1958).
1999. Schwyzer, R., *Angew. Chem.* **71,** 742 (1959).
2000. Schwyzer, R., *17th Intern. Congr. Pure Appl. Chem., Munich, 1960* Vol. II, p. 130 (1960). Butterworth, London and Washington, D. C., and Verlag Chemie, Weinheim.
2001. Schwyzer, R., *Helv. Chim. Acta* **44,** 667 (1961).
2002. Schwyzer, R., *in* "Protides of the Biological Fluids" (H. Peeters, ed.), p. 27. Elsevier, Amsterdam, 1962.
2002a. Schwyzer, R., *Pure Appl. Chem.* **6,** 265 (1963).
2002b. Schwyzer, R., *in* "Proceedings of the International Pharmacology Meeting 9th Colloquium, Bruzes, 1961," Vol. 7, p. 203. Macmillan (Pergamon), New York, 1963.
2002c. Schwyzer, R., *Ergeb. Physiol. Biol. Chem. Exptl. Pharmakol.* **53,** 41 (1963).
2002d. Schwyzer, R., *Ann. Rev. Biochem.* **33,** 259 (1964).
2003. Schwyzer, R., and Aung, T.-K., *Helv. Chim. Acta* **45,** 859 (1962).

2004. Schwyzer, R., and Carrión, J. P., *Helv. Chim. Acta* **43**, 2101 (1960).

2004a. Schwyzer, R., Carrion, J. P., Gorup, B., Nolting, H., and Aung, T.-K., *Helv. Chim. Acta* **47**, 441 (1964).

2005. Schwyzer, R., Costopanagiotis, A., and Sieber, P., *Chimia (Aarau)* **16**, 295 (1962).

2006. Schwyzer, R., Costopanagiotis, A., and Sieber, P., *Helv. Chim. Acta* **46**, 870 (1963).

2007. Schwyzer, R., and Dietrich, H., *Helv. Chim. Acta* **44**, 2003 (1961).

2008. Schwyzer, R., Feurer, M., and Iselin, B., *Helv. Chim. Acta* **38**, 83 (1955).

2009. Schwyzer, R., Feurer, M., Iselin, B., and Kägi, H., *Helv. Chim. Acta* **38**, 80 (1955).

2010. Schwyzer, R., and Gorup, B., *Helv. Chim. Acta* **41**, 2199 (1958).

2011. Schwyzer, R., and Hürlimann, C., *Helv. Chim. Acta* **37**, 155 (1954).

2012. Schwyzer, R., and Iselin, B., *Ann. Acad. Sci. Fennicae Ser. A II* **60**, 181 (1955).

2013. Schwyzer, R., Iselin, B., and Feurer, M., *Angew. Chem.* **66**, 747 (1954).

2014. Schwyzer, R., Iselin, B., and Feurer, M., *Helv. Chim. Acta* **38**, 69 (1955).

2015. Schwyzer, R., Iselin, B., Kappeler, H., Riniker, B., Rittel, W., and Zuber, H., *Chimia (Aarau)* **11**, 335 (1957).

2016. Schwyzer, R., Iselin, B., Kappeler, H., Riniker, B., Rittel, W., and Zuber, H., *Helv. Chim. Acta* **41**, 1273 (1958).

2017. Schwyzer, R., Iselin, B., Kappeler, H., Riniker, B., Rittel, W., and Zuber, H., *Helv. Chim. Acta* **41**, 1287 (1958).

2017a. Schwyzer, R., Iselin, B., Kappeler, H., Riniker, B., Rittel, W., and Zuber, H., *Helv. Chim. Acta* **46**, 1975 (1963).

2018. Schwyzer, R., Iselin, B., Rittel, W., and Sieber, P., *Helv. Chim. Acta* **39**, 872 (1956).

2019. Schwyzer, R., and Kappeler, H., *Helv. Chim. Acta* **44**, 1991 (1961).

2019a. Schwyzer, R., and Kappeler, H., *Helv. Chim. Acta* **46**, 1550 (1963).

2020. Schwyzer, R., Kappeler, H., Iselin, B., Rittel, W., and Zuber, H., *Helv. Chim. Acta* **42**, 1702 (1959).

2021. Schwyzer, R., and Li, C. H., *Nature* **182**, 1669 (1958).

2022. Schwyzer, R., and Rittel, W., *Helv. Chim. Acta* **44**, 159 (1961).

2023. Schwyzer, R., Rittel, W., and Costopanagiotis, A., *Helv. Chim. Acta* **45**, 2473 (1962).

2024. Schwyzer, R., Rittel, W., Kappeler, H., and Iselin, B., *Angew. Chem.* **72**, 915 (1960).

2025. Schwyzer, R., Rittel, W., Sieber, P., Kappeler, H., and Zuber, H., *Helv. Chim. Acta* **43**, 1130 (1960).

2026. Schwyzer, R., and Sieber, P., *Chimia (Aarau)* **10**, 265 (1956).

2027. Schwyzer, R., and Sieber, P., *Helv. Chim. Acta* **40**, 624 (1957).

2028. Schwyzer, R., and Sieber, P., *Angew. Chem.* **68**, 518 (1958).

2029. Schwyzer, R., and Sieber, P., *Helv. Chim. Acta* **41**, 1582 (1958).

2030. Schwyzer, R., and Sieber, P., *Helv. Chim. Acta* **41**, 2186 (1958).

2031. Schwyzer, R., and Sieber, P., *Helv. Chim. Acta* **41**, 2190 (1958).

2032. Schwyzer, R., and Sieber, P., *Helv. Chim. Acta* **42**, 972 (1959).

2033. Schwyzer, R., and Sieber, P., *Helv. Chim. Acta* **43**, 1910 (1960).

2034. Schwyzer, R., and Sieber, P., *Nature* **199**, 172 (1963).

2035. Schwyzer, R., Sieber, P., and Gorup, B., *Chimia (Aarau)* **12**, 90 (1958).

2036. Schwyzer, R., Sieber, P., and Kappeler, H., *Helv. Chim. Acta* **42**, 2622 (1959).

2037. Schwyzer, R., Sieber, P., and Zatskó, K., *Helv. Chim. Acta* **41**, 491 (1958).

2038. Schwyzer, R., Surbeck-Wegmann, E., and Dietrich, H., *Chimia (Aarau)* **14**, 366 (1960).

2039. Schwyzer, R., and Turrian, H., *Vitamins Hormones* **18**, 237 (1960).

2040. Sciarini, L. J., and Fruton, J. S., *J. Am. Chem. Soc.* **71**, 2940 (1949).

2040a. Scoffone, E., Marchiori, F., Tamburro, A. M., and Rocchi, R., *Gazz. Chim. Ital.* **94**, 695 (1964).

2040b. Scoffone, E., Rocchi, R., Vidali, G., Scatturin, A., and Marchiori, F., *Gazz. Chim. Ital.* **94**, 743 (1964).

2041. Semb, J., Boothe, J. H., Angier, R. B., Waller, C. W., Mowat, J. H., Hutchings, B. L., and SubbaRow, Y., *J. Am. Chem. Soc.* **71**, 2310 (1949).

2042. Semm, K., *in* "Oxytocin" (R. Caldeyro-Barcia and H. Heller, eds.), p. 336. Macmillan (Pergamon), New York, 1961.

2043. Serebryanyĭ, S. B., Jirganava, C. U., and Niepliyen, V. M., *Ukr. Khim. Zh.* **27**, 365 (1961).

2044. Seu, J. H., Smeby, R. R., and Bumpus, F. M., *J. Am. Chem. Soc.* **84**, 3883 (1962).

2045. Seu, J. H., Smeby, R. R., and Bumpus, F. M., *J. Am. Chem. Soc.* **84**, 4948 (1962).

2046. Shabarova, Z. A., Satarova, L. G., and Prokof'ev, M. A., *Dokl. Acad. Nauk SSSR* **123**, 864 (1958).

2047. Shabarova, Z. A., Sokolova, N. I., Boĭkova, L. A., and Prokof'ev, M. A., *Zh. Obshch. Khim.* **29**, 2917 (1959).

2048. Shabarova, Z. A., Sokolova, N. I., and Prokof'ev, M. A., *Dokl. Akad. Nauk SSSR* **128**, 740 (1959).

2048a. Shaltiel, S., and Patchornik, A., *J. Am. Chem. Soc.* **85**, 2799 (1963).

2049. Shankman, S., *J. Biol. Chem.* **150**, 305 (1943).

2050. Shankman, S., Gold, V., and Higa, S., *Texas Rept. Biol. Med.* **19**, 358 (1961).

2051. Shankman, S., Gold, V., Higa, S., and Squires, R., *Biochem. Biophys. Res. Commun.* **9**, 25 (1962).

2052. Shankman, S., and Higa, S., *J. Pharm. Sci.* **51**, 137 (1962).

2053. Shankman, S., Higa, S., Florsheim, H. A., Schvo, Y., and Gold, V., *Arch. Biochem. Biophys.* **86**, 204 (1960).

2054. Shankman, S., Higa, S., and Gold, V., *J. Am. Chem. Soc.* **82**, 990 (1960).

2054a. Shankman, S., Higa, S., Makineni, S., and Noll, W., *J. Med. Chem.* **6**, 746 (1963).

2055. Shankman, S., and Schvo, Y., *J. Am. Chem. Soc.* **80**, 1164 (1958).

2055a. Shchukina, L. A., and Jouse, A. L., *in* "Peptide Symposium," Oxford, 1962 (G. T. Young, ed.), p. 189. Macmillan (Pergamon), New York, 1963.

2056. Shchukina, L. A., Kara-Murza, S. N., and Gromova, G. F., *Dokl. Akad. Nauk SSSR* **136**, 1351 (1961).

2057. Shchukina, L. A., Kara-Murza, S. N., and Vdovina, R. G., *Zh. Obshch. Khim.* **29**, 340 (1959).

2057a. Shchukina, L. A., and Ravdel, G. A., Proc. 7th European Symp. on Peptides, Budapest, 1964, *Acta Chim. Acad. Sci. Hung.* (in press).

2058. Sheehan, J. C., *17th Intern. Congr. Pure Appl. Chem., 1960* Vol. II, p. 143. (1960). Butterworth, London and Washington, D. C., Chemie, Weinheim.

2059. Sheehan, J. C., and Bolhofer, W. A., *J. Am. Chem. Soc.* **72**, 2469 (1950).

2060. Sheehan, J. C., Chapman, D. W., and Roth, R. W., *J. Am. Chem. Soc.* **74**, 3822 (1952).

2061. Sheehan, J. C., Cruickshank, P. A., and Boshart, G. L., *J. Org. Chem.* **26,** 2525 (1961).
2061a. Sheehan, J. C., Drummond, P. E., Gardner, J. N., Maeda, K., Mania, D., Nakamura, S., Sen, A. K., and Stock, J. A., *J. Am. Chem. Soc.* **85,** 2867 (1963).
2062. Sheehan, J. C., and Frank, V. S., *J. Am. Chem. Soc.* **71,** 1856 (1949).
2063. Sheehan, J. C., and Frank, V. S., *J. Am. Chem. Soc.* **72,** 1312 (1950).
2064. Sheehan, J. C., Goodman, M., and Hess, G. P., *J. Am. Chem. Soc.* **78,** 1367 (1956).
2065. Sheehan, J. C., Goodman, M., and Richardson, W. L., *J. Am. Chem. Soc.* **77,** 6391 (1955).
2066. Sheehan, J. C., and Grenda, V. J., *J. Am. Chem. Soc.* **84,** 2417 (1962).
2067. Sheehan, J. C., Hasspacher, K., and Yeh, Y. L., *J. Am. Chem. Soc.* **81,** 6086 (1959).
2068. Sheehan, J. C., and Hess, G. P., *J. Am. Chem. Soc.* **77,** 1067 (1955).
2069. Sheehan, J. C., and Hlavka, J. J., *J. Org. Chem.* **21,** 439 (1956).
2070. Sheehan, J. C., and Hlavka, J. J., *J. Org. Chem.* **23,** 635 (1958).
2071. Sheehan, J. C., and McGregor, D. N., *J. Am. Chem. Soc.* **84,** 3000 (1962).
2072. Sheehan, J. C., and Richardson, W. L., *J. Am. Chem. Soc.* **76,** 6329 (1954).
2073. Sheehan, J. C., and Yang, D.-D. H., *J. Am. Chem. Soc.* **80,** 1154 (1958).
2074. Sheehan, J. C., and Yang, D.-D. H., *J. Am. Chem. Soc.* **80,** 1158 (1958).
2075. Sheehan, J. C., Zachau, H. G., and Lawson, W. B., *J. Am. Chem. Soc.* **79,** 3933 (1957).
2076. Sheehan, J. C., Zachau, H. G., and Lawson, W. B., *J. Am. Chem. Soc.* **80,** 3349 (1958).
2077. Shemyakin, M. M., *Angew. Chem.* **72,** 342 (1960).
2078. Shemyakin, M. M., *Usp. Khim.* **31,** 269 (1962).
2078a. Shemyakin, M. M., Aldanova, N. A., Vinogradova, E. I., and Feigina, M. Y., *Tetrahedron Letters* p. 1921 (1963).
2079. Shemyakin, M. M., and Antonov, V. K., *Dokl. Akad. Nauk SSSR* **129,** 349 (1959).
2079a. Shemyakin, M. M., Antonov, V. K., Shkrob, A. M., Sheinker, Y. N., and Senyavina, L. B., *Tetrahedron Letters* p. 701 (1962).
2080. Shemyakin, M. M., Chaman, E. S., Denisova, L. I., Ravdel, G. A., and Rodionov, V. Y., *Bull. Soc. Chim. France* p. 530 (1959).
2081. Shemyakin, M. M., Denisova, L. I., and Chaman, E. S., *Izv. Akad. Nauk SSSR, Otd. Khim. Nauk* p. 690 (1959).
2082. Shemyakin, M. M., Kokhlov, A. S., Kolosov, N. M., Bergelson, L. D., and Antonov, V. K., "Chemie der Antibiotika," 3rd ed. Akad. Wiss. U. S. S. R. Moscow, 1961.
2082a. Shemyakin, M. M., Ovchinnikov, Y. A., Antonov, V. K., Kiryushkin, A. A., Ivanov, V. T., Shchelokov, V. I., and Shkrob, A. M., *Tetrahedron Letters* p. 47 (1964).
2082b. Shemyakin, M. M., Ovchinnikov, Y. A., Antonov, V. K., Kiryushkin, A. A., Ivanov, V. T., Shchelokov, V. I., and Shkrob, A. M., *Izv. Akad. Nauk. SSSR* p. 2233 (1963).
2083. Shemyakin, M. M., Ovchinnikov, Y. A., Ivanov, V. T., and Kiryushkin, A. A., *Izv. Akad. Nauk SSSR Otd. Khim. Nauk* p. 1699 (1962).
2084. Shemyakin, M. M., Ovchinnikov, Y. A., Ivanov, V. T., and Kiryushkin, A. A., *Tetrahedron* **19,** 581 (1963).

2085. Shemyakin, M. M., Ovchinnikov, Y. A., Ivanov, V. T., and Kiryushkin, A. A., *Tetrahedron* **19,** 995 (1963).
2085a. Shemyakin, M. M., Ovchinnikov, Y. A., Ivanov, V. T., Kiryushkin, A. A., Zhdanov, G. L., and Ryabova, I. D., *Experientia* **19,** 566 (1963).
2085b. Shemyakin, M. M., Ovchinnikov, Y. A., Ivanov, V. T., and Kiryushkin, A. A., *Tetrahedron Letters* p. 1927 (1963).
2086. Shemyakin, M. M., Ovchinnikov, Y. A., and Kiryushkin, A. A., *Tetrahedron Letters* p. 301 (1962).
2087. Shemyakin, M. M., Ovchinnikov, Y. A., Kiryushkin, A. A., and Ivanov, V. T., *Izv. Akad. Nauk SSSR, Otd. Khim. Nauk* p. 2154 (1962).
2088. Shemyakin, M. M., Ovchinnikov, Y. A., Kiryushkin, A. A., and Ivanov, V. T., *Izv. Akad. Nauk SSSR, Otd. Khim. Nauk* p. 579 (1963).
2088a. Shemyakin, M. M., Ovchinnikov, Y. A., Kiryushkin, A. A., and Ivanov, V. T., *Izv. Akad. Nauk SSSR Otd. Khim.* p. 1148 (1963).
2089. Shemyakin, M. M., Ravdel, G. A., and Chaman, E. S., *Dokl. Akad. Nauk SSSR* **107,** 706 (1956).
2090. Shemyakin, M. M., Vinogradova, E. I., Feigina, M. Y., and Aldanova, N. A., *Tetrahedron Letters* p. 351 (1963).
2091. Shemyakin, M. M., Vinogradova, E. I., Feigina, M. Y., Aldanova, N. A., Oladkina, V. A., and Shchukina, L. A., *Dokl. Akad. Nauk SSSR* **140,** 387 (1961).
2092. Shepherd, R. G., Howard, K. S., Bell, P. H., Cacciola, A. R., Child, R. G., Davies, M. C., English, J. P., Finn, B. M., Meisenhelder, J. H., Moyer, A. W., and van der Scheer, J., *J. Am. Chem. Soc.* **78,** 5051 (1956).
2093. Shepherd, R. G., Wilson, S. D., Howard, K. S., Bell, P. H., Davis, D. S., Davis, S. B., Eigner, E. A., and Shakespeare, N. E., *J. Am. Chem. Soc.* **78,** 5067 (1956).
2094. Sheradsky, T., Knobler, Y., and Frankel, M., *J. Org. Chem.* **26,** 1482 (1961).
2095. Shiba, T., and Imai, S., *Nippon Kagaku Zasshi* **80,** 175 (1959).
2096. Shiba, T., and Imai, S., *Nippon Kagaku Zasshi* **80,** 492 (1959).
2097. Shiba, T., and Imai, S., *Nippon Kagaku Zasshi* **80,** 497 (1959).
2098. Shiba, T., Imai, S., and Kaneko, T., *Bull. Chem. Soc. Japan* **31,** 244 (1958).
2099. Shiba, T., and Kaneko, T., *Bull. Chem. Soc. Japan* **33,** 1721 (1960).
2100. Shiba, T., Yamakita, N., and Kaneko, T., *J. Inst. Polytech. Osaka City Univ. Ser. C* **5,** 144 (1956).
2101. Shibnev, V. A., Kosarenko, T. D., and Poroshin, K. T., *Izv. Akad. Nauk SSSR Otd. Khim. Nauk* p. 1500 (1960).
2102. Shields, J. E., and Carpenter, F. H., *J. Am. Chem. Soc.* **83,** 3066 (1961).
2103. Shields, J. E., McGregor, W. H., and Carpenter, F. H., *J. Org. Chem.* **26,** 1491 (1961).
2104. Shimonishi, Y., Sakakibara, S., and Akabori, S., *Bull. Chem. Soc. Japan* **35,** 1966 (1962).
2105. Shin, K. H., Sakakibara, S., Schneider, W., and Hess, G. P., *Biochem. Biophys. Res. Commun.* **8,** 288 (1962).
2106. Shingte, R., Dissertation, Freie Universität Berlin, 1960.
2107. Shizume, K., Lerner, A. B., and Fitzpatrick, T. B., *Endocrinology* **54,** 553 (1954).
2108. Shorley, P. G., and Collier, H. O. J., *Nature* **83,** 999 (1960).
2109. Shostakovskii, M. F., and Shapiro, E. S., *Izv. Akad. Nauk SSSR, Otd. Khim. Nauk* p. 517 (1961).

2110. Shostakovskii, M. F., and Shapiro, E. S., *Zh. Obshch. Khim.* **32,** 3137 (1962).
2111. Shutskever, I. E., Shibnev, V. A., Millionova, M. I., Kosarenko, T. D., and Poroshin, K. T., *Izv. Akad. Nauk SSSR, Otd. Khim. Nauk* p. 2055 (1959).
2112. Shvachkin, Y. P., Sirzewa, L. A., and Prokof'ev, M. A., *Zh. Obshch. Khim.* **30,** 2462 (1960).
2113. Sica-Blanco, Y., and Sala, N. L., *in* "Oxytocin" (R. Caldeyro-Barcia and H. Heller, eds.), p. 127. Macmillan (Pergamon), New York, 1961.
2114. Sicuteri, F., Fanciullacci, M., Franchi, G., and Michelacci, S., *Experientia* **19,** 44 (1963).
2115. Sideman, M., and Sobel, H., *Proc. Soc. Exptl. Biol. Med.* **103,** 274 (1960).
2116. Siedel, W., Sturm, K., and Geiger, R., *Chem. Ber.* **96,** 1436 (1963).
2116a. Siemion, I. Z., *Roczniki Chem.* **38,** 133 (1964).
2117. Siemion, I. Z., and Nowak, K., *Roczniki Chem.* **35,** 979 (1961).
2118. Siemion, I. Z., Nowak, K., and Kaczorowski, Z., *Roczniki Chem.* **36,** 1191 (1962).
2119. Sifferd, R. H., and du Vigneaud, V., *J. Biol. Chem.* **108,** 753 (1935).
2120. Silaev, A. B., and Stepanov, V. M., *Dokl. Akad. Nauk SSSR* **112,** 297 (1957).
2121. Silaev, A. B., Stepanov, V. M., Julikova, E. P., and Muratova, G. L., *Zh. Obshch. Khim.* **31,** 1023 (1961).
2122. Silaev, A. B., Stepanov, V. M., Julikova, E. P., and Muratova, G. L., *Zh. Obshch. Khim.* **31,** 2712 (1961).
2123. Silaev, A. B., Stepanov, V. M., Julikova, E. P., Trochko, E. V., and Levin, E. D., *Zh. Obshch. Khim.* **31,** 297 (1961).
2124. Silaev, A. B., Stepanov, V. M., and Koslov, L. V., *Zh. Obshch. Khim.* **31,** 2716 (1961).
2125. Simmonds, S., and Fruton, J. S., *J. Biol. Chem.* **174,** 705 (1948).
2126. Simmonds, S., Harris, J. I., and Fruton, J. S., *J. Biol. Chem.* **188,** 251 (1951).
2127. Sinitsyna, Z. T., and Mamiofe, S. M., *Usp. Khim.* **31,** 211 (1962).
2128. Sinn, L. G., Behrens, O. K., and Bromer, W. W., *J. Am. Chem. Soc.* **79,** 2805 (1957).
2129. Skeggs, L. T., Jr., *in* "Polypeptides which Affect Smooth Muscles and Blood Vessels" (M. Schachter, ed.), p. 106. Macmillan (Pergamon), New York, 1960.
2130. Skeggs, L. T., Jr., Kahn, J. R., Lentz, K., and Shumway, N. P., *J. Exptl. Med.* **106,** 439 (1957).
2131. Skeggs, L. T., Jr., Kahn, J. R., and Shumway, N. P., *J. Exptl. Med.* **95,** 241 (1952).
2132. Skeggs, L. T., Jr., Kahn, J. R., and Shumway, N. P., *J. Exptl. Med.* **103,** 295 (1956).
2133. Skeggs, L. T., Jr., Kahn, J. R., and Shumway, N. P., *J. Exptl. Med.* **103,** 301 (1956).
2134. Skeggs, L. T., Jr., Lentz, K. E., Kahn, J. R., and Shumway, N. P., *J. Exptl. Med.* **108,** 283 (1958).
2135. Skeggs, L. T., Jr., Lentz, K. E., Kahn, J. R., Shumway, N. P., and Woods, K. R., *J. Exptl. Med.* **104,** 193 (1956).
2136. Skeggs, L. T., Jr., Marsh, W. H., Kahn, J. R., and Shumway, N. P., *J. Exptl. Med.* **99,** 275 (1954).
2137. Skeggs, L. T., Jr., Marsh, W. H., Kahn, J. R., and Shumway, N. P., *J. Exptl. Med.* **100,** 363 (1954).
2138. Skeggs, L. T., Jr., Marsh, W. H., Kahn, J. R., and Shumway, N. P., *J. Exptl. Med.* **102,** 435 (1955).

2139. Slobin, L. I., and Carpenter, F. H., *Biochemistry* **2,** 16 (1963).
2140. Slobin, L. I., and Carpenter, F. H., *Biochemistry* **2,** 22 (1963).
2141. Smart, N. A., Young, G. T., and Williams, M. W., *J. Chem. Soc.* p. 3902 (1960).
2142. Smeby, R. R., Arakawa, K., Bumpus, F. M., and Marsh, M. M., *Biochim. Biophys. Acta* **58,** 550 (1962).
2143. Smeby, R. R., Khairallah, P. A., and Bumpus, F. M., *Cleveland Clinic Quart.* **29,** 66 (1962).
2144. Smith, C. S., and Brown, A. E., *J. Am. Chem. Soc.* **63,** 2605 (1941).
2145. Smith, E. L., *J. Biol. Chem.* **175,** 39 (1948).
2146. Smith, E. L., *J. Biol. Chem.* **176,** 21 (1948).
2147. Smith, E. L., *Advan. Enzymol.* **12,** 191 (1951).
2148. Smith, E. L., and Bergmann, M., *J. Biol. Chem.* **153,** 627 (1944).
2149. Smith, E. L., Hill, R. L., and Borman, A., *Biochim. Biophys. Acta* **29,** 207 (1958).
2150. Smith, E. L., and Polglase, W. J., *J. Biol. Chem.* **180,** 1209 (1949).
2151. Smith, E. L., and Slonim, N. B., *J. Biol. Chem.* **176,** 835 (1948).
2152. Smith, E. L., and Spackman, D. H., *J. Biol. Chem.* **212,** 271 (1955).
2153. Smith, E. L., Spackman, D. H., and Polglase, W. J., *J. Biol. Chem.* **199,** 801 (1952).
2154. Smith, M., Moffatt, J. G., and Khorana, H. G., *J. Am. Chem. Soc.* **80,** 6204 (1958).
2155. Smith, P. G. W., *J. Chem. Soc.* p. 3985 (1957).
2156. Smyth, C. N., *Brit. Med. J.* **I,** 856 (1958).
2156a. Sokolovsky, M., Sadeh, T., and Patchornik, A., *J. Am. Chem. Soc.* **86,** 1212 (1964).
2156b. Sokolovsky, M., Wilchek, M., and Patchornik, A., *J. Am. Chem. Soc.* **86,** 1202 (1964).
2157. Sokolowska, T., Kupryszewski, G., and Taschner, E., *Roczniki Chem.* **32,** 813 (1958).
2158. Sokolowska, T., Kupryszewski, G., and Taschner, E., *Bull. Acad. Polon. Sci.* **6,** 89 (1958).
2159. De Somer, P., and Van Dijck, P., *Antibiot. Chemotherapie* **5,** 632 (1955).
2160. Sondheimer, E., and Holley, R. W., *J. Am. Chem. Soc.* **76,** 2467 (1954).
2161. Sondheimer, E., and Holley, R. W., *J. Am. Chem. Soc.* **76,** 2816 (1954).
2162. Sondheimer, E., and Holley, R. W., *J. Am. Chem. Soc.* **79,** 3767 (1957).
2163. Sondheimer, E., and Semeraro, R. J., *J. Org. Chem.* **26,** 1847 (1961).
2163a. Sonnenbichler, I., Feldman, H., and Zachau, H. G., *Z. Physiol. Chem.* **334,** 283 (1963).
2164. Šorm, F., *Collection Czech. Chem. Commun.* **24,** 3169 (1959).
2165. Šorm, F., *Collection Czech. Chem. Commun.* **26,** 1174 (1961).
2166. Šorm, F., *Collection Czech. Chem. Commun.* **26,** 1180 (1961).
2167. Šorm, F., *Collection Czech. Chem. Commun.* **27,** 303 (1962).
2168. Šorm, F., *Collection Czech. Chem. Commun.* **27,** 994 (1962).
2169. Šorm, F., *Collection Czech. Chem. Commun.* **27,** 1604 (1962).
2170. Šorm, F., *Pharmazie* **17,** 257 (1962).
2171. Šorm, F., and Beránková, Z., *Collection Czech. Chem. Commun.* **26,** 2557 (1961).
2172. Šorm, F., Keil, B., Holeyšovský, V., Knesslová, V., Kostka, V., Mäsiar, P., Meloun, B., Mikeš, O., Tomášek, V., and Vaněček, J., *Collection Czech. Chem. Commun.* **22,** 1310 (1957).

2173. Šorm, F., and Rudinger, J., *Collection Czech. Chem. Commun.* **15,** 491 (1950).
2174. Sous, H., Löschner, H., Lagler, F., and Mückter, H., *Arzneimittel-Forsch.* **11,** 395 (1961).
2175. Spears, A. W., and Tieckelmann, H., *J. Org. Chem.* **26,** 1498 (1961).
2176. Spoor, H., and Ziegler, K., *Angew. Chem.* **72,** 316 (1960).
2177. Sporn, J., and Necheles, H., *Am. J. Physiol.* **187,** 634 (1956).
2178. Sprince, H., and Wolley, D. W., *J. Exptl. Med.* **80,** 213 (1944).
2179. Sprince, H., and Wolley, D. W., *J. Am. Chem. Soc.* **67,** 1734 (1945).
2180. Staab, H. A., *Angew. Chem.* **71,** 385 (1959).
2181. Staab, H. A., *Angew. Chem.* **74,** 407 (1962).
2182. Staab, H. A., Lüking, M., and Dürr, F. H., *Chem. Ber.* **95,** 1275 (1962).
2183. Staab, H. A., and Mannschreck, A., *Chem. Ber.* **95,** 1284 (1962).
2184. Staab, H. A., and Rohr, W., *Chem. Ber.* **95,** 1298 (1962).
2185. Staab, H. A., Rohr, W., and Mannschreck, A., *Angew. Chem.* **73,** 143 (1961).
2186. Staab, H. A., and Walther, G., *Ann. Chem. Liebigs* **657,** 98 (1962).
2187. Staab, H. A., and Walther, G., *Chem. Ber.* **95,** 2070 (1962).
2188. Staab, H. A., Walther, G., and Rohr, W., *Chem. Ber.* **95,** 2073 (1962).
2188a. Staab, H. A., and Wendel, K., *Chem. Ber.* **96,** 3374 (1963).
2189. Stahmann, M. A., Fruton, J. S., and Bergmann, M., *J. Biol. Chem.* **164,** 753 (1946).
2190. Stansly, P. G., Shepherd, R. G., and White, H. J., *Bull. Johns Hopkins Hosp.* **81,** 43 (1947).
2191. Stapleton, I. W., and Swan, J. M., *Australian J. Chem.* **13,** 416 (1960).
2192. Stapleton, I. W., and Swan, J. M., *Australian J. Chem.* **15,** 106 (1962).
2193. Stapleton, I. W., and Swan, J. M., *Australian J. Chem.* **15,** 570 (1962).
2194. Staub, A., Behrens, O. K., Ellis, J. T., and Kennedy, R. W., *J. Clin. Invest.* **33,** 1629 (1954).
2195. Staub, A., Sinn, L. G., and Behrens, O. K., *Science* **117,** 628 (1953).
2196. Staub, A., Sinn, L. G., and Behrens, O. K., *J. Biol. Chem.* **214,** 619 (1955).
2197. Staub, A., Springs, V., Stoll, F., and Elrick, H., *Proc. Soc. Exptl. Biol. Med.* **94,** 57 (1957).
2198. Stedman, R. J., *J. Am. Chem. Soc.* **79,** 4691 (1957).
2199. Steelman, S. L., Andersen, R. N., and McGregor, R. M., *Biochim. Biophys. Acta* **33,** 256 (1959).
2200. Steelman, S. L., and Guillemin, R., *Proc. Soc. Exptl. Biol. Med.* **101,** 600 (1959).
2201. Steelman, S. L., and Smith, W. W., *Acta Endocrinol.* **33,** 67 (1960).
2202. Stein, W. H., Moore, S., and Bergmann, M., *J. Biol. Chem.* **154,** 191 (1944).
2203. Stelakatos, G. C., Theodoropoulos, D. M., and Zervas, L., *J. Am. Chem. Soc.* **81,** 2884 (1959).
2204. Stepanov, V. M., and Silaev, A. B., *Zh. Obshch. Khim.* **31,** 3799 (1961).
2205. Stepanov, V. M., and Silaev, A. B., *Zh. Obshch. Khim.* **31,** 3804 (1961).
2206. Stepanov, V. M., and Silaev, A. B., *Zh. Obshch. Khim.* **31,** 3811 (1961).
2207. Stepanov, V. M., Silaev, A. B., and Polin, A. N., *Antibiotiki* **3,** 49 (1958).
2208. Stern, K. G., and White, A., *J. Biol. Chem.* **117,** 95 (1937).
2209. Stern, P., *Ann. N. Y. Acad. Sci.* **104,** 403 (1963).
2210. Stern, P., and Huković, S., *Med. Exptl.* **2,** 1 (1960).
2211. Stern, P., Nikulin, A., and Ferluga, J., *Arch. Intern. Pharmacodyn.* **140,** 528 (1962).

2212. Stevens, C. L., and Munk, M. E., *J. Am. Chem. Soc.* **80,** 4069 (1958).
2213. Stevens, C. M., and Watanabe, R., *J. Am. Chem. Soc.* **72,** 725 (1950).
2214. Stewart, F. H. C., *J. Org. Chem.* **25,** 1828 (1960).
2215. Stewart, I., *J. Chromatog.* **10,** 404 (1963).
2215a. Stewart, J. M., *Biochem. Pharmacol.* Suppl. 12, 180 (1963).
2215b. Stewart, J. M., and Woolley, D. W., *Biochemistry* **3,** 700 (1964).
2215c. Stewart, J. M., private communication, 1964.
2216. Stoklaska, E., and Wintersberger, E., *Arch. Exptl. Pathol. Pharmakol.* **236,** 358 (1959).
2217. Stoffel, W., Dissertation, Universität Köln, 1959.
2218. Stoffel, W., and Craig, L. C., *J. Am. Chem. Soc.* **83,** 145 (1961).
2219. Stoll, A., and Petrzilka, T., *Helv. Chim. Acta* **35,** 589 (1952).
2220. Strid, L., *Acta Chem. Scand.* **15,** 1423 (1961).
2221. Studer, R. O., *Chimia (Aarau)* **17,** 22 (1963).
2222. Studer, R. O., *Helv. Chim. Acta* **46,** 421 (1963).
2222a. Studer, R. O., *Chimia (Aarau)* (in press).
2223. Studer, R. O., and Cash, W. D., *J. Biol. Chem.* **238,** 657 (1963).
2224. Studer, R. O., and du Vigneaud, V., *J. Am. Chem. Soc.* **82,** 1499 (1960).
2225. Studer, R. O., Lergier, W., and Vogler, K., *Helv. Chim. Acta* **46,** 612 (1963).
2226. Studer, R. O., and Vogler, K., *Helv. Chim. Acta* **45,** 819 (1962).
2227. Studer, R. O., Vogler, K., and Lergier, W., *Helv. Chim. Acta* **44,** 131 (1961).
2228. Sturm, K., Geiger, R., and Siedel, W., *Chem. Ber.* **96,** 609 (1963).
2228a. Sturm, K., Geiger, R., and Siedel, W., *Chem. Ber.* **97,** 1197 (1964).
2229. Stürmer, E., and Berde, B., *Arch. Exptl. Pathol. Pharmakol.* **243,** 355 (1962).
2230. Stürmer, E., and Berde, B., *J. Pharmacol. Exptl. Therap.* **139,** 38 (1963).
2231. Stürmer, E., and Berde, B., *J. Pharmacol. Exptl. Therap.* **140,** 349 (1963).
2231a. Stürmer, E., and Berde, B., *Arch. Exptl. Pathol. Pharmakol.* **246,** 21 (1963).
2232. Stürmer, E., and Cerletti, A., *Helv. Physiol. Pharmacol. Acta* **19,** C 32 (1961).
2232a. Stürmer, E., and Cerletti, A., *Am. Heart J.* **62,** 149 (1961).
2233. Stürmer, E., and Franz, J., *Med. Exptl.* **5,** 37 (1961).
2233a. Stürmer, E., Sandrin, E., and Boissonnas, R. A., *Experientia* **20,** 303 (1964).
2234. Sumiki, Y., and Miyao, K., *J. Agr. Chem. Soc. Japan* **26,** 27 (1952).
2235. Sunahara, H., Ward, D. N., and Griffin, A. C., *J. Am. Chem. Soc.* **82,** 6017 (1960).
2236. Surbeck-Wegmann, E., Dissertation, Universität Zürich, 1961.
2237. Süs, O., *Ann. Chem. Liebigs* **572,** 96 (1951).
2238. Suzuki, K., *Tohoku Yakka Daigaku Kiyo* **4,** 117 (1957).
2239. Suzuki, K., *Tohoku Yakka Daigaku Kiyo* **4,** 127 (1957).
2240. Suzuki, K., *Tohoku Yakka Daigaku Kiyo* **4,** 135 (1957).
2241. Suzuki, S., *J. Pharm. Soc. Japan* **81,** 935 (1961).
2242. Suzuki, S., *J. Pharm. Soc. Japan* **81,** 938 (1961).
2242a. Suzuki, T., Inouye, H., Fujikawa, K., and Nagasawa, S., *J. Biochem. (Tokyo)* **54,** 173 (1963).
2242b. Suzuki, T., Inouye, H., Fujikawa, K., and Suketa, Y., *J. Biochem. (Tokyo)* **54,** 25 (1963).
2242c. Suzuki, T., Hayashi, K., and Fujikawa, K., *J. Biochem. (Tokyo)* **54,** 412 (1963).
2242d. Suzuki, T., Hayashi, K., Fujikawa, K., and Tsukamoto, K., *J. Biochem. (Tokyo)* **54,** 555 (1963).
2242e. Suzuki, T., and Fujikawa, K., *J. Biochem. (Tokyo)* **56,** 182 (1964).

2243. Swallow, D. L., and Abraham, E. P., *Biochem. J.* **70,** 364 (1958).

2244. Swallow, D. L., Lockhart, I. M., and Abraham, E. P., *Biochem. J.* **70,** 359 (1958).

2245. Swan, J. M., *Proc. Intern. Wool. Textile Res. Conf. Australia* Vol. C, p. 25 (1955). CSIRO, Victoria, Australia.

2246. Swan, J. M., *Proc. Intern. Wool. Textile Res. Conf. Australia* Vol. C, p. 175 (1955). CSIRO, Victoria, Australia.

2247. Swan, J. M., and du Vigneaud, V., *J. Am. Chem. Soc.* **76,** 3110 (1954).

2248. Sweetman, B. J., *Nature* **183,** 744 (1959).

2249. Synge, R. L. M., *Biochem. J.* **39,** 355 (1945).

2250. Synge, R. L. M., *Biochem. J.* **39,** 363 (1945).

2251. Synge, R. L. M., *Biochem. J.* **42,** 99 (1948).

2252. Synge, R. L., M., *Quart. Rev. (London)* **3,** 245 (1949).

2253. Synge, R. L. M., *Biochem. J.* **44,** 542 (1949).

2254. Tabachnik, M., Eisen, H. N., and Levine, B., *Nature* **174,** 701 (1954).

2255. Taber, W. A., and Vining, L. C., *Can. J. Microbiol.* **3,** 953 (1957).

2255a. Takita, T., *J. Antibiot. (Tokyo) Ser. A* **16,** 211 (1963).

2255b. Takita, T., and Naganawa, H., *J. Antibiot. (Tokyo) Ser. A* **16,** 246 (1963).

2255c. Takita, T., Naganawa, H., Maeda, K., and Umezawa, H., *J. Antibiot. (Tokyo) Ser. A* **17,** 90 (1964).

2256. Talbot, G., Gaudry, R., and Berlinguet, L., *Can. J. Chem.* **36,** 593 (1958).

2257. Tanaka, A., Pickering, B. T., and Li, C. H., *Arch. Biochem. Biophys.* **99,** 294 (1962).

2258. Tanaka, N., Miyairi, N., Watanabe, K., Shinjō, N., Nishimura, T., and Umezawa, H., *J. Antibiot. (Tokyo) Ser. A* **12,** 290 (1959).

2259. Tanaka, N., Shinjō, N., Miyairi, N., and Umezawa, H., *J. Antibiot. (Tokyo) Ser. A.* **11,** 127 (1958).

2260. Tanaka, N., Yamaguchi, H., and Umezawa, H., *J. Antibiot. (Tokyo) Ser. A* **15,** 33 (1962).

2261. Tanaka, N., Yamaki, H., Yamaguchi, H., and Umezawa, H., *J. Antibiot. (Tokyo) Ser. A* **15,** 28 (1962).

2262. Tanenbaum, S. W., *J. Am. Chem. Soc.* **75,** 1754 (1953).

2263. Taschner, E., *Collection Czech. Chem. Commun.* **24,** 86 (1959).

2264. Taschner, E., Biernat, J. F., Rzeszotarska, B., and Wasielewski, C., *Ann. Chem. Liebigs* **646,** 123 (1961).

2264a. Taschner, E., Biernat, J. F., and Sokolowska, T., *in* "Peptide Symposium," Oxford, 1962 (G. T. Young, ed.), p. 113. Macmillan (Pergamon), New York, 1963.

2265. Taschner, E., Chimiak, A., Bator, B., and Sokolowska, T., *Ann. Chem. Liebigs* **646,** 134 (1961).

2266. Taschner, E., Chimiak, A., Biernat, J. F., and Wasielewski, C., *Collection Czech. Chem. Commun.* **27,** 2237 (1962).

2266a. Taschner, E., Chimiak, A., Biernat, J. F., Sokolowska, T., Wasielewski, C., and Rzeszotarska, B., *in* "Peptide Symposium," Oxford, 1962 (G. T. Young, ed.), p. 109. Macmillan (Pergamon), New York, 1963.

2267. Taschner, E., and Kupryszewski, G., *Roczniki Chem.* **36,** 63 (1962).

2268. Taschner, E., Kupryszewski, G., and Liberek, B., *Roczniki Chem.* **30,** 643 (1956).

2269. Taschner, E., Kupryszewski, G., and Liberek, B., *Roczniki Chem.* **32,** 1107 (1958).

2270. Taschner, E., and Liberek, B., *Roczniki Chem.* **30,** 323 (1956).
2271. Taschner, E., and Liberek, B., *Bull. Acad. Polon. Sci., Ser. Sci. Chim.* **7,** 877 (1959).
2272. Taschner, E., Liberek, B., Wasielewski, C., and Biernat, J. F., *Angew. Chem.* **71,** 743 (1959).
2273. Taschner, E., Rzeszotarska, B., Biernat, J. F., and Grudziński, S., *Collection Czech. Chem. Commun.* **27,** 2234 (1962).
2274. Taschner, E., and Sokolowska, T., *Chimia (Aarau)* **14,** 372 (1960).
2275. Taschner, E., Sokolowska, T., Biernat, J. F., Chimiak, A., Wasielewski, C., and Rzeszotarska, B., *Ann. Chem. Liebigs* **663,** 197 (1963).
2276. Taschner, E., and Wasielewski, C., *Angew. Chem.* **71,** 743 (1959).
2277. Taschner, E., and Wasielewski, C., *Ann. Chem. Liebigs* **640,** 136 (1961).
2278. Taschner, E., and Wasielewski, C., *Ann. Chem. Liebigs* **640,** 139 (1961).
2279. Taschner, E., and Wasielewski, C., *Ann. Chem. Liebigs* **640,** 142 (1961).
2280. Taschner, E., Wasielewski, C., and Biernat, J. F., *Ann. Chem. Liebigs* **646,** 119 (1961).
2281. Taschner, E., Wasielewski, C., Biernat, J. F., and Sokolowska, T., *Chimia (Aarau)* **14,** 371 (1960).
2282. Taschner, E., Wasielewski, C., Kupryszewski, G., and Uminski, T., *Bull. Acad. Polon. Sci., Ser. Sci. Chim.* **7,** 873 (1959).
2283. Taschner, E., Wasielewski, C., Sokolowska, T., and Biernat, J. F., *Ann. Chem. Liebigs* **646,** 127 (1961).
2284. Tauber, H., *J. Am. Chem. Soc.* **71,** 2952 (1949).
2285. Tauber, H., *J. Am. Chem. Soc.* **73,** 1288 (1951).
2286. Tauber, H., *J. Am. Chem. Soc.* **73,** 4965 (1951).
2286a. Tesser, G. I., and Nivard, R. J. F., *Rec. Trav. Chim.* **83,** 53 (1964).
2287. Theobald, G. W., *J. Physiol. (London)* **141,** 30P (1958).
2288. Theobald, G. W., *in* "Oxytocin" (R. Caldeyro-Barcia and H. Heller, eds.), p. 212. Macmillan (Pergamon), New York, 1961.
2289. Theobald, G. W., *in* "Polypeptides which Affect Smooth Muscles and Blood Vessels" (M. Schachter, ed.), p. 70. Macmillan (Pergamon), New York, 1960.
2290. Theobald, J. M., Williams, M. W., and Young, G. T., *Chimia (Aarau)* **14,** 371 (1960).
2291. Theodoropoulos, D., *J. Org. Chem.* **21,** 1550 (1956).
2292. Theodoropoulos, D., *J. Org. Chem.* **23,** 140 (1958).
2293. Theodoropoulos, D., *Acta Chem. Scand.* **12,** 2043 (1958).
2294. Theodoropoulos, D., *Nature* **184,** 1634 (1959).
2295. Theodoropoulos, D., *Nature* **194,** 283 (1962).
2296. Theodoropoulos, D., Bennich, H., Fölsch, G., and Mellander, O., *Nature* **184,** 187 (1959).
2297. Theodoropoulos, D., Bennich, H., Fölsch, G., and Mellander, O., *Nature* **184,** 270 (1959).
2298. Theodoropoulos, D., and Craig, L. C., *J. Org. Chem.* **20,** 1169 (1955).
2299. Theodoropoulos, D., and Craig, L. C., *J. Org. Chem.* **21,** 1376 (1956).
2300. Theodoropoulos, D., and Fölsch, G., *Acta Chem. Scand.* **12,** 1955 (1958).
2301. Theodoropoulos, D., and Fruton, J. S., *Biochemistry* **1,** 933 (1962).
2302. Theodoropoulos, D., and Gazopoulos, J., *J. Chem. Soc.* p. 3861 (1960).
2303. Theodoropoulos, D., and Gazopoulos, J., *J. Org. Chem.* **27,** 2091 (1962).

2304. Theodoropoulos, D., Gazopoulos, J., and Souchleris, I., *J. Chem. Soc.* p. 5257 (1960).
2305. Theodoropoulos, D., Gazopoulos, J., and Souchleris, I., *Nature* **185,** 606 (1960).
2306. Theodoropoulos, D., Gazopoulos, J., and Souchleris, I., *Nature* **188,** 489 (1960).
2306a. Theodoropoulos, D., and Souchleris, I., *Biochemistry* **3,** 145 (1964).
2306b. Theodoropoulos, D., and Tsangaris, J., *J. Org. Chem.* **29,** 2272 (1964).
2307. Thiele, J., and Dent, F., *Ann. Chem. Liebigs* **302,** 258 (1898).
2308. Thompson, R. E., *J. Pharmacol. Exptl. Therap.* **80,** 373 (1944).
2309. Thomsen, V. F., *Acta Pathol. Microbiol. Scand.* **57,** 120 (1963).
2310. Thomson, W. B., *J. Physiol.* (*London*) **145,** 12P (1959).
2311. Thorn, N. A., *Acta Endocrinol.* **32,** 123 (1959).
2312. Thorn, N. A., *Acta Endocrinol.* **32,** 134 (1959).
2313. Thorn, N. A., and Milewski, B., *Proc. Soc. Exptl. Biol. Med.* **100,** 267 (1959).
2314. Tipson, R. S., *J. Org. Chem.* **21,** 1353 (1956).
2314a. Todd, L., *Ind. Chim. Belge* **27,** 1423 (1962).
2315. Toennies, G., and Callan, T. P., *J. Biol. Chem.* **129,** 481 (1939).
2316. Toennies, G., and Kolb, J. J., *J. Biol. Chem.* **140,** 131 (1941).
2317. Toennies, G., and Lavine, T. F., *J. Biol. Chem.* **113,** 571 (1936).
2318. Tomoioka, T., Soga, T., and Umezawa, S., *J. Antibiot.* (*Tokyo*) *Ser. A* **13,** 287 (1960).
2318a. Tracy, H. J., and Gregory, R. A., *Nature* **204,** 935 (1964).
2319. Trautschold, I., and Rüdel, G., *Klin. Wochschr.* **41,** 297 (1963).
2320. Trautschold, I., and Werle, E., *Z. Physiol. Chem.* **325,** 48 (1961).
2321. Tritsch, G. L., and Woolley, D. W., *J. Am. Chem. Soc.* **80,** 1490 (1958).
2322. Tritsch, G. L., and Woolley, D. W., *J. Am. Chem. Soc.* **82,** 2787 (1960).
2323. Tsakiris, A., and Bühlmann, A., *Helv. Med. Acta* **28,** 615 (1961).
2324. Tsakiris, A., and Bühlmann, A., *Deut. Med. Wochschr.* **88,** 46 (1963).
2325. Tsou, C.-I., Du, Y.-C., and Xü, G.-J., *Sci. Sinica* (*Peking*) **10,** 332 (1961).
2326. Tuppy, H., *Biochim. Biophys. Acta* **11,** 449 (1953).
2327. Tuppy, H., *Z. Naturforsch.* **12b,** 784 (1957).
2328. Tuppy, H., *Naturwissenschaften* **46,** 35 (1959).
2329. Tuppy, H., *in* "Oxytocin" (R. Caldeyro-Barcia and H. Heller, ed.), p. 315. Macmillan (Pergamon), New York, 1961.
2330. Tuppy, H., *in* "Polypeptides which Affect Smooth Muscles and Blood Vessels" (M. Schachter, ed.), p. 49. Macmillan (Pergamon), New York, 1960.
2331. Tuppy, H., and Bodo, G., *Monatsh. Chem.* **85,** 1024 (1954).
2332. Tuppy, H., and Dus, K., *Monatsh. Chem.* **89,** 407 (1958).
2333. Tuppy, H., and Kreil, G., *Monatsh. Chem.* **93,** 780 (1962).
2334. Tuppy, H., and Michl, H., *Monatsh. Chem.* **84,** 1011 (1953).
2335. Tuppy, H., and Nesvadba, H., *Monatsh. Chem.* **88,** 977 (1957).
2336. Tuppy, H., and Paléus, S., *Acta Chem. Scand.* **9,** 353 (1955).
2337. Tuppy, H., and Wintersberger, E., *Monatsh. Chem.* **91,** 1001 (1960).
2338. Turba, F., and Hetzel, H., *Biochem. Z.* **325,** 524 (1954).
2339. Turner, R. A., *J. Am. Chem. Soc.* **75,** 2388 (1953).
2340. Turner, R. A., and Schmerzler, G., *J. Am. Chem. Soc.* **76,** 949 (1954).
2341. Turrian, H., *Helv. Physiol. Pharmacol. Acta* **18,** 259 (1960).
2342. Uchio, H., and Izumiya, N., *Mem. Fac. Sci. Kyushu Univ. Ser. C* **3,** 5 (1958).

2343. Uehara, T., *Osaka Shiritsu Daigaku Igaku Zasshi* **8**, 1489 (1959).
2344. Uehara, T., *Osaka Shiritsu Daigaku Igaku Zasshi* **8**, 1499 (1959).
2344a. Ueyanag, J., Iwasaki, H., Fujino, M., Kamiya, T., and Tatsuoka, S., Symposium of the Chemical Structures of Natural Products, the University of Kyushu, 1963.
2345. Ugi, I., and Fetzer, U., *Angew. Chem.* **73**, 621 (1961).
2346. Ukita, C., and Suzuki, S., *J. Pharm. Soc. Japan* **81**, 222 (1961).
2347. Ullrich, J., Dissertation, Rheinische Friedrich Wilhelm Universität Bonn, 1958.
2347a. Unger, R. H., Eisentraut, A., McCall, M. S., and Madison, L. L. *Proc. 4th Congr. Intern. Diabetes Federation, Genf, 1961* (M. Demole, ed.), Vol. I, p. 63. Editions Médecine et Hygiène Genève, 1961.
2348. Urnes, P., and Doty, P., *Advan. Protein Chem.* **16**, 401 (1961).
2349. Utzinger, G. E., *Experientia* **16**, 136 (1960).
2350. Vajda, T., *Acta Chim. Acad. Sci. Hung.* **21**, 71 (1959).
2350a. Van de Linde, A., Kienhuis, H., Verweij, A., and Van der Holst, J. P. J., *Rec. Trav. Chim.* **80**, 1305 (1961).
2351. Vanderhaeghe, H., and Parmentier, G., *Bull. Soc. Chim. Belges* **68**, 716 (1959).
2352. Vanderhaeghe, H., and Parmentier, G., *J. Am. Chem. Soc.* **82**, 4414 (1960).
2353. Vanderhaeghe, H., Van Dijck, P., Parmentier, G., and De Somer, P., *Antibiot. Chemotherapy* **7**, 606 (1957).
2354. Van Dijck, P., Vanderhaeghe, H., and De Somer, P., *Antibiot. Chemotherapy* **7**, 625 (1957).
2355. van Dyke, H. B., *in* "Oxytocin" (R. Caldeyro-Barcia and H. Heller, eds.), p. 48. Macmillan (Pergamon), New York, 1961.
2356. van Dyke, H. B., *21st Intern. Congr. Physiol. Sci., Buenos Aires, 1959, Symp. Spec. Lectures* p. 61 (1959).
2357. van Dyke, H. B., Adamsons, K., Jr., and Engel, S. L., *Recent Progr. Hormone Res.* **11**, 1 (1955).
2358. van Dyke, H. B., Engel, S. L., and Adamsons, K., Jr., *Proc. Soc. Exptl. Biol. Med.* **91**, 484 (1956).
2359. van Orden, H. O., and Smith, E. L., *J. Biol. Chem.* **208**, 751 (1954).
2360. Van Tamelen, E. E., and Smissman, E. E., *J. Am. Chem. Soc.* **75**, 2031 (1953).
2361. Vaughan, J. R., Jr., *J. Am. Chem. Soc.* **73**, 1389 (1951).
2362. Vaughan, J. R., Jr., *J. Am. Chem. Soc.* **73**, 3547 (1951).
2363. Vaughan, J. R., Jr., *J. Am. Chem. Soc.* **74**, 6137 (1952).
2364. Vaughan, J. R., Jr., and Eichler, J. A., *J. Am. Chem. Soc.* **75**, 5556 (1953).
2365. Vaughan, J. R., Jr., and Eichler, J. A., *J. Am. Chem. Soc.* **76**, 2474 (1954).
2366. Vaughan, J. R., Jr., and Osato, R. L., *J. Am. Chem. Soc.* **73**, 5553 (1951).
2367. Vaughan, J. R., Jr., and Osato, R. L., *J. Am. Chem. Soc.* **74**, 676 (1952).
2368. Vejdělek, Z. J., *Collection Czech. Chem. Commun.* **15**, 929 (1950).
2369. Velluz, L., Amiard, G., Bartos, J., Coffinet, B., and Heymès, R., *Bull. Soc. Chim. France* p. 1464 (1956).
2370. Velluz, L., Amiard, G., and Heymès, R., *Bull. Soc. Chim. France* p. 1012 (1954).
2371. Velluz, L., Amiard, G., and Heymès, R., *Bull. Soc. Chim. France* p. 201 (1955).

2372. Velluz, L., Anatol, J., and Amiard, G., *Bull. Soc. Chim. France* p. 1449 (1954).
2373. Vernengo, M. J., and Deferrari, J. O., *Anales Asoc. Quim. Arg.* **44,** 55 (1956).
2374. Vining, L. C., and Taber, W. A., *Bacteriol. Proc.* (*Soc. Am. Bacteriologists*) p. 70 (1957).
2375. Vining, L. C., and Taber, W. A., *Can. J. Chem.* **35,** 1109 (1957).
2376. Vining, L. C., and Taber, W. A., *Can. J. Chem.* **40,** 1579 (1962).
2377. Vinkler, E., Lázař, J., and Klivényi, F., *Acta Chim. Acad. Sci. Hung.* **30,** 233 (1962).
2378. Virtanen, A. I., *Makromol. Chem.* **6,** 94 (1951).
2379. Virtanen, A. I., *Angew. Chem.* **74,** 374 (1962).
2380. Virtanen, A. I., Herkkonen, H., Hakalo, M., and Laaksonen, T., *Naturwissenschaften* **37,** 139 (1950).
2381. Virtanen, A. I., and Matikkala, E. J., *Z. Physiol. Chem.* **322,** 8 (1960).
2381a. Visser, G. H., Schattenkerk, C., Kerling, K. E. T., and Havinga, E., *Rec. Trav. Chim.* **83,** 684 (1964).
2382. Vogel, G., and Hergott, J., *Arzneimittel-Forsch.* **13,** 415 (1963).
2382a. Vogler, K., Proc. 7th European Symp. on Peptides, Budapest, 1964 *Acta Chim. Acad. Sci. Hung.* (in press).
2383. Vogler, K., and Baumgartner, H., *Helv. Chim. Acta* **35,** 1776 (1952).
2384. Vogler, K., and Chopard-dit-Jean, H., *Helv. Chim. Acta* **43,** 279 (1960).
2385. Vogler, K., Haefely, W., Hürlimann, A., Studer, R. O., Lergier, W., Strässle, R., and Berneis, K. H., *Ann. N. Y. Acad. Sci.* **104,** 378 (1963).
2386. Vogler, K., and Lanz, P., *Helv. Chim. Acta* **43,** 270 (1960).
2387. Vogler, K., Lanz, P., and Lergier, W., *Experientia* **15,** 334 (1959).
2388. Vogler, K., Lanz, P., and Lergier, W., *Helv. Chim. Acta* **45,** 561 (1962).
2389. Vogler, K., Lanz, P., Lergier, W., and Studer, R. O., *Helv. Chim. Acta* **43,** 574 (1960).
2389a. Vogler, K., Lanz, P., Quitt, P., Studer, R. O., Lergier, W., Böhni, E., and Fust, B., *Helv. Chim. Acta* **47,** 526 (1964).
2390. Vogler, K., Studer, R. O., Lanz, P., Lergier, W., and Böhni, E., *Experientia* **17,** 223 (1961).
2390a. Vogler, K., Studer, R. O., Lanz, P., Lergier, W., and Böhni, E., *Experientia* **20,** 365 (1964).
2390b. Vogler, K., Studer, R. O., Lanz, P., Lergier, W., Böhni, E., and Fust, B., *Helv. Chim. Acta* **46,** 2823 (1963).
2391. Vogler, K., Studer, R. O., and Lergier, W., *Helv. Chim. Acta* **44,** 1495 (1961).
2392. Vogler, K., Studer, R. O., Lergier, W., and Lanz, P., *Helv. Chim. Acta* **43,** 1751 (1960).
2393. Vogler, K., Würsch, J., Studer, R. O., and Lanz, P., *Chimia* (*Aarau*) **14,** 379 (1960).
2393a. Volfin, P., Chambaut, A. M., Eboué-Bonis, D., Clauser, H., Brinkhoff, O., Bremer, H., Meienhofer, J., and Zahn, H., *Nature* **203,** 408 (1964).
2394. Vollmar, A., and Dunn, M. S., *J. Org. Chem.* **25,** 387 (1960).
2395. Vollmar, A., and Dunn, M. S., *J. Org. Chem.* **26,** 4123 (1961).
2396. Vollmann, D., Dissertation, Freie Universität, Berlin, 1960.
2397. von Euler, H., and Hasselquist, H., *Arkiv Kemi* **19,** 291 (1962).
2398. von Euler, H., and Hasselquist, H., *Arkiv Kemi* **20,** 129 (1963).
2399. von Euler, U. S., *Acta Physiol. Scand.* **4,** 373 (1942).

2400. von Euler, U. S., and Gaddum, J. H., *Z. Physiol. Chem.* **72,** 74 (1931).
2401. Vuylsteke, C. A., and de Duve, C., *Arch. Intern. Pharmacodyn.* **111,** 437 (1957).
2402. Waaler, B. A., *J. Physiol. (London)* **154,** 57P (1960).
2403. Waaler, B. A., *J. Physiol. (London)* **157,** 475 (1961).
2404. Wade, R., Birnbaum, S. M., Winitz, M., Koegel, R. J., and Greenstein, J. P., *J. Am. Chem. Soc.* **79,** 648 (1947).
2405. Wade, R., Winitz, M., and Greenstein, J. P., *J. Am. Chem. Soc.* **78,** 373 (1959).
2406. Wagner, J., and Winkler, G., *Z. Anal. Chem.* **183,** 1 (1961).
2407. Wagner-Jauregg, T., O'Neill, J. J., and Summerson, W. H., *J. Am. Chem. Soc.* **73,** 5202 (1951).
2408. Waksman, S. A., and Woodruff, H. B., *Proc. Soc. Exptl. Biol. Med.* **45,** 609 (1940).
2409. Waksman, S. A., and Woodruff, H. B., *J. Bacteriol.* **42,** 231 (1941).
2410. Walaszek, E. J., *Brit. J. Pharmacol.* **12,** 223 (1957).
2411. Walaszek, E. J., and Huggins, C. G., *J. Pharmacol. Exptl. Therap.* **126,** 258 (1959).
2412. Walaszek, E. J., Bunag, R. D., and Huggins, C. G., *J. Pharmacol. Exptl. Therap.* **138,** 139 (1962).
2413. Waldschmidt-Leitz, E., and Kühn, K., *Chem. Ber.* **84,** 381 (1951).
2414. Waley, S. G., *Chem. Ind. (London)* p. 107 (1953).
2415. Waley, S. G., *J. Chem. Soc.* p. 517 (1955).
2416. Waley, S. G., *Biochem. J.* **64,** 715 (1956).
2417. Waley, S. G., *Biochem. J.* **68,** 189 (1958).
2418. Waley, S. G., and Watson, J., *J. Chem. Soc.* p. 2394 (1951).
2419. Waley, S. G., and Watson, J., *J. Chem. Soc.* p. 475 (1953).
2420. Waley, S. G., and Watson, J., *Biochem. J.* **57,** 529 (1954).
2421. Waller, J. P., and Dixon, H. B. F., *Biochem. J.* **75,** 320 (1960).
2421a. Walser, A., Haas, H. G., and Koller, F., *Helv. Med. Acta* **30,** 470 (1963).
2421b. Walser, A., and Koller, F., *Experientia* **17,** 320 (1963).
2422. Walton, E., Rodin, J. O., Stammer, C. H., and Holly, F. W., *J. Org. Chem.* **26,** 1657 (1961).
2423. Wanag, G., and Veinbergs, A., *Ber. Deut. Chem. Ges.* **75,** 1558 (1942).
2423a. Wang, Y., Hsu, J.-Z., Loh, J.-Y., Huang, K., and Huang, J.-J., *Acta Chim. Sinica* **29,** 114 (1964).
2423b. Wang, Y., Huang, C.-C., Chang, W.-C., Tu, C.-T., Wang, C.-C., Hsu, Y.-Y., Tang, Y.-S., Lu, Y.-H., Kin, S.-W., and Kung, Y.-T., *Acta Chim. Sinica* **29,** 190 (1963).
2423c. Wang, Y., Loh, H.-Y., Hsu, J.-Z., Chang, W.-C., Chen, Y.-C., Wu, C.-H., Tsai, T.-Y., Han, K.-T., and Hsu, J.-W., *Acta Chim. Sinica* **30,** 49 (1964).
2424. Watanabe, K., *J. Antibiot. (Tokyo) Ser. A* **13,** 62 (1960).
2425. Watanabe, K., *J. Antibiot. (Tokyo) Ser. A* **14,** 1 (1961).
2426. Watanabe, K., *J. Antibiot. (Tokyo) Ser. A* **14,** 14 (1961).
2427. Watanabe, K., Yonehara, H., Tanaka, N., and Umezawa, H., *J. Antibiot. (Tokyo) Ser. A* **12,** 112 (1959).
2428. Watanabe, K., Yonehara, H., Umezawa, H., and Sumiki, Y., *J. Antibiot. (Tokyo) Ser. A* **13,** 291 (1960).
2429. Warner, D. T., *Nature* **190,** 120 (1961).
2430. Warner, D. T., *J. Theoret. Biol.* **1,** 514 (1961).

2431. Wasserman, H. H., Keggi, J. J., and McKeon, J. E., *J. Am. Chem. Soc.* **83,** 4107 (1961).
2432. Wasserman, H. H., Keggi, J. J., and McKeon, J. E., *J. Am. Chem. Soc.* **84,** 2978 (1962).
2433. Weber, E. J., *Dissertation Abstr.* **22,** 3378 (1962).
2434. Weber, R., Hofer, W., Heer, W., and Brenner, M., *Helv. Chim. Acta* **44,** 2154 (1961).
2435. Webster, M. E., Clark, W. R., and Anderson, P., *Proc. Soc. Exptl. Biol. Med.* **93,** 181 (1956).
2436. Webster, M. E., and Pierce, J. V., *Proc. Soc. Exptl. Biol. Med.* **107,** 186 (1961).
2437. Webster, M. E., and Pierce, J. V., *Ann. N. Y. Acad. Sci.* **104,** 91 (1963).
2438. Weisblat, D. I., Magerlein, B. J., and Myers, D. R., *J. Am. Chem. Soc.* **75,** 3630 (1963).
2439. Weisiger, J. R., Hausmann, W., and Craig, L. C., *J. Am. Chem. Soc.* **77,** 731 (1955).
2440. Weisiger, J. R., Hausmann, W., and Craig, L. C., *J. Am. Chem. Soc.* **77,** 1323 (1955).
2441. Weitzel, G., Schneider, F., and Fretzdorff, A.-M., *Z. Physiol. Chem.* **307,** 23 (1957).
2442. Wenner, R., *Schweiz. Med. Wochschr.* **89,** 441 (1959).
2443. Werbin, H., and McLaren, A. D., *J. Am. Chem. Soc.* **73,** 501 (1951).
2444. Werle, E., *Biochem. Z.* **287,** 235 (1936).
2445. Werle, E., *Arch. Intern. Pharmacodyn.* **92,** 427 (1953).
2446. Werle, E., *in* "Polypeptides which Stimulate Plain Muscle" (J. R. Gaddum, ed.), p. 20. Livingstone, Edinburgh and London, 1955.
2447. Werle, E., *in* "Polypeptides which Affect Smooth Muscles and Blood Vessels" (M. Schachter, ed.), p. 199. Macmillan (Pergamon), New York, 1960.
2448. Werle, E., *Angew. Chem.* **73,** 689 (1961).
2449. Werle, E., *Arch. Exptl. Pathol. Pharmakol.* **245,** 254 (1963).
2450. Werle, E., and Berek, U., *Angew. Chem.* **60,** 53 (1948).
2451. Werle, E., and Berek, U., *Biochem. Z.* **320,** 136 (1950).
2452. Werle, E., and Erdös, E. G., *Arch. Exptl. Pathol. Pharmakol.* **223,** 234 (1954).
2453. Werle, E., Forell, M. M., and Maier, L., *Arch. Exptl. Pathol. Pharmakol.* **225,** 369 (1955).
2454. Werle, E., Götze, W., and Keppler, A., *Biochem. Z.* **289,** 217 (1937).
2455. Werle, E., and Grunz, M., *Biochem. Z.* **301,** 429 (1939).
2456. Werle, E., and Hambüchen, R., *Arch. Exptl. Pathol. Pharmakol.* **201,** 311 (1943).
2457. Werle, E., and Kaufmann-Boetsch, B., *Naturwissenschaften* **46,** 559 (1959).
2458. Werle, E., and Kaufmann-Boetsch, B., *Z. Physiol. Chem.* **319,** 52 (1960).
2459. Werle, E., Kehl, R., and Koebke, K., *Biochem. Z.* **320,** 372 (1950).
2460. Werle, E., Kehl, R., and Koebke, K., *Biochem. Z.* **321,** 213 (1950).
2461. Werle, E., and Markus, A., *Biochem. Z.* **296,** 275 (1938).
2462. Werle, E., and Maier, L., *Biochem. Z.* **323,** 279 (1952).
2463. Werle, E., Maier, L., and Ringelmann, E., *Naturwissenschaften* **39,** 328 (1952).
2464. Werle, E., and Preisser, F., *Klin. Wochschr.* **34,** 1042 (1956).
2465. Werle, E., and Preisser, F., *Naturwissenschaften* **46,** 376 (1959).
2466. Werle, E., and von Roden, P., *Biochem. Z.* **301,** 329 (1939).

2467. Werle, E., and Semm, K., *Arch. Gynaekol.* **187,** 449 (1956).
2468. Werle, E., and Trautschold, I., Z. *Biol.* **112,** 169 (1960).
2469. Werle, E., and Trautschold, I., *Ann. N. Y. Acad. Sci.* **104,** 117 (1963).
2470. Werle, E., Trautschold, I., and Leysath, G., Z. *Physiol. Chem.* **326,** 174 (1961).
2470a. Werle, E., Trautschold, I., and Schmal, A., Z. *Physiol. Chem.* **332,** 79 (1963).
2471. Werle, E., and Vogel, R., Z. *Vitamin-, Hormon- Fermentforsch.* **11,** 245 (1960-1961).
2472. Wessely, F., Schlögl, K., and Korger, G., *Monatsh. Chem.* **82,** 671 (1951).
2473. Weygand, F., *Chimia (Aarau)* **14,** 367 (1960).
2474. Weygand, F., *Chimia (Aarau)* **14,** 378 (1960).
2475. Weygand, F., *Angew. Chem.* **72,** 50 (1960).
2476. Weygand, F., *Bull. Soc. Chim. Biol.* **43,** 1269 (1961).
2477. Weygand, F., Prox, A., Schmidhammer, L., and König, W., *in* "Peptide Symposium," Oxford, 1962 (G. T. Young, ed.), p. 97. Macmillan (Pergamon), New York, 1963.
2478. Weygand, F., and Adermann, G., *Chem. Ber.* **93,** 2334 (1960).
2479. Weygand, F., and Csendes, E., *Angew. Chem.* **64,** 136 (1952).
2480. Weygand, F., and Geiger, R., *Chem. Ber.* **89,** 647 (1956).
2481. Weygand, F., and Geiger, R., *Chem. Ber.* **90,** 634 (1957).
2482. Weygand, F., and Geiger, R., *Chem. Ber.* **92,** 2099 (1959).
2483. Weygand, F., Geiger, R., and Glöckler, U., *Chem. Ber.* **89,** 1543 (1956).
2484. Weygand, F., Geiger, H., and Swodenk, W., *Angew. Chem.* **68,** 307 (1956).
2485. Weygand, F., and Glöckler, U., *Chem. Ber.* **89,** 653 (1956).
2486. Weygand, F., and Hunger, K., Z. *Naturforsch.* **13b,** 50 (1958).
2487. Weygand, F., and Hunger, K., *Chem. Ber.* **95,** 1 (1962).
2488. Weygand, F., and Hunger, K., *Chem. Ber.* **95,** 7 (1962).
2489. Weygand, F., and Kaelicke, J., *Chem. Ber.* **95,** 1031 (1962).
2490. Weygand, F., Klinke, P., and Eigen, I., *Chem. Ber.* **90,** 1886 (1957).
2491. Weygand, F., Klipping, G., and Palm, D., *Chem. Ber.* **93,** 2619 (1960).
2492. Weygand, F., Kolb, B., Prox, A., Tilak, M. A., and Tomida I., Z. *Physiol. Chem.* **322,** 38 (1960).
2493. Weygand, F., and Leising, E., *Chem. Ber.* **87,** 248 (1954).
2494. Weygand, F., Prox, A., Jorgensen, E. C., Axén, R., and Kirchner, P., Z. *Naturforsch.* **18b,** 93 (1963).
2495. Weygand, F., Prox, A., Schmidhammer, L., and König, W., *Angew. Chem.* **75,** 282 (1963).
2496. Weygand, F., and Reiher, M., *Chem. Ber.* **88,** 26 (1955).
2497. Weygand, F., and Rinno, H., *Chem. Ber.* **92,** 517 (1959).
2498. Weygand, F., and Röpsch, A., *Chem. Ber.* **92,** 2095 (1959).
2499. Weygand, F., and Steglich, W., Z. *Naturforsch.* **14b,** 472 (1959).
2500. Weygand, F., and Steglich, W., *Chem. Ber.* **92,** 313 (1959).
2501. Weygand, F., and Steglich, W., *Chem. Ber.* **93,** 2983 (1960).
2502. Weygand, F., and Steglich, W., *Angew. Chem.* **73,** 757 (1961).
2503. Weygand, F., and Swodenk, W., *Chem. Ber.* **90,** 639 (1957).
2504. Weygand, F., and Swodenk, W., *Chem. Ber.* **93,** 1693 (1960).
2505. Weygand, F., and Zumach, G., Z. *Naturforsch.* **17b,** 807 (1962).
2506. White, H. J., Alverson, C. L., Baker, M. J., and Jackson, E. R., *Ann. N. Y. Acad. Sci.* **51,** 879 (1949).
2507. White, J., *J. Biol. Chem.* **102,** 249 (1933).

2508. White, J., *J. Biol. Chem.* **106,** 141 (1934).
2508a. White, J. E., and Engel, F. L., *J. Clin. Invest.* **37,** 1556 (1958).
2509. White, W. F., *J. Am. Chem. Soc.* **75,** 503 (1953).
2510. White, W. F., *J. Am. Chem. Soc.* **75,** 4877 (1953).
2511. White, W. F., *J. Am. Chem. Soc.* **76,** 4194 (1954).
2512. White, W. F., *J. Am. Chem. Soc.* **77,** 4691 (1955).
2513. White, W. F., and Gross, A. M., *J. Am. Chem. Soc.* **79,** 1141 (1957).
2514. White, W. F., and Landmann, W. A., *J. Am. Chem. Soc.* **77,** 771 (1955).
2515. White, W. F., and Landmann, W. A., *J. Am. Chem. Soc.* **77,** 1711 (1955).
2516. White, W. F., and Peters, R. L., *J. Am. Chem. Soc.* **78,** 4181 (1956).
2517. Wieczorek, H., Diplomarbeit, Freie Universität Berlin, 1960.
2518. Wiegershausen, B., Rauch, J., and Häussler, R., Vortrag auf der 4. Arbeitstagung der Industrie: und Hochschulpharmakologen, Wernigerode/Harz, 1961.
2518a. Wiegershausen, B., Stopp, G., and Eichstädt, M., *Acta Biol. Med. Ger.* **12,** 443 (1964).
2519. Wieland, T., *Angew. Chem.* **69,** 44 (1957).
2520. Wieland, T., *Angew. Chem.* **71,** 417 (1959).
2521. Wieland, T., *Collection Czech. Chem. Commun.* **24,** 6 (1959).
2522. Wieland, T., *Angew. Chem.* **72,** 892 (1960).
2523. Wieland, T., *Helv. Chim. Acta* **44,** 920 (1961).
2523a. Wieland, T., *Pure Appl. Chem.* **6,** 339 (1963).
2524. Wieland, T., and Bernhard, H., *Ann. Chem. Liebigs* **572,** 190 (1951).
2525. Wieland, T., and Bernhard, H., *Ann. Chem. Liebigs* **582,** 218 (1953).
2526. Wieland, T., and Boehringer, W., *Ann. Chem. Liebigs* **635,** 178 (1960).
2527. Wieland, T., and Bokelmann, E., *Ann. Chem. Liebigs* **576,** 20 (1952).
2528. Wieland, T., Bokelmann, E., Bauer, L., Lang, H. U., and Lau, H., *Ann. Chem. Liebigs* **583,** 129 (1953).
2529. Wieland, T., and Determann, H., *Angew. Chem.* **75,** 209 (1963).
2530. Wieland, T., Determann, H., and Albrecht, E., *Ann. Chem. Liebigs* **633,** 185 (1960).
2531. Wieland, T., Freter, K., and Gross, E., *Ann. Chem. Liebigs* **626,** 154 (1959).
2532. Wieland, T., and Heinke, B., *Ann. Chem. Liebigs* **599,** 70 (1956).
2533. Wieland, T., and Heinke, B., *Ann. Chem. Liebigs* **615,** 184 (1958).
2534. Wieland, T., Heinke, B., Vogler, K., and Morimoto, H., *Ann. Chem. Liebigs* **655,** 189 (1962).
2535. Wieland, T., and Höfer, A., *Ann. Chem. Liebigs* **619,** 35 (1958).
2536. Wieland, T., and Jaenicke, F., *Ann. Chem. Liebigs* **613,** 95 (1958).
2537. Wieland, T., Kern, W., and Sehring, R., *Ann. Chem. Liebigs* **569,** 117 (1950).
2538. Wieland, T., and Köppe, H., *Ann. Chem. Liebigs* **581,** 1 (1953).
2539. Wieland, T., and Krantz, H., *Chem. Ber.* **91,** 2619 (1958).
2540. Wieland, T., Lang, H. U., and Liebsch, D., *Ann. Chem. Liebigs* **597,** 227 (1955).
2541. Wieland, T., and Mannes, K., *Angew. Chem.* **69,** 389 (1957).
2542. Wieland, T., Mannes, K., and Schöpf, A., *Ann. Chem. Liebigs* **617,** 152 (1958).
2543. Wieland, T., and Mohr, H., *Ann. Chem. Liebigs* **599,** 222 (1956).
2544. Wieland, T., Niemann, E., and Pfleiderer, G., *Angew. Chem.* **68,** 305 (1956).
2545. Wieland, T., and Ohly, K. W., *Ann. Chem. Liebigs* **605,** 179 (1957).

2546. Wieland, T., and Sarges, R., *Ann. Chem. Liebigs* **658,** 181 (1962).
2547. Wieland, T., and Schäfer, W., *Ann. Chem. Liebigs* **576,** 104 (1952).
2548. Wieland, T., Schäfer, W., and Bockelmann, E., *Ann. Chem. Liebigs* **573,** 99 (1951).
2549. Wieland, T., and Schnabel, H. W., *Ann. Chem. Liebigs* **657,** 218 (1962).
2550. Wieland, T., and Schnabel, H. W., *Ann. Chem. Liebigs* **657,** 225 (1962).
2551. Wieland, T., and Schön, W., *Ann. Chem. Liebigs* **593,** 157 (1953).
2552. Wieland, T., and Schöpf, A., *Ann. Chem. Liebigs* **626,** 174 (1959).
2553. Wieland, T., and Sehring, R., *Ann. Chem. Liebigs* **569,** 122 (1950).
2554. Wieland, T., and Stimming, D., *Ann. Chem. Liebigs* **579,** 97 (1953).
2555. Wieland, T., and Urbach, H., *Ann. Chem. Liebigs* **613,** 84 (1958).
2556. Wieland, T., and Vogler, K., *Angew. Chem.* **73,** 435 (1961).
2557. Wieland, T., and Vogler, K., *Angew. Chem.* **74,** 904 (1962).
2558. Wieland, T., and Weiberg, O., *Ann. Chem. Liebigs* **607,** 168 (1957).
2559. Wieland, T., and Weidenmüller, H. L., *Ann. Chem. Liebigs* **597,** 111 (1955).
2560. Wieland, T., and Wieland, O., *Pharmacol. Rev.* **11,** 87 (1959).
2561. Wiemer, B., Dissertation, Freie Universität Berlin, 1960.
2561a. Wilchek, M., and Patchornik, A., *J. Org. Chem.* **28,** 1874 (1963).
2562. Wilken, D. R., *Dissertation Abstr.* **21,** 3254 (1961).
2563. Wilkinson, J. R., and Young, G. T., *J. Chem. Soc.* p. 2626 (1959).
2564. Wilkinson, S., *Nature* **164,** 622 (1949).
2565. Wilkinson, S., *J. Chem. Soc.* p. 104 (1951).
2565a. Wilkinson, S., and Lowe, L. A., *Nature* **202,** 1211 (1964).
2565b. Wilkinson, S., and Lowe, L. A., *Nature* **204,** 185 (1964).
2565c. Wilkinson, S., and Lowe, L. A., *Nature* **204,** 993 (1964).
2566. Williams, M. W., and Young, G. T., *J. Chem. Soc.* p. 881 (1963).
2566a. Williams, M. W., and Young, G. T., *in* "Peptide Symposium," Oxford, 1962 (G. T. Young, ed.), p. 119. Macmillan (Pergamon), New York, 1963.
2567. Winitz, M., Bloch-Frankenthal, L., Izumiya N., Birnbaum, S. M., Baker, C. G., and Greenstein, J. P., *J. Am. Chem. Soc.* **78,** 2423 (1956).
2568. Winitz, M., and Fruton, J. S., *J. Am. Chem. Soc.* **75,** 3041 (1953).
2569. Wintersberger, E., Tuppy, H., and Stoklaska, E., *Monatsh. Chem.* **91,** 577 (1960).
2570. Winterstein, A., Hegedüs, B., Fust, B., Böhni, E., and Studer, A., *Helv. Chim. Acta* **39,** 229 (1956).
2571. Wilson, S., and Dixon, G. H., *Nature* **191,** 876 (1961).
2572. Wilson, S., Dixon, G. H., and Wardlaw, A. C., *Biochim. Biophys. Acta* **62,** 483 (1962).
2573. Wiseblatt, L., Wilson, L., and McConnell, W. B., *Can. J. Chem.* **33,** 1295 (1955).
2574. Witkop, B., *Experientia* **12,** 372 (1956).
2575. Witte, S., Schricker, K. T., and Schmid, E., *Arzneimittel-Forsch.* **11,** 619 (1961).
2576. Wolf, D. E., Valiant, J., Peck, R. L., and Folkers, K., *J. Am. Chem. Soc.* **74,** 2002 (1952).
2577. Wolman, Y., Gallop, P. M., and Patchornik, A., *J. Am. Chem. Soc.* **83,** 1263 (1961).
2578. Wolman, Y., Gallop, P. M., Patchornik, A., and Berger, A., *J. Am. Chem. Soc.* **84,** 1889 (1962).
2579. Wood, J. L., and du Vigneaud, V., *J. Biol. Chem.* **130,** 109 (1939).

2580. Woodward, R. B., and Olofson, R. A., *J. Am. Chem. Soc.* **83,** 1007 (1961).
2581. Woodward, R. B., Olofson, R. A., and Mayer, H., *J. Am. Chem. Soc.* **83,** 1010 (1961).
2582. Woolley, D. W., *J. Biol. Chem.* **166,** 783 (1946).
2583. Woolley, D. W., *J. Biol. Chem.* **172,** 71 (1948).
2584. Woolley, D. W., and Merrifield, R. B., *J. Am. Chem. Soc.* **76,** 316 (1954).
2585. Woolley, D. W., and Merrifield, R. B., *Science* **128,** 238 (1958).
2586. Woolley, D. W., and Merrifield, R. B., *Ann. N. Y. Acad. Sci.* **104,** 161 (1963).
2587. Woolley, D. W., Merrifield, R. B., Ressler, C., and du Vigneaud, V., *Proc. Soc. Exptl. Biol. Med.* **89,** 669 (1955).
2588. Work, E., *J. Gen. Microbiol.* **25,** 167 (1961).
2589. Work, E., Birnbaum, S. M., Winitz, M., and Greenstein, J. P., *J. Am. Chem. Soc.* **77,** 1916 (1955).
2590. Wright, P. H., *Brit. Med. Bull.* **16,** 219 (1960).
2591. Wright, W. W., and Welch, H., *Antibiot. Ann.* p. 61 (1959-1960).
2592. Wünsch, E., Dissertation, Ludwig-Maximilian-Universität München, 1955.
2593. Wünsch, E., *Angew. Chem.* **71,** 743 (1959).
2594. Wünsch, E., *Collection Czech. Chem. Commun.* **24,** 60 (1959).
2595. Wünsch, E., *in* "Peptide Symposium," Oxford, 1962 (G. T. Young, ed.), p. 89. Macmillan (Pergamon), New York, 1963.
2595a. Wünsch, E., *Z. Physiol. Chem.* **332,** 288 (1963).
2595b. Wünsch, E., *Z. Physiol. Chem.* **332,** 295 (1963).
2596. Wünsch, E., Fries, G., and Zwick, A., *Chem. Ber.* **91,** 542 (1958).
2597. Wünsch, E., and Fürst, G., *Z. Physiol. Chem.* **329,** 109 (1962).
2597a. Wünsch, E., and Heidrich, H.-G., *Z. Physiol. Chem.* **332,** 300 (1963).
2597b. Wünsch, E., Heidrich, H.-G., and Grassmann, W., *Chem. Ber.* **97,** 1818 (*1964*).
2597c. Wünsch, E., and Jentsch, J., *Chem. Ber.* **97,** 2490 (1964).
2597d. Wünsch, E., and Wendlberger, G., *Chem. Ber.* **97,** 2504 (1964).
2597e. Wünsch, E., and Zwick, A., *Chem. Ber.* **97,** 2497 (1964).
2597f. Wünsch, E., Wendlberger, G., and Jentsch, J., *Chem. Ber.* **97,** 3298 (1964).
2597g. Wünsch, E., and Zwick, A., *Chem. Ber.* **97,** 3305 (1964).
2597h. Wünsch, E., and Zwick, A., *Chem. Ber.* **97,** 3312 (1964).
2598. Wünsch, E., and Zwick, A., *Z. Physiol. Chem.* **328,** 235 (1962).
2598a. Wünsch, E., and Zwick, A., *Z. Physiol. Chem.* **333,** 108 (1963).
2598b. Würz, H., Tanaka, A., and Fruton, J. S., *Biochemistry* **1,** 19 (1962).
2599. Yalow, R. S., and Berson, S. A., *J. Clin. Invest.* **39,** 1157 (1960).
2600. Yalow, R. S., and Berson, S. A., *Diabetes* **10,** 339 (1961).
2601. Yamamoto, M., Kotaki, A., Okuyama, T., and Satake, K., *J. Biochem.* (*Tokyo*) **48,** 84 (1960).
2602. Yamashita, K., and Yashiro, M., *J. Agr. Chem. Soc. Japan* **28,** 674 (1954).
2603. Yamashita, T., *Nippon Kagaku Zasshi* **81,** 801 (1960).
2604. Yamashita, T., *J. Biochem.* (*Tokyo*) **48,** 651 (1960).
2605. Yamashita, T., *J. Biochem.* (*Tokyo*) **48,** 846 (1960).
2606. Yamashita, T., and Izumiya, N., *J. Biochem.* (*Tokyo*) **46,** 991 (1959).
2607. Young, F. G., *Proc. Roy. Soc.* **B157,** 1 (1962).
2607a. Young, F. G., *in* "Comparative Endocrinology" (U. S. von Euler and H. Heller, eds.), Vol. 1, p. 371. Academic Press, New York, 1963.
2608. Young, G. T., *Angew. Chem.* **71,** 742 (1959).

2609. Young, G. T., *Collection Czech. Chem. Commun.* **24**, 114 (1959).
2610. Young, J. D., and Carpenter, F. H., *J. Biol. Chem.* **236**, 743 (1961).
2611. Young, R. W., Wood, K. H., Joyce, R. J., and Anderson, G. W., *J. Am. Chem. Soc.* **78**, 2126 (1956).
2612. Zachau, H. G., *Chem. Ber.* **93**, 1822 (1960).
2613. Zachau, H. G., and Karau, W., *Chem. Ber.* **93**, 1830 (1960).
2614. Zahn, H., *Chimia (Aarau)* **15**, 378 (1961).
2615. Zahn, H., and Determann, H., *Chem. Ber.* **90**, 2176 (1957).
2616. Zahn, H., and Diehl, J. F., *Angew. Chem.* **69**, 135 (1957).
2617. Zahn, H., and Diehl, J. F., *Z. Naturforsch.* **12b**, 85 (1957).
2618. Zahn, H., Diehl, J. F., and Haas, W. L., *Z. Naturforsch.* **16b**, 149 (1961).
2619. Zahn, H., and Fahnenstich, R., *Ann. Chem. Liebigs* **663**, 184 (1963).
2620. Zahn, H., and Falkenburg, H. R., *Ann. Chem. Liebigs* **636**, 117 (1960).
2621. Zahn, H., and Kessler, H., *Makromol. Chem.* **27**, 218 (1958).
2622. Zahn, H., and Kunde, J., *Angew. Chem.* **69**, 713 (1957).
2623. Zahn, H., and Kunde, J., *Chem. Ber.* **94**, 2470 (1961).
2624. Zahn, H., Kunde, J., and Heidemann, G., *Makromol. Chem.* **43**, 220 (1961).
2625. Zahn, H., Kunde, J., and Zabel, R., *Ann. Chem. Liebigs* **663**, 177 (1963).
2626. Zahn, H., and La France, N. H., *Z. Physiol. Chem.* **319**, 143 (1960).
2627. Zahn, H., and La France, N. H., *Ann. Chem. Liebigs* **630**, 26 (1960).
2628. Zahn, H., and La France, N. H., *Ann. Chem. Liebigs* **630**, 37 (1960).
2628a. Zahn, H., Meienhofer, J., and Klostermeyer, H., *Z. Naturforsch.* **19b**, 110 (1964).
2628b. Zahn, H., Meienhofer, J., and Schnabel, E., Proc. 7th European Symp. on Peptides, Budapest, 1964 *Acta Chim. Acad. Sci. Hung.* (in press).
2629. Zahn, H., and Meissner, A., *Ann. Chem. Liebigs* **636**, 132 (1960).
2630. Zahn, H., and Messerknecht, H. E., *Z. Physiol. Chem.* **315**, 228 (1959).
2631. Zahn, H., and Osterloh, F., *Ann. Chem. Liebigs* **595**, 237 (1955).
2632. Zahn, H., and Osterloh, F., *Proc. Intern. Wool Textile Res. Conf. Australia* Vol. C., p. C 144 (1955). CSIRO, Victoria, Australia.
2633. Zahn, H., and Otten, H. G., *Ann. Chem. Liebigs* **653**, 139 (1962).
2633a. Zahn, H., and Pätzold, W., *Chem. Ber.* **96**, 2566 (1963).
2634. Zahn, H., and Pieper, W., *Chem. Ber.* **95**, 1069 (1962).
2635. Zahn, H., and Schade, F., *Chem. Ber.* **94**, 843 (1961).
2636. Zahn, H., and Schnabel, E., *Ann. Chem. Liebigs* **604**, 62 (1957).
2637. Zahn, H., and Schnabel, E., *Ann. Chem. Liebigs* **605**, 212 (1957).
2638. Zahn, H., and Schüssler, H., *Ann. Chem. Liebigs* **641**, 175 (1961).
2639. Zahn, H., and Siepmann, E., *Biochem. Z.* **335**, 303 (1961).
2640. Zahn, H., and Umlauf, E., *Z. Physiol. Chem.* **297**, 127 (1954).
2641. Zahn, H., and Zabel, R., *Ann. Chem. Liebigs* **659**, 163 (1962).
2642. Zahn, H., and Ziegler, K., *Ann. Chem. Liebigs* **610**, 132 (1957).
2643. Zahn, H., Zuber, H., Ditscher, W., Wegerle, D., and Meienhofer, J., *Chem. Ber.* **89**, 407 (1956).
2644. Zahn, H., and Zürn, L., *Chem. Ber.* **91**, 1359 (1958).
2645. Zahn, H., and Zürn, L., *Ann. Chem. Liebigs* **613**, 76 (1959).
2646. Zaoral, M., *Angew. Chem.* **71**, 743 (1959).
2647. Zaoral, M., *Collection Czech. Chem. Commun.* **24**, 1314 (1959).
2648. Zaoral, M., *Collection Czech. Chem. Commun.* **27**, 1273 (1962).
2649. Zaoral, M., and Arnold, Z., *Tetrahedron Letters* p. 9 (1960).

2650. Zaoral, M., Pliška, V., Řezábek, K., and Šorm, F., *Collection Czech. Chem. Commun.* **28,** 746 (1963).
2651. Zaoral, M., Pliška, V., Řezábek, K., and Šorm, F., *Collection Czech. Chem. Commun.* **28,** 747 (1963).
2652. Zaoral, M., and Rudinger, J., *Collection Czech. Chem. Commun.* **20,** 1183 (1955).
2653. Zaoral, M., and Rudinger, J., *Proc. Chem. Soc.* p. 176 (1957).
2654. Zaoral, M., and Rudinger, J., *Collection Czech. Chem. Commun.* **24,** 1993 (1959).
2655. Zaoral, M., and Rudinger, J., *Collection Czech. Chem. Commun.* **26,** 2316 (1961).
2656. Zaoral, M., Rudinger, J., and Šorm, F., *Collection Czech. Chem. Commun.* **18,** 530 (1953).
2656a. Zhenodarova, S. M., and Morozova, E. A., *Vest. Mosk. Univ. Ser. II Khim.* **15,** 31 (1960).
2657. Zepf, K., and Berg, H., *Pharmazie* **14,** 396 (1959).
2658. Zervas, L., *Collection Czech. Chem. Commun.* **27,** 2242 (1962).
2659. Zervas, L., Benoiton, L., Weiss, E., Winitz, M., and Greenstein, J. P., *J. Am. Chem. Soc.* **81,** 1729 (1959).
2659a. Zervas, L., Borovas, D., and Gazis, E., *J. Am. Chem. Soc.* **85,** 3660 (1963).
2660. Zervas, L., and Dilaris, I., *J. Am. Chem. Soc.* **77,** 5354 (1955).
2661. Zervas, L., and Katsoyannis, P. G., *J. Am. Chem. Soc.* **77,** 5351 (1955).
2662. Zervas, L., Otani, T. T., Winitz, M., and Greenstein, J. P., *Arch. Biochem. Biophys.* **75,** 290 (1958).
2663. Zervas, L., Otani, T. T., Winitz, M., and Greenstein, J. P., *J. Am. Chem. Soc.* **81,** 2878 (1959).
2664. Zervas, L., and Photaki, I., *Chimia (Aarau)* **14,** 375 (1960).
2665. Zervas, L., and Photaki, I., *J. Am. Chem. Soc.* **84,** 3887 (1962).
2666. Zervas, L., Photaki, I., Cosmatos, A., and Ghelis, N., *in* "Peptide Symposium," Oxford, 1962 (G. T. Young, ed.), p. 27. Macmillan (Pergamon), New York, 1963.
2667. Zervas, L., Photaki, I., and Ghelis, N., *J. Am. Chem. Soc.* **85,** 1337 (1963).
2668. Zervas, L., and Theodoropoulos, D. M., *J. Am. Chem. Soc.* **78,** 1359 (1956).
2669. Zervas, L., Winitz, M., and Greenstein, J. P., *Arch. Biochem. Biophys.* **65,** 573 (1956).
2670. Zervas, L., Winitz, M., and Greenstein, J. P., *J. Org. Chem.* **22,** 1515 (1957).
2671. Zervas, L., Winitz, M., and Greenstein, J. P., *J. Am. Chem. Soc.* **83,** 3300 (1961).
2672. Zetler, G., *Arch. Exptl. Pathol. Pharmakol.* **242,** 330 (1961).
2673. Zetler, G., *Arch. Exptl. Pathol. Pharmakol.* **243,** 360 (1962).
2674. Zetler, G., *Arch. Exptl. Pathol. Pharmakol.* **245,** 263 (1963).
2674a. Zhuze, A. L., Jošt, K., Kasafírek, E., and Rudinger, J., *Coll. Czech. Chem. Commun.* **29,** 2648 (1964).
2675. Ziegler, K., and Spoor, H., *Biochim. Biophys. Acta* **33,** 138 (1959).
2676. Zilkha, A., and Liwschitz, Y., *J. Chem. Soc.* p. 4397 (1957).
2677. Zomzely, C., Marco, G., and Emery, E., *Anal. Chem.* **34,** 1414 (1962).
2678. Zuber, H., *Chimia (Aarau)* **14,** 405 (1960).
2679. Zuber, H., *Ann. N. Y. Acad. Sci.* **104,** 391 (1963).
2680. Zuber, H., and Jaques, R., *Helv. Chim. Acta* **43,** 1128 (1960).
2681. Zuber, H., and Jaques, R., *Angew. Chem.* **74,** 216 (1962).

2682. Zyl, G., van, van Tamelen, E. E., and Zuidema, G. D., *J. Am. Chem. Soc.* **73**, 1765 (1951).
2683. Gaddum, J. R. (ed.), "Polypeptides which Stimulate Plain Muscle," 140 pp. Livingstone, Edinburgh and London, 1955.
2684. Schachter, M. (ed.), "Polypeptides which Affect Smooth Muscles and Blood Vessels," 336 pp. Macmillan (Pergamon), New York, 1960.
2685. Stern, P. (ed.), "Symposium on Substance P," 143 pp. Oslobodenji Sarajewo, 1961.
2686. Caldeyro-Barcia, R. and Heller, H. (eds.), "Oxytocin," Proceedings of an International Symposium held in Montevido 1959, 443 pp. Macmillan (Pergamon), New York, 1961.
2687. Bradykinin and vasodilating polypeptides. Papers of the 1st International Pharmacological Meeting, *Biochem. Pharmacol.* **10**, 94 pp. (1962).
2688. Wood, J. E., and Ahlquist, R. P. (eds.), Symposium on Angiotensin *Circulation* **25**, 167–270 (1962).
2689. Structure and function of biologically active peptides: Bradykinin, Kallidin and Congeners. *Ann. N. Y. Acad. Sci.* **104**, (1963).
2690. Verhandlungen der Deutschen Pharmakologischen Gesellschaft 27. Tagung, *Arch. Exptl. Pathol. Pharmakol.* **245**, (1) 141–288 (1963).
2691. Schwyzer, R., CIBA AG. Basel, D. A. S. 1033, 214 (Nov. 11, 1953).
2692. Schwyzer, R., Iselin, B., and Feurer, M., CIBA AG. Basel, D. A. S. 1,060,380 (Feb. 4, 1955).
2693. Schwyzer R., Iselin, B., and Feurer, M., CIBA AG. Basel, D. A. S. 1,085,881 (Oct. 1, 1956).
2694. Schwyzer, R., Iselin, B., Kappeler, H., and Rittel, W., CIBA AG. Basel, Schweiz. Patent 356,122 (1956).
2695. Schwyzer, R., Iselin, B., Rittel, W., and Sieber, P., CIBA AG. Basel, D. A. S. 1,090,663 (Jan. 19, 1957).
2696. Schwyzer, R., Iselin, B., Kappeler, H., and Rittel, W., CIBA AG. Basel, D. A. S. 1,121,059 (Sept. 2, 1957).
2697. Schwyzer, R., Iselin, B., Kappeler, H., Rittel, W., and Riniker, B., CIBA AG. Basel, D. A. S., 1,125,942 (Sept. 2, 1957).
2698. CIBA AG. Basel, Oesterr. Patent 202,276 (Dec. 18, 1957).
2699. CIBA AG. Basel, British Patent 836,726 (Dec. 19, 1957).
2700. CIBA AG. Basel, Oesterr. Anm. A 6067/58 (Aug. 29, 1958).
2701. CIBA AG. Basel, Franz. Patent 1,271,027 (April 20, 1959).
2702. Schwyzer, R., CIBA AG. Basel, D. A. S. 1,112,525 (April 22, 1959).
2703. Schwyzer, R., and Kappeler, H., CIBA AG. Basel, D. A. S. 1,108,232 (June 24, 1959).
2704. CIBA AG. Basel, Dän. Patent 91,765 (Aug. 6, 1959).
2705. Schwyzer, R., CIBA AG. Basel, D. A. S. 1,121,620 (Aug. 27, 1959).
2706. Schwyzer, R., Iselin, B., Kappeler, H., Rittel, W., and Zuber, H., CIBA AG. Basel, South Africa Patent 60 1792 (April 25, 1960).
2707. Schwyzer, R., Iselin, B., Kappeler, H., Rittel, W., and Riniker, B., CIBA AG. Basel, South Africa Patent 61 617 (June 30, 1961).
2708. Kappeler, H., and Schwyzer, R., CIBA AG. Basel, South Africa Patent 62 1755 (April 25, 1962).
2709. Herrick, E. C., and Todd, C. W., E. I. du Pont de Nemours and Co., Wilmington, Delaware, U. S. Patent 2,723,973 (April 29, 1954).

2710. Fields, M., and Steven, M. A., E. I. du Pont de Nemours and Co., Wilmington Delaware, D. A. S. 1,143,815 (March 1, 1957).
2711. Weygand, F., Steglich, W., Prox, A., and Kaelicke, J., Farbwerke Hoechst, AG. D. A. S. 1,114,200 (May 15, 1961).
2712. Farbwerke Hoechst, AG. Niederl. Patent 265,516 (June 2, 1961).
2713. N. V. Organon (Holland) Holländ. Patent 149,558 (Oct. 7, 1959).
2714. N. V. Organon (Holland) Franz. Patent 1,305,345 (Oct. 7, 1960).
2715. Boissonnas, R. A., Sandoz AG. Basel, D. A. S. 1,080,113 (May 4, 1959).
2715a. Boissonnas, R. A., and Sandrin, E., Sandoz AG. Franz. Patent 1,329,839 (1963) (July 12, 1962).
2715b. Boissonnas, R. A., and Sandrin, E., Sandoz AG. Franz. Patent 1,329,840 (1963) (July 12, 1962).
2715c. Chillemi, F., Sicieta Farmaceutica Italia, Franz. Patent 1,337,576 (1963) (Oct. 2, 1962).
2716. UCLAF, British Patent 826,925 (July 10, 1956).
2717. Velluz, L., Amiard, G., Bartos, J., Goffinet, B., and Heymès, R., UCLAF, Paris, D. B. Patent 1,038,053 (March 22, 1957).
2718. Losse, G., and Zönnchen, W., VEB Berlin-Chemie, Berlin-Adlershof, D. A. S. 1,117,134 (June 6, 1960).
2719. Geipel, H., and Reichwald, W., VEB-Berlin-Chemie, Berlin-Adlershof, D. A. S. 1,133,396 (March 2, 1961).

AUTHOR INDEX

Numbers in parentheses are reference numbers and indicate that an author's work is referred to although his name is not cited in the text. Numbers in italic show the pages on which the complete references are listed.

E

L

O

P

S

T

SUBJECT INDEX

C

D

I

K

L

R

S

U

V

W

X